Haldane, Frankland & Hofmeyr
Three Accounts of Captivity and Escape from Pretoria

Captain Aylmer Haldane.

Lt. Thomas Frankland.

Rev. Adrian Hofmeyr.

Armoured Train Disaster by Thomas Frankland.

Officers Prison at the State Model School.

First person accounts assembled herein:

Captain Aylmer Haldane, DSO: How We Escaped From Pretoria.

Rev. Adrian Hofmeyr: The Story of My Captivity During the Transvaal War 1899-1900.

Lieutenant Thomas Hugh Colville Frankland: Held by the Enemy: Extracts from the Journal of Lieutenant Thomas Hugh Colville Frankland, Royal Dublin Fusiliers, lately prisoner of war at Pretoria.

Volume 1 of the series, *The Last Horse War*

Series compiled, annotated, illustrated and expanded by Payton Fireman, JD.

ISBN: 978-1-7320124-3-1

Cover and Title Page Photo Credits:[1,2,3,4,5,6]

[1] Please see the Bibliography and License Links sections regarding the use of material in this volume.

Contents

Series Introduction

When this material was originally published all of the subjects and content were current events. Everyone saw the pictures and read the books. In the present, this saga has become disparate and scattered amongst various bookshelves and servers. The interconnected stories no longer are. Here they are re-assembled into a narrative of the time and place where the events took place. There is an avalanche of material that I am not able to include. Space considerations have been paramount in order to provide the most vital, trenchant and relevant material.

The series title, 'The Last Horse War', recognizes that the Boer War of 1899-1902 occurred just prior to the practical development of the internal combustion engine automobile. The Ford Model T would not be introduced until 1908 and before that time the mass production of automobiles did not occur. In all my research I have not identified one automobile in use during the conflict. I have been able to identify the use of steam traction engines;[2] but there is nothing to suggest that they played any role other than for transport. Fifteen years later, during the First World War, the use of trucks and automobiles had become ubiquitous and by its conclusion armored fighting vehicles, e.g. Tanks, had eliminated cavalry as a means of conducting large scale offensive military operations. In this conflict, particularly in the Orange Free State campaign and the succeeding last two years of the war, large scale cavalry operations were the primary means of projecting military force.

Traction Engines crossing a drift.[7] Traction Engine stuck in a soft spot.[8]

Over four hundred thousand British troops were deployed to this theatre of war and an equal number of horses from every corner of the Empire and beyond.

Remount Camp at Stellenbosch. "The heavy mortality amongst the horses and mules during the campaign put the British forces to considerable inconvenience, and large numbers were purchased in North and South America, Spain and Italy."[9] Loading Horses onto Trains.[10]

"Gift of the Indian Princes." Indian horses arriving in Cape Town."[11] "Boer Commando assembled." [12]

[2] A traction engine is a steam-powered tractor used to move heavy loads on roads, plough ground or to provide power at a chosen location. They are sometimes called road locomotives to distinguish them from railway locomotives. Traction engine - Wikipedia

Army Corps shipping horses.[13]

Landing mules for transport.[14]

Troops of one of the British mounted regiments collecting forage for their horses.[15] Officers of the Lothians and Berwickshire Yeomanry tending their horses.[16]

Ninth Lancers at Modder River.[17] Troops of one of the British mounted regiments watering their horses while on the move.[18]

With the 1888 invention of Roll Film[19] and its subsequent distribution by Kodak; the word 'Kodaking'[20] was coined to described going out and about to take photographs with a portable camera. This term of art was in use as late as the 1940s;[21] oftentimes with the ubiquitous Kodak Brownie Camera.

This conflict coincides with the widespread availability for both professionals and amateurs to engage in this means of spontaneous photography. One also sees the word 'snap-shooting'[22] used. This was made possible by because the speed of the film[23] was now fast enough that shutter speeds could be shortened to catch actions rather than simply poses. Most single speed cameras, generally had shutter speeds of 1/35 to 1/50 of a second.

It also gave rise to more extensive photojournalism.[24] I have made use of Navy & Army Illustrated,[25] Black & White (Magazine)[26] as well as other illustrated periodicals, to document this series. Also, during this period, printed books were able to incorporate far more pictures because advances in printing technology allowed high speed book presses to print photographs along with text.

Before the disaster, November 15th. [1899] Men of the Dublin Fusiliers embarking on the armoured train. In the [train] truck on the left is…Rene' Bull, snap-shooting.[27] [The man on the left standing on the railroad truck with a hat on and looking down into his camera.] "The Misses Eloff. Granddaughters of President Kruger." [With portable camera.][28]

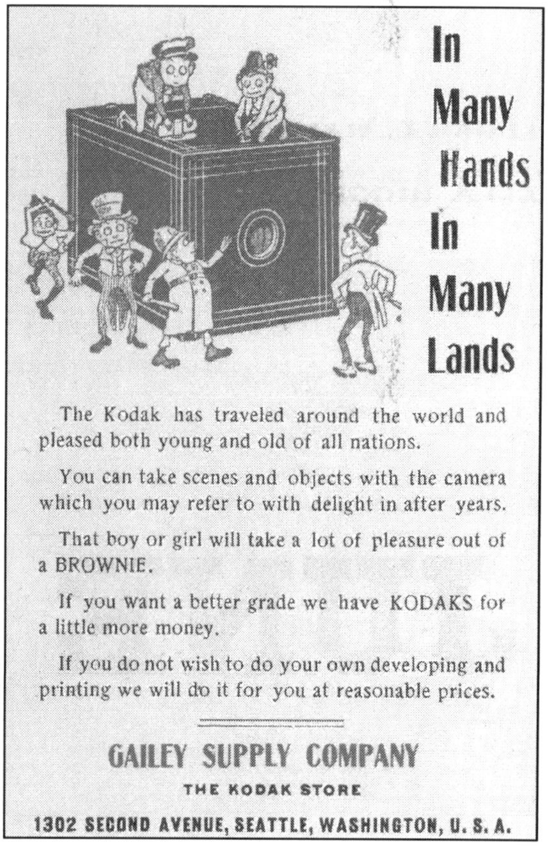

This is a model No. 4 Cartridge Eastman Kodak Camera. This It was used by the Scots Guards during the conflict.[29] It may be the same model that Ms. Eloff is holding. A Gailey Supply Company Advertisement.[30]

Document Discussion

The reason this series is composed of a collection of first person accounts is that my background as an attorney encourages me to cite original sources and follow the best evidence rule.[3] In that context, the original authors are the best witnesses of their experiences and I see no reason why they need to have their tales retold by someone other than themselves. It's their story, and they did a fine job of setting down their recollections.

These recollections are then augmented by the knowledge, facts, and circumstances of the other memorialists who participated, oftentimes in the same noteworthy events. To some degree they are synoptic, as each has their own perspective, and ancillary facts included in individual accounts. I have also done my best to fill in gaps in their knowledge that are now, in hindsight, available to be added. I have also made a large number of probably glaring mistakes regarding identifications and locations that despite my best efforts I am still unaware of at the time of publication.

I've focused my efforts on augmenting and contextualizing their narratives; supplementing, annotating, illustrating as well as filling in gaps in their knowledge that their circumstances prevented them from knowing at the time. The goal has been to update a number of dated popular narratives into comprehensible windows to another time and place, since the passage of time has obscured the social and political landscape that was as familiar to readers of the narrators' time as our own is to us. Many of the cultural and demographic references have fallen out of common use and knowledge, and many locations have changed their names as well. To get the most out their works, one needs to be familiar with the personages and nomenclature then current.

There is a story arc for the series that extends well beyond the actual narratives as the participants continue to interact for decades afterwards. They exchange letters, ask favors, and write about each other during the ensuing decades.

I hope the annotations and illustrations I have added will allow readers without a working knowledge of this time period to fully enjoy the narrators' writing and observations.

Please see the South African Background infra, for an introduction to the time period, the events leading up to the conflict and the first engagements before our protagonists arrive.

This series is built upon the ability to search large Word documents and downloaded PDF's for words and images that have been previously processed into useable form. As the work progressed I could not remember where I'd last seen that image, person, or place name. London to Ladysmith is a 118 megabyte Word document and the other volumes are not far behind it. Maps, images and individuals are reused where the lines intersect with the other narratives. The same search ability holds true for the hundreds of downloaded PDF's from the Internet Archive and Hathitrust, along with many, many other sources. Any cue can send me off on a hour's long search for that map, place or image but it's easier than it used to be. Using the resources I've assembled it is possible to continue this interconnected journey of images, memoirs and historical title searching somewhat akin to genealogy to further explore and extrapolate these events into the War and Peace in Africa that it was.

As to the time frame of the conflict, it is an intimate overview of the political, military and social zeitgeist of the age which could be surmised as the high tide of Victorian Manifest Destiny; During this phase of the Empire it was a regular occurrence to direct military force to be applied to effect political or social change. This particular conflict, was intensively documented, narrated and studied at the time but has subsequently taken a back seat to more recent attempts at this recurring theme.[4]

Also, the events and social movements that engendered this conflict have been obscured by the subsequent cataclysms of the First and Second World Wars. De-colonialism and the Cold War further deprecated this saga and kept it in the shadow of history.

[3] The best evidence rule applies to documents, positing that an original is superior to a copy. My application of this concept conflates this rule with the rule regarding Hearsay that in broad brush terms states that first person accounts are more accurate than second hand one's.

[4] For instance, the American conflicts in Vietnam and Afghanistan, with better results for, Japan, Korea, the Philippines. Iraq?

Text and Spelling

I've kept the text and spelling as intact as possible and sparingly inserted a few bracketed words for clarity. Oftentimes different writers will spell the same name or place differently. This is far too common to try and harmonize globally. Sometimes I will [bracket] an inserted word or two for context. I have also mostly retained the rather idiomatic spelling and hyphenation of compound words that are now handled differently. Original footnotes are enclosed in brackets [ ].

I deprecate italics on aesthetic grounds generally and in particular when using Times New Roman as a font. I trust that the titles of books, articles and newspapers will be recognizable to the reader. Citations are numerous but eclectic as to their form, hopefully not as to their content.

I have taken the liberty of dividing up some of the longer paragraphs. Very rarely there are certain offensive words and phrases and I have declined to serve as a censor.

Photographs

Regardless of whether there are quotation marks included in the photo captions, I have almost universally used the original captions. Sometimes they may not be accurate as per the original editor's license to ascribe them to different military units or different battles. For example:

"Dorset Yeomanry Entering Johannesburg"[31]

"The City [London] Volunteers Entering Johannesburg."[32]

It's been a learning curve to decide when and where to put in an illustration; particularly with Churchill and Deneys Reitz. They are so good at descriptive prose that I do not want to interrupt them by placing a photograph in the middle of their work. This conflicts mightily with page layout and sometimes I may get it wrong.

There are duplicates of pictures, maps and footnotes in the various sections and volumes based relevance to the topic at hand and space available. My desire was the each section should illustrate the topic under discussion and to some extent to be able to stand on its own as a separate work.[5]

[5] Sometimes the picture quality I've been able to find is very low, the balancing test employed is to weigh the probative value of the image against the quality of the reproduction. Some high value pictures accordingly have very poor quality. I have tried to refrain from using artists renderings of events except where they have artistic value and also where there is no other visual means of representing the event.

Research & Style Guide

Almost everyone who is mentioned by first and last name or even just unit, rank and last name has been researched for additional information that if located, is recited in footnotes. Either that or they are so famous that no citation is really necessary. Those without footnotes associated with them have simply not been identifiable with the resources I have been able to devote to the search.[6] Some are specifically noted as regretfully not individually identifiable.[7]

I occasionally interject text quoted or otherwise into the story and these interjections are often identified by vertical lines on each side of the text.

I realize that there are more footnotes than truly necessary. Please feel free to read right past and through them. Many are destined for readers in decades to come when they may prove more useful than to us who are closer to the events described.[8] Some are solely for my own edification. .[9]

I reference Wikipedia[10] to identify a large number of topics. This is a deliberate choice since, they are already satisfactory for an introduction to the person, place or event, and once created are likely to be permanent and will continue to improve over time.

Each volume contains an individual if somewhat abbreviated bibliography focusing on texts that are most relevant to that that volume. For the full bibliography please see Volume 5 – A Century of Wrong.

Space Limitations

In order to keep each of the volumes under ~350-375 pages, some useful material such as chronologies, tables maps and a full bibliography have been allocated to other volumes in the series on a space available basis.

In the same vein, interior space for text is at a premium and I have sometimes chosen to end chapters and begin subsequent ones on the same page rather than take up room in the book with a blank page.

Acknowledgements

I would like to thank my collaborator Joe Ervin for all the support he has provided to my efforts. Ms. Rae Jean Sielen and her staff at Populore Publishing[11] have assisted with manuscript repair before I publish. All, and I do mean all, errors and idiosyncrasies in formatting and footnoting are mine. I am particularly fortunate to have contacted the Churchill Archive Centre[12] and obtained the assistance of the archivist, Ms. Katharine Thomson, for whom I tried mightily to solve my own research problems but who always resolved them most handedly when I could not. I have also been fortunate to have made the acquaintance of Dr. David Biggins, and Neville Constantine along with Rob Davidson and all of the members of the community that supports angloboerwar.com,[13] which is a comprehensive repository and forum of information on this time period.

[6] My primary source for individuals who do not have Wikipedia pages has been https://www.angloboerwar.com/name-search.

[7] For those who are footnoted, I've made a concerted effort to include an identifying footnote associated with the first time that individual is mentioned. Sometimes I've had to defer their annotated reference because the notable event is being discussed by the author, is revealed in the annotation. This always makes it more difficult to go back and identify the person or event but is done to not interrupt the author's train of thought or steal his thunder.

[8] Footnotes and endnotes both have Arabic numerals. This was unavoidable since using Roman numerals for endnotes became unwieldly. In order to make some distinction I have used different fonts and font sizes to distinguish the notation type. It becomes easier to follow once the number sequencing diverges farther along into the book.

[9] Everything about series has been a heuristic endeavour to simply find a solution to the problem at hand along the way. This is quintessentially a first edition, with all the flaws that moniker entails. Hopefully, a second edition will correct some of its flaws.

[10] Wikipedia, the free encyclopedia

[11] Populore Publishing Company | LinkedIn

[12] The Churchill Archives Centre – Churchill College (cam.ac.uk)

[13] Anglo Boer War - Home

3 Page Introduction

The Steamship Age

By the 1880s the age of sail had largely passed and it was possible to transport large amounts of supplies, troops and ammunition by steamship, in relatively short order to project British Imperial power across the globe on an industrial scale. This was the first time that British Colonial troops from Canada and Australia, among other colonies, were transported en-mass to fight on behalf of the Empire. They were employed to fight and defeat two independent Republics in order to destroy their sovereignty and then continued to maintain that hegemony until South Africa left the British Commonwealth in 1961.

The example set by this logistical achievement; to transport a half million man army to a foreign continent proved to be a trial run for what would occur fifteen years later in 1914 at the onset of the First World War.

Weapons, Transportation and Tactics

As demonstrated in the United States Civil War of 1861-1865 the use of railroads was effective in bringing large quantities of supplies and troops to the vicinity of engagements but Southern Africa is a vast country and could only be navigated fully by mobile units. Without motorized transport battles would be fought in a classical manner by infantry and cavalry without mechanization.

The lessons of the United States Civil War regarding the firepower available to repeating arms[14] had not been incorporated into British military doctrine. In the interim, the lethality of combat zones had increased dramatically with the introduction of cased cartridges containing cordite and nitrocellulose based smokeless power, employed in magazine fed bolt action rifles like the Mauser and Lee-Enfield.

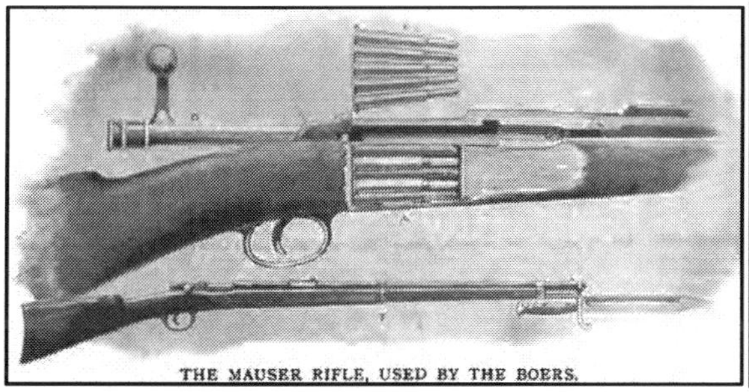

The Mauser Rifle, Used By the Boers.[33]

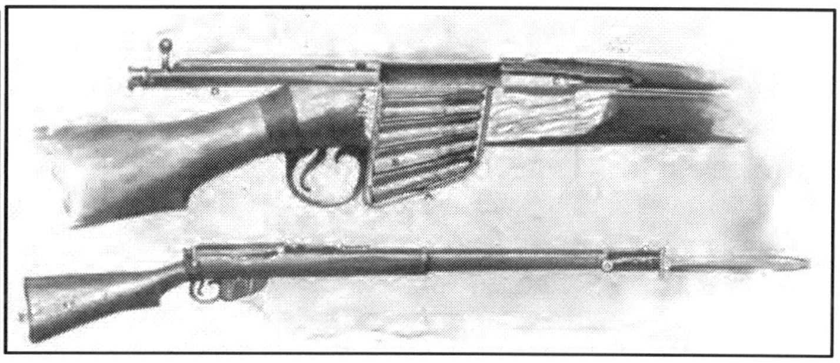

The Lee-Enfield Rifle, Used By The British Troops.[34]

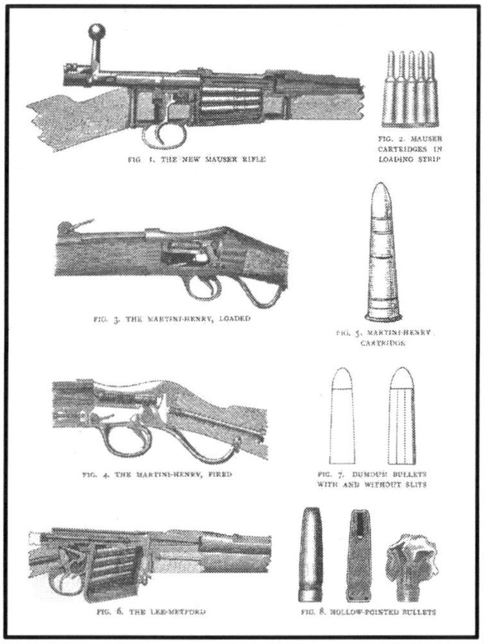

Various contemporary rifles.[35] [36]

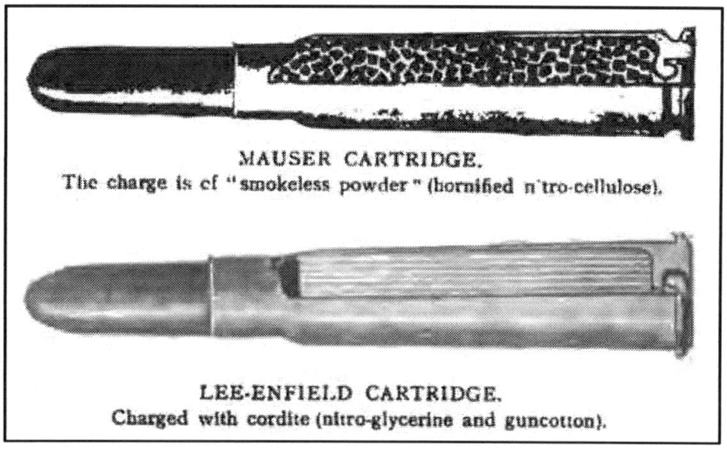

Difference between British and German smokeless powder.[37]

[14] See the: Henry rifle - Wikipedia

These high velocity weapons extended the average effective battlefield range from roughly 200 yards using muzzle loading black power rifles, to over 1,000. The Boer armies were well aware of the effectiveness of their small arms both in their range and rounds per minute per man during infantry combat. In the British army marksmanship was not as high a priority. Their infantry were still drilled in essentially massed volley fire rather than individual aiming at targets like their Boer opponents.

These doctrinal differences were also evident between the commanders on each side and the British were consistently at a disadvantage while the Boer tactics contributed significantly to the success of their armies in the early stages of the conflict.

It will be worthwhile to keep in mind that the outlook of these soldiers towards battle is still imbued with an 18th and 19th century romanticism about combat. Also, the British Regular Troops, were a professional army, it was their job to do these things the way they had been trained to.

On many occasions infantry units are asked to close 800 yards of open ground in the face of withering rifle fire before fixing their bayonets for the final charge. The initial pyric victories achieved with high casualties attest to the discipline and valor necessary to conduct these assaults. Portable machine guns available in the First World War will increase the lethality of direct assaults but they will still be employed. e.g. the Somme.

It is clear that all the British narrators have received a classical education. They quote Latin phrases with ease. They write in French. Their tales are replete with references to Shakespeare and Dickens. They are serving their county at the apogee of their Queen's Empire, in the age of Rudyard Kipling. Their courage is as willing and strong as that displayed in Tennyson's "The Charge of the Light Brigade."[15]

As successful and admirable as this courageous outlook will prove in this conflict; it will come at a high cost of mortality. None of this exuberant desire for feats of valor will be tempered by their commanders who strategically and tactically put them in the position where; "fix your bayonets, and die like men." was one of only two remaining options.[16]

It is also possible to recognize that 20th and 21st century armies and their warriors, while just as willing and just as courageous, do not emphasize the same romantic, heroic ideal regarding warfare and particularly of individual combat. The modern battlefield is just too lethal to engage in daring charges to try and carry the day. Modern military operations are more corporate. Danger is to be faced and heroic deeds may occur but the Forlorn Hope[17] is no longer rationally considered as a tactical option in modern professional military doctrine. These soldiers, on both sides, are entering this conflict with 19th century ideas about the lethality of the encounters they will be confronted with. In Ian' Hamilton's March, Churchill, with Romanesque sardonicism, can sincerely express, 'Qui vivra verra,' [Who(ever) lives will see.] about the campaign yet to be decided.

For the Boer armies the call to battle was more Old Testament. Starting from the Cape they had been forced to trek from place to place for over fifty years, longer than the Israelites wandered in the wilderness. Their goal was to find a refuge where they could thrive along with their own culture, pretty much in isolation. They did not treat African peoples well and would not incorporate them into their society but in large part, like many religious bands looking for a place to settle, they just wanted to be separate and apart from other societies. Much like the Pilgrims to New England and the American Amish.

The Dutch Reformed Church,[18] was the dominant religious denomination of the Boer Republics. It had its roots in Calvinist doctrine; one that emphasized the sovereignty of God and the authority of the Bible. It also employs the concept of covenant to describe the way God enters into fellowship with people in history. The concept of

[15] See: Charge of the Light Brigade - Wikipedia

[16] At the conclusion of the Battle of Nicholson's Nek: "They were absolutely hemmed in, and what had commenced as a skirmish seemed about to become a butchery. The grim order was passed round—" Faugh-a-Ballaghs, (Faugh an Beallach is a battle cry of Irish origin, meaning "clear the way"). fix your bayonets and die like men!" There was the clatter of steel, the moment of suspense, and then the "Cease Fire" sounded. [by bugle] Again, and again it sounded, but the Irish Fusiliers were loath to accept the call, and continued firing for many minutes. Then it was unconditional surrender and the men laid down their arms ...' Col. Carleton and Walter Adye were both among the 954 officers and men to be taken prisoner that day. It is said that the officers of the Royal Irish Rifles were 'so exasperated at the exhibition of the white flag that they set to work and smashed their swords rather than give them up."

[17] A forlorn hope is a band of soldiers or other combatants chosen to take the vanguard in a military operation, such as a suicidal assault through the kill zone of a defended position, or the first men to climb a scaling ladder against a defended fortification, where the risk of casualties is high. The term comes from the Dutch verloren hoop, literally 'lost heap.' Forlorn hope - Wikipedia

[18] Dutch Reformed Church - Wikipedia

covenant is so prominent in Reformed theology that Reformed theology as a whole is sometimes called "covenant theology." Amongst the Pilgrims in America the dividing line of theology was queried with the question as to whether one was under a covenant of grace or a covenant of works with the Calvinist's being under the predestination based covenant of grace.[19] This doctrine became a crucial part of Afrikaner nationalism in South Africa.[20]

In effect they had indeed lived out a second Israelite trek into a promised land, and one filled with gold and diamonds as well. To their lights they had defeated and displaced their opponents by divine providence just as the Israelites had the Amalekites.[21] They intended to hold their land against all comers and at all costs.

Both of these warrior cultures will meet in the last large scale cavalry and infantry war ever fought. The British are fresh off of their victory over the Mahdist army[22] eliminating one of the last enclaves of the slave trade in Soudan. The British unfortunately prevailed their industrial way of war and traditional tactics that they thought it would be 'walk over' to defeat the Boers and extend their Empire.

The Boers went to war like Gideon and the British emulated Wellington. Both cultures were imbued with the ethos of European horse warrior culture with all adventure, romance and the nobleness that concept entails.

A man on a horse, on both sides of this conflict, is heir to Alexander, Saladin, and Bonaparte. Their infantry is heir to the 300 at Thermopylae, Caesars legions and Cromwell's Roundheads. One gets the sense that, however dangerous, war is an adventure, something chivalrous and noble. The vanquished are congratulated on their courage and tended to. Modern war is very different: A soldier driving an armoured vehicle may be king of the road but it is a stretch to see oneself as a knight errant.[23]

Cartography

Many place name spellings have changed over the years since the original publications. I have cited the modern spellings where appropriate and left the respective authors' spelling as originally published. It is sometimes difficult to know what cartographic feature a writer is referring to. As the most egregious instance I've come across, the Elands River may refer to several rivers in South Africa:[24] The same holds true for the Crocodile River, Buffalo River and the Klip River amongst others. For Instance: "This article is about the river of the Crocodile River (East) system. For the tributary of the Crocodile River (West), see Elands River (North West). For other places with the same name, see Elands River (disambiguation)."

I have been regularly searching for a town on Google Earth, only to end up on the other side of the African continent because the original town name has been changed and there's now another one somewhere else. Two: Klipstapel's, two Wonderkop(s). Definitely two 'Sand River'(s) etc..[25] [26]

[19] A second covenant, called the covenant of grace, is said to have been made immediately following Adam and Eve's sin. In it, God graciously offers salvation from death on condition of faith in God. This covenant is administered in different ways throughout the Old and New Testaments, but retains the substance of being free of a requirement of perfect obedience. Calvinism - Wikipedia

[20] Reformed Churches in South Africa - Wikipedia

[21] Amalek - Wikipedia

[22] Mahdist War - Wikipedia

[23] Knight Rider notwithstanding. "Knight Rider is an American action crime drama television series. (1982-86) The show stars David Hasselhoff as Michael Knight, a sleek and modern crime fighter assisted by KITT, an advanced, artificially intelligent, self-aware, and nearly indestructible car. Knight Rider (1982 TV series) - Wikipedia

[24] Elands River - Wikipedia Disambiguation.

[25] To quote Kurt Vonnegut.

[26] See: List of Rivers in South Africa: List of rivers of South Africa - Wikipedia
List of settlements in South Africa: List of settlements in South Africa - Wikipedia
List of Commandos by Country: Anglo Boer War - Boer forces

Maps & Tables

Multiple maps are provided of various areas. Each has details or nomenclature that are not reproduced in others.

SA Political

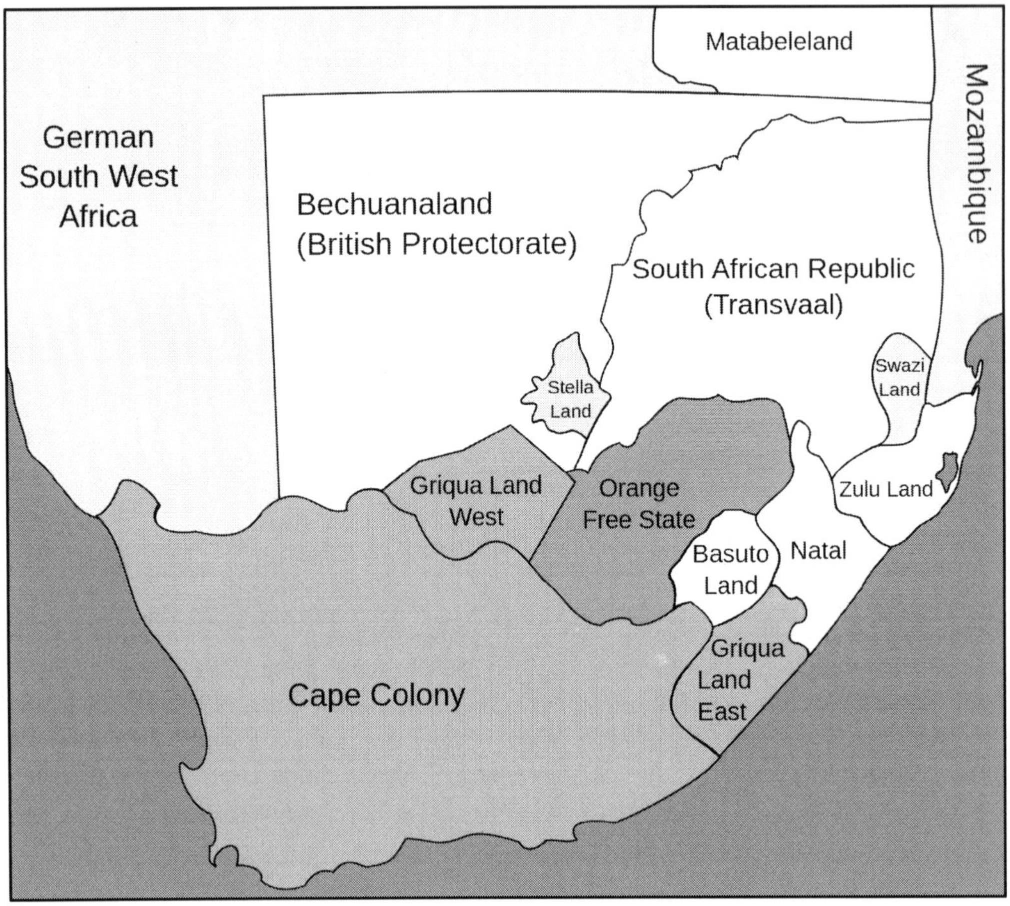

South Africa 1885.[38]

Relative Size

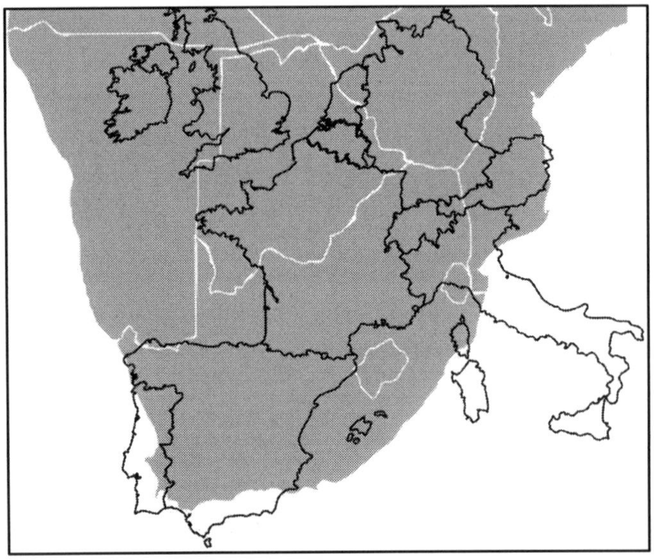

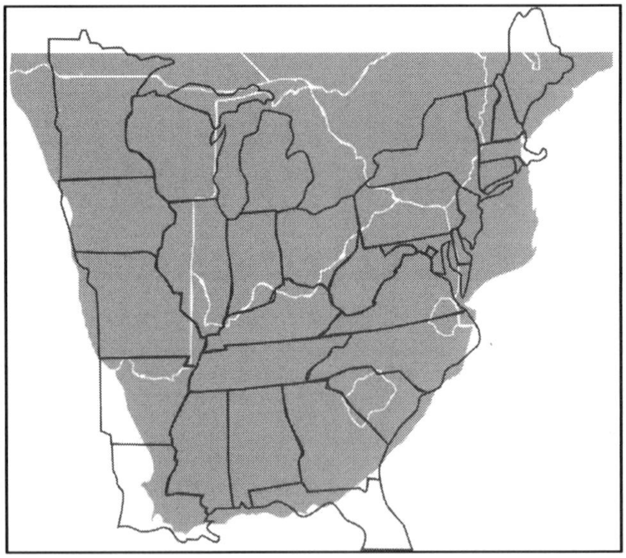

Relative size of Southern Africa compared to Europe and North America.[27]

[27] Adapted from an original appearing in Hiley and Hassell's, The Mobile Boer, Grafton Press, New York, (c. 1902) "The above map, comparing South Africa with Europe was published by "C.I." as long ago as Jan 6th, 1900."

Theatre of Conflict

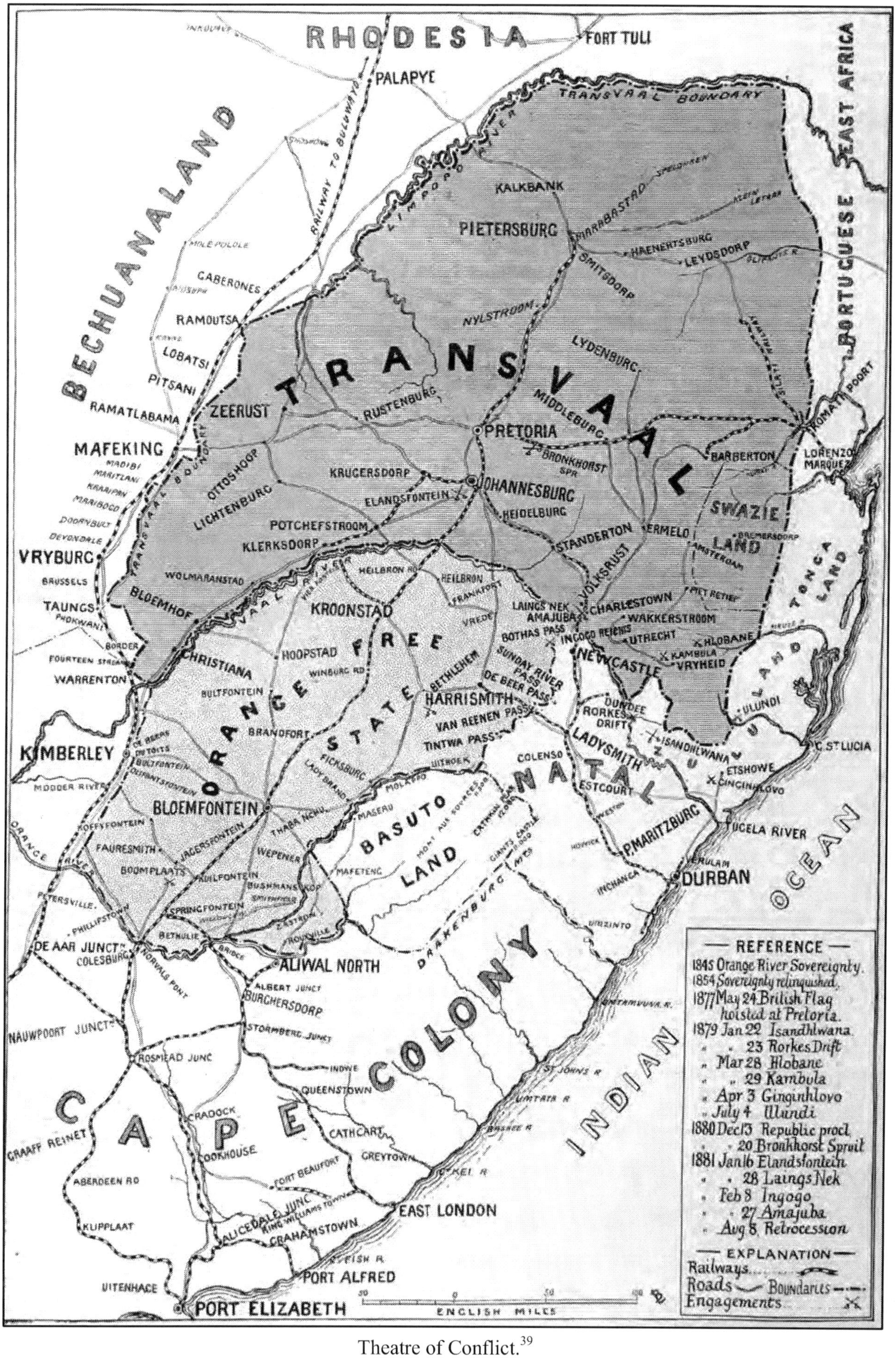

Theatre of Conflict.[39]

Tables of Distances

Table of Distances	Miles	Kilometers
By Sea		
London to Capetown	6,181	9,947
Capetown to Port Elizabeth	428	689
Port Elizabeth to East London	131	211
East London to Durban	253	407
Durban to Delagoa Bay	300	483
Capetown to Durban	812	1,307
By Rail	Miles	Kilometers
Capetown to De Aar Junction	501	806
Capetown to Kimberley	647	1,041
Capetown to Vryburg	774	1,246
Capetown to Mafeking	870	1,400
Capetown to Bulawayo	1,361	2,190
Capetown to Naauwpoort	570	917
Capetown to Norval's Pont	628	1,011
Capetown to Bloemfontein	750	1,207
Capetown to Johannesburg	1,014	1,632
Capetown to Pretoria	1,040	1,674
Port Elizabeth to Naauwpoort	270	435
Port Elizabeth to Bloemfontein	450	724
Port Elizabeth to Pretoria	740	1,191
Durban to P'Maritzburg	70	113
Durban to Ladysmith	189	304
Durban to Harrismith	249	401
Durban to Laing's Nek	301	484
Durban to Johannesburg	483	777
Durban to Pretoria	511	822
Lorenzo Marques to Komati Poort	58	93
Lorenzo Marques to Johannesburg	395	636
Larenzo Marques to Johannesburg	395	636

Table of Distances by Sea and Rail.[28]

[28] The Boer war: its causes and its interest to Canadians by Emerson Bristol Biggar. 3rd.Ed. 1899. "John Noble's 'Handbook of the Cape and South Africa for 1893,' gave the area and white and colored population of South Africa, as [shown]."

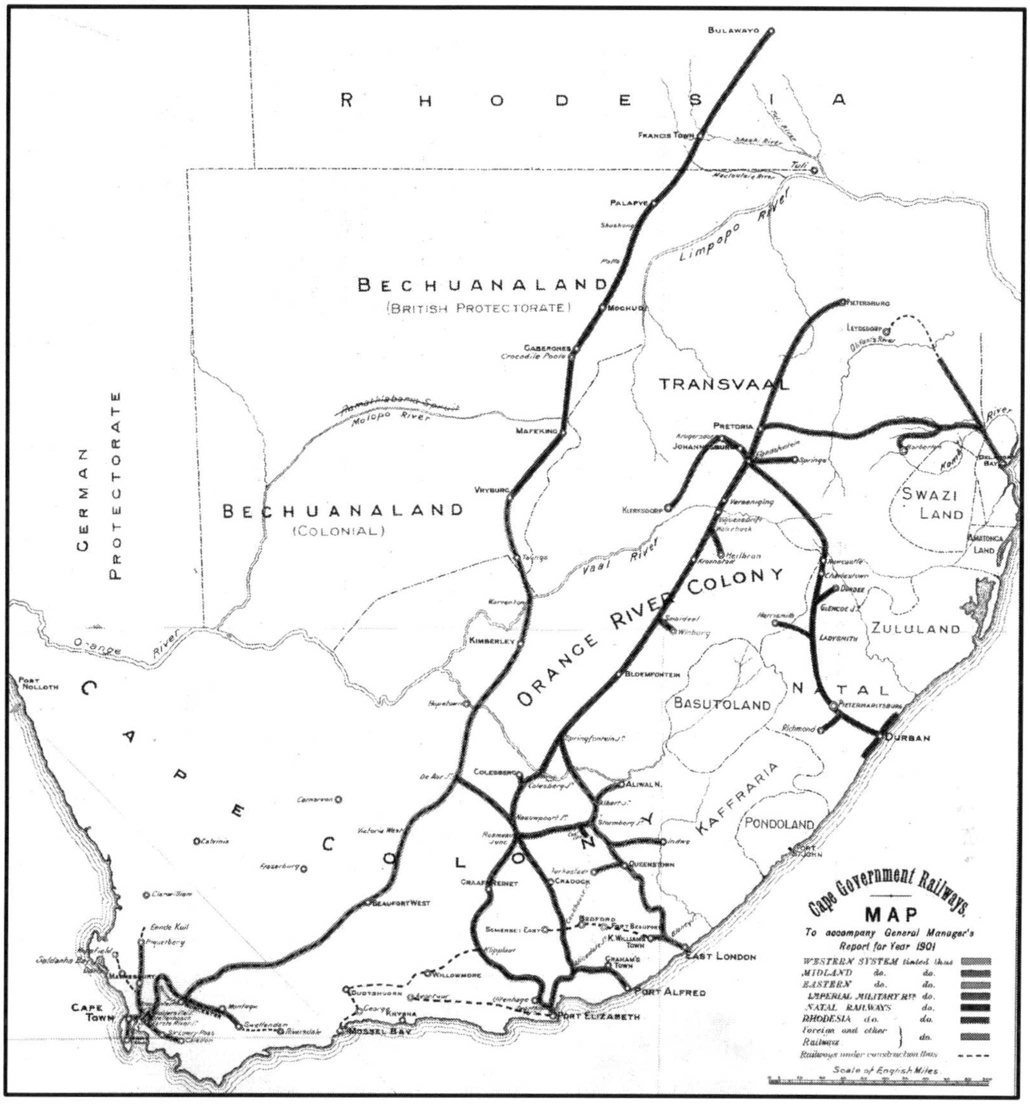

South African Railway System 1901.[40]

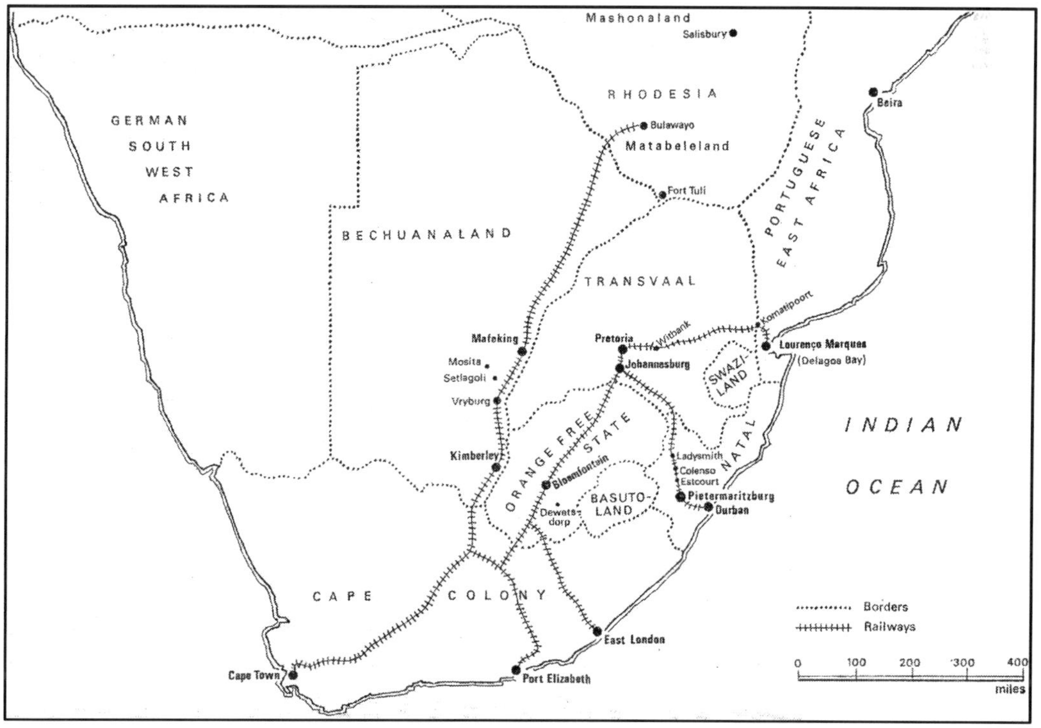

South Africa Railroad Network circa 1899.[41]

General Maps

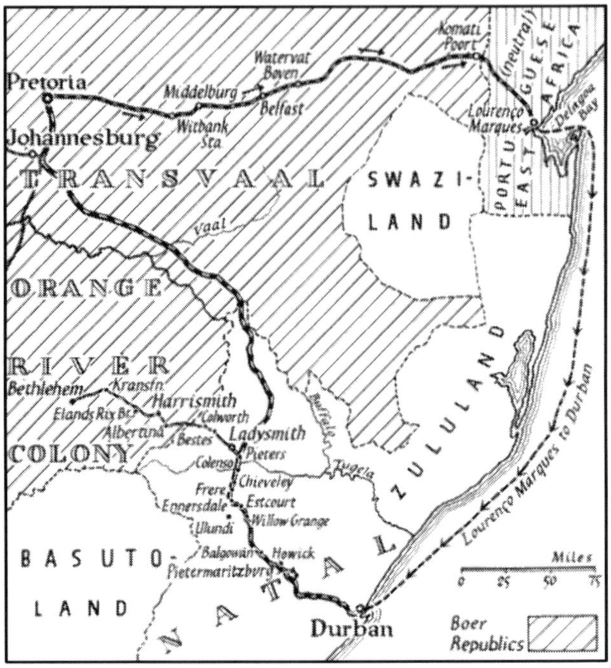

General Map of SA showing route taken by Churchill and also by Haldane, Le Mesurier & Brockie.[42] A second version of the territory.[43]

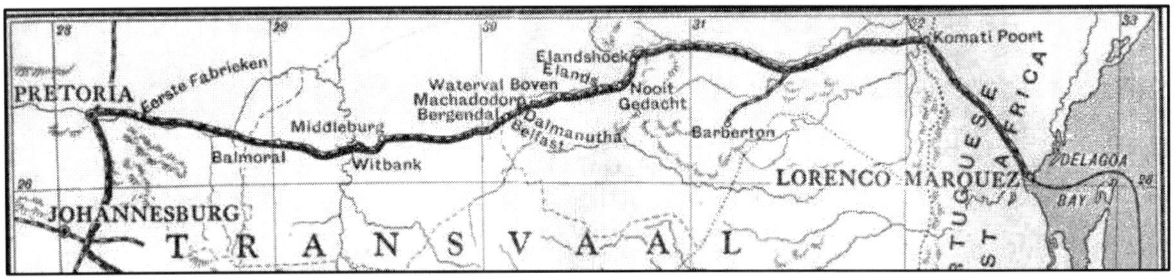

General Map of SA showing route taken by Churchill and also by Haldane, Le Mesurier & Brockie.[44]

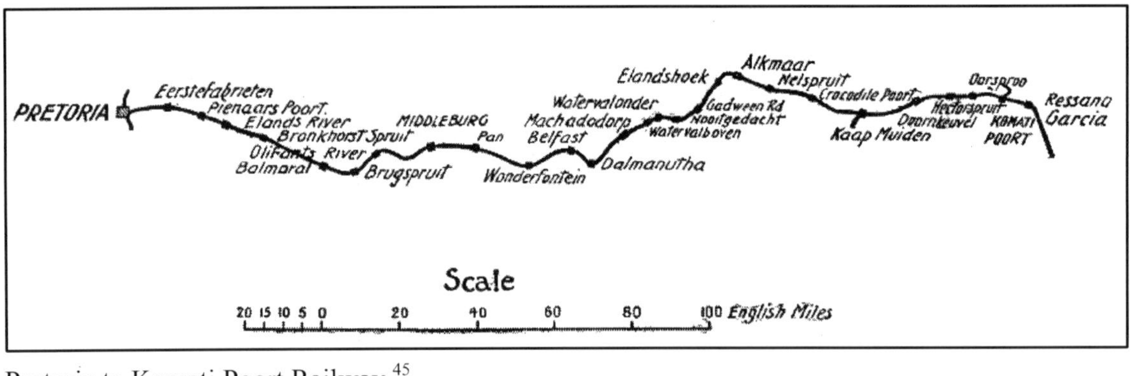

Pretoria to Komati Poort Railway.[45]

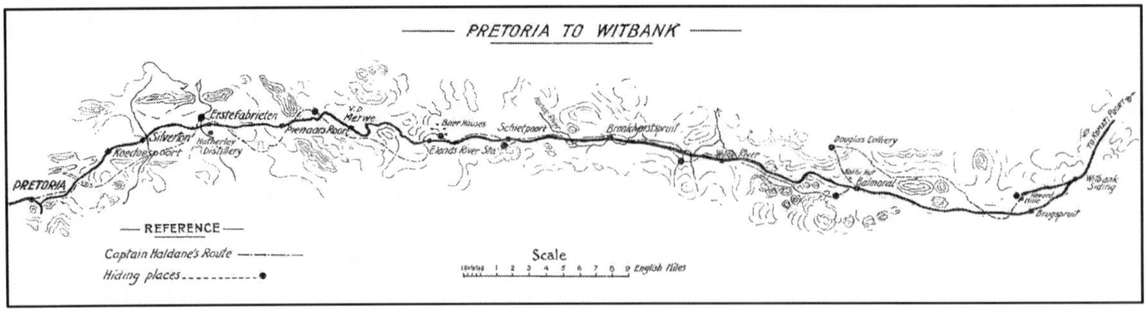

Pretoria to Whitbank[46]

Northern Natal 1

Northern Natal.[47] Note Volksrust and Majuba at top.

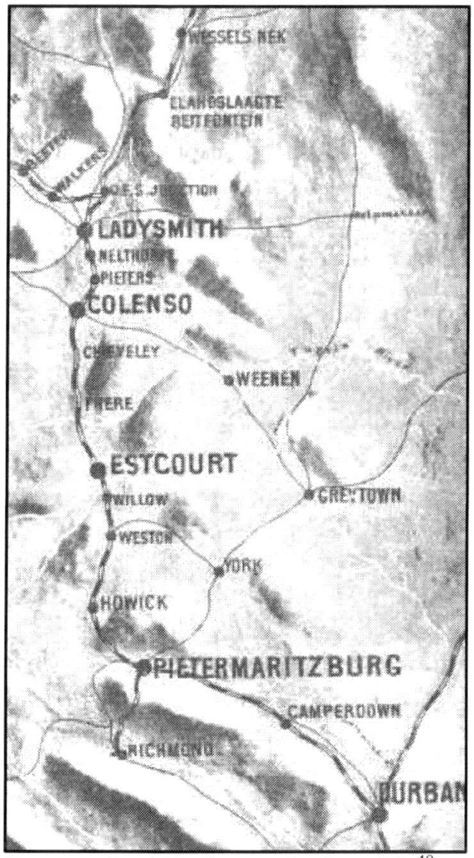

Country from Durban to Ladysmith.[48]

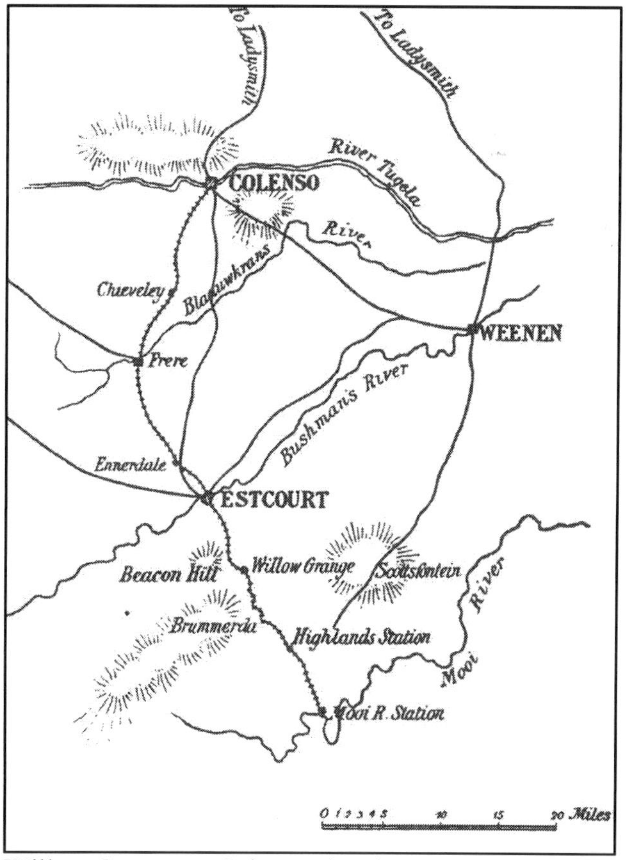

Willow Grange to Colenso showing Joubert's Invasion of Natal.[49]

Glossary

Assegai	A light spear or javelin
Bad	A spring
Berg	Mountain (the plural is formed by adding en)
Biltong	Dried meat
Bloemfontein	Flower fountain
Boer	Literally a farmer
Bosch	Thicket or bush
Boschveldt	An open plain covered with bush
Burg	A town
Burgher	A male citizen with full political rights
Bushveld	Low country, in contrast to high plains, the high veld
Commando	A Boer military force of any size.
Donga	Ravine, a gully; the bank of a river.
Dorp	A village
Drift	A crossing point of A river
Fontein	Fountain or spring, as Bloemfontein (fountain of flowers)
Hoek	A corner; or secluded valley,also used for pass and ravine
Inspan	To harness up Outspan or uitspan, to unharness or untether
Karroo	Desert land covered with scrubby plants
Klip	Rock
Kloof	A valley or ravine
Kop or Kopje	A hill or peak
Kraal	Collection of native huts; also a cattle enclosure
Krantz	A cliff or precipice
Laager	A camp and for defense; often made by A circle of wagons
Laagte	Low land or valley, as Elandslaagte (Eland valley)
Landdrost	Local governor
Modder	Mud
Nek	Neck, applied to a depression between two mountains
Outspan	To unharness
Pan	Bed of a dried-up salt marsh
Poort	A passage between mountains
Rand	Edge, Margin, Also Highland;
Rooinek	Term of contempt applied to British by Boers
Sangar	A stone brestwork
Schantze	A heap of stones to protect A soldier against rifle fire
Sluit	A ditch, An artificial water course.
Span	A team of animals usually horses or oxen
Spruit	A small river, sometimes dry
Stad	A town or city
Transvaal	The territory across the Vaal River
Trek	To journey
Uitlander	An outlander or foreigner, usually British
Unspan	To unharness, to stop
Vaal	Valley
Veld or Veldt	(Literally a field), an extent of country
Voortrekkers	Pioneers
Witwatersrand	The White-water–highlands - See Rand
Zarp	Transvaal Police (Zuid Afrikaansche-Republiek-Politie)

Glossary Sources.[50 51 52]

Historical Background

Eurocentric Introduction

I am including a few Eurocentric observations as a summary.[29]

In the 17th Century (~1650) the Dutch controlled a trading network that encompassed the entire known world. The economic and naval competition that this engendered, primarily with England resulted in a century long struggle for preeminence that the British eventually won by virtue of the British adoption of the technology available as a result of the Industrial Revolution and the Dutch not being able to catch up quickly enough.

For the Dutch, despite starting out with a more lucrative overseas trading network, better organization in the form of corporations, extensive wind power manufacturing processes, and more advanced technology in the form of cargo/warships, they were overhauled by their English competitors during a seventy-five-year conflict involving four major wars.[30]

During this period the English adopted Dutch methods, eventually even to the point of co-opting their head of state, William of Orange into their redefined monarchical/parliamentary system during the Glorious Revolution of 1688. By adopting the Dutch empire as an appendage of the British, the Dutch were relegated to being a second-tier empire throughout the remainder of the Colonial period. One factor in this was the failure of the Dutch to incorporate the advances made possible by the Industrial Revolution; particularly the application of Steam Power to industrial processes of manufacturing. It had been plentiful wind power had enabled the Dutch to establish a lead in manufacturing efficiency and as a society they continued using this method of powering their economy for too long since they did not have ready access to the coal necessary to power steam engines.

South African Background

Readers who are familiar with South African history can easily skip this background summary. I came to this topic without a native understanding of the history of the region and it took some time to assimilate enough information to obtain a working knowledge of the region's history. I felt that other Eurocentric readers might benefit from an introduction to the historical background of South Africa.

The Second Boer War,[31] 1899-1902, was the culminating conflict between the British Empire and the two Dutch speaking Boer republics: The Orange Free State and the South African Republic (Transvaal).

European settlement in South Africa began with the Dutch establishing a trading post and waystation near the Cape of Good Hope in 1652. By occupying this strategic point, the sea journey to and from India and the Far East was secured. This colony grew and also changed hands several times between then and the end of the Napoleonic Wars when it was ceded to the British in 1814.

The Dutch residents did not assimilate with their new British administration. British colonies were expected to be self-supporting through internal taxation and this was novel to the Dutch. One of the primary causes of friction was the policy of the British authorities towards slavery in South Africa. In 1828, the Cape governor declared that all native inhabitants but slaves were to have the rights of "citizens", in respect of security and property ownership, on parity with the settlers. In 1828, the Cape Governor declared that all native inhabitants were to have equal treatment under the law for all, regardless of race. In 1834, the British went further and abolished slavery in the Empire altogether. The Boers opposed this. This resentment along with that associated with taxation, culminated in a mass migration of substantial numbers of the Boers into the hitherto unexplored frontier, to get beyond the control of British rule. The migration became known as the Great Trek[32] into the hinterland of South Africa in order to maintain their independence and conduct their own affairs.

This was not a move in pursuit of immediate greater material wealth or safety. It was no gold rush. They abandoned their farms, businesses and most of their possessions when they departed. They left behind the security, schools, infrastructure, and business opportunities that were already extant in the colony.

[29] My Eurocentric Introduction ballooned into a fifty page behemoth that I dare not impose on my readers at the outset. It became an indictment so universal in scope and stern in its explication with respect to both Western and non-Western civilizations that it will be moved to the back of Volume 9.

[30] See: First Anglo-Dutch War - Wikipedia Second Anglo-Dutch War - Wikipedia Third Anglo-Dutch War - Wikipedia Fourth Anglo-Dutch War - Wikipedia.

[31] Second Boer War - Wikipedia

[32] Great Trek - Wikipedia

"A Traveler's Difficulty: This unwieldy wagon is crossing the ford and has not been well managed. The driver does not know which way to make his turn without throwing his wagon into the stream. The oxen stand comfortably in the cool water waiting for the command to 'trek.'"[53] Also: "Dutch Boers Outspanned: The span of oxen has been taken out and placed within the inclosure for the night. The Boer travelers have lit their fire and are cooking their evening meal. They are tall, straight, powerful men, accustomed to life in the open air, to physical exposure, and to some sleep within the wagon, and some on the ground beneath it with their guns always within reach."

Throughout the remainder of the 19th century various republics and other political entities both European and Native engaged in numerous wars, incursions and annexations.[33]

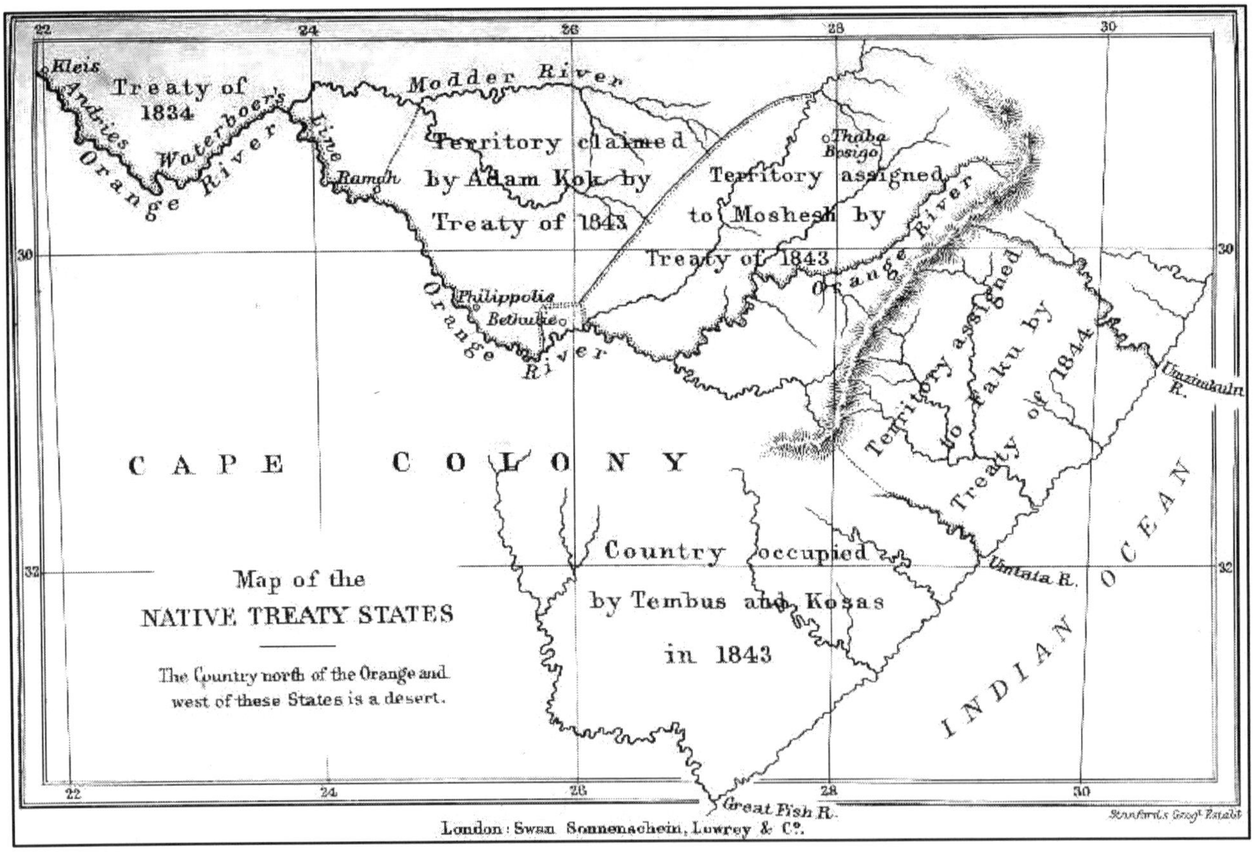

Native Treaty States.[54]

[33] See for example: The Ninth Frontier War (1877–79) Mgolombane Sandile - Wikipedia.

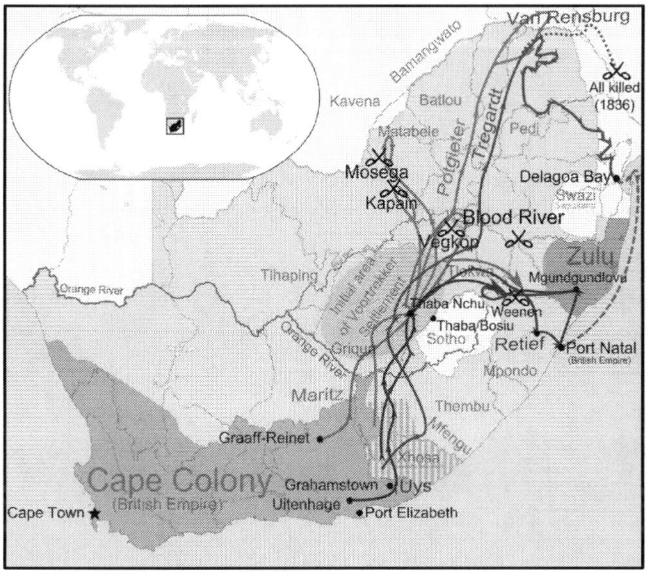

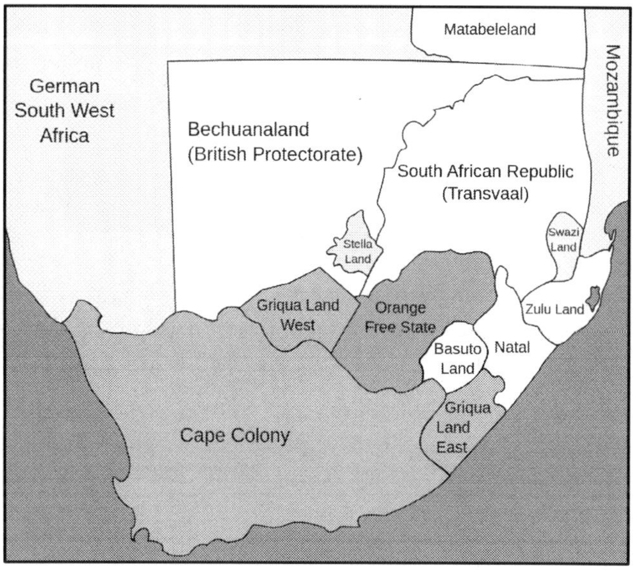

Great Trek Map[34] charting the routes of the largest trekking parties during the first wave of the Great Trek (1835-1840) along with key battles and events. South Africa 1885.[55]

These resulted in a large body of associated treaties and conventions attempting to define the sovereignty of the various states and their relations to one another.[35] One of note is the Natalia Republic,[36] which was founded by the Boers in the interior of Natal from 1839 until 1845. The British, who controlled Port Natal (Durban), did not want to lose this strategic asset or allow the Boers to have access to a seaport and so annexed the territory for the Empire. Following the policy first set forth in the Cape Colony, the conditions of government for Natal were similar to those of the Cape Colony with as much local control as possible but also including the fundamental condition, "that there should not be in the eye of the law any distinction or disqualification whatever, founded on mere difference of colour, origin, language or creed."

Many of the Boers who would not acknowledge British rule trekked once more over the mountains into what became the Orange Free State and Transvaal provinces to seek their freedom and independence. At the end of 1843 there were not more than 500 Dutch families left in Natal.

The Orange Free State obtained its independence from the British in 1854 with the signing of the Orange River Convention.[37] After the discovery of diamonds near Kimberly in 1867, the Free State, the Native Griqua People[38] and the British reached a compromise allowing the British, in 1876, to annex Griqualand West in consideration of a £90,000 payment to the Free State "in full satisfaction of all claims which it considers it may possess to Griqualand West."

The South African Republic[39] (Transvaal or ZAR) was recognized by the British as an independent republic by virtue of the Sand River Convention[40] in 1852. The treaty contained the following provisions:

1. The British government guarantees and grants the emigrant farmers across the Vaal river the right to govern themselves, according to their own laws free from any and all British interference and that the British Government wishes to promote peace, free trade and friendly intercourse with the new country.
2. Britain disclaims any and all alliances with coloured nations to the North of the Vaal river.

[34] Solid Blue: Louis Tregardt's route (1833 to 1838) Dashed Line: Survivors of Tregardt's trek evacuated by sea, 1839 Dots: Van Rensburg's route, after it separated from Tregardt's Red: Hendrik Potgieter's trek, campaign and scouting routes. Green: Gerrit Maritz's route. Burgandy: Piet Retief's route, including missions with his entourage. Blue: Piet Uys's route. Afrikaans: Kaart van die eerste uittog van die Groot Trek (1835-1840), met die roetes van die grootste trekpartye van die Voortrekkers aangedui. Gebaseer op 'n kaart in "The Great Trek" (1985) deur B.P.J. Erasmus en die "Illustrated History of South Africa: The Real Story" deur Reader's Digest (1988), bladsy 115. Author Discott File:Great Trek map full.png - Wikimedia Commons.

[35] Military history of South Africa - Wikipedia
[36] Natalia Republic - Wikipedia
[37] Orange River Convention - Wikipedia (Bloemfontein Convention)
[38] Griqua people - Wikipedia
[39] South African Republic - Wikipedia
[40] Sand River Convention - Wikipedia

3. No slavery [is to] be practised in the country to the North of the Vaal river.

4. Ammunition and arms crossing over the border from the South of the Vaal river shall require a certificate from a British Magistrate, and that no arms or munitions be supplied to the natives by either the British or by the people of the new country.

5. Criminals will be exchanged between the British and the new country and that summonses for witnesses from both sides of the river be backed up by the magistrates at both sides of the river.

6. It is agreed that certificates of marriages will be recognised on both sides of the river.

7. It is agreed that any and all people now residing in British land but being in possession of land in the new country shall have the free right to sell the property and to move freely over the Vaal river.

In 1874 Lord Carnarvon,[41] Secretary of State for the Colonies, who had successfully brought about federation in Canada in 1867, thought that a similar scheme might work in South Africa.[42] Carnarvon, in an attempt to extend British influence in 1875, approached the Boer states of the Orange Free State and the Transvaal Republic and tried to organize a federation of the British and Boer territories, but the Boer leaders turned him down.

Also in 1875, the Transvaal government under the presidency of Thomas Burgers[43] began to plan a railway line linking the Transvaal to the sea by building eastwards towards Delagoa Bay, which was not under British control. While claimed as part of the Transvaal because of the incorporation of the Lydenburg Republic[44] and the Republic of Utrecht in 1860,[45] this area was also occupied by the Pedi (Bapedi) people[46] who occupied the Limpopo district[47] on the Transvaal's northeastern border.

In addition to the railway issue, alluvial gold had been found in the Lydenburg district in 1873 and tensions increased after the Pedi king Sekhukhune[48] refused to pay taxes to the Transvaal government. The Transvaal declared war in May 1876, the outcome of which was that the Transvaal's attack failed, the government became insolvent and president Burgers left the country while Paul Kruger became the head of the government.

Although Sekhukhune made peace,[49] he was not defeated and this fact, together with the shaky financial position of the Transvaal, gave Sir Theophilus Shepstone[50] the pretext he required to annex the republic as the Transvaal, a British colony, in April of 1877.[51]

At this time the Zulu nation[52] was also an intact political entity and a formidable military force in the region under its king, Cetshwayo.[53]

The Transvaal Boers objected to the annexation, but as long as the Pedi and Zulu military power existed, found themselves between two threats; they feared that if they took up arms to resist the British annexation actively, the Zulus and Pedi would take the opportunity to attack.

Also in 1877, Carnarvon chose Henry Bartle Frere[54] as High Commissioner for Southern Africa[55] with a brief attempt to implement the policy of federation.

Shepstone in his capacity as the Governor and British Secretary for Native Affairs in Natal railed against the disruptive effect of allowing Cetshwayo's regime to remain in place. Frere collaborated in this warmongering

[41] Henry Herbert, 4th Earl of Carnarvon - Wikipedia
[42] Anglo-Zulu War - Wikipedia
[43] Thomas François Burgers - Wikipedia
[44] Lydenburg - Wikipedia
[45] Utrecht, KwaZulu-Natal - Wikipedia
[46] Pedi people - Wikipedia
[47] Limpopo - Wikipedia
[48] Sekhukhune - Wikipedia
[49] There was some closure when on February 16th, 1877, the two parties, mediated by Alexander Merensky, signed a peace treaty.
[50] Theophilus Shepstone - Wikipedia
[51] Transvaal Colony - Wikipedia
[52] Zulu Kingdom - Wikipedia
[53] Cetshwayo - Wikipedia
[54] Henry Bartle Frere - Wikipedia
[55] The British office of high commissioner for Southern Africa was responsible for governing British possessions in Southern Africa, latterly the protectorates Basutoland (now Lesotho), the Bechuanaland Protectorate (now Botswana) and Swaziland (now Eswatini), as well as for relations with autonomous governments in the area. The office was combined with that of Governor of Cape Colony from 1847 to 1901, with that of the governor of Transvaal Colony 1901 to 1910, and with that of Governor-General of South Africa from 1910 to 1931. https://en.wikipedia.org/wiki/High_Commissioner_for_Southern_Africa

since he had been promised the position of British Governor-General of a federated southern African dominion, if he was successful.

Together they escalated a small series of border boundary disputes and cross border incidents into a justification for war. Cetshwayo did not wish to fight the British and worked to resolve disputes as they arose. His only defender in Natal with some authority was the bishop of Natal, John Colenso.[56] Bishop Colenso's concern about the misleading information that was being provided to the Colonial Secretary in London by Shepstone and Frere prompted him to speak out on behalf of the Zulus against Boer oppression and official encroachments.

British Prime Minister Benjamin Disraeli's Tory administration in London (1874 to April 1880) did not want a war with the Zulus. "The fact is," wrote Sir Michael Hicks Beach,[57] who would replace Carnarvon as Secretary of State for the Colonies, in November 1878, "that matters in Eastern Europe and India ... were so serious an aspect that we cannot have a Zulu war in addition to other greater and too possible troubles."

However, by this time Frere's plans were well advanced. He had asked for additional forces for defense of Natal and they were granted with the specific caveat that they should not to be used for offensive action.[58]

Frere took deliberate advantage of the length of time required for correspondence to pass between South Africa and London to conceal his intentions from the government or at least defer giving them the necessary information until it was too late for them to act.

In December of 1878 Frere presented Cetshwayo with an ultimatum that the Zulu army be disbanded and the Zulus accept a British resident. This was unacceptable to the Zulus as it effectively meant that Cetshwayo would have lost his throne. Cetshwayo ignored the ultimatum.

Frere ordered Lord Chelmsford[59] to invade Zululand. On January 11th, 1879, British troops crossed the Tugela River; and on January 22nd the disaster of the battle of Isandlwana[60] occurred. Shortly thereafter the result of the battle was reported to London and that was enough for the House of Commons to demand that Frere be recalled; however, in a strange compromise he was censured but asked to stay on.

After the failure of the first invasion of Zululand, it was not politically possible for the British to simply sue for peace and return to the situation ante bellum. Reinforcements were sent and a second invasion was undertaken in June of 1879.

Garnet Wolseley[61] was sent to South Africa to replace Chelmsford in command of the forces in the Zulu War, and as governor of Natal and the Transvaal, and as High Commissioner of South Africa. Wolseley had cabled Chelmsford ordering him not to undertake any serious actions:

"Concentrate your force immediately and keep it concentrated. Undertake no serious operations with detached bodies of troops. Acknowledge receipt of this message at once and flash back your latest moves. I am astonished at not hearing from you."

Chelmsford, knowing he was to be replaced, dodged Wolseley's orders to stand fast until he arrived. Chelmsford knew that his only chance to redeem his reputation was to avenge the failure of his first invasion by defeating the Zulus in battle. Cetshwayo attempted to negotiate peace but his envoys and gifts were rejected.

[56] John Colenso - Wikipedia

[57] Michael Hicks Beach, 1st Earl St Aldwyn - Wikipedia

[58] In January 1879 Hicks Beach wrote to Bartle Frere:

"I may observe that the communications which had previously been received from you had not entirely prepared them (Her Majesty's Government) for the course which you have deemed it necessary to take. The representations made by Lord Chelmsford and yourself last autumn as to the urgent need of strengthening Her Majesty's forces in South Africa were based upon the imminent danger of an invasion of Natal by the Zulus, and the inadequate means at that time at your disposal for meeting it. In order to afford protection to the lives and property of the colonists, the reinforcements asked for were supplied, and, in informing you of the decision of Her Majesty's Government, I took the opportunity of impressing upon you the importance of using every effort to avoid war. But the terms which you have dictated to the Zulu king, however necessary to relieve the colony in future from an impending and increasing danger, are evidently such as he may not improbably refuse, even at the risk of war; and I regret that the necessity for immediate action should have appeared to you so imperative as to preclude you from incurring the delay which would have been involved in consulting Her Majesty's Government upon a subject of so much importance as the terms which Cetshwayo should be required to accept before those terms were actually presented to the Zulu king." Colenso, Frances E. (1880). History of the Zulu War and Its Origin. Assisted by Edward Durnford. London: Chapman & Hall. Pp 260-262 Anglo-Zulu War - Wikipedia. Footnote # 27.

[59] Frederic Thesiger, 2nd Baron Chelmsford - Wikipedia

[60] Battle of Isandlwana - Wikipedia

[61] Garnet Wolseley, 1st Viscount Wolseley - Wikipedia

At the very time Wolseley was riding north from Durban, Chelmsford was preparing to engage the enemy. Wolseley's efforts to reach the front in time were in vain.

The British and Zulu forces clashed on July 4th, 1879, at the battle of Ulundi[62] during which the British army broke the military power of the Zulu nation by defeating the main Zulu army and immediately afterwards capturing and burning the royal kraal of oNdini. Cetshwayo was captured in late August and exiled to London, where he would remain for three years.

I have gone into detail regarding this period of South African history because we will see this pattern of local governors and commanders provoking conflict, exceeding their authority, then asking for more troops for "defense" and subsequently using those troops to provoke a war repeated by Alfred Milner and George White during the run-up to the Second Boer War.

Although the British had first condemned the Transvaal War against the Pedi king Sekhukhune, after the defeat of the Zulus, Wolseley engaged in the Second Sekhukhune War; during which an army of 2,000 British soldiers and Boers, and 10,000 Swazis[63] defeated Sekhukhune's forces and in December of 1879 he was captured and imprisoned but later released after the First Boer War when the Transvaal regained its independence.[64]

In late April of 1880 the United Kingdom held a general election and the Conservative ministry of Benjamin Disraeli was defeated by the Liberal party leading to the second ministry[65] of William Gladstone.[66]

Gladstone was convinced that Disraeli's policies had alienated the Boers but did not appear to be making any effort to reverse or alleviate the annexation. Representative government would be restored but within the framework of the Empire itself and not as an independent country.

Major-General Sir George Pomeroy Colley,[67] took over as Governor of Natal, Transvaal, and High Commissioner of South Africa as well as Military Commander from Wolseley in July 1880. While Colley knew many of the senior Boers, he did not travel to the Transvaal and instead he relied on reports from the Administrator, Sir Owen Lanyon,[68] who had no understanding of the Boer mood or capability.

With the defeat of the Zulus, and the Pedi, the Transvaal Boers were able to give voice to the growing resentment against the 1877 British annexation of the Transvaal and complained that it had been a violation of the Sand River Convention of 1852, and the Bloemfontein Convention of 1854.

On December 13th, 1880, the members of the last Volksraad were summoned to a meeting at Paardekraal, where authority was placed in the hands of a triumvirate comprising Paul Kruger,[69] Piet Joubert[70] and Marthinus W. Pretorius.[71] They declared Heidelberg their seat of government and hoisted the Vierkleur flag there on December 16th.[72]

[62] Battle of Ulundi - Wikipedia

[63] The Swati are a Bantu ethnic group in Southern Africa, predominantly inhabiting Eswatini, South Africa's Mpumalanga province, and Mozambique. Swazi people - Wikipedia

[64] On 3 August 1881, the Pretoria Convention was signed, which stipulated in Article 23 that Sekhukhune would be released. On 13 August 1882, Sekhukhune was murdered by his half-brother Mampuru II, who claimed to be the lawful king. Mampuru was captured by the Boers, tried for murder, and hanged in Pretoria on 21 November 1883.

[65] Premierships of William Ewart Gladstone - Wikipedia Foreign policy of William Ewart Gladstone - Wikipedia

[66] William Ewart Gladstone - Wikipedia

[67] George Pomeroy Colley - Wikipedia

[68] Owen Lanyon - Wikipedia

[69] Paul Kruger - Wikipedia

[70] Piet Joubert - Wikipedia

[71] Marthinus Wessel Pretorius - Wikipedia

[72] First Boer War - Wikipedia And: South African Wars (1879–1915) - Wikipedia

Annual Demonstration at Paardekraal Monument.[73]

The Paardekraal Monument enclosing original cairn.[56]

Gathering at the Paardekraal Monument.[57]

Gathering at the Paardekraal Monument.[58]

The fighting opened with an irregular attack, after giving short notice under flag of truce, to a British regiment on the march at Bronkhorstspruit.[74]

The British were subsequently defeated in several more engagements including Laing's Nek,[75] Schuinshoogte,[76] and on February 27th, 1881 at Majuba Hill,[77] where General Colley was among the many fallen soldiers. While the British would in other circumstances have sent more troops and continued fighting, Gladstone concluded that Great Britain had no strategic interest in the colony, nor did the territory possess any particularly valuable resources, and, too, that any further attempt to control the colony politically would require substantial troop reinforcements, and it was likely that the war would be costly, messy, and protracted. Unwilling to get bogged down in a distant war, the British government asked for a truce.

Sir Evelyn Wood signed an armistice and a peace treaty with Paul Kruger in March of 1881 and there was a final peace treaty, the Pretoria Convention;[78] in it the British agreed to complete Boer self-government in the Transvaal under British suzerainty.

[73] Image from:War in SA, Harding 1899. The Paardekraal Monument is situated in Krugersdorp. It commemorates the original site of the vow made by Transvaal Boers prior to the First Boer War, when they vowed to regain their independence. The original monument was a cairn made of thousands of stones representing the gathered burgers' vow. This official monument enclosed the original stone cairn and was built in 1890 by the South African Republic. Paardekraal Monument - Wikipedia

[74] At the first battle at Bronkhorstspruit on 20 December 1880, Lieutenant-Colonel Philip Anstruther and 120 men of the 94th Foot (Connaught Rangers) were killed or wounded by Boer fire within minutes of the first shots. Boer losses totaled two killed and five wounded. This mainly Irish regiment was marching westward toward Pretoria, led by Lieutenant-Colonel Anstruther, when halted by a Boer commando group. They were halted when they approached a small stream called the Bronkhorstspruit, 38 miles from Pretoria. Its leader, Commandant Frans Joubert, (brother of General Piet Joubert), ordered Anstruther and the column to turn back, stating that the territory was now again a Boer Republic and therefore any further advance by the British would be deemed an act of war. Anstruther refused and ordered that ammunition be distributed. The Boers opened fire and the ambushed British troops were annihilated. In the ensuing engagement, the column lost 56 men dead and 92 wounded. With the majority of his troops dead or wounded, the dying Anstruther ordered surrender. Battle of Bronkhorstspruit - Wikipedia and First Boer War - Wikipedia

[75] Battle of Laing's Nek - Wikipedia

[76] Battle of Schuinshoogte - Wikipedia

[77] Battle of Majuba Hill - Wikipedia

[78] Pretoria Convention - Wikipedia and: Pretoria Convention - Wikisource, the free online library

Paul Kruger, Evelyn Wood, and Redvers Buller at the second conference between the British and Boers in March of 1881 at O'Neils Farm at the foot of Majuba.[59]

The Boers accepted the Queen's nominal rule and British control over external relations, African affairs, and native districts. This Convention ending the conflict was signed in Pretoria on August 3rd, 1881, subject to ratification by the Volksraad within three months. The Volksraad first raised objections to a number of the clauses of the treaty, but did eventually ratify the version signed in Pretoria, after Britain refused any further concessions or changes to the treaty.

The Pretoria Convention was superseded in 1884 by the London Convention,[79] which provided for full independence and self-government although still with British control of foreign relations.[80] In addition the London Convention stipulated that the South African Republic had the right to enter into a treaty with the Orange Free State without approval from the British. Any other treaty with any other nation would require approval from the British.

Practically simultaneously Basutoland, modern Lesotho,[81] was able to maintain its independence. In the 1860s and 1870s many of the residents had gone to work on constructing railroads and in the diamond mines near Kimberley. With their earnings they had purchased guns for defense of their flocks and homes. When the Cape Colony passed a "Peace Preservation Act" as a means of forcing disarmament, they refused. The resulting conflict known as the Basuto Gun War[82] followed and lasted from September 1880 to April 1881. Despite the expenditure of considerable effort, the British were unsuccessful in enforcing their policy and the conflict ended by treaty with the Basuto maintaining their independence from the Cape Colony as a High Commission Territory so as to deter any aggression from the Orange Free State. This meant that Basutoland was not incorporated into the Union of South Africa in 1910, facilitating Lesotho's ascension to independence in 1966.

The Boers[83] operated a primarily rural economy. For almost fifty years they had sought and fought tenaciously to establish their independence, and it seemed that this result had been achieved. Then in 1886 a large deposit of gold was discovered in the Witwatersrand[84] range of hills, which forms part of the central rebound peak of the Vredefort crater.[85] The asteroid that created this structure struck the Earth two billion years ago and exposed rocks that had been laid down almost a billion years earlier when Archean rivers from the north and west formed fan deltas, with many braided channels, before flowing into the "Witwatersrand Sea" to the south.

[79] London Convention (1884) - Wikipedia

[80] Suzerainty is the rights and obligations of a person, state or other polity who controls the foreign policy and relations of a tributary state, while allowing the tributary state to have internal autonomy. Suzerainty - Wikipedia

[81] Lesotho - Wikipedia

[82] Basuto Gun War - Wikipedia

[83] One translation of Boer is directly to 'Farmer."

[84] "Witwatersrand" (literally "white water ridge" – a watershed) Witwatersrand - Wikipedia

[85] Vredefort crater - Wikipedia

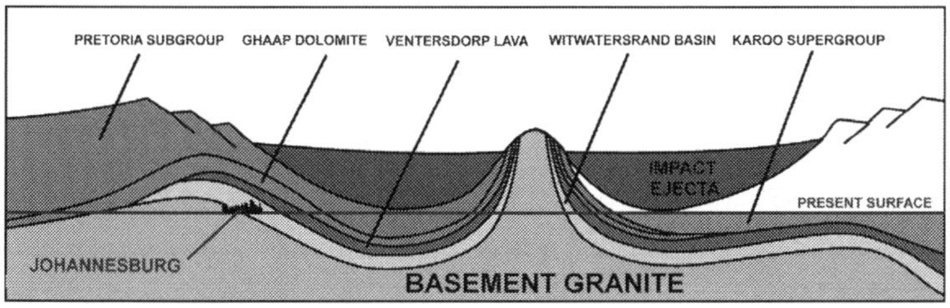

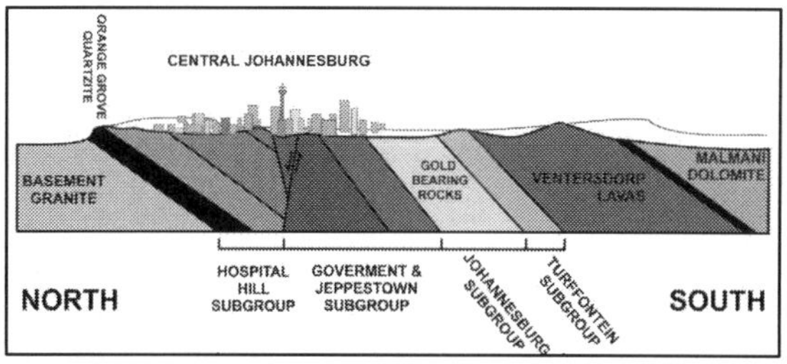

Vredefort crater cross section[60] Witwatersrand Super group[61]

As a result Johannesburg was founded in 1886, starting with Ferreiraskamp[86]

Ferreiraskamp in 1886.[62] The mine of Ignatius Ferreira in 1886.[87]

Within ten years it had grown to become a city of over 100,000 residents.

First Train to reach Johannesburg, September 15th 1892.[63] The Rand Club, Johannesburg.[64]

Morning market in Johannesburg.[65] Johannesburg, 1896. Commissioner Street.[66]

The resulting gold rush drew tens of thousands of people into the Transvaal from all over the world, and a large proportion of the immigrants were British citizens. The surface deposits of gold were fully exploited relatively quickly, and it became apparent that deep mines would be required to extract the remaining ore. This shift in production methods required massive amounts of capital and the most sophisticated machinery available at the time as well as qualified miners and engineers to operate it.

[86] Ferreirasdorp - Wikipedia

[87] Ignatius Ferreira - Wikipedia

Almost all of this capital and expertise came from outside the Transvaal. Collectively these immigrants were referred to as Uitlanders,[88] literally outlanders. The Boers did not consider these immigrants to be permanent residents of the Transvaal but temporary workers who would work the gold fields and when the gold was fully extracted, would then leave with their money, less taxes paid.

Head-Gear of the Van Rhyn Gold Mine.[67]

Crushing battery of the Van Rhyn gold mine.[68]

Gold Mines at Johannesburg, 1899.[69]

Crushing Battery at a Johannesburg mine.[70]

To prevent their republic from being substantially altered politically by the influx of new residents and possibly being changed radically in terms of its constitution, principals and alliances, the Boers established a residency requirement of fourteen years before full citizenship would be granted.

Initially the Transvaal legislature, the Volksraad,[89] was a unicameral body. In 1890 it was divided into two chambers of 24 members each in order to keep Boer control over state matters while still giving Uitlanders (foreigners) a say in local affairs. Immigrants could now apply for citizenship after 5 years of living in the Transvaal. After a further two years, they were permitted to vote for the "Second Volksraad" which was granted legislative powers in the fields of mining, road construction, copyright and certain commercial affairs, all subject to ratification by the "First Volksraad."[90]

Full citizenship carried with it some relief from taxes but also the obligation to serve in the army as well as other responsibilities. During the last phase of the negotiations with the British, Paul Kruger offered to reduce the residency requirement of fourteen years as shown in the accompanying cartoon, in its case to nine years. Subsequently offers were made down to seven and finally the Transvaal made and offer of five years. By that time events had gone so far that it proved to be too late.

[88] Uitlander - Wikipedia

[89] Volksraad - Wikipedia

[90] See this Excellent Article too: Second Anglo-Boer War - 1899 - 1902 | South African History Online (sahistory.org.za) No credited author but probably Dr. David Biggins.

President Kruger (to the British colonial): "There see what I am doing for you. Ain't you very grateful. Who is your true friend now?"[91] In the background is Joseph Chamberlin[92] looking stern and not satisfied.

Some placer gold had been extracted prior to the discovery of the Witwatersrand reserves and the following table shows just how dramatic an effect this discovery had on the production of gold in the Transvaal between 1884 and 1897, climbing to 25% of world output of gold.

Year	Value in £	% Change
1884	10,096	
1885	6,010	-40.5%
1886	34,710	577.5%
1887	169,401	488.0%
1888	967,416	571.1%
1889	1,490,568	54.1%
1890	1,866,645	25.2%
1891	2,924,305	56.7%
1892	4,541,071	55.3%
1893	5,480,498	20.7%
1894	7,667,152	39.9%
1895	8,569,555	11.8%
1896	8,603,821	0.4%
1897	11,476,260	33.4%
Total	53,807,508	

Gold Production in Transvaal.[93]

Gold Output 1894	Value £	%
United States	9,000,000	32.1%
Australasia	8,000,000	28.6%
South Africa	7,000,000	25.0%
Russia (1892)	4,000,000	14.3%
Total	28,000,000	100.0%

World Gold Production 1894.[94]

In addition, while under British control, the diamond mines near Kimberley were adjacent to the Orange Free State and potentially vulnerable to capture and annexation by the Orange Free State. During the Second Boer War the town was besieged for four months.[95]

[91] Slow and Sure - JM Staniforth - Uitlander - Wikipedia

[92] Joseph Chamberlain - Wikipedia

[93] William Harding War in South Africa and the Dark Continent from Savagery to Civilization 1899 Dominion Company (The strange story of a weird world from the earliest ages to the present, including the war with the Boers) War in South Africa and the Dark Continent from Savagery to Civilization - Google Books

[94] South Africa and the Transvaal War, Louis Creswicke, Vol 1 Edinburgh: T. C. & E. C. Jack, 1900 Page 129 https://archive.org/details/southafricatrans01cres.

[95] See: Siege of Kimberley - Wikipedia.

Mine	Year	Loads Hoisted	Loads Washed	Carats of Diamonds	Carats per Load	£ Sold For	Shillings per Carat	Cost per Load. S.	Dividend Paid £	% Equal to.
Debeers & Kimberley	1889	944,706	712,263	914,121	1.28	901,818	19	9	188,329	20.9%
Debeers & Kimberley	1890	2,192,226	1,251,245	1,450,605	1.16	2,330,179	32	8	789,682	33.9%
Debeers & Kimberley	1891	1,978,153	2,029,588	2,020,515	1.00	2,974,670	29	8	789,791	26.6%
Debeers & Kimberley	1892	3,338,553	3,239,134	3,035,481	0.94	3,931,542	25	7	1,382,134	35.2%
Debeers & Kimberley	1893	3,090,183	2,108,626	2,229,805	1.06	3,239,389	29	6	987,238	30.5%
Debeers & Kimberley	1894	2,999,431	2,577,460	2,308,463	0.90	2,820,172	24	6	987,238	35.0%
Debeers & Kimberley	1895	2,525,717	2,854,817	2,435,541	0.85	3,105,957	25	6	987,238	31.8%
Debeers & Kimberley	1896	2,698,109	2,597,026	2,363,437	0.91	3,165,382	26	7	1,579,582	49.9%
Debeers & Kimberley	1897	2,515,889	3,011,288	2,769,422	0.92	3,722,099	26	7	1,579,582	42.4%
Debeers, Kimberley	1898	3,332,688	3,259,692	2,603,250	0.80	3,451,214	26	6	1,489,573	43.2%
Premier Mine	1898	1,146,984	691,722	189,356	0.27	196,659	20	2	90,009	45.8%
Totals		26,762,639	24,332,861	22,319,996		29,839,081			10,850,396	36.4%

Diamond Production.[96]

Open Diamond Mine in Kimberly in 1899.[71]

"Trolley lines for individual claims."[72]

Arriving at worksite via trolley.[73]

Sifting the gravel for diamonds.[74]

[96] Adapted from: South Africa and the Transvaal War, Louis Creswicke, Vol 1 Edinburgh: T. C. & E. C. Jack, 1900 Page 135 https://archive.org/details/southafricatrans01cres.

Main shaft of the Kimberley Mine.[75] Diamond washing machinery in Kimberly.[76]

This phenomenal concentration of mineral wealth attracted the attention of many foreign powers many of whom were engaged in what became known as the "Scramble for Africa."[97] Germany occupied and colonized what is now Namibia from 1884 until 1915 and were right next door to British Bechuanaland and the Cape Colony.[98]

Cecil Rhodes as the Colossus of Rhodes.[99]

Given the history I have outlined, it is little wonder that both the Transvaal and the Orange Free State governments used a large proportion of their revenues to purchase modern armaments for themselves and their

[97] See: Scramble for Africa - Wikipedia.

[98] See The Berlin Conference. The Berlin Conference of 1884–1885, also known as the Congo Conference (German: Kongokonferenz) or West Africa Conference (Westafrika-Konferenz), regulated European colonization and trade in Africa during the New Imperialism period and coincided with Germany's sudden emergence as an imperial power. The conference was organized by Otto von Bismarck, the first chancellor of Germany. The conference contributed to ushering in a period of heightened colonial activity by European powers, which eliminated or overrode most existing forms of African autonomy and self-governance. Berlin Conference - Wikipedia See also: The Treaty of Alcáçovas, 1479. It is one of the first international documents formally outlining the principle that European powers are empowered to divide the rest of the world into "spheres of influence" and colonise the territories located within such spheres, and that any indigenous peoples living there need not be asked for consent. This remained a generally accepted principle in the ideology and practice of European powers up to the 20th century decolonization. The Treaty of Alcáçovas could be regarded as the ancestor of many later international treaties and instruments based on the same basic principle. Treaty of Alcáçovas - Wikipedia See also: Treaty of Tordesillas, 1494, which divided the newly discovered lands outside Europe between the Portuguese Empire and the Spanish Empire (Crown of Castile), along a meridian 370 leagues west of the Cape Verde islands, off the west coast of Africa. Treaty of Tordesillas - Wikipedia

[99] Rhodes is depicted promoting his Cape to Cairo Rail and Telegraph project. File:Punch Rhodes Colossus.png - Wikipedia

citizen militias. Every citizen, Burgher, between the ages of 16 and 60 was required to serve in the militia army and to come equipped with a horse and rifle when called up.

Citizens of the Transvaal.[77] A Boer commando passing through the streets of Johannesburg.[78]

Men of the Pretoria Commando.[79] Boer family of sharpshooters at Johannesburg.[80] Three Generations of Boers Now Fighting.[81]

"Adjutant Roos and Gunners on the Tugela."[82] Lieutenant and Adjutant Carl Roos.[83]

Johannesburg Volunteers.[84] Practicing Running Shots.[85]

For the British, the idea of incorporating the two Boer republics into a federation of South African republics was returned to repeatedly and economic competition was a part of the strategy of limiting the power of the Transvaal Republics.

In September and October 1895 the Drifts Crisis[100] between the Cape Colony and the Transvaal occurred. The British Cape colony under the leadership of Cecil Rhodes had finished building a railway line to Johannesburg and tried to get as much of the Transvaal's railway traffic by reducing its rates. It was aware that the Transvaal's Delagoa Bay line was almost complete. The Transvaal government increased the rates on the part of the railway that ran through the Transvaal once it had crossed the Vaal River into the Transvaal.

In answer to this goods, were taken to the Vaal River by train, and then taken the remaining 50 miles (80 km) to Johannesburg by wagon to avoid paying the higher prices in the Transvaal.

Kruger responded by closing the drifts to all goods traffic. There was a great deal of anger toward this protectionism and exclusion, not only in the Cape Colony but also in the Orange Free State and among Afrikaners.

The British government demanded that Kruger open the drifts and used the situation to involve itself directly in Transvaal affairs. Eventually, the Transvaal Government was forced to relent, and the crisis passed.

In late 1895, Cecil Rhodes,[101] Prime Minister of the Cape Colony, attempted to overthrow the Transvaal government by sponsoring an incursion of the country that was to be coordinated with a local uprising by the members of the Reform Committee[102] of resident British workers. It should be noted that the lack of voting rights was exacerbated by significant taxation and inefficient administration of social services such as schools. This policy of taxation without representation, for a long residence time before citizenship could be obtained, was a legitimate and significant disability.

Rhodes enlisted Leander Star Jameson[103] to mount a raid with his militia of Bechuanaland and Rhodesian (Matabeleland) policemen[104] to invade the Transvaal. The raid failed disastrously. Jameson's troops tried to cut communication lines to Pretoria, but cut the wrong lines. This meant that the Transvaal government knew the raiders were on their way before and the party was intercepted, defeated and captured well before reaching Johannesburg, but the British government was also implicated in the arrangements as well.

Cecil Rhodes was forced to resign as Prime Minister of Cape Colony in 1896 due to his involvement in planning and assisting in the raid. For conspiring with Jameson, the members of the Reform Committee (Transvaal), including Cecil's brother, Colonel Frank Rhodes, and John Hays Hammond, were jailed, found guilty of high treason, and sentenced to death by hanging. This sentence was later commuted to 15 years' imprisonment, and in June 1896, all surviving members of the committee were released on payment of stiff fines.

100 Drifts Crisis - Wikipedia
101 Cecil Rhodes - Wikipedia
102 Johannesburg Reform Committee - Wikipedia
103 Jameson Raid - Wikipedia
104 Who had participated in the First Matabele War in 1893-1894. First Matabele War - Wikipedia

Jameson himself and many other prisoners were returned to London for trial and conviction, not for treason, but for acting as mercenaries without permission.[105][106] The Transvaal government received considerable compensation from the British South Africa Company as well.[107]

Though the raid was ineffective and no uprising took place, it did alert the Boer government to just how much jeopardy they were in and caused them to redouble their arms purchase efforts and military preparations.

An unintended consequence of this attempted coup was that it removed to prison and exile the military force that had previously been stationed in Matabeleland,[108] and this left the whole territory vulnerable. Seizing on this weakness, and discontent with the British South Africa Company who administered the province, the Ndebele[109] and Shona[110] people revolted during March 1896 in the Second Matabele War.[111] Hundreds of European settlers were killed within the first few weeks of the revolt, and many more would die over the next year and a half. With few troops to support them, the settlers had to quickly build a laager around the town of Bulawayo[112] to defend themselves and used cavalry to rescue outlying settlers and harass the enemy. The British were required to send troops and months passed before the British forces were strong enough to break the sieges and defend the major settlements, and war raged on until October of 1897 when the Ndebele and Shona would finally lay down their arms.

Under the Unionist Government elected 1895,[113] Joseph Chamberlain[114] began serving as Secretary of State for the Colonies in June. He had foreknowledge of and acquiesced to the planning of the Jameson Raid later that year but managed to avoid escape culpability and kept his post. Both he and his chosen representative in South Africa, Alfred Milner,[115] were of the opinion that the increasing wealth of the two Boer republics would eventually shift the balance of power in the region to the point where any future union of Southern African states would be as a Boer-dominated republic outside the British Empire.

Both men were of the opinion that a "forward" policy of precipitating a conflict was required and that they would use the civil and voting rights of the Uitlanders as the means of doing so.

When Paul Kruger was reelected to his fourth term as President in 1898 by an overwhelming majority of the voters, it was clear that the British idea of reforming the Republic by using strategic patience in the hope that the electorate would alter by the inclusion of more Anglo friendly voters was not going to be effective and no substantial electoral reforms would be undertaken by the current administration.

During late 1898 Anglo-German relations warmed with Berlin disavowing any interest in the Transvaal, and this opened the way for Milner and Chamberlain to take a firmer line in pressing their demands.

The usual incidents of friction and cause celèbre were trumpeted by the press. A Boer policeman was acquitted of manslaughter for killing a British subject and the South African League, a new Uitlander movement, prepared two petitions, each with more than 20,000 signatures, that appealed to Queen Victoria for intervention against the Transvaal government, which they called inefficient, corrupt, and oppressive. Other Uitlanders produced a counter-petition in which about as many affirmed their satisfaction with Kruger's government.

At the insistence of Orange Free State president Martinus Theunis Steyn,[116] a conference was held in the capital, Bloemfontein[117] in late May of 1899. Milner took a hardline approach with the following set of demands:

1. Enactment of a law that would immediately give Uitlanders enfranchisement and the right to vote.
2. Use of the English language instead of Dutch in the Volksraad (Transvaal Parliament).
3. All laws of the Volksraad would need to be approved by the British Parliament.

[105] Privateer, Filibuster(s) or Freebooters. See: Filibuster (military) - Wikipedia
[106] Foreign Enlistment Act 1870 - Wikipedia
[107] British South Africa Company - Wikipedia
[108] Matabeleland - Wikipedia
[109] Northern Ndebele people - Wikipedia
[110] Shona people - Wikipedia
[111] The territory was suffering from a drought, locust plagues, and the cattle disease rinderpest all ravaging the country at the time. Second Matabele War - Wikipedia
[112] Bulawayo - Wikipedia
[113] Unionist government, 1895–1905 - Wikipedia
[114] Joseph Chamberlain - Wikipedia
[115] Alfred Milner, 1st Viscount Milner - Wikipedia
[116] Martinus Theunis Steyn - Wikipedia
[117] Bloemfontein Conference - Wikipedia

Sir Alfred Milner, High Commissioner for South Africa.[86] Marthinus Steyn, President of the Orange Free State.[87] Paul Kruger, President of the Transvaal Republic.[88]

At one point in the conference Kruger famously remarked, almost in tears, "It is our country you want."

Milner backed off to some degree, only wanting full voting rights after five years' residence, and increased legislative representation for the new burghers. Kruger was willing to reduce the period of Uitlander enfranchisement from the present fourteen years to seven years, and offered naturalization after two years' residence, retroactive, and full franchise after five more, non-retroactive years.

At this point Milner was concerned that he might gain such concessions that he would lose the pretext for war, and despite encouragement from Chamberlain for him to continue the talks, Milner walked out of the conference on June 5th, and no resolution concerning the fate of the Uitlanders was reached.

Back in Pretoria Kruger introduced a draft law to give the mining regions four more seats in each Volksraad and to fix a seven-year residency period for voting rights. Jan Hendrik Hofmeyr[118] persuaded Kruger to make this fully retroactive but Milner still deemed this insufficient.

Chamberlain issued an ultimatum in September 1899 in which he insisted on five years without conditions, or else the British would "formulate their own proposals for a final settlement." British public opinion had now swung in favor of war, and by the beginning of October 1899 nearly 20,000 British troops were based in the Cape and Natal, with thousands more en-route.

Aware of the re-deployment of British troops from elsewhere in the Empire to South Africa, Kruger and his government surmised that from a military standpoint the Boers' only chance was to invoke Grotius's[119] recognition of a state's right of self-defense to include the right to forestall an attack forcibly by invasion of their adversary's territory. They reasoned that if they waited for the Empire to assemble all resources at its disposal, the two small Boer republics would stand no chance of successfully repelling an invasion.

Orange Free State President Steyn was anxious that they not be seen as the aggressors and insisted they delay until there was absolutely no hope of peace. On October 9th, 1899, he changed his mind and his government's policy and informed Kruger that he also now thought war unavoidable. That same afternoon, the Transvaal government handed the British envoy an ultimatum advising him that all disputes between the Orange Free State and the Transvaal should be settled by arbitration (by implication that if the ultimatum was accepted they would not attack) and that if Britain did not withdraw all troops from the border within 48 hours, a state of war would exist. The British government considered the conditions impossible and informed Kruger of this on October 11th, 1899. The start of the Second Boer War was announced in Pretoria that day.

[118] Jan Hendrik Hofmeyr (Onze Jan) - Wikipedia
[119] Hugo Grotius - Wikipedia

"England and the War in South Africa"

"War and Capitalism, or the Transmutation of Human Blood into Gold."[89]

Prime Minister Salisbury, Joseph Chamberlain, and Cecil Rhodes are shown working their war into gold still.

L to R. Erasmus, Reitz, Schalk Burger, General Kock.

Bottom: Joubert, Kruger, Cronje.

The Executive Council of the Boer Government at the start of the war.[90]

Early Campaign Summary

One day after war was declared, on October 12th the Boers entered Bechuanaland[120] and fought the battle of Kraaipan.[121] They attacked and captured the British garrison and railway siding 37 miles (60 kilometers) southwest of Mafeking. The siege of Mafeking[122] began the next day and lasted until May 17th, 1900. The siege of Kimberley[123] began a day later and lasted from October 14th, 1899, until February 15th, 1900. In neither case were the Boers able to force the garrisons to surrender before they were relieved.

The armoured train captured at Kraaipan.[91 92]

Botha Commando. "The guns which the Boers captured from us a Kraaipan have been used against gallant little Mafeking." [124]

Boer Artillery unit operating near Mafeking. [125]

[120] Botswana - Wikipedia Formerly: Bechuanaland Protectorate - Wikipedia

[121] Battle of Kraaipan - Wikipedia

[122] Siege of Mafeking - Wikipedia

[123] Siege of Kimberley - Wikipedia

[124] PF: Note the heliograph set up on the right. Black & White Weekly Vol 1-2 1899-1900.

[125] Black & White Weekly Vol 1-2 1899-1900 "The guns which the Boers captured from us a Kraaipan have been used against gallant little Mafeking." PF: Note the heliograph on a tripod on the right-hand side.

General Redvers Buller[126] was placed in overall command of the British forces to be sent to South Africa.[127] He had fought in the Cape Frontier Wars, the Anglo Zulu War, and the First Boer War and knew a great deal about South Africa and his opponents.

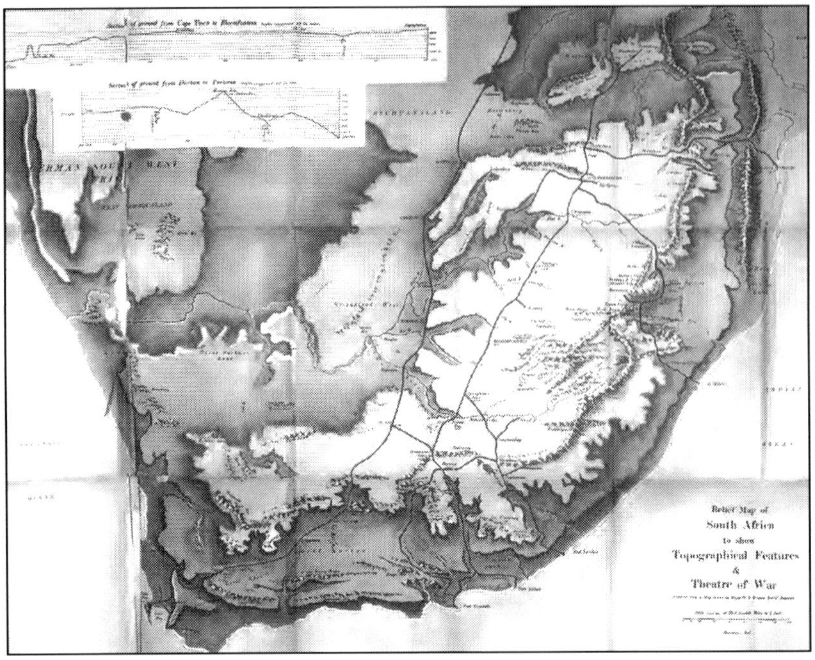

Rt. Hon. Sir Redvers Henry Buller, V.C.[93] Topographical Map of South Africa showing the High Veld[128] of the Kaapvaal Craton.[129]

Buller's strategy was to start from the Cape, move north and get his army up onto the High Veld of the Kaapvaal Craton to push north, taking the Orange Free State out of the war by capturing the capital Bloemfontein and then continue using the existing rail lines to move towards Pretoria and Johannesburg. While the distance was shorter through Natal, attacking from that direction would involve fighting in the mountainous terrain of Northern Natal, which favored defense.

Buller realized that the northernmost line in Natal, with open country, where British forces could effectively operate and that could be defended, was the line of the Tugela River that ran west to east through Colenso. Beyond that the country was surrounded on three sides by the Transvaal and Orange Free State, in terrain that was not defensible with the forces available at this early stage of the war. The unavoidable consequence was that parts of Natal would be under Boer control until additional forces were available or Buller closed in from the southwest.

Buller and a large proportion of his forces were still in England when the war began. We now turn to examine the actions of the local commanders to see how Buller's strategy was thwarted.

In May of 1899 Sir William Penn Symons[130] was appointed as the General Officer in command of forces in South Africa. He had about 10,000 troops under his command spread out from the Cape to Natal. By September this was increased by another 10,000, and General George White[131] was appointed in overall command.

By the time General White arrived in Cape Town, Penn Symons had deployed his Natal force to Ladysmith 15 miles north of Colenso on the Tugela River, in a location where the approaches were elevated on both side of the railway line. He had also pushed a full brigade [~4,000 soldiers][132] another 35 miles north beyond the Biggersberg to Dundee.

[126] Redvers Buller - Wikipedia
[127] Natal Field Force - Wikipedia
[128] A type of wide open rural landscape Veld - Wikipedia
[129] Geology of South Africa - Wikipedia
[130] Penn Symons - Wikipedia
[131] George White (British Army officer) - Wikipedia
[132] British Divisions generally consisted of three Brigades, each with 4,000 soldiers in four battalions of ~1,000 plus support troops.

Drakensburg Mountains.[133] [94]

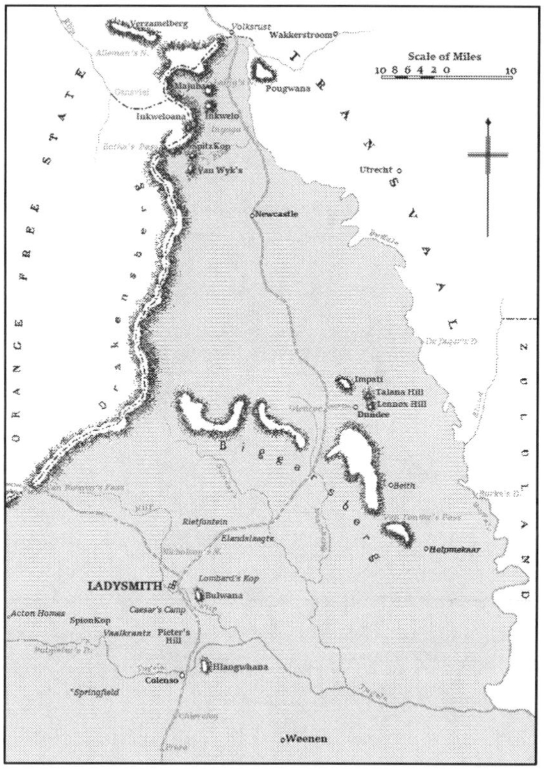

Northern Natal. Note Drakensburg bordering the Orange Free State.[95]

After being placed in overall command and before departing England, General Buller had instructed General White not to concentrate his forces north of the Tugela River because he would be surrounded on three sides by the Transvaal and the position would not be defensible with the forces at his disposal.

When General White arrived, he wanted to recall the forces at Dundee back to Ladysmith if not all the way to Colenso. However, there was a good deal of jingoistic pre-war optimism that the conflict would result in a one-sided victory for the British. Of course the local commanders wanted to succeed in order to enhance their reputations. There was also political pressure from the governor of Natal, Walter Hely-Hutchinson.[134] Natal wanted to defend as much territory as possible so as not to abandon the citizens and towns of northern Natal to foreign occupation for the duration of the war. There was already a huge exodus of refugees that Natal was dealing with and more would strain the colony further.

"Johannesburg Uitlanders, fleeing from the expected war, fighting for seats in open trucks."[96]

Refugees.[97]

From their point of view they did not ask the British Army to show up in order to retreat. Also, the northern area of Natal was a coal mining region and coal was a strategic asset needed for the war effort. Coal was mined at

[133] The Drakensberg is the eastern portion of the Great Escarpment, which encloses the central Southern African plateau. The Great Escarpment reaches its greatest elevation – 2,000 to 3,482 meters (6,562 to 11,424 feet) within the border region of South Africa and Lesotho. Drakensberg - Wikipedia

[134] Walter Hely-Hutchinson - Wikipedia

Dundee; in fact, the large town 35 miles north of Dundee is called "Newcastle" after the famous coal mining town in England. This position was way too exposed to attack by the Boers and was clearly being left out of the planning. The bold move that White could have taken was to move farther north up into the passes of the Drakensburg Mountains themselves so that he could block them effectively.

That would have been exceeding his authority, and given the considerations mentioned General White could convince himself that he would compromise by abandoning Newcastle but not order Penn Symons to retreat from Dundee. Since Penn Symons was still in Dundee it followed that White could not retreat from Ladysmith. On Friday October 13th, the day after the war began, Newcastle was abandoned, and it was occupied by the Boers on Sunday the 15th.[135]

Even so, this half measure of retreat would not be enough. The decision to allow tactical considerations to override strategic ones would prove to be a grave mistake and affect the entire course of the war. The strategic purpose of garrisoning Natal with reinforcements was to protect Durban and its harbor from falling under Boer control, not to try to win the war before substantially more reinforcements arrived.

Buller left England on Saturday, October 14th. We will join Captain Haldane's narrative less than a month later. By the time Buller arrived in Cape Town on Tuesday, October 31st, the Boers had moved quickly to exploit their local numerical superiority.

Talana Hill

With the passes into Natal open and undefended, by the night of Thursday the 19th, General Lucas Meyer[136] had moved his force of approximately 4,000 men 35 miles south towards Dundee and climbed up to the top of Talana Hill. On the morning of Friday, October 20th, he was able to begin shelling the garrison. In the resulting Battle of Talana Hill[137] [98](Glencoe/Dundee) General Penn Symons was mortally wounded while bravely but recklessly exposing himself to enemy fire in order to motivate his troops. General Yule assumed command, and the British did successfully storm the hill in the face of withering magazine fed, Mauser bolt-action rifle fire. The Boers retreated on horseback and the British troops were refused permission to use their rifles to fire upon them as they departed.

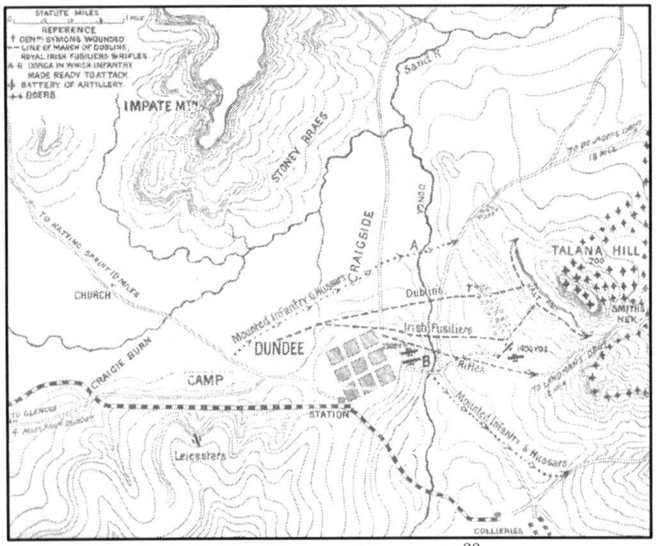

Battle of Talana Hill (Glencoe/Dundee).[99]

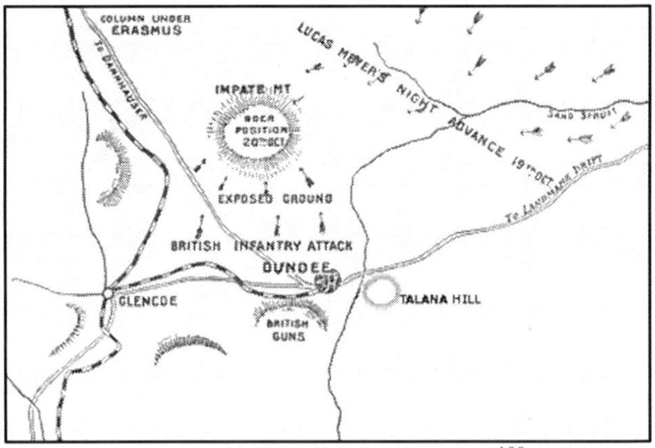

Battle of Talana Hill (Glencoe/Dundee).[100]

[135] Boer War Chronology - Adapted from Creswicke, South Africa and the Transvaal War Vols. 1-7
[136] Lucas Johannes Meyer - Wikipedia
[137] Battle of Talana Hill - Wikipedia

British troops assaulting Talana Hill.[101]

British troops crossing a river near Talana Hill.[102]

The British artillery had performed well but there were also friendly fire incidents where they had killed their own comrades by shelling after the British had gained the summit. During the Boer retreat the artillery did not fire on the enemy for fear of hitting their own troops again. At the beginning of the battle, anticipating victory in storming the hill, the cavalry had been detached to cut off the expected retreat. This failed as well because a large number were surrounded by a second force of 4,000 men under General Erasmus, men who came from another direction, late to the battle but just in time to capture the cavalry. See: Colonel Moller's[138] surrender after Talana Hill.

Boers watching the fight at Dundee.[103] Boer Commandants reading a message from British officers after the fight at Dundee.[104]

[138] Bernhard Drysdale Moller Lieutenant Colonel QSA. Severe Wound, Van Wyks Vlei, 21 Aug 00. Placed on Half-Pay, 01 Sep 00. Source: QSA medal rolls 18th (Victoria Mary, Prince of Wales's Own) Hussars. The regiment was at Glencoe under General Penn Symons when war broke out, and took part in the battle of Talana Hill (see 1st Leicestershire Regiment). In Sir G. White's dispatch of 2nd November 1899, para 12, he says: "Turning now to our cavalry, the 18th Hussars received orders at 5.40 am to get round the enemy's right flank and be ready to cut off his retreat. They were accompanied by a portion of the Mounted Infantry and a machine-gun. Making a wide turning movement, they gained the eastern side of Talana Hill. Here Lieutenant Colonel Moller halted with one squadron 18th Hussars, the machine-gun, and Mounted Infantry, sending his two other squadrons farther to the east. These two latter squadrons took part in the pursuit of the enemy, who retreated eastward, but Lieutenant Colonel Moller and the troops with him appear, so far as can be ascertained, to have pursued in a northerly direction, to have come in contact with superior forces not previously engaged, and to have been surrounded and forced to surrender while endeavouring to return to camp round the north of Impati Mountain" The Times historian (Leo Amery) states that Colonel Moller arrived at a strong defensible position from which he could have stampeded the whole of the Boers' ponies and commanded their line of retreat and that Major Knox begged to be allowed to fire on the ponies, but instead was ordered to advance with two squadrons right in the rear of the Boer position. This and the rest of Colonel Moller's proceedings are inexplicable. After exhibiting great rashness he seems to have become unnerved. Knox with difficulty got back with his two squadrons. The others were taken prisoners. In his evidence before the court of inquiry, Colonel Moller gave his reasons for taking the road he did, but Captain Lonsdale of the 2nd Dublin Fusiliers Mounted Infantry stated that he had informed the colonel he was taking the wrong road. The officers and men were exonerated, but Roberts did not allow Colonel Moller to rejoin his regiment.

In fact the Boers had made a serious mistake of their own. If General Meyer had waited one day, General Erasmus would have arrived and the combined force of about 7,000 highly motivated burghers would have outnumbered the British garrison by 2 to 1 at least.

On the 19th, just before the battle of Talana Hill, the telegraph wires between Ladysmith and Dundee had been cut and General White learned that General Johannes Kock,[139] in command of the Johannesburg commando, was third force in the area that had occupied the train station at Elandslaagte.

General and Mrs. Lucas Meyer.[105] Commandant Erasmus.[106] "The first prisoners of war, captured near Dundee."[107]

Sir William Penn Symons.[108 140] J. H. Yule.[109 141] George White.[110 142]

Casualties In Action	Killed	Wounded		Missing and Prisoners			Forces Engaged		
	Officers	Men	Officers	Men	Officers	Men	Total	British	Boer	UK Loss
Talana Hill /Dundee	8	42	11	84	25	306	476	3,400	7,000	14.0%

Talana Hill / Dundee / Glencoe Casualties.[143]

[139] Johannes Hermanus Michiel Kock - Wikipedia

[140] Penn Symons - Wikipedia

[141] J. H. Yule Brigadier General Officer Commanding near Dundee. MID LG: 8 February 1901.

[142] Sir George Stuart White, (6 July 1835 – 24 June 1912) He was commander of the forces in Natal at the opening of the Second Boer War and fought at the Battle of Elandslaagte in October 1899. He commanded the garrison at the siege of Ladysmith: although instructed by General Sir Redvers Buller to try and break out or surrender the garrison he responded "I hold Ladysmith for the Queen" and held out for another four months before being relieved in February 1900. George White (British Army officer)-Wikipedia

[143] Doyle, Arthur Conan. The Great Boer War: Seventh Impression 1900 (September) London, Smith, Elder & Co. Pages 535-536. And With The Flag to Pretoria Vol 1 Preface page vi (Casualties) Forces engaged from various sources as a general guide, not exact figures.

Grave at The Foot of Talana Hill Behind Smith's farm. [111]

Elandslaagte

It was not until White learned that the Boers had retreated after the battle of Talana Hill[144] that he dispatched General John French[145] with about 2,000 men, including Captain Haldane, on the afternoon of the 20th, to recapture Elandslaagte. This force arrived shortly before dawn on Saturday, October 21st. General French and his forces enjoyed numerical superiority of around 3:1, indicating that he faced about 700 Boer combatants.[146]

While three batteries of British field guns bombarded the Boer position, the Devonshire Regiment advanced frontally in open order. The main attack was commanded by Colonel Ian Hamilton.[147] It moved around the Boers' left flank. The sky had steadily been growing dark with thunderclouds, and as the British made their assault, the storm burst. In the pouring rain and with poor visibility, the British infantry had to face a barbed-wire farm fence, in which several men were entangled and shot. Nevertheless, they cut the wire or broke it down, and occupied the main part of the Boer position.

Elandslaagte and train.[112]

Gordon Highlanders advancing at Elandslaagte.[113]

[144] He was probably notified by heliograph or maybe a rider was sent his way. Heliograph - Wikipedia
[145] John French, 1st Earl of Ypres - Wikipedia
[146] Battle of Elandslaagte - Wikipedia
[147] Ian Hamilton (British Army officer) - Wikipedia

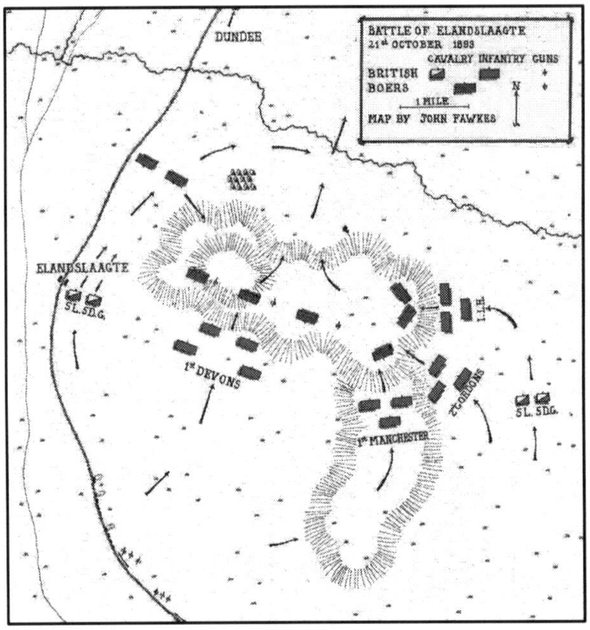

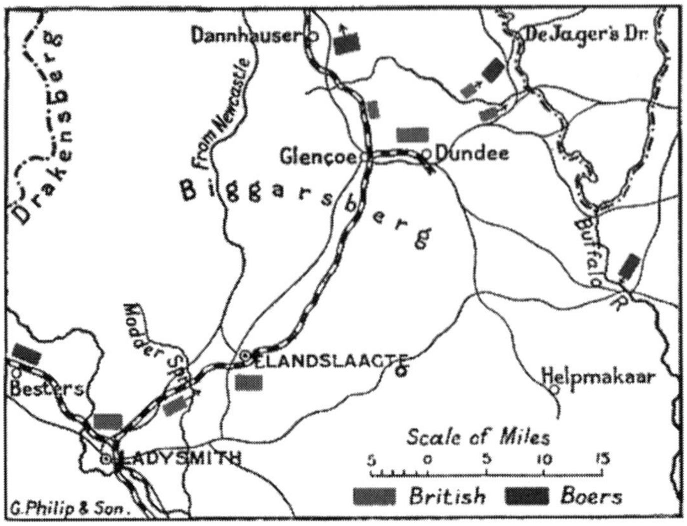

Battle of Elandslaagte.[114] See Also: Battle of Elandslaagte.[115] North Natal Area.[116]

Dr. ELSBURGER Col. A. H. SCHIEL JUDGE KOCK COUNT ALBEDYLL THE LATE GENERAL J. H. M. KOCK COUNT ZEPPELIN FIELD CORNET POTGIETER

General Kock and his staff.[117] Kock and Staff identified.[118 148]

Just as the main Boer line began to wave white flags of surrender and the British ceased fire, General Kock led a counterattack, dressed in a top hat and his Sunday best. His party drove back the British in confusion, but they were rallied by the infantry commander, Ian Hamilton, and Kock and his assault force were killed or severely wounded. Kock was captured and died a week later while a prisoner in Ladysmith.

[148] "General Kock And His Staff : The Heroes Of Elandslaagte , 21 October 1899. J. H. M. Kock , the gallant old Transvaaler , Member of the Executive Raad , died at Ladysmith , on the 31st of October 1899, of inflammation of the lungs contracted on the field of battle, where he was wounded and lay long in the pouring rain before the English ambulance found him . His illness was increased by sorrow at his defeat. His son, Judge Kock, fell into the hands of the enemy, but was exchanged by after his father's death. General White Count Zeppelin was killed after firing off all his ammunition and eventually defending himself to the last against the charging Lancers with his whip. Colonel A. H. Schiel was till the end of the war a prisoner at St. Helena. Albedyll succeeded him in his command. With the exception of Potgieter and the lieutenant of the Transvaal Artillery , who did not take part in the battle, the others are all members of the German Corps, of which Dr. Elsburger stood at the head of the ambulance." Heroes of the Boer War. Review of Reviews, 1903.

Landscape of Elandslaagte.[119] "Elandslaagte. The flat ground where the 5th Lancers charged the retreating Boers. [Note Coal mine shaft and works. Northern Natal is underlain by coal seams which is one reason the British were prepared to defend it rather than retreating to the plains of Natal.][120]

"The German Corps. Before the Battle the of Elandslaagte (October 21st, 1899). The corps lost many killed and wounded in this battle."[121]

The Hollander Corps leaving Pretoria for the front.[149] [150]

The Dutch Corps Monument.[122] [151]

[149] Departure of The Hollander Corps from Pretoria 6 October 1899. "There was not a foreigner in South Africa, with the exception of the British, who did not take side with the Boers when war had once become inevitable. The Hollanders were the first to form themselves into a Corps, and they rendered invaluable services during the war." Image: Heroes of the Boer War. Review of Reviews.

[150] The Battle of Elandslaagte (10/21/1899) was also notable for being the first and last battle of the volunteer Hollanderkorps. The Hollanderkorps was a group of ca. 150 Dutch volunteers which had been established a mere month earlier. During the battle the Hollanderkorps suffered 9 fatalities, including Herman Coster, along with fellow officer Cars Geerts de Jonge and seven soldiers: P.J. van den Broek, H. van Cittert, J.A. Lepeltak Kieft, Jan Moora, J.Th. Rummeling, M. Schaink, and F.W. Wagner. A further 35 of the Hollanders were taken prisoner by the British. Among the prisoners was Willem Frederik Móndriaan (brother of the Dutch artist Piet Mondrian) who was wounded in the battle. Although he was able to crawl away from the battlefield he was taken prisoner by the British shortly after. He was later sent to Saint Helena as a prisoner of war, returning to South Africa in 1903. Cornelis Vincent 'Cor' van Gogh, the brother of the Dutch artist Vincent van Gogh, also fought in the battle where he was wounded and taken prisoner. He died shortly after. The Boer generals were deeply unhappy with the Hollanderkorps' performance, and it was disbanded after the battle though several hundred Dutch volunteers continued to fight in Boer regiments. The names of the deceased Hollanders, including Coster, were inscribed at a monument at the location of the battle. The monument was destroyed by vandals in 2014. Battle of Elandslaagte - Wikipedia

After this last counterattack the remaining Boers mounted their horses and began to retreat. A squadron of lancers and another of dragoons were able to get in amongst them with lances and sabers, inflicting severe casualties.

With possession of the field, the way was now clear for the force at Dundee to retreat towards Ladysmith using the railroad or at least the main road. However, General White was concerned that an additional force from the Orange Free State was advancing from the West and ordered his units to retreat in the dark towards Ladysmith. By giving up the advantage won in this battle, the force in Dundee was again isolated.

Searching for wounded at night.[123]

Removing wounded after the Battle of Elandslaagte.[124]

Carrying the wounded from the battlefield of Elandslaagte.[125] [126] "Mrs. Otto Krantz, the wife of a hunter. Accompanied her husband to Natal and fought by his side at Elandslaagte, also in the battles on the Tugela, and after wars in the Free State. She is by no means the only Boer woman who has carried a rifle in the war, and used it."[127] And.[128]

Elandslaagte bridge; damaged during the war.[129]

Battlefield of Elandslaagte.[130]

[151] The Dutch Corps Monument is a provincial heritage site in Kliprivier in the KwaZulu-Natal district of South Africa. It was designed by Gerard Moerdijk. [a] The monument was vandalized in 2014. In 1982, it was described in the Government Gazette as
This memorial, which was designed by G. Moerdyk of Pretoria, was built by the firm Barker and Nel of Volksrust and was erected by the citizens of The Netherlands in memory of members of the Dutch Corps who died during the battle of Elandslaagte on 21 October 1899. Dutch Corps Monument - Wikipedia

At this time and throughout most of the war, the parties did not do battle on Sunday, nor did they engage in night battles.[152] General Yule in Dundee realized that he needed to retreat and he could not use the main road or railroad to do so. On Sunday, October 22nd, the remaining Dundee garrison was forced to leave the wounded behind, including General Penn Symons, in order to begin a forced cross-country march, at night, to retreat towards Ladysmith. It would take him four days until Thursday the 26th for him to reach his goal. Initially the Boers did not realize the retreat had occurred, and after they occupied Dundee there was indiscipline during which the Boer forces began drinking and looting Dundee. It was some time before military order could be restored and all spirits in the town destroyed by order of the commander. Yule's force was later harassed but not attacked in force during the retreat, and the column made it safely through the defile of Van Tonder's Nek, which was the most dangerous pass that could have been blocked by even a small party of opponents.

Casualties In Action	Killed		Wounded		Missing and Prisoners			Forces Engaged		
	Officers	Men	Officers	Men	Officers	Men	Total	British	Boer	UK Loss
Elandslaagte	5	50	30	169	0	4	258	3,500	2,000	7.4%

Charge of the 5th Lancers. Trumpeter Shurlock is firing his revolver.[131] Yule's Retreat from Dundee.[132]

It should also be noted that during the early stages of the war, when the Boers were repeatedly defeating the British forces, they did not harass them unduly while they retreated. At Colenso, Spion Kop and Vaal Krantz British forces were compelled to retreat back across the Tugela River and the Boers made no concerted effort to shell or snipe at them while they did so. It has been said that this was the last of the chivalrous wars and first of the modern.

While General Yule and his column were making their way back towards Ladysmith, General White realized that once clear of the Biggarsberg mountain pass the four-mile-long column would have to turn west in order to approach Ladysmith and this would bring it closer to Elandslaagte, which had been re-occupied by the Free State commando.

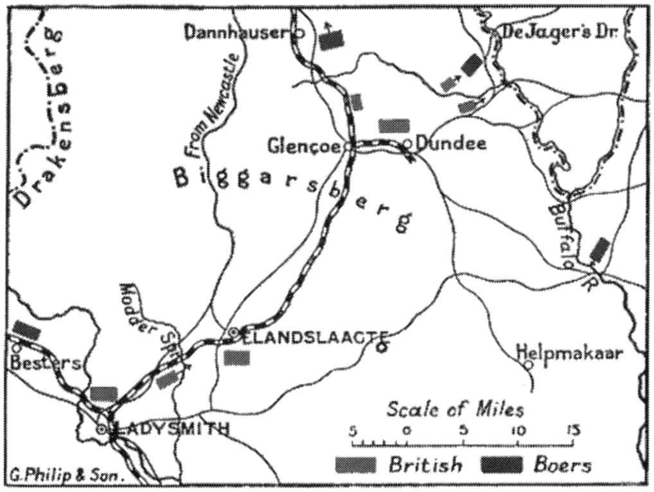

North Natal Area.[133] This action is at Modder Spruit. (Reitfontein/ Intintanyone).

[152] See Times History Vol 2 p. 215 for one time this almost happened but was called off by the British at the last moment.

In order to provide protection for Yule's forces White set out from Ladysmith with artillery and infantry that would occupy the attention of the Boers and also block them from moving westward across the railroad and towards Yule's column. In order to bring his mountain battery guns close enough, 1,500 yards, so it could effectively shell the tops of the ridge where the Boers had their artillery, it was necessary to place infantry in front of those guns so they could not be easily assaulted by Boer cavalry or infantry. Consequently, the infantry engaged in a long-range duel of rifle fire while the artillery shelled the mountain tops. Due to some confusion one regiment, the Gloucester, began advancing beyond the ridge that was providing cover and was immediately subjected to withering fire; six were killed and another forty wounded.

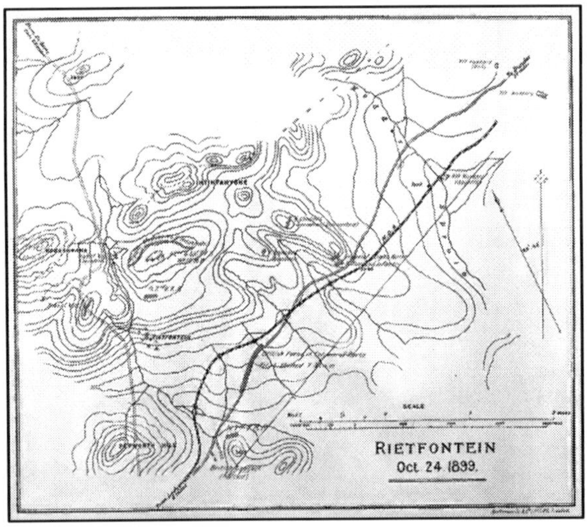

Rietfontein[134] or Battle of Tintwa Inyoni (Intintanyone).

Heliograph in use.[135]

Casualties In Action	Killed		Wounded		Missing and Prisoners			Forces Engaged		
	Officers	Men	Officers	Men	Officers	Men	Total	British	Boer	UK Loss
Rietfontein (N. Cape)	1	11	6	98	0	2	118	5,100	7,000	2.3%

At noon, after about four hours of engagement, White received a heliograph message from Yule's column identifying its position, and with that information determined that Boers were no longer a threat to the column. He had the infantry retire under cover of the artillery and began to withdraw towards Ladysmith.

What White did not realize was that this Boer force was simply a screen for General Joubert's[153] army in order to keep the British from reoccupying Elandslaagte until Joubert's large force could come down from Dundee/Glencoe. Alternatively, White could have simply monitored the Boers on the ridges and if they did come down off of the mountain tops, he could have engaged them on the flat ground where the Boers would have no cover and the British army was trained to fight.

If it was just a military demonstration to protect Yule that White was attempting, then there was no need to engage. If the position was worth taking, then he should have made the effort to do so.[154] Certainly the experience of the Gloucester regiment in attempting to advance told the leadership that a coordinated attack would be required for any assault. Also, White knew that, in total, he was facing combined forces that were very much larger than his own. He decided to retire to Ladysmith and wait for Yule's column to arrive before taking further action.

On the 25th, General Joubert's forces effectively joined up with those of the Orange Free State and were moving consistently towards Ladysmith. While Penn Symons and General White had both won their tactical engagements and General White had concentrated his forces, both engagements had failed to achieve any strategic purpose in that the British forces had failed to keep the Boer forces separated so they could be dealt with individually.

On Thursday, October 26th, Yule's column arrived at Ladysmith. The march had been exhausting, and the last 30 miles or so had been made under dreadful conditions of rain and forced marching. The troops had not been allowed to rest and arrived in no shape to engage in any offensive military action.

[153] Piet Joubert - Wikipedia

[154] Amery, Times History Vol 2 p. 206.

Ladysmith, Natal.[136]

Observation balloon at Ladysmith.[137]

The permanent camp at Ladysmith, Natal, and second page of the panorama photograph.[138] PF: Too much infrastructure and supplies for such a vulnerable outpost. No wonder they didn't abandon it.

By Saturday the 28th the Boers had cut the waterlines feeding Ladysmith, although the British still had access to the Klip River that ran through town. On Sunday the 29th, they began putting large caliber artillery, the Creusot "Long Tom," within view of Ladysmith on Pepworth Hill, roughly 4 miles (6.4 km) northeast of the town.

Long Tom showing breech block, shell, and powder bag, along with ramrod/swab and water bucket.[155] Boer Long Tom gun in use at Vaal Krantz, February 5th, 1900.[139] Note shell in loader's hands and bag of powder on the shoulder of assistant. This may be the same crew but different location.

[155] The 155 mm Creusot Long Tom was a French siege gun (artillery piece) manufactured by Schneider et Cie in Le Creusot, France, and used by the Boers in the Second Boer War as field guns. Four guns, along with 4,000 common shells, 4,000 shrapnel shells, and 800 case shot were purchased by the Transvaal in 1897. 155 mm Creusot Long Tom - Wikipedia Image Source: 155 mm Creusot Long Tom - Wikipedia File:Long tom on platform at Mafeking.jpg - Wikimedia Commons

Cronje and lieutenants around one of the Creusot guns. General Cronje near gun, with whip pointing down. [140]
(L to R) Commandant Steenkamp, General Cronje and Commandant Van der Merwe identified. [141]

"An attack upon the Boers on Pepworth Hill for the 29th had been discussed on Saturday but put off, partly from an unwillingness to attack the Boers on a Sunday … ."[142] It was decided that a general action would take place on Monday the 30th.

This battle has been referred to by several names. There were two distinct actions each of which have their own name. It is now generally known as the Battle of Ladysmith,[156] but it is also referred to as Lombard's Kop & Farquhar's Farm, and there is also the distinct action that takes place near Nicholson's Nek.

General White's plan was to continue using the same tactics that had been used at Talana Hill and Elandslaagte, an artillery bombardment followed by a frontal and a flanking infantry attack.

Also, like Talana Hill, White planned for a column to proceed into the rear of the enemy in order to block reinforcements and also prevent the retreat of the Boer forces that might flee if the main attack was successful. This action developed into that of Nicholson's Nek.

[156] Battle of Ladysmith - Wikipedia Also Known as Farquhar's Farm and Nicholson's Nek.

Ladysmith & Nicholson's Nek/Lombard's Kop

During the night of October 29th, three separate forces left Ladysmith to prepare for the general engagement.

The plan that General White implemented was for Colonel Grimwood[157] to lead a force in order to attack the Boer left, which lay to the east of Ladysmith. Boer forces had been seen on Long Hill and that was to be the point attacked. The attack was to begin from the base of the southern kopje, Lombard's Kop, which was thought to be unoccupied. This proved to be an erroneous assumption. During the night of the 29th the Boers had withdrawn from Long Hill, probably to sleep in their laagers behind the hill. However there were Boer forces farther south on top of Lombard's Kop facing Grimwood's right flank. These forces comprised the Johannesburg commando now led by Louis Botha,[158] in place of Lucas Meyer.

General Lucas Meyer General Louis Botha and Commandant D. J. E. Erasmus[143] "A Boer Commando passing through the streets of Johannesburg"[144]

Johannesburg Vrijwilliger Corps.[145] Burgomaster Schutte Of Johannesburg Witnessing The Departure of the Burghers : 4 November 1899.[146]

The third force was led by Col. Frank Robert Carleton to be sent out as the blocking force behind the enemy lines. This was the same strategy that was employed at Talana hill where Col. Moller's force was cut off and captured.

During the night the first force under Grimwood marched out of Ladysmith and towards the base of Lombard's Kop at Farquhar's Farm where he faced north towards Long Hill and waited for dawn.

[157] Colonel G. G. Grimwood King's Royal Rifle Corps QSA DoL Provisional list of recipients Source: Ladysmith Siege Account and Medal Roll Anglo Boer War - Name search

[158] Louis Botha - Wikipedia

Col. Grimwood and officers of 1st and 2nd King's Royal Rifles.[147] Colonel Ian Hamilton.[148]

A second force led by Ian Hamilton was to march out directly towards Pepworth Hill for an assault supported by artillery.

British troops on the march.[149] Royal Lancasters January 19th, 1900.[150]

He expected that his artillery detachment would bombard the summit of Long Hill, just to the north of him and then his force would assault that position. He also expected that General French and his cavalry would be supporting his right flank.

[During the night march] Colonel Coxhead's[159] division of artillery, which was in the middle of Grimwood's column, and which Grimwood understood was to accompany him to some position south of Long Hill, turned aside during the march in order to take up the position behind the kopje known afterwards as Flag Hill, east of Limit Hill … Not only did the artillery leave the column in the darkness without Grimwood's knowledge, but it carried away with it the Liverpools, Dublin Fusiliers, and the mounted infantry companies of the 60th and Leicesters. The whole incident was a very bad piece of staff work on somebody's part, and gravely prejudiced the chances of the day. The three leading battalions marched on in blissful ignorance and, shortly before daybreak, formed up behind a couple of low-lying rocky kopjes 1,800 yards south of Long Hill.[160]

With daylight Grimwood found that part of his force, including his two batteries of guns, were missing and French's cavalry was not positioned to support him. Without cavalry the right flank and rear of the infantry line was completely in the air. Grimwood also realized that the Boers were no longer on Long Hill. They had moved during the night so that they were positioned on his right flank in and on the hills he was moving across.

As Grimwood's men moved forward the Boers opened a heavy fire into their right flank, taking the British troops by surprise and causing considerable confusion. Instead of conducting an attack along the ridge and rolling up the Boer flank northwards, the British found themselves in a desperate struggle to maintain their position on the hill under a withering fire from hidden Boer riflemen to the east.

159 Colonel J. A. Coxhead Commanding Brigade Division Royal Artillery, Elandslaagte and Reitfontein; Siege of Ladysmith, slightly wounded. Entered Service 1872; 1898. Staff service: ADC to Captain-General and Governor-in-Chief, Jamaica, 1883-87. War service: Boer War, 1899-1900. Anglo Boer War - Name search

160 Times History Vol 2 p. 222

With his right flank already under attack there was nothing to be gained by continuing to move northwards. This would only have allowed the enemy to work still further round the rear of the brigade. Colonel Grimwood decided to remain where he was.

At the start of the day, the artillery being employed was directing its fire onto the top of Long Hill as had been ordered, but those trenches were unoccupied. It took several hours before they were repositioned and redirecting their fire in the direction that Grimwood was actually fighting.

The brigade was gradually swung round, so as to face east instead of north, and was eventually extended in a long irregular line from within rifle range of Farquhar's Farm to some distance along the low ridges that stretch down northwards from Lombard's Kop.

However, the opportunity to roll up the Boer left flank was lost because the column was fully engaged and General French's cavalry had been in the wrong position to clear the flank before so many Boer reinforcements had arrived.

Hamilton's brigade had meanwhile marched forward shortly before daybreak to take up its position behind Limit Hill. Rifle and artillery fire were directed towards the Pepworth Hill position, but any attempt at moving across open ground was met with withering rifle fire. Without some success from Grimwood's efforts it was not possible to frontally assault the Pepworth Hill position.

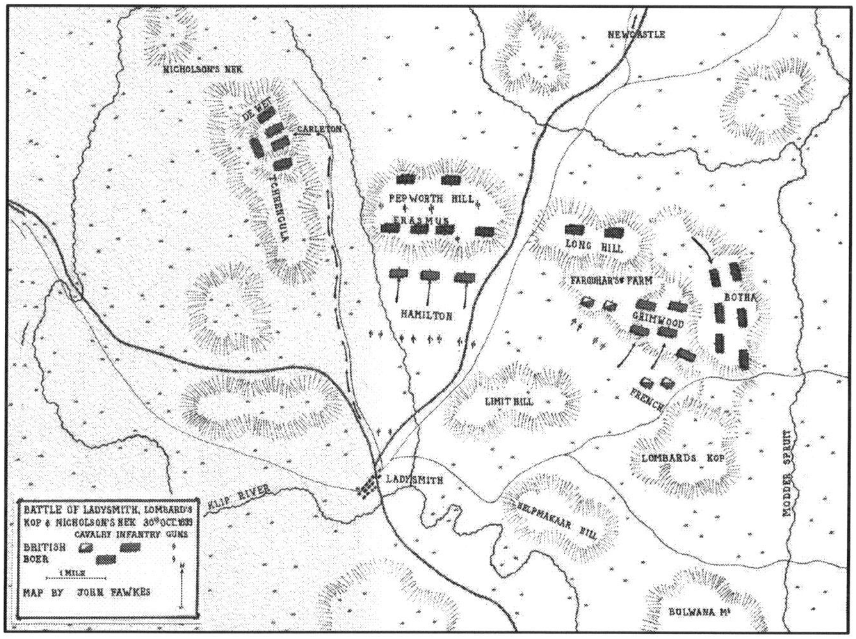

Ladysmith Battle Map.[151] Samuel Begg painted Melton Prior sketching the Battle of Nicholson's Nek. [152]

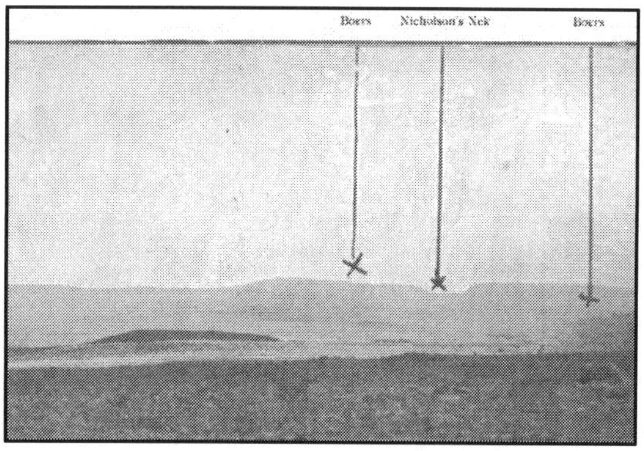

Nicholson's Nek, scene of De Wet's first victory where twelve hundred British were taken prisoners.[153] Area of Lombard's Kop Battle on October 30th. [154]

Just after daylight the Long Tom gun on Pepworth Hill began lobbing shells 8,500 yards into Ladysmith. They did not cause much damage but the novelty of it caused an increasing amount of civilian panic. White received scattered reports from mule drivers who had retreated after the stampede down the mountain and some of the gunners who had tried to chase after their guns that Carleton's force, at Nicholson's Nek, was in trouble. (See infra for that part of the battle.)

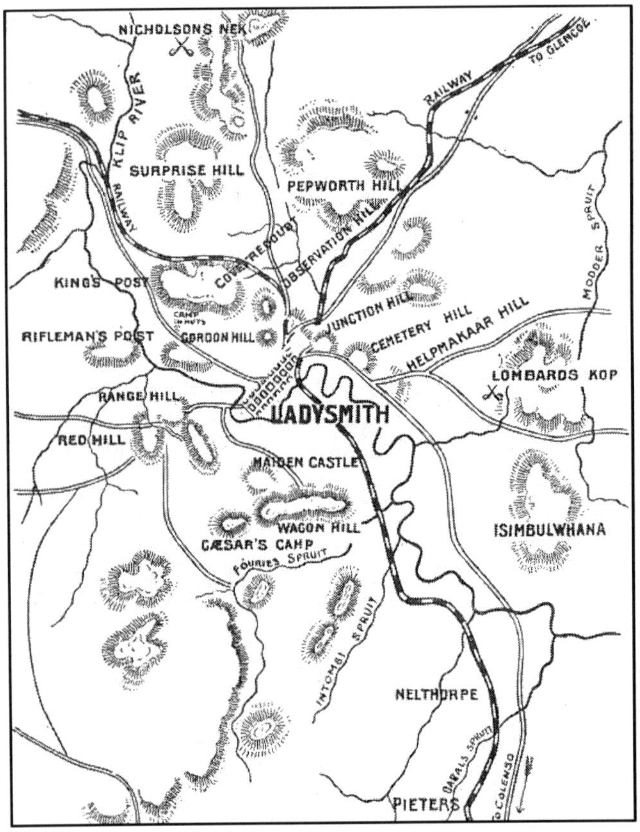

Ladysmith showing landmarks in detail.[155]

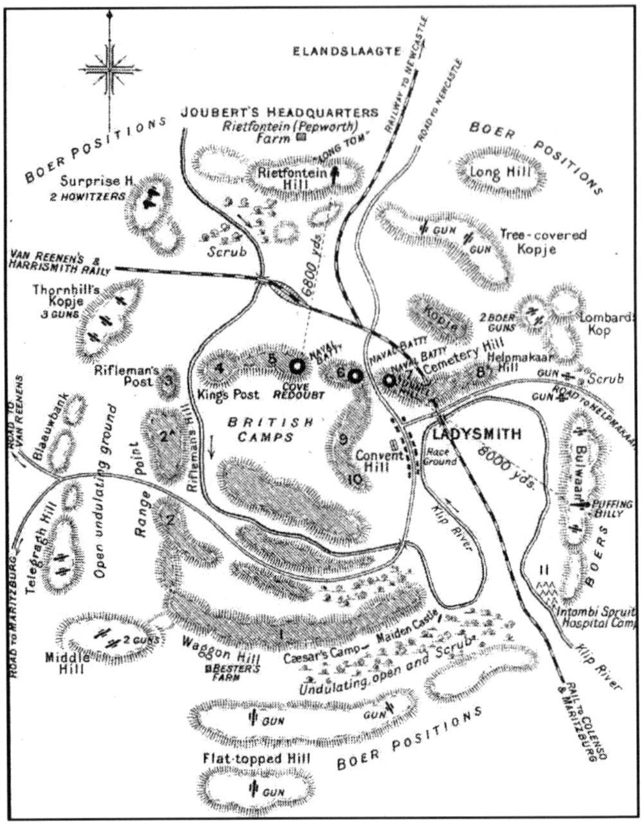

Ladysmith Map.[156]

In sum, with the attacks of both Hamilton and Grimwood stalled out and not able to take the hills, and the steady fire of Long Tom, together with some knowledge that Carleton's force was not successful either, as well as rumors that more forces from the Orange Free State Boers might join the attack, White ordered his commanders to disengage and retire into Ladysmith. By noon the forces were in retreat towards the town. Carleton's force would hold on for another two hours before surrendering as we will see.

The one bright spot was that during the engagement:

> The naval brigade from the HMS *Powerful* had reached Ladysmith by train about 9.30 AM to find the shells from "Long Tom" raining down on the town. Their 4.7 inch guns were unloaded and hitched to bullock teams and marched out to Limit Hill. This was too close to the enemy and they were ordered to retreat. Just as they started to retreat Long Tom was turned on to their position and the second shell broke the carriage of the leading gun, seriously wounding the crew. The other guns retreated and took up a position on the open ground just north of Ladysmith, and at about 12.30 PM opened fire on Pepworth at a range of 6,500 yards. In a very few rounds they effectively silenced Long Tom and the rest of the battery on Pepworth. To this point the British had been outranged by the Boer artillery. [The effect on morale] of the arrival of artillery with a range equal to that of the enemy's guns of position was significant. It went far to counteract the feeling of panic which had pervaded the town since it was apparent that the battle was not going successfully.[161]

General White must have known this artillery unit was on its way. If he had waited until its actual arrival before setting his plan in motion it would have balanced the scales and given him a better chance to prevail. In fact the Admiralty had refused White's initial requests for heavy guns because the captains were reluctant to take them off their ships and repurpose them for land warfare.

[161] Adapted from Times History Vol 2 p. 236

4.7 Inch Naval Gun on improvised mounting (From HMS *Terrible* with notation "Who sup with me require a devil of a long spoon.")

"Naval 4.7 Inch Gun."[162]

Example of a 15-pounder field gun. The largest gun the British had before the arrival of the naval guns.[157]

Devonshire Regiment facing Pepworth Hill, firing from behind boulders, which provided for an effective cover.[158]

Nicholson's Nek

Like at Talana Hill the plan of battle included a column to be sent out separately to try and cut off the retreat or reinforcement of the Boer forces. Also like Talana Hill, this tactic would be unsuccessful and result in over 800 British soldiers being captured. This action came to be known as Nicholson's Nek since that was the intended target for emplacement before the expected ambush but the actual battle was fought on Tchrengula.

Col. Frank Robert Carleton's column was late in setting out on the night of the 29th, accompanied by the 10th Mountain Battery, which required 200 mules to carry all the equipment and ammunition. The order of march was that the Royal Irish Fusiliers led the column followed by the Mountain Battery with the Gloucester regiment in the rear.

He was able to march virtually undetected through the Boer lines at Pepworth Hill, towards Nicholson's Nek, but Carleton was concerned he was not moving fast enough and that he would be caught on the road in the valley well short of his destination of Nicholson's Nek by sunrise and decided to take his force up the hill on his left, Tchrengula or Hog's Back.

[162] See: Navy & Army Illustrated Vol 10 Photo by Cribb "Not only did Captain (Percy) Scott's ingenuity enable the long 12-pounder to be brought into use for fieldwork, but he also adapted a carriage by which the "four-point-seven" and 6-in guns could be utilised with troops in the field. This gun has a calibre and range equal to the heavy Creusot guns with which the Boers had supplied themselves."

No.10 Mountain Battery On The March In Natal, October, 1899.[159]

Close up: British Mountain Battery.[160]

Going out.[161]

While his troops were struggling up the hillside, which became progressively steep, some rocks were dislodged that rolled down amongst the mules and horses; alternatively, several mounted Boers rode down the hill and through the column, firing as they went, with the result that the artillery mules bolted back down the track through the Gloucesters, who thought they were under attack and fired on the mules. As the artillery mules raced through the column they also stampeded the mules carrying the small arms ammunition reserve and disappeared into the night.

Kopje climbing at its most extreme.[162]

Fighting on a Kopje in South Africa.[163]

The Gloucesters re-formed with some difficulty and continued on up the hill. The force arrived on top around 3 AM. The troops occupied positions around what they thought was the summit of the hill and piled up stone sangars for protection. With daylight it became apparent that Carleton's force had occupied a plateau overlooked by a number of higher peaks. The Boers, led by Christian de Wet, assembled on these points and opened fire on the British troops. The British, deprived of their artillery and extra ammunition, tried to hold their ground as best they could. This battle went on from daylight at 5 AM until around 2 PM when a small outlying detachment of British soldiers began waving white flags. Learning of this, Colonel Carlteon, seeing no help coming from Ladysmith or other reinforcements over the course of seven hours of battle, came to the conclusion that the only recourse left to his troops was a bayonet charge into rifle and machine-gun fire and that, while gallant, it would be suicidal. Accordingly, he ordered his bugler to sound "cease fire" and surrendered his force. This was later criticized because some of his forces had been defending portions of the line that were not under as severe attack and others wanted to fix bayonets and "die like men."

The total casualties at Nicholson's Nek were about 46 men killed, and 130 men and 8 officers wounded. The Glo'sters suffered most heavily, losing 33 men killed, and 5 officers and about 75 men wounded, or about 27 percent of their strength on the hill.

In the moment of their triumph the Boers behaved with the same courtesy and unaffected kindheartedness which they had shown at Majuba, and which they displayed after most of their victories. Though exultant, they were not insulting. They fetched water and blankets for the wounded, and treated the prisoners with every consideration. The latter, about 850 altogether, including many of the lightly wounded, after being disarmed, were marched down to Joubert's laager. That same evening they were driven in wagons to Waschbank Station, where they entrained for Pretoria.

No one can read Mr. Carter's vivid narrative of the fight on Majuba (Hill) and then read the account of Nicholson's Nek without feeling that the two are but slightly different variants of the same story. On the British side there is the same failure to occupy advanced positions from which the enemy could have been kept at arm's length, the same crowding of men at points that were never assailed, the same inability to modify the dispositions to meet the actual attack.[163] Refer to: Colonel Carleton's surrender at Nicholson's Nek.[164]

Casualties In Action	Killed		Wounded		Missing and Prisoners			Forces Engaged		
	Officers	Men	Officers	Men	Officers	Men	Total	British	Boer	UK Loss
Ladysmith & Nicholson's Nek	6	58	9	244	43	925	1,285	10,000	20,000	12.9%

[163] Times History Vol 2 p. 254 PF: This mistake will be repeated again at the battle of Spion Kop.

[164] Carleton Frank Robert Crofton Lieutenant Colonel QSA. QSA known to exist Source: List of QSAs with the clasp Talana from the Talana book 1st Battalion QSA roll (Princess Victoria's) Royal Irish Fusiliers.

At the battle of Ladysmith, General White sent a detachment consisting of the 1st Battalion, the Royal Irish Fusiliers and half of the 1st Battalion, the Gloucestershire Regiment, with Number 10 Mountain Battery (equipped with RML 2.5-inch Mountain Guns) to capture a pass known as Nicholson's Nek which lay about 3 miles (4.8 km) to the northwest of Pepworth Hill. The force was commanded by Lieutenant-Colonel Carleton of the Royal Irish Fusiliers. White intended that by seizing the pass, this detachment would prevent the Boer force from the Orange Free State reinforcing the Transvaal Boers on Pepworth Hill, and also prevent the defeated Boers retreating directly north. Carleton's men began preparing sangars on what they thought was the summit of Tchrengula. In fact the mile-long summit was divided into two peaks, and in the dark, the British had occupied only the southern, lower summit.

Christiaan De Wet quickly grasped the situation, and led some men on horseback to gallop to the northern end of the hill, where they dismounted and climbed the hill to occupy the undefended northern summit. From there, the Boers pressed forward along the summit, which was strewn with rocks that gave individual Boer riflemen ample cover. The British sangars were exposed and obvious targets to the Boers, and gave inadequate protection. Like the artillery, the British infantry relied on drill: firing volleys on the command of an officer. By the time the order had been given, the intended target was safe behind cover, while the British soldiers were exposed to fire. The British infantry resisted for several hours, but the numbers of wounded and killed increased and ammunition ran out. Finally, as the main British force could be seen retreating into Ladysmith, Carleton gave the order to surrender. A British officer waved a white flag at the same time. Some of the British troops had seen no reason for the surrender and felt themselves humiliated by the order. Eight hundred soldiers were taken prisoner.

British Prisoners from Nicholson Kop Arriving at Pretoria.[164] British Prisoners leaving Pretoria for Waterval.[165]

General Piet J. Joubert, the Boer commander, came under heavy criticism from his men for permitting the British troops under Hamilton and Grimwood to withdraw into Ladysmith without a more vigorous pursuit. When urged to order a general assault on the retreating British, Joubert is reputed to have said, "When God holds out a finger don't take the whole hand."[165]

General Joubert.[166].   General Joubert and staff at Modder spruit.[167]

A Stone-Wall Defense.[168] Captured British 2.5. in. Mountain Gun.[169] One of the 10th Mountain Battery's 2.5-inch Mountain Guns captured from the British at Nicholson's Nek on 30 October 1899.

[165] He also unilaterally granted a truce for the following day in order to collect wounded and soldiers killed in action.

British forces retreating into Ladysmith after the battle.[170]

It quickly became apparent that siege was inevitable. After allowing the British to retreat relatively unmolested, the Boers did not launch a direct assault on the town. The British reorganized and constructed a defensive perimeter. The garrison would end up besieged for four full months, until February 28th of 1900 when it was relieved by General Buller on his fourth attempt to break the Boer lines on the Tugela River.[166]

Although General French, the cavalry commander, pointed out that cavalry were unlikely to be of much use in a besieged town, White refused him permission to break out. White was trying to preserve the illusion that he still commanded a field force of combined arms, artillery, cavalry, and infantry. The cavalry regiments were not allowed to escape the siege and join Buller to the south of the Tugela River. Instead they languished inactive in Ladysmith for the months of the siege. Finally the garrison ate the horses.

General French must have put in for a transfer because on November 2nd he was ordered to leave Ladysmith along with Douglas Haig.[167] They escaped while under fire, on the last train to leave Ladysmith before the siege began. The Boers tore up the track shortly after their train passed.

"An armoured train at Ladysmith, off to meet the Boers."[168] Defence of Ladysmith. Armoured Train Manned by "Terrible's" Marines Under Captain Mullins, R.M.L.I.

[166] 1. Battle of Colenso - Wikipedia 2. Battle of Spion Kop - Wikipedia 3. Battle of Vaal Krantz - Wikipedia 4. Battle of the Tugela Heights - Wikipedia

[167] Douglas Haig, 1st Earl Haig - Wikipedia

[168] Black & White Weekly Vol 1-2 1899-1900 PF: Patrols were also conducted southwards until such time as Ladysmith was completely

Ian Hamilton remained in Ladysmith for the duration of the siege during which he played an instrumental role in fighting off the assault on the garrison during the battle of Wagon Hill.[169] After the relief of Ladysmith, he was given a command under Lord Roberts who was conducting a campaign up through the Orange Free State. The account of Hamilton's participation in this campaign is chronicled by Winston Churchill in his book *Ian Hamilton's March*.[170]

"The battle of Ladysmith was the first engagement on a large scale between British troops and Dutch burghers, and the first in which the two military systems were fairly matched against each other. And it showed conclusively that in the open field, 12,000 British troops were not a match for an equal number of Boers."[171]

The day after the Battle of Ladysmith, Sir Redvers Buller landed at Cape Town, almost a thousand miles away to the news that the battle had been lost and the garrison was about to be cut off. This thwarted the entire plan of the campaign that Buller had hoped to pursue.

The British could not ignore 12,000 troops under siege and leave them to their fate. They had to be relieved and quickly. Rather than fighting on the rolling country of the Orange Free State, General Buller would have to concentrate his efforts to fighting in the mountains of Northern Natal against an entrenched enemy rather than on the open plains of the Orange Free State. The British would continue to lose battles in northern Natal for four more months before Ladysmith could be relieved.

The vast majority of the forces in Natal were concentrated in the north and were trapped in Ladysmith. After the siege began there was little in the way of military force to prevent the Boer armies from following the train tracks all the way down to Durban. They did conduct raids into this area but chose to concentrate their efforts on breaking the siege and occupying Ladysmith rather than continue their invasion. The fact that the garrison held out against this siege did serve to dissuade the large-scale invasion of Natal.

Conclusion

The first engagements of the conflict saw the Boer armies invade Natal and the Cape Colony, and inflict a series of significant defeats on British forces. The strategy of preemptive invasion was based upon the fact that the Boer republics could not wait for an invasion of their territory since the British Empire could postpone military action until it had mobilized the overwhelming forces at its disposal. This calculated risk of attempting to turn the tables, such that the British would rather sue for peace than undergo the arduous process of reclaiming by conflict the territory occupied by the Boer armies, was not successful. The invasion strategy was also based upon a jingoistic optimism and assumption that people of Dutch descent in the adjoining colonies would join the war on the side of the Boers and that the possibility existed that the British could be evicted from South Africa entirely.

Both of these assumptions proved to be untenable.

surrounded.

169 The Battle of Wagon Hill, January 5th and 6th, 1900. The Boers around Ladysmith were also growing weak from lack of forage. With little action, many fighters took unauthorized leave or brought their families into the siege encampments. Eventually, with the Tugela in flood, preventing Buller from giving any support, some younger leaders persuaded Joubert to order a storming attempt on the night of 5 January 1900, before another relief attempt could be made. The British line south of Ladysmith ran along a ridge known as the Platrand. Under Ian Hamilton, they had constructed a line of forts, sangars and entrenchments on the reverse slope of the Platrand, of which the Boers were unaware. In the early hours of 6 January 1900, Boer storming parties under General C. J. de Villiers began climbing Wagon Hill and Caesar's Camp. They were spotted and engaged by British working parties who were emplacing some guns. The Boers captured the edge of both features, but could not advance further. British counterattacks also failed. At noon, de Villiers made another attack on Wagon Hill. Some exhausted defenders panicked and fled, but Hamilton led reserves to the spot and recaptured some empty gun pits. Late in the afternoon, a terrific rainstorm broke, and the Boers withdrew under cover of it. The British suffered 175 killed and 249 wounded. 52 dead Boers were left in the British positions, but their total casualties were not recorded. Siege of Ladysmith - Wikipedia See Also: Platrand Monument. Johannesburg Light Horse Regiment - Wikipedia

170 Winston Spencer Churchill, Ian Hamilton's March, 1900 The Copp, Clark Company, Ltd. Toronto Ian Hamilton's March (fadedpage.com) Also here: Ian Hamilton's March by Winston Churchill - Free Ebook (gutenberg.org)

Ian Hamilton's march : Churchill, Winston, 1874-1965 : Free Download, Borrow, and Streaming : Internet Archive

171 Times History Vol 2 p. 255

The Boer Lilliputian and British Gulliver.[171]

President Kruger stands on top of John Bull who has been tied down with straps representing the battles won and sieges imposed by the Boer forces. Dundee, Glencoe, Elandslaagte, Modder River, Spion Kop. Colenso is not mentioned by name but only indicated by a strap titled Tugela. Ladysmith, Belmont, and Mafeking are also straps encumbering the giant. A broken sword inscribed "British Prestige" lies on the ground.

The British assumption that their disciplined and professional army would quickly defeat the Boer militia army also proved to be untenable. They eventually defeated the Boer armies in the field and occupied the capitals of both Boer republics, after which they presumed that the Boers would sue for peace to revert to colony status within the Empire. The Boers did not capitulate, and the war dragged on as a guerilla insurgency for over two more years until June of 1902. This part of the conflict resulted in the creation of the first "concentration camps" for Boer civilians, and the death toll in these camps was significant.[172]

After the war the British granted self-governing status to both colonies in 1906, and after a short period of adjustment finally realized its goal of federating the region in 1910 with the creation of the Union of South Africa.[173]

Just after Ladysmith is cut off from reinforcement, Captain Haldane returns to Estcourt after being wounded during the Battle of Elandslaagte on the 21st of October. So, his convalescence was about two weeks. He is still recovering when he returns to duty.

Captain Haldane has a very exciting tale to tell about his capture, imprisonment and escape from a Boer prisoner of war camp. He is also a very good writer and raconteur. Captain Haldane's narrative deals with the period from the outbreak of hostilities in October of 1899 until April of 1900.

[172] Second Boer War concentration camps - Wikipedia
[173] Union of South Africa - Wikipedia

How We Escaped from Pretoria

By: Captain Aylmer Haldane, DSO
2nd Battalion. Gordon Highlanders

"abiit, excessit, evasit, erupit"[174]

Frontispiece Photo, First Edition.[172] Captain Aylmer Haldane DSO.[175] Frontispiece Photo, Revised Edition. Aylmer Haldane, signed photograph taken around 1900.[176]

[174] *He has left, absconded, escaped and disappeared*: abiit, excessit, evasit, erupit: Or, *he has left, gone off, escaped, broken away.* (Cicero, said of Catiline's flight.)

[175] General Sir James Aylmer Lowthorpe Haldane, GCMG, KCB, DSO (17 November 1862 – 19 April 1950) was a Scottish British Army General. Haldane was appointed a staff captain in the Intelligence Section at the War Office on 27 June 1901, promoted to major on 23 July 1902, and received the brevet rank of lieutenant colonel on the following day. He was military attaché with the Imperial Japanese Army from July 1904 to September 1905 during the Russo-Japanese War, and accompanied Japanese forces into Manchuria. Following promotion to brevet colonel, Haldane was appointed Companion of the Bath on 16 March 1906 and granted the rank of colonel on 29 October 1906. From 1906 to 1909, he served as assistant director of military intelligence. On 1 October 1909, Haldane was promoted to temporary brigadier-general and in 1910 become commander of 10th Infantry Brigade. Haldane fought in World War I initially as General Officer Commanding 3rd Division, then part of the British Expeditionary Force. He was given command of 6th Army Corps in France in 1916. After WWI, Haldane was appointed General Officer Commanding Mesopotamia in 1920 and remained in that post until 1922, in the process putting down a revolt with "utmost severity." He retired in 1925. Aylmer Haldane-Wikipedia

[176] PF: While similar, these two photographs were taken during different sittings. In the frontispiece photo for the 1901 edition he is wearing four medals. In the second signed copy he is only wearing three, including his DSO; which had been awarded by the time he published his account in Blackwoods Magazine in August of 1900. File:Aylmer Haldane 002.jpg-Wikimedia Commons

Athenaeum Book Review

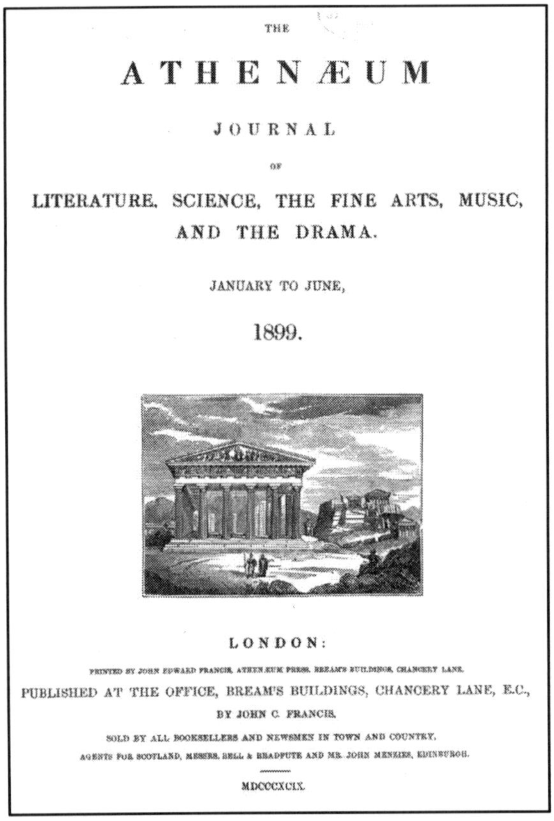
THE

ATHENÆUM

JOURNAL

OF

LITERATURE, SCIENCE, THE FINE ARTS, MUSIC, AND THE DRAMA.

JANUARY TO JUNE,

1899.

LONDON:

PRINTED BY JOHN EDWARD FRANCIS, ATHENÆUM PRESS, BREAM'S BUILDINGS, CHANCERY LANE.

PUBLISHED AT THE OFFICE, BREAM'S BUILDINGS, CHANCERY LANE, E.C.,

BY JOHN C. FRANCIS.

SOLD BY ALL BOOKSELLERS AND NEWSMEN IN TOWN AND COUNTRY.

AGENTS FOR SCOTLAND, MESSRS. BELL & BRADFUTE AND MR. JOHN MENZIES, EDINBURGH.

MDCCCXCIX.

Sample Cover of the Athenaeum.[177]

> **Capt. Haldane's thrilling narrative of his escape from Pretoria, which has been appearing in *Blackwood's*, is now published by Messrs. Blackwood & Sons under the title *How We Escaped from Pretoria*. We can strongly recommend this little paper-covered book as the best work of adventure to which the war has as yet given rise.**
>
> **A less admirable production, but one which is also bright and readable, is Lord Rosslyn's *Twice Captured*, also published by Messrs. Blackwood & Sons, a record of somewhat harum-scarum adventure, put together without much literary art.**

Capt. Haldane's thrilling narrative of his escape from Pretoria, which has been appearing in Blackwood's, is now published by Messrs. Blackwood & Sons under the title How We Escaped from Pretoria. We can strongly recommend this little paper-covered book as the best work of adventure to which the war has as yet given rise.

A less admirable production, but one which is also bright and readable, is Lord Rosslyn's Twice Captured, also published by Messrs. Blackwood & Sons, a record of somewhat harum-scarum adventure, put together without much literary art.

Citation to the review.[178]

[177] The Athenæum was a British literary magazine published from 1828 to 1921. The Athenaeum (British magazine) - Wikipedia
Internet Archive Catalog Record: https://archive.org/details/texts?tab=collection&query=%22The+Athen%C3%A6um%22&sort=date
HathiTrust Catalog Record: Catalog Record: The Athenaeum | HathiTrust Digital Library

[178] Ian's March & Haldane Review: The Athenaeum-No. 3808, Oct. 20th, 1900 p. 507. U. Illinois: #523 - The Athenaeum 1900:2. - Full View | HathiTrust Digital Library

Blackwood's Edition

BLACKWOOD'S

Edinburgh

MAGAZINE.

109571

VOL. CLXX.

JULY—DECEMBER 1901.

NEW YORK:
THE LEONARD SCOTT PUBLICATION CO.,
7 AND 9 WARREN STREET.

Publishers of
NINETEENTH CENTURY, CONTEMPORARY REVIEW, WESTMINSTER REVIEW,
FORTNIGHTLY REVIEW, BLACKWOOD'S MAGAZINE,
QUARTERLY AND EDINBURGH REVIEWS.

1901.

BLACKWOOD'S

EDINBURGH MAGAZINE.

No. MXVIII. AUGUST 1900. VOL. CLXVIII.

HOW WE ESCAPED FROM PRETORIA.

BY

CAPTAIN AYLMER HALDANE, D.S.O.,

2ND BATT. GORDON HIGHLANDERS.

As those of my readers who chanced to see the brief accounts of our escape from Pretoria which have appeared in the English and other papers may possibly observe some slight discrepancy between those narratives and the one in the following pages, I take this opportunity of explaining the cause thereof. In the press account I was unable to take the public entirely into my confidence, since on the night of the seventh day after leaving Pretoria we met with very material assistance from our own countrymen, the fact of which could in no way be disclosed without placing the lives and property of those who unhesitatingly risked both in the most imminent peril. Until the war approached a successful termination I could not give a full and complete account of our adventures, and at the same time tender our heartfelt thanks to those without whose timely aid we should have had the greatest difficulty in reaching the Portuguese frontier.

LADYSMITH, *May* 1900.

I.—CHIEVELEY TO PRETORIA.

After the fight of the 21st October at Elandslaagte I was sent to Pietermaritzburg together with other wounded, there to be cared for and made fit to resume my place at the front. But during the last days of October and the first

VOL. CLXVIII.—NO. MXVIII. M

Citation to Article(s)[179]

Book Cover

[179] How We Escaped From Pretoria in Two Parts in: Blackwood's Magazine - Wikipedia
August Issue: Blackwood's Magazine 1900-08: Vol 168 Iss 1018 : Free Download, Borrow, and Streaming : Internet Archive PDF of August Issue: Blackwood's Magazine 1900-08: Vol 168 Iss 1018 (archive.org)
September Issue: Blackwood's Magazine 1900-09: Vol 168 Iss 1019 : Free Download, Borrow, and Streaming : Internet Archive PDF of September Issue: Blackwood's Magazine 1900-09: Vol 168 Iss 1019 (archive.org)

Note to Revised Edition.

Encouraged by the favourable reception of the first account of my adventures and those of my companions, and at the desire of many friends, I have been induced to prepare a fuller narrative of our escape from Pretoria. The original articles on the subject, published in Blackwood's Magazine,[180] were written very shortly after the events they recorded. Since that time I have recalled many incidents then over looked, and have been placed in possession of facts of which I was then unaware. The countless questions to which I have been subjected may have been difficult to bear with perfect equanimity, but they have at least had the merit of stimulating memory and drawing attention to those points which were in the first instance not fully described. In the present edition it has been my endeavor to omit nothing which may present features of interest.

My acknowledgment and thanks are due to Major the Hon. Robert Carnegie of my regiment for the drawings-executed by him from my description-represented, Hiding under floor of school sketch, Hiding in river sketch, Hiding in grass sketch, and Sketch inside native dwelling. and to Messrs. Harmsworth for .... Majuba Hill From the South, Pretoria, Model School (back view) and Emergency Ration.

London, 15th November 1901.

Note to First Edition.

As those of my readers who chanced to see the brief accounts of our escape from Pretoria which have appeared in the English and other papers may possibly observe some slight discrepancy between those narratives and the one in the following pages, I take this opportunity of explaining the cause thereof.

In the press account I was unable to take the public entirely into my confidence, since on the night of the seventh day after leaving Pretoria we met with very material assistance from our own countrymen, the fact of which could in no way be disclosed without placing the lives and property of those who unhesitatingly risked both in the most imminent peril.

Until the war approached a successful termination I could not give a full and complete account of our adventures, and at the same time tender our heartfelt thanks to those without whose timely aid we should have had the greatest difficulty in reaching the Portuguese frontier.

Ladysmith, May 1900.

[180] How We Escaped From Pretoria in Two Parts in: Blackwood's Magazine - Wikipedia
August Issue: Blackwood's Magazine 1900-08: Vol 168 Iss 1018 : Free Download, Borrow, and Streaming : Internet Archive
September Issue: Blackwood's Magazine 1900-09: Vol 168 Iss 1019 : Free Download, Borrow, and Streaming : Internet Archive

I. Chieveley to Pretoria

"The most unkindest cut of all"[181]

After the fight of the 21st October 1899 at Elandslaagte[182] I was sent to Pietermaritzburg together with other wounded,[183] there to be cared for and made fit to resume my place at the front. But during the last days of October and the first of November events rapidly developed; and Ladysmith becoming surrounded by the Boer main army,[184] I found myself cut off from my regiment,[185] which formed part of the beleaguered garrison. As soon as I could walk I obtained permission to proceed to Estcourt, the farthest point north held by the relieving force, whence I hoped to have an opportunity of passing through the Boer lines with despatches, and so rejoin my regiment.

At Estcourt I was attached for duty to the 2nd battalion of the Royal Dublin Fusiliers, one of those battalions which had taken part in the dashing capture of Talana Hill[186] and the retreat to Ladysmith, whence it had been sent to Colenso and later to its present quarters.

Panorama of Estcourt. 1 Camp of Dublin Fusiliers. 2 Camp of Durban Light Infantry "X" Camp of 'A' squadron of Imperial Light Horse.[173]

"A war view of the gallant Dublin Fusiliers (the 'Fighting Devils') at Estcourt."[174]

[181] Shakespeare, *Julius Caesar*, When Antony calls Brutus's stabbing of Caesar "the most unkindest cut of all," he is playing on two senses of "unkind"— "inhumane" and "unnatural."

[182] The Battle of Elandslaagte (21 October 1899) was a battle of the Second Boer War, and one of the few clear-cut tactical victories won by the British during the early stages of the conflict. However, the British force retreated afterwards, throwing away their advantage. Battle of Elandslaagte-Wikipedia

[183] PF: I believe Captain Haldane was wounded in the leg but whether by gunfire, shrapnel or other cause is not known.

[184] After the battle of Lombards Kop on October 30th 1899, General White's 'Field Force' retreated to Ladysmith proper and the last train south towards Colenso got through under heavy fire on November 5th. After that the Boers tore up the tracks and the city and garrison were under siege. Battle of Ladysmith - Wikipedia

[185] He was a Captain in the Gordon Highlanders. Gordon Highlanders-Wikipedia

[186] The Battle of Talana Hill, also known as the Battle of Glencoe, was the first major clash of the Second Boer War. A frontal attack by British infantry supported by artillery drove Boers from a hilltop position, but the British suffered heavy casualties in the process, including the mortal wounding of their commanding general Sir William Penn Symons. Battle of Talana Hill-Wikipedia

Several days passed unbroken by any event worth recording. Day after day, generally at the same hour, the armoured train, or (as it was named by the men in camp) "Wilson's death-trap," used to press forth unattended beyond the line of outposts, heralding, by agonised gasps and puffs, and clouds of smoke and steam, its advent to the far-sighted, long-hearing Boer.

PF: Note that no door is provided for troops to enter or exit in order to engage any hostile forces.[175]

A second view of the train boarding, on November 5th, the day Rene Bull accompanied the reconnaissance; titled "Before the disaster, November 15th.[176] [1899] Men of the Dublin Fusiliers embarking on the armoured train. In the [train] truck on the left is Rene Bull,[187] snap-shooting." [The man standing up in the railroad truck with a hat on and looking down into what is presumably his camera.] Loading a truck with rails and tools.[177] for repairing the line broken up by the Boers between Colenso and Pietermaritzburg. The truck was afterwards attached to the armoured train.

[187] René Bull was a British illustrator and photographer. He was born in Dublin on 11 December 1872 to a French mother and an English father. He went to Paris to study engineering, but embarked on an artistic career after meeting and taking drawing lessons. He went to South Africa to record the Boer War until the relief of Ladysmith in March 1900. As he was wounded in 1900, Bull was invalided out. René Bull - Wikipedia

Firing on the Boers. Reconnoitering in Armoured train near Colenso.[178] "In touch with the Boers near Colenso"[188]

"A reconnaissance from Estcourt. Asking a Zulu if he has seen any Boers about."[189 190] Trench at Estcourt.[179]

Daily, too, did it return in safety to the siding whence it had sallied forth at daybreak on its fruitless mission.

How relieved the occupants looked when they climbed over its plated sides and congratulated themselves that their turn to form the freight of this moribund engine of war would not come round again for at least some days.

Mine, alas! arrived on the 15th of November.

On the preceding day there had been an alarm in camp. Corps stood to their arms and forthwith occupied the various defences and localities that they were detailed to hold. That night I received orders to reconnoitre on the following day with the armoured train.

I am not going to recount the events of that day, which already have been portrayed by an abler pen than mine.[191]

[188] Black & White Weekly Vol 1-2 1899-1900 Photo by "Our Special Correspondent" PF: this must be early on since it is an open carriage, even more dangerous.

[189] Black & White Weekly Vol 1-2 1899-1900 Photo by "Our Special Correspondent" René Bull. The man in the hat, with moustache, appears to be Bennet Burleigh. He was a friend of René Bull's, and features in several of René Bull's photos. Burleigh was the Correspondent for the *Daily Telegraph* and wrote "The Natal Campaign."

[190] Bennet Graham Burley (1840 – June 17, 1914) was a Scottish-born pirate, Confederate spy and journalist. Later in life, he changed his surname to Burleigh and became a celebrated war correspondent for London's *The Daily Telegraph*. Bennet Burleigh - Wikipedia File:Bennet Burleigh 1898.jpg - Wikipedia The Natal campaign : Burleigh, Bennet, d. 1914 : Free Download, Borrow, and Streaming : Internet Archive

[191] Captain Haldane is referring to Churchill's account which was previously published in his work *London to Ladysmith via Pretoria*.

Major-General H. J. T. Hildyard,[180] commanded the troops at Estcourt. Colonel Long. [181] Haldane receives his orders from Colonel Long.

After Action Report

Captain Haldane wrote the following report while a prisoner of war in Pretoria. It was transmitted with the permission of the Transvaal authorities to Haldane's superiors while he was in their custody.

Captain Haldane's Report of Action on November 15th, 1899[192]

From: Captain Aylmer Haldane 2nd Battalion Gordon Highlanders

To: The Chief of Staff Natal Field Force.

[From] Pretoria 30th November 1899[193]

Sir:

I have the honour to forward the following report, concerning the capture of an armoured train, which left Estcourt under my command on the 15th November 1899.

1. On the night of the 14th, November, I received verbal instructions from Colonel Long, R.A.[194] commanding at Estcourt, to take a 9 lb. muzzle loading naval gun, and detachment H. M. S. Tartar, 1 company Royal Dublin Fusiliers and 1 company Durban Light Infantry—and with the – armoured train reconnoitre in the direction of Colenso, proceeding with caution and keeping out of range of the enemy's guns.

[192] PF: With the assistance of Ms. Katharine Thomson an archivist with the Churchill Archives Centre, she was able to locate Captain Haldane's handwritten after action report. I was able to transcribe a large portion of the text but I would like to credit Ms. Thomson for her invaluable assistance in transcribing the remainder and correcting my portion as well. This transcription has used full words where abbreviations occur in the original. Some punctuation has been added and text that has been inserted for clarity has been bracketed with square brackets [ ]. Source Repository: Churchill Archives Centre. Reference Code: CHAR 1 23 18-19 Title: Report of Action on November 15th, 1899 by Captain Aylmer Haldane. Unbeknownst to us at the time, Randolph Churchill had transcribed this report and included it in volume 1 of his official biography (p. 447-450 of Houghton Mifflin 1966 Ed.) and again in the companion volume 1 Part 2 (1896-1900) (p. 1064-1067) This Crown Copyright material is public domain in the UK 50 years after the first publication, (2017).

[193] Dr. J. W. B. Gunning is the censor at the State Model School where Captain Haldane is detained. It is remarkable that this report was able to be transmitted two weeks after the event. Churchill was also permitted to transmit his report dated Monday November 20, 1899, titled The Fate of the Armoured Train as well. [Full name of Gunning per, Brian Davis, Churchills in Africa p. 201]

[194] C. J. Long: Colonel Long (Royal Horse Artillery) is a gunner of much distinguished service, and it is impossible not to admire the gallantry that led him into the crucial mistake made in the handling of the guns at Colenso on December 15. Upon that unfortunate occasion Colonel Long was dangerously wounded. It was intended that he should come into action covered by the 6th Brigade, but the 14th and 66th Batteries were led into the open, where they were subjected to a fire in which it was impossible for men to live. Ten of the twelve guns were later captured by the enemy. Anglo Boer War - Name search

2. The train which was made up and manned as follows: -

[1] Ordinary [railroad] truck gun detachment - H.M.S. Tartar line guard.

[2] Armoured [truck] 3/4 company Royal Dublin Fusiliers.[195]

[3] Engine

[4] Tender

[5& 6] 2 Armoured trucks 1/4 company Royal Dublin Fusiliers. 1 company Durban Light Infantry.

[7] Breakdown gang truck and line guard.

Left Estcourt at 5:10 AM, and on arrival at Frere station at 6:20 AM I reported [by telegraph] that no enemy had been seen.[196]

Second Battalion Royal Fusiliers arriving in Cape Town.[182]

Cape Town Harbor.[183]

Onboard an Armoured Train.[197]

Armoured Train passing Frere Station.[184]

Here, I met a party of 8 men of the Natal Mounted police,[198] from whom I learnt that their patrols were in advance [of the train] and reconnoitering towards Chieveley, whither I proceeded after a halt of 15 minutes.[199]

[195] In 1930, when Churchill diagrams his recollections of the event he includes two armoured trucks for a total of eight cars on the train.

[196] PF: It was raining on the day that the reconnaissance took place. Leo Amery, Churchill's tent mate, wanted to go but when Winston shook him at 5 AM he declined to go because it was raining and the train never left early; usually around 8 AM. This day, Haldane was determined to get the reconnaissance accomplished as soon as possible and left on time.

[197] These are Royal Dublin Fusiliers but not on the day that the ambush took place. It was raining that day and the train left the station at 5:10 AM. Credit: Neville Constantine, Neville C. at Angloboerwar.com

[198] See: Natal Mounted Police - Wikipedia

[199] Subsequently Haldane will opine that it is Churchill's encouragement that affected Haldane's better judgement to venture beyond

On reaching there [Chieveley] at 7:10 AM, I received a telegraph message from Estcourt (enclosure marked "A") and in acknowledging it stated that a party of about 50 Boers and 3 waggons was visible, moving South on the West side of the railway. I at once retired towards Frere and on rounding the spur of a hill which commanded the line, the enemy suddenly opened on us at 600 yards with artillery, one shell striking the leading truck.

7.5 cm Krupp Field Gun.[200]

Front View of 7.5cm Krupp Gun.[201]

Boer Pom-Pom, Or Vickers-Maxim Gun.[202] Note the belt fed ammunition held by the man behind the gun. ZAR Norden Maxim Mafeking.[185]

ZAR Staats Artillery using Maxim Pom Pom.[186] Maxim Automatic Machine-Gun (the "Pom Pom")[187] And the Boer Maxim-Nordenfeldt,[188]

Frere. This contemporary account indicates that Haldane had additional information that the patrols were "in advance" and "reconnoitering towards Chieveley" to support his decision to proceed.

[200] The Oranje Vrijstaat Artillerie Corps made use of the "export model" of the German horse artillery's 8-cm leichtes Feldgeschütz Material C/73. Officially it was denoted as: 7.5cm Krupp: Feldkanone L/27, but during the Boer War it became known as the "Free State Krupp". Courtesy of Neville Constantine: Anglo Boer War - Artillery and Ammunition - Page 26 - Boer War Forum

[201] Courtesy of Neville Constantine: Anglo Boer War - Artillery and Ammunition - Page 13 - Boer War Forum

[202] This gun was in action at Alleman's Nek and elsewhere. Buller's Campaign with the Natal Field Force 1902

3. The driver put on full steam and ran down the gradient at a rapid rate and almost immediately on reaching the level, 3/4 of a mile from Frere, the break down gang truck and 2 armoured trucks, which preceded the engine were derailed, one of the latter standing partly across the track. The enemy, lying in ambush had allowed the train to pass towards Chieveley, and had then placed a stone on the line, which the guard, probably owing to the shell fire had neither seen nor reported.

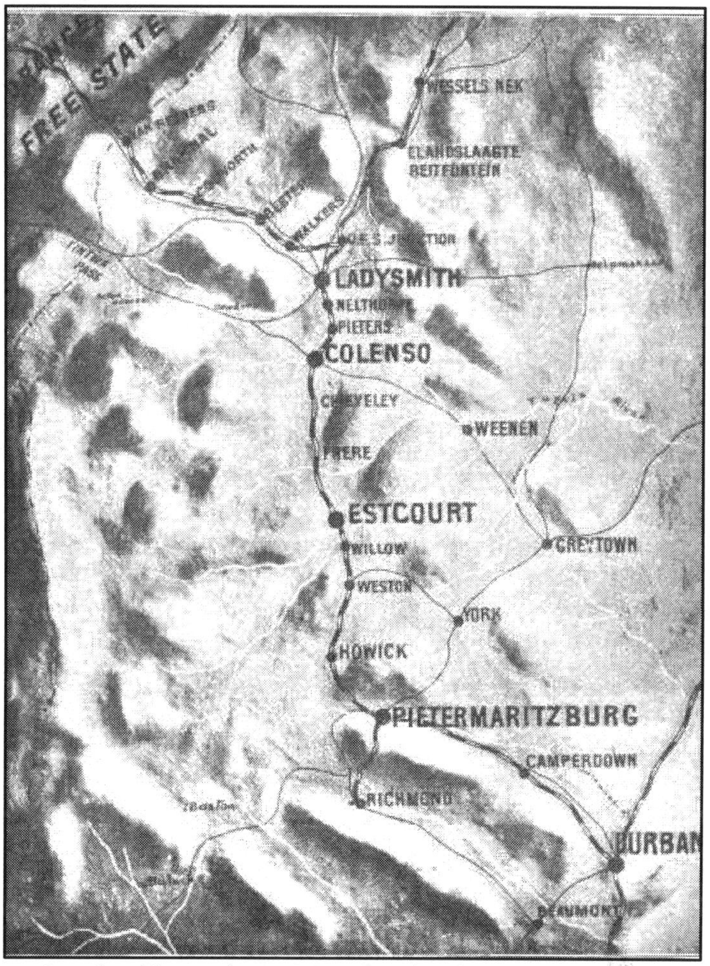

Close-up map. Google Image © 2023 Image Landsat/Copernicus. Map showing larger area including Ladysmith.[189]

Reported to be place where the stones were laid on the track.[203] 1st Royal Dublin Fusiliers fighting from inside the train.[190]

[203] This is perhaps staged put purports to show the stones used to throw the train off the tracks. Credit: Neville Constantine, Neville C. at Angloboerwar.com

The Armoured Train Disaster.[204]

Sketch of the train after its return to Estcourt.[191]

4. Mr. Winston S. Churchill, who was with me in the truck next [to] the gun truck offered me his services and knowing how thoroughly I could rely on him, I gladly accepted them and undertook to keep down the enemy's fire; while he endeavoured to clear the line. Our gun came into action at 900 yards, but after 4 rounds was struck by a shell and knocked over: I reached the gun detachment with the armoured truck whence I kept up a continuous fire on the enemy's guns, considerably disconcerting their aim as I was afterwards informed and killing 2 and wounding 4 men. The Boers meantime maintained a hot fire with rifles three 15 lb. Creusot guns and a maxim shell gun, but as the shells struck the engine and trucks obliquely the armoured plating was only occasionally penetrated.

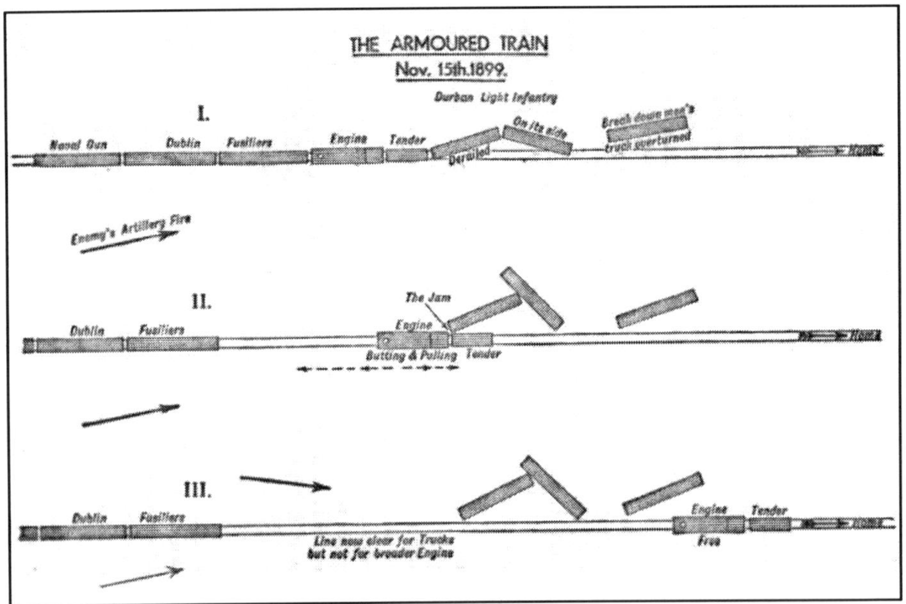

Diagram of the armoured train.[205]

For an hour efforts to clear the line were unsuccessful as the trucks were heavy and jammed together and the break down gang were not to be found, but Mr. Churchill with indomitable perseverance continued his difficult task and about 8.30 A.M. the engine forced its way past the obstructing truck, which however again fell forward some inches across the line. Shortly before this, the coupling which fastened the armoured and gun trucks to the engine had been broken by a shell and when the engine cleared the wreckage, they were left behind.

[204] A sketch by Mr. Thomas Frankland, an officer in the train, forwarded by Mr. Churchill from the States School Prison at Pretoria to the Editor of The Graphic with the comment: "The enclosed sketch seems to me to be a very accurate presentation of what actually occurred." By permission of the "Daily Graphic," From: Winston Spencer Churchill by A. MacCallum Scott, Methuen, London 1905

[205] Image Source: WSC, My Early Life. Churchill has included 8 cars in the train where Haldane only recites 7 in his report.

"Disaster strikes the armoured train on November 15th, 1899."[192]

Perhaps it was possible to again remove the wreckage and re-couple the trucks to the engine, but as the cab of the latter was now crammed with wounded who would have been scalded had a shot struck the boiler as the pipe of the reserve water tank was torn open and the water rushing out, as the front of the engine was in flames and because I apprehended lest the enemy seeing the engine free should again tamper with the rails I resolved to allow it to retreat out of range towards Frere and withdrawing the men from the trucks made a run for some houses 800 yards distant where I had hopes of making a further resistance.

"The attack on the armoured train."[193]

As the enemy had not relaxed his artillery and musketry fire and there was absolutely no cover, the men became considerably scattered along the line and in a formation ill-adapted to offer resistance. To my disgust, I saw 2 men 200 yards in front of me holding up white handkerchiefs. I shouted and ran towards them, but at this moment the Boers, who unperceived by us had left the hill on our right rear and seeing the signal had stopped their long range fire, galloped among the retreating soldiers—who remained uncertain what to do—and called upon them to surrender. Under these circumstances, I, 1st Lieutenant Frankland, 2nd Dublin Fusiliers and 50 men were captured, the remainder making for their retreat across the railway bridge to Frere station; whence I understand they and the engine reached the British lines at Estcourt and safety.

5. I mention here with some diffidence, owing to the unfortunate termination of this reconnaissance, to record my acknowledgement of the services of the following:

1st Lieutenant Frankland of 2nd Dublin Fusiliers a gallant young officer who carried out my orders with coolness.

A[ble] seaman E. Read[206] H.M.S. Tartar, L[ance] Corporal L. Connell,[207] P[rivates] Phoenix[208] and Kavanagh,[209] 2nd Dublin Fusiliers, who notwithstanding that they were repeatedly knocked down by the concussion of the shells striking the armoured trucks, continued steadily to fire until ordered to cease.

Captain Wylie[210] and his men of the Durban Light Infantry who ably assisted in covering the working party engaged in clearing the line and who were much exposed to the enemy's fire. I regret I am unable to give exact details of any deserving special mention but I beg to suggest Captain Wylie be called upon for a report.

Mr. Winston S. Churchill, whose valuable services have already been detailed in the text of this report and whom, owing to the urgency of the case, I formally placed on duty. I would point out that while engaged in the work of saving the engine, for which he is mainly responsible, he was continually exposed to the full fire of the enemy. I cannot speak too highly of his gallant conduct.

The driver of the engine, who though wounded remained at his post.

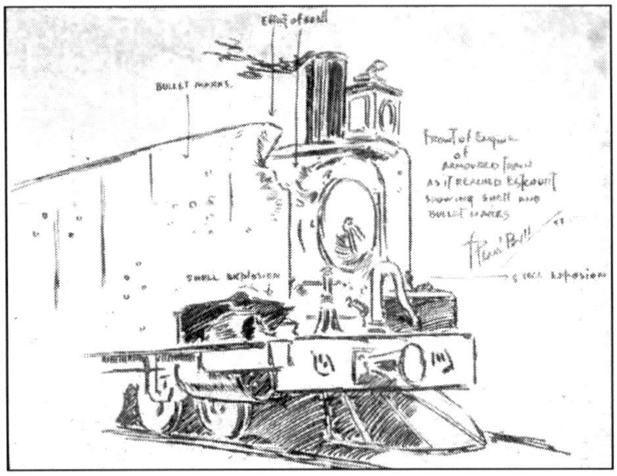

Charles Wagner and Alexander Stewart.[211] The Albert Medal.[194] Sketch of the train after its return to Estcourt.[195]

6. I understand that the Boers buried 4 of our killed and that they have 13 wounded in hospital. Of the 52 taken prisoner, 7 including Mr. Churchill are slightly wounded.

7. I should like to be accorded an opportunity of making a report in detail of the many defects of the armoured train and the dangers and disabilities under which it lies in an unsupported reconnaissance, but the popular opinion of this flimsy military machine may best be shown by the fact that in the camp it was known as Wilson's deathtrap.[212]

8. I attach list of prisoners, rough sketch and copy of telegram referred to in para[graph] 2

I have the honour to be
Your obedient servant
Aylmer Haldane, Captain
2nd battalion Gordon Highlanders

[206] E. J. Read. Able Seaman QSA (3) Nat DH Belf. Ref: 187.682. Prisoner Frere 15 Nov 99 Source: QSA medal rolls. Anglo Boer War - Name search

[207] No person named L. Connel is listed in Royal Dublin Fusiliers. There is a J. Connel in the 1st Battalion. Anglo Boer War - Name search

[208] Two individuals are listed in the Royal Dublin Fusiliers. M. Phoenix and T. Phoenix. Both sources are the QSA and KSA medal rolls.

[209] Two individuals are listed in 1st Battalion P. Kavanagh and T. Kavanagh. Both sources are the QSA and KSA medal rolls.

[210] James Wylie No. 28 Lieutenant Colonel Source: Nominal roll in WO127 Durban Light Infantry. Anglo Boer War - Name search "Sixty men of the Durban Light Infantry formed part of the personnel of the armoured train which at this time patrolled daily from Estcourt to Colenso. On 15th November a rail was removed or twisted, and the train was attacked; 2 men of the Durban regiment were killed, Captain J. Wylie and 15 were wounded, and 19 were taken prisoners, of whom 8 were wounded, 1 mortally. A squadron of Carbineers and one of Imperial Light Horse came out to the help of the armoured train. These reinforcements drove back the enemy, killing 3." See: Durban Light Infantry article @ https://www.angloboerwar.com/?option=com_content&view=article&id=326

[211] Mr. Wagner and Mr. Alexander Stewart were awarded the Albert Medal, for Lifesaving first class and second class respectively, in 1910 when Churchill was Home Secretary. Sandys p. 67. See Albert Medal for Lifesaving - Wikipedia Image: Alexander M. Stewart.

[212] There is apparently no individual or firm named Wilson who was responsible for the construction of the armored train. It was simply an appellation that was given to the contraption and it stuck.

Aftermath

"Out to the relief of the wrecked train, November 15th, The Imperial Light Horse crossing the bridge to the north of Estcourt which the Boers are threatening to blow up."[196]

Securing the site.[197] The overturned ganger truck.[198] Because the engine was reversing back towards Estcourt. This was the lead car of the train as it struck the stones laid on the track.

Overturned Railroad Ganger Truck.[199] The Men in the Background are Gathered at the Gravesite. Wrecked Armoured Train.[200] Effects of the Boer Shells on the Wrecked Armoured Train. 'The Plates were Torn Like Paper.'[201] And Close-up of Damage Done to Train.[202]

The Grave of those killed in the armoured train disaster. (Before 1904)[203] Gravesite in 1978.[204]

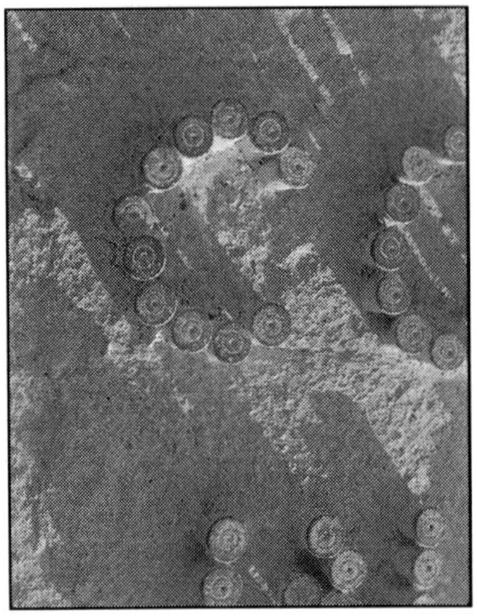

Grave of fallen Royal Dublin Fusiliers.[205] Rifle shell casings used to spell words on grave.[206]

Private J. Birney.[213] Private J. McGuire.[214] Private M. Balfe.[215]

Buller's Thoughts

General Sir Redvers Buller to Field Marshal Lord Roberts.

January 19th 1900.[216] Spearman's Hill

Sir,

I have the honour to forward herewith Captain Aylmer Haldane's report respecting the capture of the armoured train near Frere on the 15th November 1899.

Captain Haldane was unfortunate as this was the first engagement he was in after his recovery from a wound received at Elandslaagte.

From all accounts which I have received I believe that Captain Haldane's action on this occasion was very creditable to him, and that he acted in trying circumstances with great judgement and coolness.

I have the honour to be, Sir, Your obedient servant.

Redvers Buller General Officer Commanding-in-Chief, Natal

213 Birney J. No. 5031 Private 2nd Battalion Killed in action 15-11-1899 Place: Frere. Armoured Train Site Source: In Memoriam by S Watt. Also QSA (1) Source: List of QSAs with the clasp Talana from the Talana book.

214 McGuire E / J No. 5861 Private 2nd Battalion Demise: Killed in action 15-11-1899 Place: Frere, Armoured Train Site Source: In Memoriam by S Watt.

215 Balfe M. No. 5795 Private 2nd Battalion Demise: Killed in action 15-11-1899 Place: Frere, Armoured Train Site Source: In Memoriam by S Watt. See also: Balfe M. No. 5792 Private 2nd Btn. Killed at Estcourt. 12 Nov 1899. Source: Natal Field Force Casualty Roll, page 196 line 3 Also QSA (1) Source: List of QSAs with the clasp Talana from the Talana book.

216 OBC Vol 1 Part 2 p. 1068.

Roberts' Observations

Lord Roberts to Lord Lansdowne.[217]

February 15th 1900.[218] Army Head-Quarters South Africa

Camp Wegdraai Drift

Confidential

My Lord,

I have the honour to submit for your Lordship's information, despatch No MS/294, dated 19th January 1900, from General Sir Redvers Buller, enclosing a report by Captain J. A. L. Haldane, DSO, Gordon Highlanders, on the capture of an armoured train near Frere on the 15th November 1899.

I concur in the remarks made by Sir Redvers Buller.

I may add that, though armoured trains may be useful as an adjunct to the defence of a fortress, or possibly against an uncivilized enemy, I regard their employment in civilized warfare for purposes of reconnaissance as distinctly dangerous to the troops occupying them, and very rarely advantageous. The information which an armoured train is sent out to acquire can, as a rule, be obtained just as accurately and with infinitely less risk, by a few intelligent mounted Officers or men.

I have the honour to be My lord, Your Lordships' most obedient Servant

Roberts, Field-Marshal, Commanding-in-Chief, South Africa

Armoured Train Leaving Durban. [207] Nicknamed 'Wilson's Death Trap' by the troops. Apparently for no particular reason. There was no 'Wilson' associated with its design.

[217] Henry Charles Keith Petty-Fitzmaurice, 5th Marquess of Lansdowne, (14 January 1845 – 3 June 1927), was a British statesman who served successively as Governor General of Canada, Viceroy of India, Secretary of State for War and Secretary of State for Foreign Affairs. In 1917, during the First World War, he wrote the "Lansdowne letter", advocating in vain a compromise peace. He had the distinction of having held senior positions in Liberal and Conservative Party governments.

After the Unionist victory in the general election of October 1900, Salisbury reorganised his cabinet, gave up the post of Foreign Secretary and appointed Lansdowne to replace him. Lansdowne remained at the Foreign Office under Salisbury's successor Arthur Balfour. Henry Petty-Fitzmaurice, 5th Marquess of Lansdowne - Wikipedia

[218] OBC Vol 1 Part 2 p. 1068.

Resume Haldane Narrative

I will take up the tale at the point where I became a captive in the hands of the Boers-for it is my purpose in these pages to treat mainly of my captivity and escape, not to launch into matters to which full justice already has been done.[219]

The morning of my capture-which took place at 8.50 A.M. after defending the armoured train for an hour and a half was raw and damp: mist hung heavily on the surrounding hills, and a small drizzling rain, which from time to time exerted itself until it attained to the magnitude of a shower, rapidly penetrated our garments, and added to the misery and discomfort of our unenviable position. Our escort politely begged us not to hurry, saying that there were lots of time, and that although they were unable themselves to spare us either food or clothing, we should find on reaching Pretoria all we required-nay, even that there we should be provided with the usual games with which the British officer delights to exercise himself. From their description, all that could delight the heart of man or prisoner awaited us in the Boer capital.

No doubt they meant kindly, very kindly; but no promised land, no vista of untold delight, as seen through our eyes at that moment, could have made one's pulse beat even a shade faster. Some deep and dreary dungeon, or, still more so, an *oubliette*,[220] seemed consonant with my feelings.

Meanwhile, trudging along the muddy veldt road, we came upon a number of Boers who were cleaning their rifles after being engaged with us and with them was a party of Staats Artillery officers.

With one of the latter, the officer in command, who was neatly dressed in cord coat and breeches, faced with blue, I had some conversation. He began by asking me why we had not surrendered at once, and congratulating us on the defence of the train, but lamenting that his guns had not been better laid, in which case, according to his reckoning, our shrift would have been a short one.

He mentioned that the Boers had fully resolved to capture the train and its luckless occupants, but that they had had sufficient patience to wait until the psychological moment, or perhaps that which coincided best with their own convenience, had arrived.

At a later period I ascertained that General Joubert had personally superintended the arrangements for the derailment of the train, and in addition had given directions as to where the guns and men who engaged us were to be posted. He then went on to say that his three heavy guns had each fired an average of thirty rounds, and that (as we well knew) he had employed a Vickers-Maxim gun against us, a machine which had since received the *sobriquet* of "Pom-pom."

Passing on from here, we were ushered into the midst of the camp of a large commando which was snugly ensconced behind some low hills not far from the commanding ground overlooking the spot where the Boers had wrecked our train.

Here Churchill's papers were examined and taken to General Joubert,[221] whom, I regret to say, we were not allowed to see; and great excitement was displayed on its becoming known that the real live son of a lord was amongst the prisoners.

A crowd of rough looking, and for the most part bearded, fellows soon gathered round us, and while Churchill's case was under the consideration of the commandant-general, a running fire of questions on various matters connected with the struggle which had so recently begun was maintained.

It was plain to me now that we were in the thick of a strong force which was on its way southwards. According to my own computation, I should have put down the numbers seen as from 3,000 to 4,000 men; but the Boers themselves stated that there were 6,500. As events afterwards proved, our unfortunate reconnaissance had encountered that portion of the Boer army which had been detached from the force engaged in besieging Ladysmith, and which was about to make what resulted in an abortive attempt to reach Durban.

After a brief halt, during which General Joubert had come to the decision that Churchill, although a non-combatant, should not be then and there released, but should accompany the other prisoners to Pretoria, we were ordered to move on. Doubtless the Boer commander was well aware of the valuable assistance Churchill had

[219] He is referring to Churchill's account in *London to Ladysmith via Pretoria.*

[220] A secret dungeon with access only through a trapdoor in its ceiling.

[221] Piet Jacobus Joubert (20 January 1831 – 28 March 1900), was Commandant-General of the South African Republic from 1880 to 1900. He also served as Vice-President to Paul Kruger from 1881-1883. Piet Joubert-Wikipedia

given in saving the engine from capture, and this, added to the fact that he must have mentally noted the strength of the commando, offered a strong argument against release.

Boers on campaign during the Boer War.[208] General Joubert With His Son-In-Law, Malan, and His Secretary, Bracht, In His Tent Before Ladysmith.[209]

The country about here is not interesting, although it was destined to become so ere a month had passed, and we ourselves were not in a mood to appreciate the beauties of nature, more especially as the inner man was beginning to cry out that he was being neglected.

Somewhat weary after a tramp of sixteen miles, and further fatigued by the exhausting experience of the morning, we at length came within sight of the village of Colenso.

It presented every appearance of some unusual condition of affairs,—windows smashed, doors unhinged, furniture and crockery scattered broadcast over the grass grown streets-in short, every indication of a hostile occupation.

Colenso Railway Bridge looking South from the Boer side of the river. Note tents in open area and people on island in the river. Colenso Bridge sometime after the battle of Colenso which took place on December 15th, 1899.[210] Note the change in the river's flow from dry to rainy season.

Feeling considerable anxiety as to what reports regarding our fate might have travelled back to Estcourt with the engine, we were much relieved to find here two Boer war correspondents, who, after assuring us that that portion of the train had escaped, undertook to telegraph to England at the earliest opportunity the fact that we were alive.

Night was now falling, and we were hurried on to a large goods-shed contiguous to the main platform of the insignificant railway-station. We were directed to accept this as our lodging for the night, and expectations were held out that shortly an ox would be killed, and we should have our share of what promised to be a somewhat tough and un-toothsome meal. This promise was not long in the fulfilment. The beast of burden was slain, and almost as soon as the proverbial Indian dâk-bungalow chicken finds its way from the compound to the pot, we were busily engaged in cooking scraps of meat on sticks held over a fire, and looking forward to enjoying what are called in the East *kababs*.

The wind was blowing, and the rain falling, as we crouched round the wood fires each and all intent on the rapid preparation of our frugal meal. The warmth of the burning sticks and the discomfort of burnt fingers rapidly did their share in appeasing our hunger, and we betook ourselves to our temporary prison, where, selecting a corner apart from the men, we made a bed by spreading on the concrete floor the contents of some compressed forage bales. Burying ourselves in the short dry hay, we huddled close together, endeavoring thereby to retain such warmth as still remained in our weary limbs.

The building in which we were congregated contained two wooden doors, of which one was securely fastened with large nails; while the other, guarded by a double sentry on the outside, was closed upon us shortly after nightfall. In the corrugated iron roof of the shed was a small skylight. A glance round on first entering it had raised hopes of a possible escape during the night, which as regards climatic conditions was singularly favourable. To have fashioned some equivalent of a rope, mounted by means of the rafters, removed the glass of the skylight, reached the exterior of the roof and thence dropping to the ground, on that side where we supposed no watch was kept, covered the thirty miles which lay between us and Estcourt, seems no difficult affair when regarded by the light of events which were yet to come. But though, when I look back on this lost chance, it seems like child's play compared with the difficulties which later opportunities presented. To our weary bodies and fagged-out brains it appeared to offer little chance of success, and we were only too glad to procrastinate, and to console ourselves with the thought that on the night of the morrow we should be within range of the guns of Ladysmith, and possibly in a far better position to give our captors the slip.

Having for the time dissolved the committee on ways and means of evasion, we turned our thoughts towards slumber, and made such dispositions as were most likely to lure "nature's soft nurse" to our aid and so "steep our senses in forgetfulness." But nature's soft nurse must have had an important engagement elsewhere; for she peremptorily declined to say anything to my case, and I was fain to lie awake through the long night, the events of the past twelve hours coursing through my brain, chased by the vain imaginings of what still lay before us.

Stuffiness reigned supreme, the only door of the building being fast closed by the guard, who, no doubt, considered that we within were not unenviably situated compared with those without. Outside, the pitter-patter of the rain resounded monotonously on the metal roof. Inside, some fifty mortals lay, for the most part, wrapped in heavy slumber, no sound save the rhythmic snore of some weary warrior, or the unintelligible jargon from the lips of some sleeper busy in his dreams, breaking in upon the profound silence. But, hark! what means that sound of deep and manly voices, now swelling, now dying away, without, in the cheerless night![222]

For a moment one almost forgot where one was—amid the pious God-fearing Boers.

We strain the ear and soon distinguish the tune of the "Old Hundredth,"[223] followed again by other and familiar chants. Such, indeed, is the custom of the enemy. Nightly, in every camp, the commandant assembles all his men whom the call of duty does not take elsewhere, and before retiring to rest a short time is devoted to prayer and praise.

Honesty compels me to show the reverse of this picture. These very men, who are almost as full of apt texts as the great Cromwell's Ironsides, are far from acting up to the high religious sentiments that they profess. They are singularly untruthful, eminently boastful, lamentably immoral, and their ideas of *meum* and *tuum*[224] would do credit to a London pickpocket. My statement regarding their characteristics may seem sweeping; nevertheless it is true, and "Never trust a Dutchman" is a byword among the British in South Africa.

But the Boers, who here have no enemy immediately in front of them, and consist solely of our escort, have now retired to rest, and we turn over and hope for morning.

At length it dawns, and we are told to bestir ourselves and prepare for the road. Munching a cold *kabab*, washed down by a draught of water from the engine-pump, we follow our captors in melancholy procession to the banks of the river. Here we are invited to wade across-an unnecessary proceeding, considering the close proximity of the foot-bridge; but the Tugela river before the rains is only a paltry obstacle, and our company is soon assembled on the farther bank. We plod along steadily and somewhat silently, until we reach the summit of the hill on which is a Boer post overlooking Pieter's station. Here we halt, and make short work of a tin of bully

[222] I believe this is a quote from Shakespeare's Macbeth.

[223] "Old Hundredth" is a hymn tune in long metre, from second edition of the Genevan Psalter. It is one of the best-known melodies in many occidental Christian musical traditions. The tune is usually attributed to the French composer Louis Bourgeois (c. 1510 – c. 1560). Although the tune was first associated with Psalm 134 in the Genevan Psalter, the melody receives its current name from an association with the 100th Psalm, in a translation by William Kethe entitled "All People that on Earth do Dwell." Old 100th-Wikipedia

[224] Mine and thine, or mine and yours.

beef which, with a scrap of biltong,[225] some kindly rustic presents to us, and in return for which I make him a present of my Sam-Browne sword belt,[226] foreseeing that it is not likely to be of any use for some time to come.

"Siege of Ladysmith: British prisoners at Mooder Spruit and Boer lady visitors to the laager."[211]

Away to the north-west we can clearly distinguish the captive balloon hovering like some strange phantom of mythology over what assuredly must be Ladysmith.[227] Ever and anon[228] the deep boom of the Boer great gun reaches our ears, followed by the responsive and angry roar of our Naval Artillery. Sometimes a great cloud of white smoke discloses to us where Long Tom lies so cunningly hid.

War balloon at Ladysmith.[212]

Preparing the telegraph apparatus for an observation balloon.[213]

Observation Balloon[214]

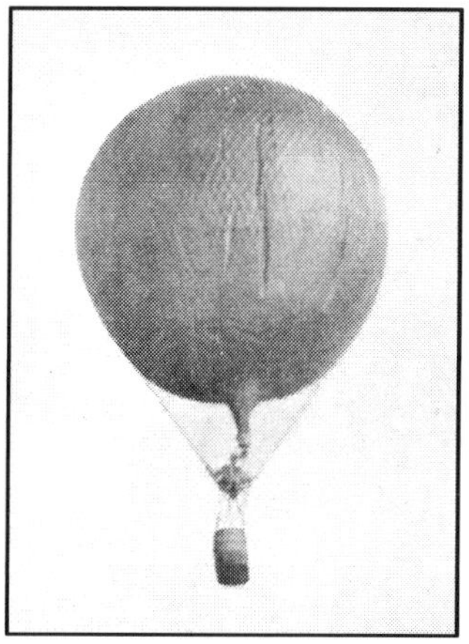

Military Balloon[215]

Observation Balloon[216]

[225] *Biltong* is a form of dried, cured meat that originated in Southern African countries. Biltong-Wikipedia

[226] The Sam Browne belt is a wide belt, usually leather, supported by a narrower strap passing diagonally over the right shoulder Sam Browne belt-Wikipedia.

[227] The siege itself would not begin until November 2nd and would end on February 28th, 1900. Siege of Ladysmith-Wikipedia

[228] Soon, short while, presently.

The Balloon at Ladysmith.[217]

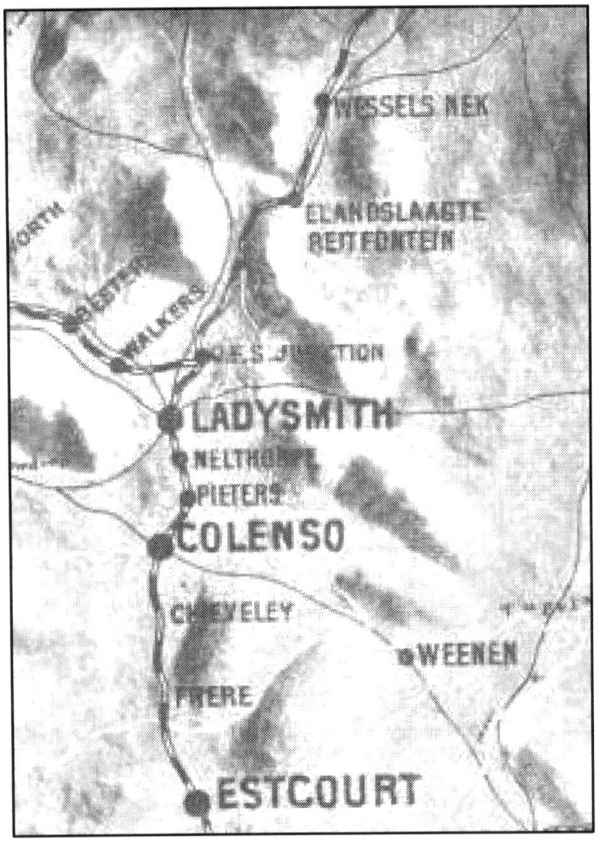

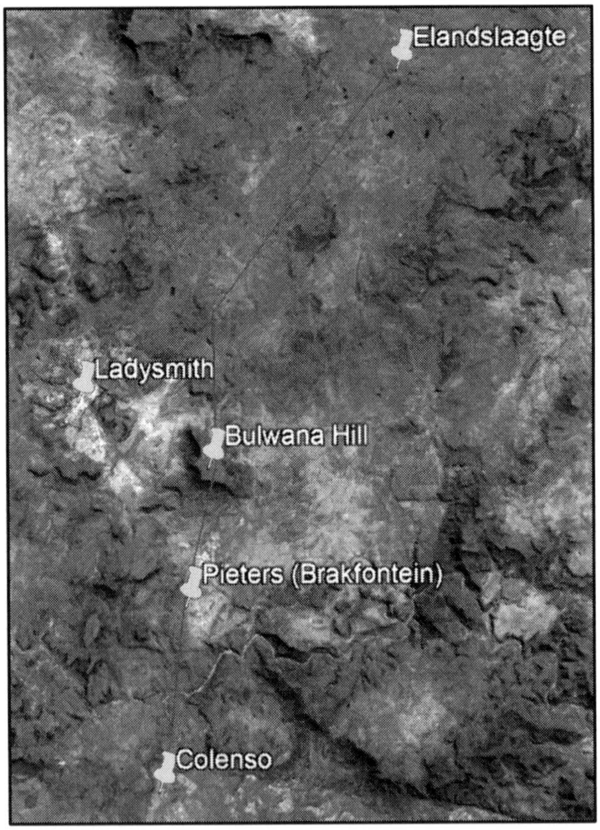

Map showing country from Durban to Ladysmith.[218] It is ~12 miles from Colenso to Ladysmith and another 12 to Elandslaagte. The older map does show a train junction just north of Ladysmith. This may be what Haldane is referring to as Modder Spruit. And: Google Image © 2023 Image Landsat/Copernicus.

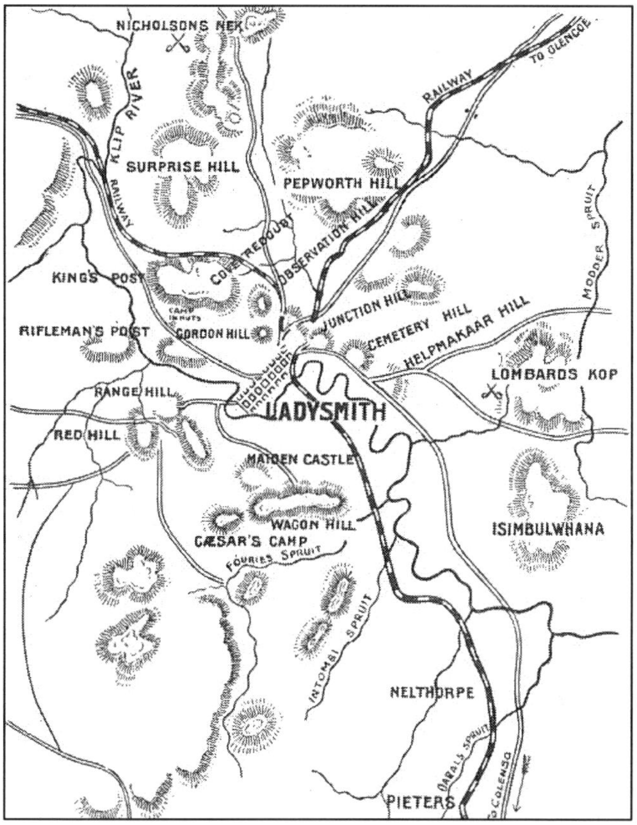

Ladysmith showing landmarks in detail.[219]

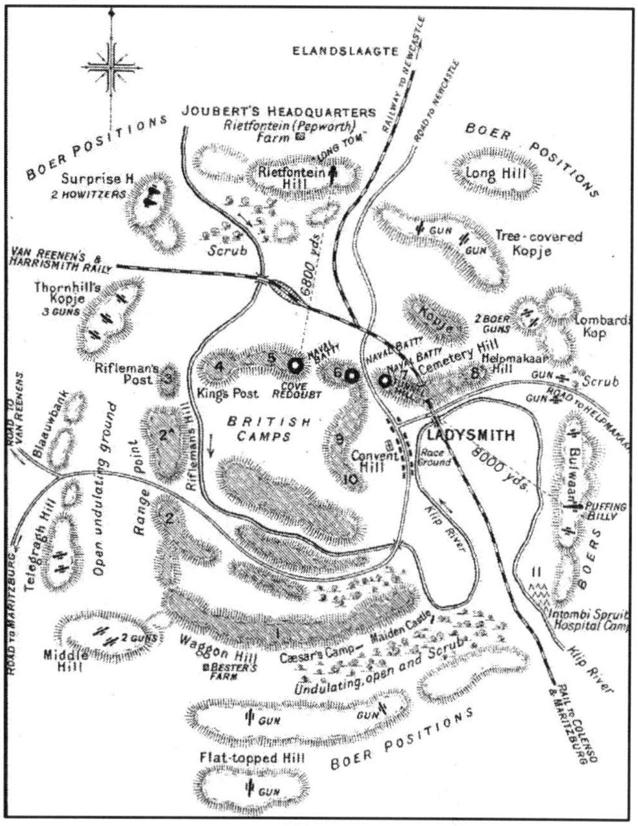

Ladysmith Map Detail. [220]

Ladysmith, Natal.[221]

Ladysmith, 1900. Town Hall with damaged clock tower.[222]

We descend the rocky side of the hill, and crossing the railway strike a much-worn wagon-track, which we are told leads beyond the Klip river[229] to our 'night's halting-place. A couple of miles over a grassy plain brings us to the drift, a fairly deep one, reaching nearly to our waists. Over we go, with difficulty keeping our feet on the slippery stones, and a short half-hour brings us to the end of that day's tramp.

Seating ourselves under the welcome shade of a stunted oak, we are soon surrounded by every living soul from the adjacent camp, which in our eyes more resembles a gypsy settlement than a military post.

[229] The Klip River is a main tributary of the Tugela River Klip River (KwaZulu-Natal)-Wikipedia.

General Joubert's camp at Glencoe.[223]

Boer Camp at Elandslaagte.[224]

A Boer Encampment "Waggons Massed on the Veldt."[225]

A typical Boer camp.[226]

The old commandant, Davel, soon appeared on the scene, and in broken English invited us to share his tent, secure from the sun's rays, which had become oppressive.

We acquiesce, and following him to the brow of the hill, enter the welcome shade he proffers.

I have the most vivid and lasting recollection of the attention and kindness which this old Dutch farmer lavished upon us.

Throughout the journey to the Boer metropolis we met with nothing but the greatest consideration; but this old commander not only gave us of the best he could, and at that time the Boer commissariat arrangements were in their infancy, but insisted on lending us some of his blankets, provided us with candles, and with his own hands brought us coffee on the following morning. He presented Churchill with a blanket, and we really felt quite sorry to leave his camp.

We passed many hours in his tent, which seemed to be the focus of movement in the camp.

All who came to the post he commanded seemed to drop in, give him the time of day, and, after a few minutes' conversation, vanish. During the afternoon two nephews of General Joubert belonging to the Staats Artillery returned to the camp from their tour of duty on Long Tom, and pressed on us the contents of their cigarette-cases.

Churchill, needless to say, afforded all the greatest source of interest. With much ability, coupled with quick repartee, he defended the justice of the war; but it struck me that these plainspoken, ignorant farmers, who based their arguments on the capitalists and Mr. Chamberlain, declaring that but for these two ruling factors matters would never have come to blows, somewhat shook his faith, and certainly gained his sympathy.

Knowing nothing of the Boers personally, beyond an experience of twenty-four hours, I for the time being dismissed from my thoughts the overwhelming importance of the question of British supremacy in South Africa, and felt inclined to sympathise with these rude tillers of the soil in their plaint[230] that the arbitrament of arms had been forced upon them.

A few months have made it clear to me how insidiously the country Boer has been taught to believe as gospel those specious arguments which are everywhere on the tip of his tongue. He firmly believes that the war is one of capitalists, and that we wish to occupy his country, and perhaps he is not very wide of the mark now. Had this simple-minded agriculturist known the true state of affairs, I question greatly whether even President Kruger, backed by the Psalms of David, could have kept him in the field for half as many months as he has. Leave the country Boer to himself,—and there is plenty of room for him,—let him lead the pastoral existence for which

[230] An accusation; a charge; a complaint; a lamentation.

alone he is suited, let him live, die, and be buried, as is his custom, on his own farm, and I take leave to think that to him it is not a matter of stupendous importance whether he be under the Dutch flag or the English.

I have been told by those who have a good claim to know that no mean proportion of the rural population will welcome our advent, which is dreaded solely by the rapacious officials and overbearing underlings of the most corrupt and venal State of latter days.

But I have wandered far afield. Later in the day a member of the Honourable Artillery Company, an Irishman, paid us a visit, and told us that he formed one of the gun-crew serving Long Tom on Umbulwana Hill, and that being a burgher of the Transvaal he had been forced to take up arms for his adopted country.

Creusot Boer Field Artillery.[231] with directions. The artillerists with shells (left) and shrapnel shell (right).[227]

He assured us that the naval guns in Ladysmith, though they made good practice, had not yet succeeded in searching out the weapon which he and others served. Long Tom, it seems, was cunningly hidden in a bomb-proof casemate, built of sand-bags that protected both him and his attendants. As this was early in the history of the siege, we suggested that when his turn came round for laying the gun, he would do well to direct the sights so that the projectile might fall where it would do no harm. He chaffingly consented, and three weeks later, being granted short leave of absence, paid us a visit, bringing with him a large quantity of cigarettes.

The natural question, "When do you mean to take Ladysmith?" being put, all with one accord affirmed, with the utmost confidence, that they would do so shortly, but that they were in no violent hurry to carry out the operation. *Delenda est Carthago*[232] was easier said than done, and here it never passed beyond the stage of threat. On my pointing out to the anxious listeners that in their place we should not sit down and rest contented with looking at it, hoping the wall would come to us and so save our having to go to the wall, like Mahomet, they had a reply ready to hand. "Were the commandant-general" (meaning General Joubert), one of them insubordinately answered, "to order me to go and attack Ladysmith to-morrow, I should refuse point-blank, and my comrades would do likewise: most of us have wives and children, and we don't want to be killed! To meet the case, someone suggested that the young unmarried men might be suitably employed in the assault; but this did not appear to find favour with any young man present.

Another topic which, under present circumstances, did not lend itself to lengthened discussion, and which seemed to be on the tongue of every Boer we met, evidently arose from some offensive remarks published at one time or another in the English newspapers.

It may be that our captors were not above laying themselves out to be complimented, or were honestly offended at the picture of their characteristics as depicted in the journals of their enemy. What the exact condition of their minds may have been was beyond my knowledge and penetration, but their thoughts were speedily translated into the frequently reiterated questions— "Do *you* think we are uncivilised? Are we then such barbarians as you English say, and is this your own experience of us?" I will leave the reader to supply the answer which at once and without hesitation came from our lips, and which, though doubtless anticipated, was none the less agreeable to our hearers.

[231] Text per Rob Davidson: Here's a photo of Lt Heinrich Grothaus of the ZAR Staatsartillerie, at the siege of Ladysmith. The field gun is a 75mm Krupp QF. The medic on the right is Dr. Elsberger of the German corps. Note the tarpaulin over the epaulement under the muzzle of the Krupp. I understand this is to prevent a cloud of dust from rising up when the gun is fired, revealing its position. Image Per Neville Chamberlain: Anglo Boer War - Artillery and Ammunition - Page 15 - Boer War Forum.

[232] "Carthage must be destroyed."

The Boers were then on their best behaviour; and indeed, with rare exceptions, wherever they have been met in organized bodies and under leaders, they have acted towards their prisoners in a humane and considerate manner.

Another point regarding which they seemed to have doubts and preconceived ideas impossible to dispel, was that of the numbers of men of which the British army was said to be composed. The more ignorant among them had doubtless been bred up in the fallacy that our army had been annihilated at Majuba,[233] and that it could be possible to find as many as 200,000 *rooineks*[234] to fight them seemed utterly beyond their power of comprehension. A few among the more intelligent ones gave indications that their mental outlook in some measure grasped the magnitude of the operations, which were then only in their infancy; but even these appeared to think that they had now in front of them the entire British army, and that reinforcements which might come hereafter would but be of inferior quality, and consequently more easily overcome. A great part of the day was passed in futile attempts to convince our captors of the fatal error that had been made in declaring war, and when night came I felt glad to be relieved both of their presence and their arguments.

Our hopes of escape, apart from the feeling that it would be rather mean to fly from our hospitable commandant and launch him in a sea of trouble, were on this night, as on the previous, doomed to disappointment.

We were located in a small tent between another in which were the men captured with us and a waggon, underneath which a party of Boers had taken up their quarters, moving in and out during the night with disturbing frequency. Seated at the tent door were two Boers with loaded rifles, and a similar precaution prevented exit at the other end. The moon was full, or nearly so, and a small object at a distance of several hundred paces would have been visible and no difficult target for a Boer marksman.

Under these circumstances the matter had to be faced philosophically, and having decided to defer our hopes of escape till later, we retired to rest, nor did we hear the nightly strains of music which no doubt arose later.

Daybreak saw us once more afoot and wending our way northwards, the men following in our wake. The day was fine, and all seemed so calm and peaceful—for at so early an hour the daily great gun duel had not begun—that it was hard to believe that the campaign was proceeding, still harder to realise that we were really prisoners of war.

We were told that when we reached the railway-station of Modder Spruit,[235] [Churchill indicates they boarded at Elandslaagte] we should find a train waiting to take us to Pretoria, and also that we should obtain some food, which some of us sorely needed. Dropping down the far side of the neck, which we had climbed on the previous day, we passed behind Umbulwana Hill and soon after crossed the Modder Spruit, then on behind Lombards Kop and over a grassy rocky plain. During our march we passed more than one large camp, all being well sheltered, by choice of site, from the iron hail of the Ladysmith guns; and in the distance towards Pepworth's farm and Surprise Hill I saw and mentally noted the positions of several other smaller encampments.

All these camps appeared to be connected with each other by the electric telegraph, and as our little company threaded its way through some of them, the occupants crowded round us like bees.

The thought crossed my mind that if only somehow one could convey the knowledge of the enemy's position gained, on the march from Pieter's to Modder Spruit, to the garrison of Ladysmith, how useful might it not be.

But the iron heel of fate crushed any hopes of effecting such a purpose, and our near approach to the railway brought one's thoughts back to the stern reality of the situation.

We reached the station at 9.30 A.M., and shortly after were invited to enter a first-class carriage, the men being accommodated with covered trucks. Here our party was augmented by the arrival of Sergeant-Major A. (Adam) Brockie,[236] Imperial Light Horse,[237] who had been captured whilst swimming the Klip river on the previous

[233] The Battle of Majuba Hill on 27 February 1881 was the final and decisive battle of the First Boer War that was a resounding victory for the Boers. Battle of Majuba Hill-Wikipedia

[234] An Englishman, originally specifically a British soldier during the Boer wars, but by the early 20th century a general colloquial term, commonly derogatory, for anyone with English connections.

[235] Spruit: A small headstream; a stream flowing through a village, dry in summer. It is not this town which is west of Pretoria. Moddersspruit is a town in Bojanala District Municipality in the North West province of South Africa. Moddersspruit-Wikipedia

[236] Four Entries for Adam Brockie are located at Anglo Boer War - Name search My supposition is that they all refer to the same individual. He may have been promoted because Haldane states infra that he is later tasked as a recruiting officer in Johannesburg. Brockie Adam No. 489 Sergeant QSA (5) Source: List of QSAs with the clasp Elandslaagte Imperial Light Horse.

morning, on his way back to Ladysmith from a scouting expedition through the Boer camps. Brockie had been sent in charge of the Boer prisoners taken at Elandslaagte to Durban, whence he had returned to Estcourt.

Adam Brockie.[228]

After a few days at that place he had obtained permission to attempt to rejoin his regiment, then besieged in Ladysmith, taking with him despatches for Sir George White.

After he had related to me his adventures in passing through the Boer lines, I decided that from his knowledge of the country and its inhabitants, and his acquaintance with the Dutch and Kaffir [238]languages, be would be a useful companion with whom to attempt an escape. He whispered to me that he had given out to his captors that he was a lieutenant in the Natal Carabineers, and we undertook to support his deception.

His object in this misrepresentation, to further which he had removed the regimental badges from his slouch hat and shoulder straps, was to avoid being lodged in Pretoria jail, whence escape would have been no easy matter. He knew that such would be his fate should it transpire to what corps he actually belonged, for the members of the Imperial Light Horse are held in peculiar aversion by the Boers. The reason is not far to seek. The majority of those serving in this corps, which has so greatly distinguished itself during the war, are residents of Johannesburg, and many had taken a prominent part in the Jameson Raid, some actually having served on the Reform Committee.[239]

Before leaving this station we met with the only insulting language experienced throughout our journey, and this from an Irishman. I was glad to observe that the men of the Royal Dublin Fusiliers treated him with a silent scorn which his cowardly conduct merited.

By way of escort we were provided with two Boers, who travelled in the carriage as far as Volksrust, where they were relieved by two others.

Brockie A. Lieutenant QSA (5) Source: List of QSAs with the clasp Elandslaagte, Johannesburg Mounted Rifles. Brockie A. Lieutenant QSA (5) CC Eland DoL Tr SA01. QSA (5). Source: Ladysmith Siege Account and Medal Roll Johannesburg Mounted Rifles. Brockie Adam Corporal BSACM Mashonaland 1897 (0). Source: BSACM rolls British South Africa Police Roll The South African National Archives website shows evidence that Brockie died in 1904 as there is a deceased estate file for him. He died in the Transvaal. http://www.national.archsrch.gov.za/sm300cv/smws/sm30ddf0?2021092602480064684B63&DN=00000003

[237] Johannesburg Light Horse Regiment - Wikipedia

[238] During the Dutch and English Colonial Period in South Africa the word Kaffir was used to describe monotheistic peoples (Nguni ethnic groups in particular) of South Africa, who were not of a Christian or of an Islamic religious background, without derogatory connotations, until the early twentieth century. It appears in many historical accounts by anthropologists, missionaries and other observers, as well as in academic writings. See Also: Kaffraria and British Kaffraria (Queen Adelaide's Province) were historical regions along the coast north or Port Elizabeth. See: File:Eastern Frontier, Cape of Good Hope, ca 1835.png - Wikimedia Commons Subsequently it became a racial slur. Kaffir (racial term) - Wikipedia See also the British military unit Kaffrarian rifles which were white and black units which were still organized as late as 1947. East London, 1 March 1947. King George VI inspecting 'Kaffrarian Rifles' guard of honour. - Atom site for DRISA And: Kaffrarian Rifles AKA: Buffalo Volunteer Rifles Buffalo Volunteer Rifles - Wikipedia Also: a historical region in Afghanistan and the setting of most of Rudyard Kipling's famous 1888 novella "The Man Who Would Be King." Kafiristan - Wikipedia And: an Arabic and Islamic term which, in the Islamic tradition, refers to a person who disbelieves in God as per Islam. Kafir - Wikipedia PF Note: I have no authority to change the wording used by original authors. There are instances where a pejorative or admiring modifier is used. While this is racist in and of itself, to identify someone by their complexion, in this narrative, identifiers such as Dutch, Boer, English, Uitlander, Indian, and many others are used as means of identification sometimes with pejorative or admiring modifiers.

[239] This pro-British organization had helped precipitate the Jameson Raid, been disbanded and reformed later to lobby for British interests in the Transvaal. Having been residents of the Transvaal, enlisting in enemy forces was viewed as a betrayal.

The younger of the two[240] former confided to me, with youthful glee, that he had fired over one hundred rounds at us on the 15th,[241] and had got a week's leave to visit his *vrouw*,[242] while his elder comrade, who spoke but little English, and resembled somewhat a Scottish gamekeeper, was described to us as the man who put the stone on the line which was the cause of the derailment of the armoured train. He had been awarded a fortnight's leave for his successful feat, and not some costly war decoration.

The Boers have a strange and objectionable custom of firing from the trains when in motion, when any game they may chance to see is within range. This, I believe, is prohibited by their laws, but neither that nor the fact of its being the close season had any deterrent effect on our fellow-travelers.

Passing Elandslaagte,[243] which from its name might be taken to suggest the proximity of one species of antelope, a herd of buck passed us in Indian file some four hundred yards distant. Pop, pop, went the Mausers, the bullets knocking up the dust fairly near the terrified animals. They galloped off apparently untouched, and I was glad to see that the vaunted marksmanship was at fault. An equally unsuccessful attempt at useless slaughter was repeated the following day in the Transvaal.

We passed many trains bearing living freight of men and horses to the front, and at every station on the line numbers of people of both sexes crowded round the carriage windows, some expressing disappointment to find that I was not arrayed in the garb of old Gaul.

Departure of Pretoria Commando for active service, October 1899.[229]

Boers departing for the front.[230]

A Burgher's farewell to his family.[231]

A Boer commando in fighting trim.[244]

[240] Churchill relates that: "I gradually fell into conversation with this man. His name, he told me, was [H.G.] Spaarwater, which he pronounced Spare-water." When they parted at Volksrust: "I gave … [Mr.] Spaarwater, a little slip of paper setting forth that [he] had shown kindness and courtesy to British prisoners of war, and personally requesting anyone into whose hands the papers might come to treat them well, should they themselves be taken by the Imperial forces."
"The Bearer, H. G. Spaarwater has been very kind to me and British Officers captured in the Estcourt Armoured train. I shall be personally grateful to anyone who may be able to do him any service should he himself be taken prisoner." (Sandy's p. 76 and image courtesy of Martha Bam on p. 77)

[241] When the armoured train was attacked.

[242] Borrowed from Dutch *vrouw*. Doublet of *frow* and *frau*.

[243] Elandslaagte is farming and coal-mining centre some 26 km north-east of Ladysmith. Afrikaans for 'elands flat', it was the scene of the Battle of Elandslaagte, one of the first battles of the Second Anglo-Boer War, on 21 October 1899. Elandslaagte - Wikipedia

[244] Anglo Boer War - Images from 'On the Transvaal border' published by W H Gill, 1900 - Boer War Forum Credit Dr. David Biggins.

Fat and forward *vrouws* [245] intimated to us that they would like to be the happy possessors, and no doubt wearers, of our regimental buttons and badges; but we were ungallant enough to repel their unwelcome advances.

At length, shortly before nightfall, we came in view of the solid mass presented by Majuba Hill,—for me a spot of the greatest interest, my regiment having left on the top, in the action of the 27th February 1881, ninety-six killed and wounded out of one hundred and twenty.

Majuba Hill from the south.[232 246]

Summit of Majuba.[233]

I had never thought to see this hill in such fashion, and I mentally registered a vow one, too, which seven months later was fulfilled—that I should yet live to gaze at it, going northwards with my [Gordons] regiment.

Through the tunnel which pierces Laing's Nek there was an opportunity to spring forth from the window, like Charles Peace[247] the burglar, and make for the open veldt, but we were now full a hundred miles from Ladysmith, the country between us and it infested with Boers, and ourselves clad in uniform.

The Boer guards, too, were singularly watchful, and a glance out of the carriage window displayed to view on right and left more than one bearded face casting its eyes along the train ready for any eventuality that might occur.

Drakensberg Railway Switchbacks.[234]

Laing's nek Tunnel damaged later in the conflict.[235]

245 Borrowed from Dutch *vrouw*. Doublet of *frow* and *frau*.

246 The Battle of Majuba Hill on 27 February 1881 was the final and decisive battle of the First Boer War that was a resounding victory for the Boers. Major General Sir George Pomeroy Colley occupied the summit of the hill on the night of 26–27 February 1881.
The Boers believed that he might have been attempting to outflank their positions at Laing's Nek. The hill was not considered to be scalable by the Boers for military purposes and so it may have been Colley's attempt to emphasize British power and strike fear into the Boer camp. The battle is considered by some to have been one of the "most humiliating" defeats suffered by the British in their military history. Battle of Majuba Hill-Wikipedia

247 Charles Peace (14 May 1832 – 25 February 1879) was an English burglar and murderer, who embarked on a life of crime after being maimed in an industrial accident as a boy. Charles Peace-Wikipedia

Express Coach Office between Johannesburg and Newcastle.[236]

Charlestown,[248] the last British settlement in North Natal, was passed, and at Volksrust,[249] the first station inside the Transvaal border, our escort was changed, and an inspector of police having joined us, Sergeant-Major Brockie was subjected to a close examination, with a view to find out whether he had resided long in Johannesburg, and if he were a burgher. As he was talking somewhat freely about his acquaintance with that city, a young Boer from Swaziland standing at the window whispered to me that if I did not want to get him into trouble I had better warn him to put a guard on his tongue. I think that this little incident shows how well disposed towards us the people were in many cases. I do not think that this predilection altogether begins and ends with us soldiers, though undoubtedly a large majority look upon us as blindly carrying out our orders without being much concerned with what we are fighting about.

At this station, where we halted for half an hour, a rough-looking fellow came to the window and seemed to be on the point of making himself offensive to us, but the escort quickly ordered him about his business. A loquacious Frenchman pressed to the front and began talking to us in his own language, but the police-inspector, not understanding what was passing, requested us to cease. This same inspector, who accompanied us for several stations, was, like many of his countrymen, of a highly suspicious nature; for while Brockie and I were whispering together regarding the best means of escaping from Pretoria, he kept his lynx-like eyes bent on us, and frequently interrupted our conversation, saying, "I know what you are talking about," and it is probable that he did.

The night passed as nights generally do pass in a train where there are too many occupants of the carriage for all to recline, and the morning came and reminded us that we had before us the ordeal of arriving at Pretoria. Of breakfast at Heidelberg[250] I have no very distinct recollection, save that the coffee was indifferent and the butter very rancid; but these trifles were borne with complacency by our escort, who, I think, appreciated the full-flavoured lubricant to which we took exception.

[248] Charlestown is a small town situated at the top of Laing's Nek pass in KwaZulu-Natal, South Africa, between Newcastle and Volksrust. It started out as an important railway station and customs post between Natal and Transvaal in 1891 until the Union of South Africa came into being in 1910, and customs tariffs were abolished. Charlestown, KwaZulu-Natal - Wikipedia

[249] Volksrust is the first town inside the Transvaal. Presently it is a town in the Mpumalanga province of South Africa near the KwaZulu-Natal provincial border, some 240 km southeast of Johannesburg, 53 km north of Newcastle and 80 km southeast of Standerton. Volksrust - Wikipedia

[250] Heidelberg began in 1862 as a trading station built by a German, Heinrich Julius Ueckermann. A town was laid out around the store and named after Ueckermann's alma mater. In 1866, the District of Heidelberg was created from the eastern portion of the Potchefstroom district with its own landdroost (magistrate) having been laid out as a churchplace in 1865. Heidelberg is some 50 kilometers south-east of Johannesburg, close to the Mpumalanga border. Heidelberg, Gauteng - Wikipedia

Heidelburg (Population 2,500) is situated in the south-eastern corner of the Witwatersrand Basin on the main line to Johannesburg. [237]

The country was ugly, flat, and devoid of trees, and the train rolled along at a steady irritating pace, until by the hilly nature of the ground it was easy to tell that we were approaching the Boer capital. First a fort appeared frowning down upon us from a wooded hill on our right, then another on a similar feature to our left. Passing down the gorge between these two mute guardians of the city on this, its southern side, we clashed over some points and knew it was the junction of the railway to Delagoa.

A few minutes more and we were steaming alongside the platform to the terminus, the cynosure of all eyes; for a goodly crowd had collected to view the latest captures by the brave burghers at the front. Our journey to the metropolis of the South African Republic was at an end, and nothing remained but to employ one's wits in devising the best means of completing the circular tour so unwillingly commenced. For that purpose, however, I recognised that a return ticket and a personally conducted journey were equally unnecessary and undesirable.

II. Life at the Model School

"All ye who enter here leave hope behind."

Perhaps a few words regarding the seat of Government (or shall I say Misgovernment?) of the Transvaal, in which four weary months were to be passed, will not be out of place at this stage of the narrative.

Pretoria, named after Pretorius, the first governor of the South African Republic, lies on the northern slope of the Aapies, a small tributary of the Crocodile river. The town is regularly laid out in parallelograms, the sides which form them running nearly due north and south, and east and west, a not inconsiderable advantage to those who may desire to find their way out of the city in some particular direction.

The streets, which are of uniform width, are bordered with magnificent willows, a tree in which the Dutchman delights, but which gives the whole town somewhat the appearance of a great necropolis. The town is brilliantly lighted by electricity, and the water-supply is excellent. The temperature varies from 89° Fahrenheit during summer to about 42° Fahrenheit in winter. Six forts of modem type encircle the town, constructed on a series of heights which serve to conceal its existence until the near vicinity is reached. At the foot of Signal Hill, to the south of the town, still stands "Jess's" cottage, described in Rider Haggard's[251] novel.

Pretoria.[238] The Dutch Reformed church steeple is left and "Raadzaal" is right. A street in Pretoria.[239]

Transvaal Government Buildings and Dutch Reformed Church.[240] Muster of Town Burghers in Pretoria, November 1st, 1899.[241]

[251] Sir Henry Rider Haggard KBE (22 June 1856 – 14 May 1925) was an English writer of adventure fiction set in exotic locations, predominantly Africa, and a pioneer of the lost world literary genre. Notably King Solomon's Mines. H. Rider Haggard-Wikipedia Jess (novel)-Wikipedia The book is named after Jess, one of the two orphaned nieces of a farmer in the Transvaal.

Supply Preparations – Despacthing Stores from Pretoria, October 1899.[242]

Market Street in South Pretoria.[243]

The country round Pretoria is said to possess few natural beauties, but of these I am unable to speak with authority, having only seen a portion by moonlight, at a time too when my attention was fully occupied with matters which will be referred to later on.

A few minutes after the train came to a standstill at Pretoria we were requested to alight. This was no sooner done than necks craned forward to obtain a glimpse of the new arrivals, and the camera fiend, ever lying in wait for the unwary, began the operation of transferring our features to his film. Pressing my hat over my brows, I endeavoured to put my physiognomy as far as possible into the shade.

Prisoners arriving in Pretoria November 18th, 1899.[244]

November 18th in Pretoria[245] [246] Winston Churchill and apparently Captain Haldane with his face obscured by his helmet. The figure is wearing an officer's sash but no Sam Browne belt. Haldane is listed as 6 foot 1 inch in his wanted poster. It is likely that Adam Brockie would also be with the 'officers' at the front of the column (Far left, although according to Haldane he went to prison for a day of interrogation). PF Note: Churchill is wearing the same "Irish Fusilier" (Dublin Fusilier) cap referred to supra, that was tossed to him by a Boer during his march through the rain on November 15th. Also, the consensus is that the young 'officer' standing at the front next to Churchill is not Thomas Frankland. It may be Chudleigh, or Lieutenant Grimshaw or Lieutenant Southey. In the background are other soldiers of 1st Royal Dublin Fusiliers all arriving in Pretoria on November 18th, 1899.

Pretoria November 18th, 1899.[247]

A lane was cleared through the living mass, and, joined by an officer of the Natal Carabineers[252] (a prisoner from Harrismith)[253], who was privily informed of Brockie's deception, we were formed up in a small party; the

[252] In September 1899, the Natal Carbineers were mobilized for active service in the British campaign to subdue the Boer republics of the Transvaal and Orange Free State. The Regiment served until October 1900, when the Natal Volunteer Forces were demobilized. Ingobamakhosi Carbineers - Wikipedia

[253] The Harrismith commando was initially sent to occupy mountain passes of the Drakensberg between the Orange Free State and the British Colony of Natal. English speakers of Harrismith were sent to the Oliviershoek Pass or serve in the Town Guard. Harrismith Commando - Wikipedia

men were in a larger[sic].[254] To my dismay, Brockie—with whom, during our journey, I had come to an agreement that we would spare the Boers in Pretoria the trouble of entertaining us for any length of time—was led off to prison by two plain-clothes policemen, while Churchill, conducted by a burly, evil-looking police official, left us to join the men. I remonstrated with this Jack-in-office—who, it is some satisfaction to know, is now in Ceylon in a similar position to ours[255] at that time—and pointed out to him that a war correspondent ranked as an officer, and further informed him who this particular correspondent was. All the reply vouchsafed was, "We know and care nothing for your lords and ladies here."

I disdained further conversation with such a person. Like criminals we were marched off under a strong escort of police, but after a few yards I perceived a portly and still more ill-favoured looking person, who turned out to be a certain Malan,[256] field cornet,[257] of Pretoria. Accosting him, I repeated my explanations, and was rewarded by Churchill joining us a few moments later. Sometime afterwards we heard incidentally that when the first party of officers and men captured reached Pretoria,[258] they were marched all-round the town, a trophy for the inhabitants to see. The Boers, however, were greatly disappointed that they were not wearing the red coats as in 1881, and complained bitterly that wearing khaki was not fair, it was "so much more difficult to see." The men were marched to the racecourse, on the outskirts of the town, and, after passing along a few unfrequented streets, we arrived in front of the Staats Model School.

On the verandah I noticed several acquaintances, and once inside the gate we were beset by inquiries as to how we were captured, what the latest news was, and where Buller and his army were. At this early stage of the campaign there was little to tell, and, tearing myself away, I speedily found someone who explained to me the routine of the establishment, and where one would sleep, and feed, and so on.

I regret that I must weary my readers with a somewhat minute description of the building and its surroundings, as it was from here that I made more than one unsuccessful, and at length a successful, attempt to escape. A reference to the accompanying plan will here and later help to elucidate matters.

The Staats Model School, a substantial single-storeyed red brick edifice, is built at the corner of one of the parallelograms into which the town is subdivided, its length running approximately north and south.

Staats Model School.[248]

Prisoners Arriving Staats School.[249]

[254] Perhaps misspelling of Laager.

[255] He had been transported as a prisoner of war to the camps in Ceylon. Over 5,000 Boers were interned in Ceylon (Sri Lanka) during the Boer War. The Ceylon camps were the second set of overseas POW camps after those in St. Helena had filled. http://libcom.org/library/pow-camps-ceylon-during-boer-war.

[256] Field Cornet Malan was a grandchild of President Kruger. (Sandys p. 82) This was not exceptional because Kruger had 17 children and over 100 grandchildren. He was captured in August of 1901 and exiled to Ceylon. (Haldane)

[257] The term was used for a civilian official in a local government district (drostdy) of the Cape Colony, acting as and invested with the authority of a military officer and empowered to act as a magistrate. The office had its origins in the position of veldwachtmeester in the Dutch Cape colony, and was regarded as being equivalent to a sergeant. The British administration enhanced its importance with the term field cornet, making it equivalent to an officer's rank. Field cornet-Wikipedia

[258] Probably the ~600 men under Col. Moller who were captured at the battle of Talana Hill/Glencoe.

Model School (back view). The hospital is in the foreground. The houses and gardens as well as the overhanging trees that provided were available for the first escape attempt are just visible in the lower left.[250]

It is, as its name implies, an educational establishment for the youth of Pretoria, and is divided into a number of schoolrooms and lecture-halls. It contains in all sixteen rooms in the body of the building, including two at each end. A long central passage runs almost throughout its length, terminated by the end rooms, and across this is the passage from the front entrance to the door into the yard or playground.

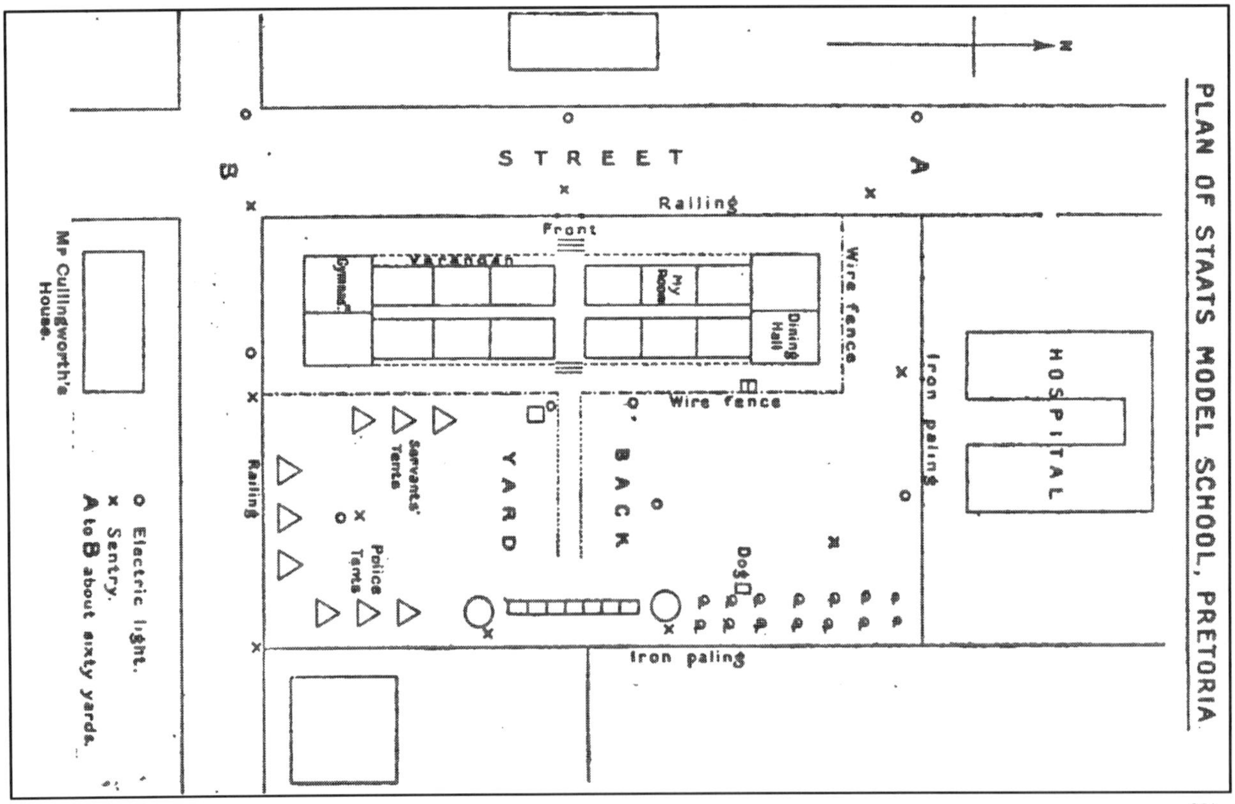

Haldane's Plan of Staats Model School showing Cullingworth's house on left and hospital on right side.[251]

On both sides of the building is a verandah which extends along the exterior of the six central rooms, and this again is overlapped by the four end rooms. One of these end rooms is fitted up as a gymnasium, and another was used by us as a fives-court.[259] Outside, a railing breast-high ran round the west and south sides, the two remaining sides being enclosed by a corrugated iron paling[260] six feet and a half high. In addition to these, a wire netting ten feet high ran parallel to the paling but close to the building, and through this an opening immediately facing the back door led to the grass-covered back-yard. Across the yard, in which were the tents of our soldier servants and the police guard, were some low buildings, which at the time of our arrival were connected with the corrugated iron paling previously referred to. A double row of trees ran close to the eastern paling.

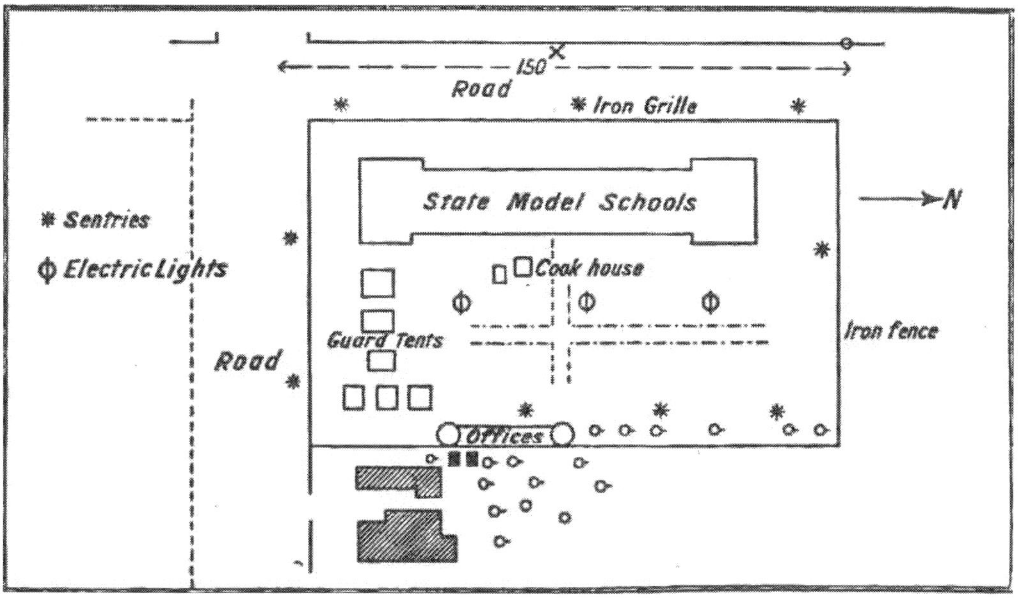

State Model School Plan.[252]

At night this yard was lighted by four electric lights. On the two sides of the building which did not look upon the street were private houses in gardens, that on the north side being used as a Red Cross hospital, with a door of communication in the iron paling which opened into the schoolyard.

The houses across the streets were occupied. Our guard at this time consisted of twenty-seven men and three corporals of the South African Republican Police, which furnished nine sentries in reliefs of four hours. They were armed with Lee-Metford rifles[261] —and revolvers, and carried whistles.

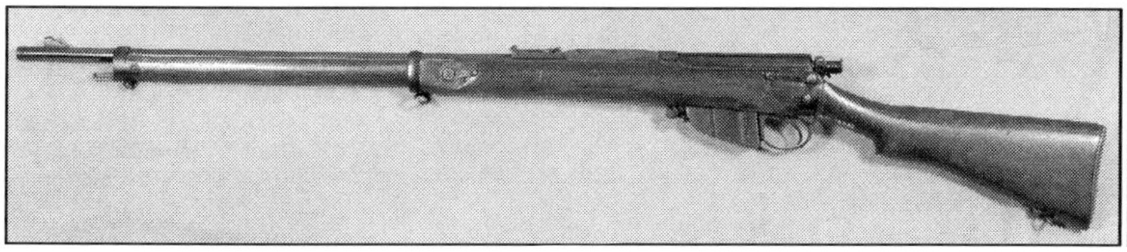

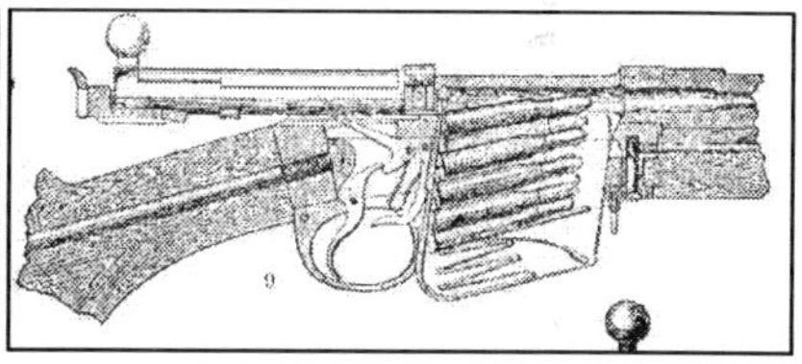

Lee-Metford Mk II.[262] The Lee-Metford action representing a modern bolt action design.[253]

The rifles were not loaded, but each sentry wore a bandolier full of cartridges. Five of these sentries were posted outside the enclosure and the remainder within, and the public were forbidden to approach nearer to us than the opposite side of the street. No vehicle, horse, or bicycle was permitted to pass the school; and even dogs were driven away lest they should be the bearers of messages.

The town was guarded, principally for the protection of property and for the arrest of suspicious persons, by special constables, who were posted in every street, armed with revolvers, and in some cases accompanied by

[259] Fives is an English sport believed to derive from the same origins as many racquet sports. In fives, a ball is propelled against the walls of a 3- or 4-sided special court, using a gloved or bare hand as though it were a racquet, similar to hand-pelota. Fives-Wikipedia

[260] A pointed stick used to make a fence. Also a fence made of such sticks. In this case a corrugated iron form.

[261] The Lee–Metford rifle (a.k.a. Magazine Lee–Metford), was a bolt action British army service rifle, combining James Paris Lee's rear-locking bolt system and detachable magazine with an innovative seven groove rifled barrel designed by William Ellis Metford. It replaced the Martini–Henry rifle in 1888, Lee–Metford - Wikipedia

[262] Rifle Lee-Metford Mk II. Caliber .303 British. From the collections of Armémuseum (Swedish Army Museum), Stockholm, Sweden.

their canine friends. These mounted duty at 6 or 7 P.M., according to the hour of sunset, and after 10 P.M. were entitled to demand passes from persons walking in the town.

In the vicinity of the Model School they appeared to be exceedingly alert and vigilant, frequently signaling to each other with their whistles, and closely scanning the passers-by. Beyond them, again, were mounted patrols, who by day and night watched the main roads into the town, preventing the ingress and egress of persons unprovided with permits. No one not in possession of a passport could travel by rail, which effectually barred that means of exit to us.

The prisoners living at the school at this time consisted of the officers of the cavalry and mounted infantry captured at Dundee on the 20th October those taken at Nicholson's Nek, and some stray prisoners taken at other times.

In charge of us were two officials—one a Boer commandant [Opperman],[263] the other a Hollander gentleman,[264] Dr. Gunning,[265] who at the same time discharged the double functions of manager of the Zoological Gardens and Curator of the Museum.

Central Hall of The State Museum At Pretoria.[266] Lion at Staats Museum.[267] Kruger & Lion at his house.[254]

The Fountain In Burgers Park, Pretoria.[268]

The Club, Pretoria.[255]

[263] "Commandant [R.W. L.] Opperman, who was directly responsible for our safe custody, was in times of peace a Landdrost or Justice. He was too fat to go and fight, but he was an honest and patriotic Boer, who would have gladly taken an active part in the war. He firmly believed that the Republics would win, and when, as sometimes happened, bad news reached Pretoria, Opperman looked a picture of misery, and would come to us and speak of his resolve to shoot his wife and children and perish in the defence of the capital." (WSC London to Ladysmith.)

[264] In this and other narratives the term Hollander refers to recent Dutch immigrants to the Transvaal, who were recruited for their technical and educational skills, many of whom have prominent places in the Government and are often associated with the "War Party." See Hofmeyr Chapter 2: "I knew that that particular Hollander element which swayed the Government counsels, was a particularly narrow-minded one, vengeful and spiteful an element which believed that it could only prove its true love for the Transvaal by exhibiting excessive bigotry of zeal against its enemies—by out-Heroding Herod."

[265] Dr. J. W. B. Gunning is the administrator and the censor at the State Model School where Captain Haldane is detained. Dr. J. W. B. Gunning, per Brian Davis, Churchills in Africa p. 201.

[266] Heroes of the Boer War. Review of Reviews, 1903. Note what appear to be Coelacanths among the displays. Coelacanth - Wikipedia

[267] The Boer "Tower of London." A Visit To The Staats Museum, Pretoria. By Rev. John C. Harris, Of Johannesburg. The Strand Magazine, Volume XIX April 1900 No. 112 p. 46. Cecil Rhodes gave a lion to the Zoological Garden and Gunning accepted it. Kruger was so angry that he demanded that the animal be returned. An alternative was found whereby the animal was shot and stuffed.

[268] After Pretoria: The Guerilla War Vol 3 August, 1900. "In the distance is the house of Mr. S. Heys. valued at £35,000. Roberts

The commandant was not exactly a pattern of every virtue, but the doctor, whose wife was an Englishwoman, was uniformly courteous, and did all that he could to make our confined existence as bearable as possible. Our daily routine reminded one of school, without work or common task to fill the hours which dragged so slowly by. Fortunately, however, we were allowed to join the Staats Library, from which a fair selection of books to suit most tastes could be obtained. This was indeed a boon to most of us.

The morning succeeding our arrival, Brockie, who had been successful in keeping up the illusion that he was an officer, and had evaded the pitfalls which the Pretoria Landdrost[269] had prepared for him in his cross-examination, arrived at the school. He told us that he had been so fortunate as to find among the officials at the prison a Johannesburg acquaintance, who had recommended him to conceal the fact of his knowing the Dutch language. He was thereby enabled to prepare beforehand his reply to each question, as the officials discussed in his presence and in their own tongue the interrogatory which they proposed to put to him.

To my surprise, for otherwise escape would have been made much more difficult, we were allowed to wear plain clothes, and were even provided with a suit and some other necessaries on arrival. The small party which arrived at the same time as I did were provided each with a ready-made suit of hideous mustard-colour, quite unsuitable for passing through Pretoria unnoticed.

We managed, however, by representing that the garments issued to us were not sufficiently large, to procure others of a more sombre and less conspicuous hue. An article of dress the acquisition of which was hedged in with difficulties was a suitable headdress. The authorities declined to provide us with or allow us to buy the slouch hat so universally worn in South Africa. Fortunately I had been given one by a railway employee at the time of the armoured train affair, having lost my helmet in that accident; and Churchill, shortly before his escape, borrowed that of the Reverend Adrian Hofmeyr.

Our meals were simple but sufficient. The Transvaal Government gave us a free daily ration of 1/2 lb. of beef, tinned or fresh, some bread, tea, and potatoes. As this was inadequate, a mess committee undertook the thankless task of feeding us at the rate of about 2S. 6d.[270] per diem. Those who desired could purchase luxuries, and indeed almost anything but firearms, through a local dealer.

Space for exercise was limited, for beyond the small yard in which some played rounders[271] and quoits,[272] we were restricted to walking round and round the building, reminding one of life on board ship. A few of the more energetic spirits determined that their incarceration should not cause a deterioration of muscle, and formed a class to practise Sandow's[273] exercises. At one time the feminine recreation provided by the skipping-rope became the fashion, and the inhabitants of Pretoria must have wondered indeed to see youths not long out of their teens and men of middle age nimbly hopping over the swinging cord. Inside we had the gymnasium and a room which was eventually used as a fives-court.[274]

Chess, bridge-whist, "patience,"[275] and even halma,[276] helped to pass the evenings and wet days. Those who had friends in hospital next door were permitted to visit them, and by this means we occasionally obtained some news of the outside world. The daily papers published in Pretoria and Johannesburg also were to be obtained; but their contents, coupled with the information which our jailors volunteered, were of a nature to depress one's spirits—always provided one was foolish enough to place any credence in them.

occupied it for six months In 1900. The park is named after the Rev. T. F. Burgers a progressive and philanthropic clergyman of the Dutch Reformed Church who was elected president of the Transvaal in 1872. This recreation ground covers an area of about two acres, is situated some three-quarters of a mile away from Church Square."

[269] Landdrost: From South African Dutch *landdrost*, from *land* + *drost* (plural *landdrosts*) (historical) A type of magistrate in South Africa. Landdrost - Wikipedia

[270] Two shillings and six pence a day.

[271] Rounders is a bat-and-ball game played between two teams. It involves hitting a small, hard, leather-cased ball with a round-ended wooden, plastic, or metal bat. The players score by running around the four bases on the field. Played in England since Tudor times, it is referenced in 1744 in the children's book *A Little Pretty Pocket-Book* where it was called Base-Ball. Rounders-Wikipedia

[272] Quoits is a traditional game which involves the throwing of metal, rope or rubber rings over a set distance, usually to land over or near a spike (sometimes called a hob, mott, or pin). The sport of quoits encompasses several distinct variations. Quoits-Wikipedia

[273] Eugen Sandow (born Friedrich Wilhelm Müller, 2 April 1867 – 14 October 1925) was a Prussian bodybuilder and showman. Eugen Sandow-Wikipedia

[274] Fives (historically known as hand-tennis) is an English handball sport, similar to the game of Handball, Fives - Wikipedia

[275] Solitaire. Patience (game)-Wikipedia

[276] Halma (from the Greek word meaning "jump") is a strategy board game invented in 1883 or 1884 by George Howard Monks, an American thoracic surgeon at Harvard Medical School. His inspiration was the English game Hoppity which was devised in 1854. Halma-Wikipedia

I must not omit to say here that beer, on payment, was allowed, and that one might walk round the buildings as late as one liked, and sleep on the verandah. The inhabitants of Pretoria took considerable interest in us, and some amusing expedients were adopted to procure our photographs.

One day a cab drove up at an hour when, like wild beasts in confinement, we were walking round and round our cage: the window nearest to us was draped with black cloth, in which a small hole, pierced the size of a camera lens, was visible. This vehicle remained some time, after which the occupant, having got all he wanted, drove off.

Detectives were frequently to be seen watching us as we took our exercise, and taking down notes as to our appearance, gait, and so on. That their observations were by no means perfunctory, but attained to some degree of accuracy, will be apparent as my tale proceeds. Sunday, however, was the day of the week when we were most favoured. Towards sunset the youth and beauty (I can't say rank and fashion) of the town passed and repassed our scholastic retreat, some gazing on us with looks of sympathy, others as if they would be glad to have a shot at us through the railings.

I must not forget one young lady in particular, evidently of a deeply sympathetic nature, who, I heard, was named after one of those flowers which in heathen mythology was adopted as a disguise by some forlorn maiden pursued by the enamoured Jupiter. Beyond glances shot over the railings at twenty yards, she was quite safe from the inmates of our dwelling. It was said, and I believe truly, that after Churchill's escape, when every English person in Pretoria was suspected of complicity, she was arrested by the unchivalrous Dutch, and though she could have rendered him no assistance whatever, was fined £25.

Looking over the railings, beyond the hideous and exasperating landscape of armed police, much of the town could not be seen. To the southward were the fortified hills through which we had passed at the time of our arrival, and between us and them ran the railway to Delagoa Bay. Being summer, the foliage of the trees effectually prevented an extended view, which might have given us a better idea of our surroundings.

The interest which Churchill's arrival had at first provoked had by no means abated, and he received many visitors with whom animated discussions on the all-absorbing topic of the war were engaged in. Some hope of release was held out to him; but the impression which had gone forth (an incorrect one) that he had borne arms against the Republic during the armoured train incident made that somewhat problematical.

The thought of evading the vigilance of the guard and making my way back to the front had scarcely for an instant been absent from my mind, and I was fully resolved to effect it at the earliest favourable opportunity.

An attempt to bribe the sentries to look the other way, while one passed the barrier, met with no success; and although some months later I renewed it, offering a sentry who was well disposed towards us £100, he was not to be tampered with. He said that if I got out of the building I could never get out of the country, there were so many patrols and other precautions to prevent the escape of prisoners, and that in any case the sentries on his right hand and left would see, and not be silent. Another difficulty which presented itself, even had one succeeded in gaining over our amiable policeman, was the impossibility of being sure that he would be on duty at the best place of exit on the most suitable occasion.

Ultimately we came to the decision that outside assistance in getting out of the guarded enclosures was unattainable, and that we must depend on our own devices.

Various expedients whereby we might gain our freedom were considered, but none of these appeared easy when worked out in detail, and most of them exceedingly difficult and uncertain.

From the dealer [Bosscher],[277] with whom, when the attention of the commandant could be temporarily distracted, a few moments' conversation could occasionally be snatched, we learnt that forged passports might possibly be obtained—for a price. Armed with these, we might succeed in quitting the country by rail; but our ignorance of the language, and the chance of identification by detectives at the station, did not recommend this venture to me.

[277] Churchill refers to him as Mr. Boshof. Frankland calls him Boscher.

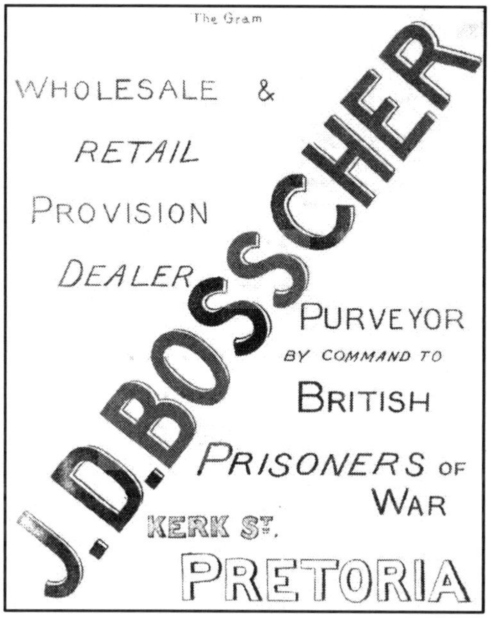

Cover page and an advertisement Mr. Bosscher took out in the prisoners' newsletter 'The Gram.' [278] [279] [256]

The plan which Sergeant-Major Brockie and I at length evolved was to climb out of the enclosure at the place where the low buildings at the back of the yard were situated. The spot chosen was one on which the electric light did not shine; and provided the sentry, who generally stood close to this place, happened to move a few paces from it, no great difficulty was to be anticipated. Only one sentry could possibly see anyone climbing over, and if his back were turned, provided he heard nothing, his eyes would certainly be of no avail.

Once outside the yard, creeping through the adjacent gardens we could gain the streets, then the exterior of the town, and, following the railroad to Delagoa Bay, either board the 11.10 P.M. train (a portion of which consisted of trucks) at some steep gradient, or wait for it at Eerste Fabriken Station, thirteen miles distant. Travelling with it till close on daybreak, which would bring us near Balmoral[280] (the centre of a coal district), we would hide all day, and at nightfall search for a coal-truck in the adjacent sidings.

By this means, if successful, the Portuguese frontier might be passed. It is perhaps desirable that I should explain that the night train from Pretoria went no farther than Middelburg, which is a few stations east of Balmoral; but by the timetable published in the papers we knew that even if we failed to find a coal-truck we should be able to board a train at night at Waterfall Boven, which place would have to be reached on foot.

The information regarding trains, special police, and other details was gradually accumulated, and early in December the plan was ripe for execution.

On the 7th of that month [December] two of our soldier servants belonging to the 18th Hussars succeeded in breaking out and escaping. They climbed over the offices at the back of the yard during the dinner-hour, a time at which the vigilance of the sentries was somewhat relaxed. I believe the police found out that they had gone, but thinking that they would not be missed said nothing about it to the commandant. The two soldiers, we heard afterwards, were caught at no great distance from Pretoria, and after being identified were lodged in jail.

The escape of these men made one feel that no time was to be lost, and all that was required was a slack and unobservant sentry, one who would during the dinner-hour move a few paces from his post and provide the necessary opportunity.

[278] PF: Mr. Bosscher was the only advertiser, even though he need not have done so, since he still had the concession. See Hofmeyr, for particulars, and Rosslyn. The magazine did not begin publishing until May 1900 well after the prisoners had been transferred to a purpose-built prison outside of Pretoria.

[279] "The Gram" A social Magazine Founded by British Prisoners of War in Pretoria, Facsimile Edition Eyre & Spottiswoode, Her Majesty's Printers, London #1 - The Gram ; a social magazine founded by British prisoners ... (no.1-3). - Full View | HathiTrust Digital Library Scanned from the University of Minnesota Copy.

[280] Balmoral is a town in the Emalahleni Local Municipality in the Mpumalanga province of South Africa. The village was established as a railway station of the Oosterlijn from Pretoria to Maputo in 1894. The village and railway station were erected at the farm Eenzaamheid. For decades the village was nothing more than a railway stop with a primary school for the surrounding farms. During the Second Boer War the British built a concentration camp near the railway station. Balmoral, Mpumalanga - Wikipedia.

Up to this time Churchill, as I have said, had had some hopes of release, but on the 9th of December he told me they were dissipated, and knowing that Sergeant-Major Brockie and I intended to escape, he suggested coming with us. We consented, though the certainty that he would be missed within a few hours lessened the chances of success.

As for Brockie and me, we knew that our absence would hardly be noticed, since at this time no roll-calls were held and no apparent record kept of the number of prisoners at the Model School.

We made a half-hearted attempt on the 11th [December], when matters were unfavourable,[281] and on the next night [December 12th] decided to go if possible but as the moon would rise about 7.45 P.M. and illuminate the dark spot where we meant to break out, the opportunity would, we knew, be a fleeting one. "There is many a slip," as the proverb has it, and Brockie and I were doomed to disappointment. Churchill saw his chance and took it, but when I tried to follow, the sentry, who now obstinately stuck to his post, saw me, and to go was impossible.

I will not dwell longer on this unsuccessful attempt, for on the principle that:

"Things without remedy should be without regard: what's done is done."[282]

The only course was to await events and adopt some other plan. We knew now, however, that escape by our first scheme was no longer practicable; and this was amply proved by the events of the following day, which I will leave to the next chapter.[283]

Scaling the Prison Wall.[257]

Churchill at the top of the wall looking at the guards.[258]

[281] Haldane Letter from 1935. See OBC Vol 1 Part 2 p. 1103. The night of the 11th December came, and we decided to go if possible. About 6.50 - ten minutes before the dinner hour - Churchill and I crossed over the back yard to the latrine in company with several officers, who knew what was intended.
They were to leave one by one, and it was hoped that the sentry, thinking that all had gone would quit his post and move up the line of trees. He did not, however, display any signs of emerging from the corner in which he stood, and, after waiting what seemed to be a quarter of an hour, though it could not have been so much, we whispered together and decided that to go then was impossible.

[282] Lady Macbeth says, "Things without all remedy should be without regard: what's done is done" to Macbeth at the beginning of Act 3 Scene 2 when they are talking about the continual plague of thought and fear that killing King Duncan has brought to them.

[283] PF: There was extended controversy about Winston taking the initiative to go over the wall when the opportunity arose. He was subsequently accused of leaving his comrades in the lurch. Winston had seized the moment while Haldane and Brockie had retired to get dinner before the kitchen closed since their initial reconnaissance did not look hopeful for success that evening. Subsequently Winston did wait an hour in the dark, next to the outside wall, for his comrades to return, discover he was not around, and go to the fence. Haldane's effort to scale the wall was seen, after which he told Winston it was "All up" and that Winston was on his own.

III. Revolving Plans of Escape

"Hope springs eternal in the human breast."[284]

In order to conceal the sudden departure of the special war correspondent of the Morning Post, and obtain for him as long a start as possible, we made up a dummy figure which, placed in his bed, had such a natural appearance that early on the morning of the 13th it was invited to accept a cup of coffee by a soldier servant. No reply being vouchsafed, the beverage was placed on a chair. But our attempts to defer the evil hour of discovery were speedily frustrated; for a barber[285] of the town, who, watched by a policeman, plied his trade on certain days in the school, came by appointment at 8 A.M., and failing to find his client, roused suspicion in the mind of his escort. These suspicions were conveyed to Dr. Gunning, who came himself to see what was the matter.

I had endeavoured to get rid of the barber by telling him that he was not required that morning, but, unfortunately, he was an inquisitive, persistent fellow, who was unwilling to depart before earning his expected fee. Dr. Gunning came and made inquiries regarding Churchill's whereabouts, but obtained no information. At 9.30 A.M., when the commandant returned, a roll was called, and each prisoner's presence carefully noted. The fact of there being an absentee became apparent.

I knew that by this time, if Churchill had caught the night train, he should be at least fifty miles from Pretoria, and well out of the reach of any patrols which might be despatched thence. Steps were now undertaken for his capture; his description was telegraphed everywhere; and so great was the Government's annoyance at his escape and its desire to capture him that the members thought of nothing else for some days, and the whole State machinery came to a standstill. Several policemen who were supposed to have been bribed were sent to the front and replaced by others; many Englishmen's houses were searched; an English nurse in the hospital next door, falsely accused of helping him, was put over the border; and more than one arrest of persons supposed to answer to his description was made.

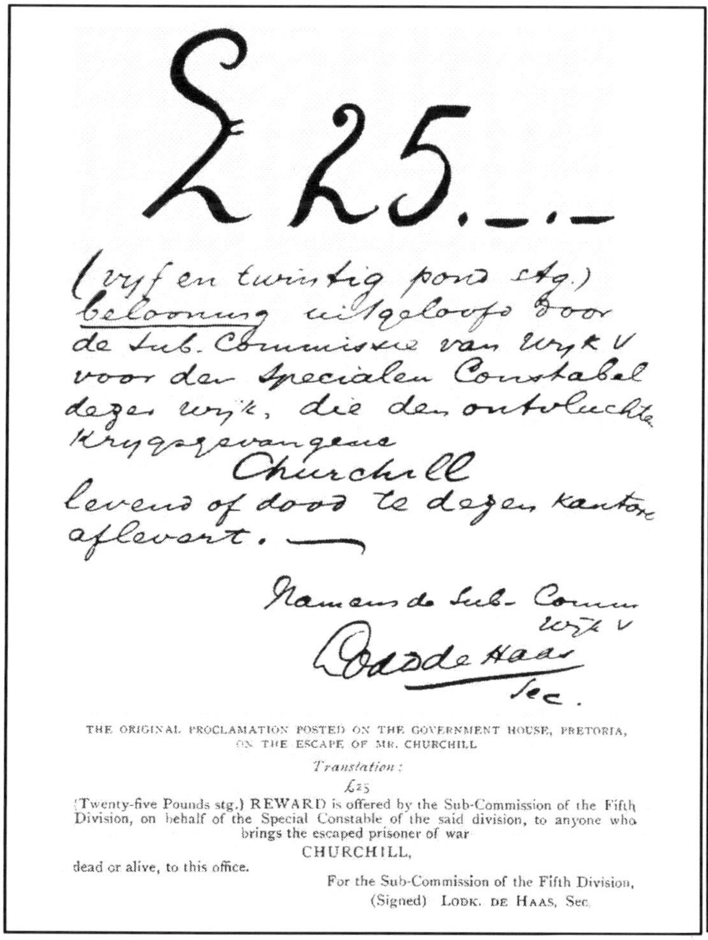
£25.-.-

(vijf en twintig pond stg.)
belooning uitgeloofd door
de Sub-Commissie van Wijk V
voor den Specialen Constabel
dezer wijk, die den ontvluchte
Krijgsgevangene
Churchill
levend of dood te dezen kantore
aflevert. —

Namens de Sub-Comm.
Wijk V
Lodk de Haas
Sec.

THE ORIGINAL PROCLAMATION POSTED ON THE GOVERNMENT HOUSE, PRETORIA, ON THE ESCAPE OF MR. CHURCHILL

Translation:

£25

(Twenty-five Pounds stg.) REWARD is offered by the Sub-Commission of the Fifth Division, on behalf of the Special Constable of the said division, to anyone who brings the escaped prisoner of war

CHURCHILL,

dead or alive, to this office.

For the Sub-Commission of the Fifth Division,
(Signed) Lodk. de Haas, Sec.

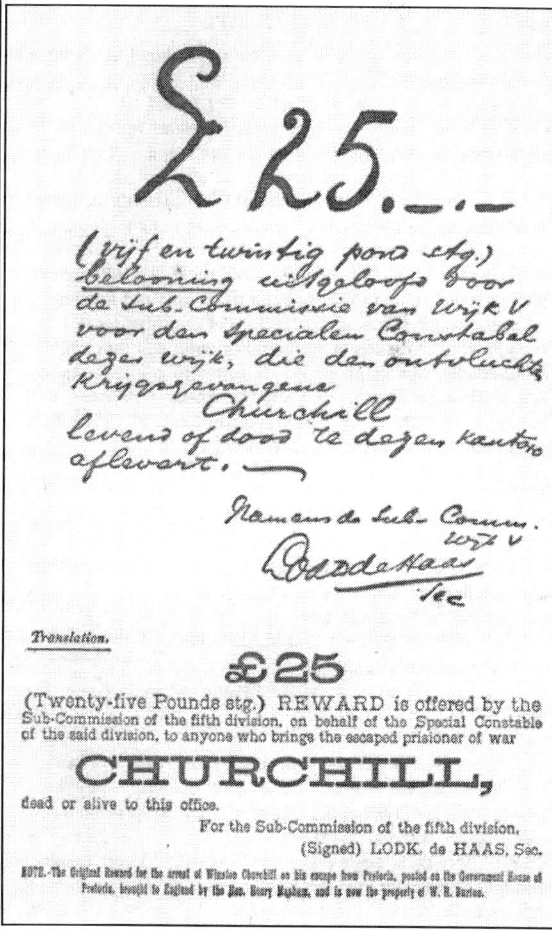
£25.-.-

(vijf en twintig pond stg.)
belooning uitgeloofd door
de Sub-Commissie van Wijk V
voor den Specialen Constabel
dezer wijk, die den ontvluchte
Krijgsgevangene
Churchill
levend of dood te dezen kantore
aflevert. —

Namens de Sub-Comm.
Wijk V
Lodk de Haas
Sec.

Translation.

£25

(Twenty-five Pounds stg.) REWARD is offered by the Sub-Commission of the fifth division, on behalf of the Special Constable of the said division, to anyone who brings the escaped prisioner of war

CHURCHILL,

dead or alive to this office.

For the Sub-Commission of the fifth division.
(Signed) LODK. de HAAS, Sec.

NOTE.-The Original Reward for the arrest of Winston Churchill on his escape from Pretoria, posted on the Government House at Pretoria, brought to England by the Hon. Henry Masham, and is now the property of W. R. Burton.

Reward Poster (dead or alive, to this office.) first published in 1905.[259] Citation continued on next page.

[284] Alexander Pope.

[285] Henri Adelaar (Sandys p. 100).

Churchill purchased a copy in 1904 and later reset the type for republication in 'My Early Life.' [260]. The Dutch text appears to be the original posted at the Government house in Pretoria and tracks the English text at the bottom of each closely, if not exactly.

Winston left a letter under his pillow.[286] The holographic original has survived. He later commented: "I arranged that this letter, which I took great pleasure in writing, should be left on my bed, and discovered so soon as my flight was known."[287]

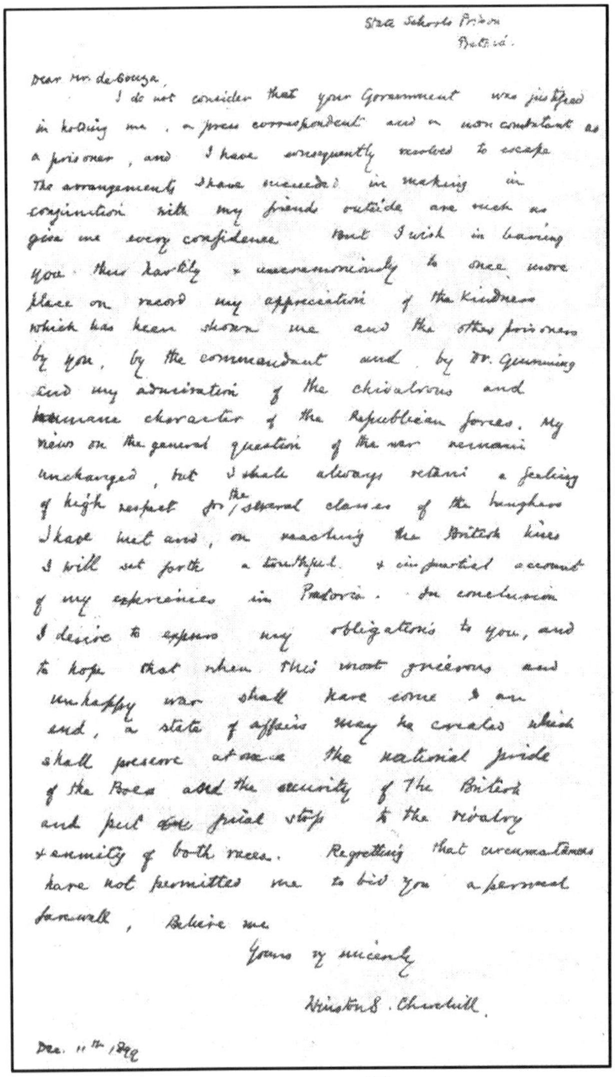

State Schools Prison
Pretoria.

Dear Mr. de Souza,
I do not consider that your Government was justified in holding me, a press correspondent and a non combatant as a prisoner, and I have consequently resolved to escape. The arrangements I have succeeded in making in conjunction with my friends outside are such as give me every confidence. But I wish in leaving you thus hastily & unceremoniously to once more place on record my appreciation of the kindness which has been shown me and the other prisoners by you, by the commandant and by Dr. Gunning and my admiration of the chivalrous and humane character of the Republican forces. My views on the general question of the war remain unchanged, but I shall always retain a feeling of high respect for the several classes of the burghers I have met and, on reaching the British lines I will set forth a truthful & impartial account of my experiences in Pretoria. In conclusion I desire to express my obligations to you, and to hope that when this most grievous and unhappy war shall have come to an end, a state of affairs may be created which shall preserve at once the national pride of the Boers and the security of the British and put a final stop to the rivalry & enmity of both races. Regretting that circumstances have not permitted me to bid you a personal farewell, Believe me

Yours very sincerely

Winston S. Churchill.

Dec. 11th 1899

Post Escape Letter 12/11/1899.[261]

State Schools Prison, Pretoria.

Dear Mr. de Souza,—I do not consider that your Government was justified in holding me, a press correspondent and a non-combatant, as a prisoner, and I have consequently resolved to escape. The arrangements I have succeeded in making with my friends outside are such as to give me every confidence. But I wish, in leaving you thus hastily and unceremoniously, to once more place on record my appreciation of the kindness which has been shown me and the other prisoners by you, by the commandant, and by Dr. Gunning, and my admiration of the chivalrous and humane character of the Republican forces. My views on the general question of the war remain unchanged, but I shall always retain a feeling of high respect for the several classes of the Burghers I have met, and on reaching the British lines I will set forth a truthful and impartial account of my experiences in Pretoria. In conclusion, I desire to express my obligations to you, and to hope that when this most grievous and unhappy war shall have come to an end, a state of affairs may be created which shall preserve the national pride of the Boers and the security of the British, and put a final stop to the rivalry and enmity of both races. Regretting that circumstances have not permitted me to bid you a personal farewell, believe me, yours very sincerely

Winston Churchill.

December 11th 1899

[286] Text from several sources but see Sandys P. 94 "As a touch of humor, he had written the letters 'p.p.c. (*pour prendre congé*, to take my leave) on the envelope." The letter is dated December 11th, but the attempt to escape that evening was not successful. When he did escape on the 12th, his bed was made up by Haldane to look like a sleeping person and the ruse was successful. The letter must have been nearby or given to de Souza by another prisoner.

[287] WSC London to Ladysmith.

Louis de Souza.[262]

Louis de Sousa and his spouse Marie.[263]

Next day [December 13th] Mr. De Souza, Under Secretary of State for War, a Portuguese gentleman, told me that General Joubert had telegraphed from before Ladysmith to the effect that he was satisfied that Mr. Winston Churchill had not borne arms against the Republic, and that he approved of his unconditional release. Mr. De Souza added that his Government was now indifferent as to whether he was found or not, as in the former case he would be released.[288]

Joubert Telegram

In 1902 Aylmer would obtain a copy of the following telegram by General Joubert and forward it to Winston with his compliments. See the section Haldane Forwards Telegram in the epilogue, infra.

> [December 12th, 1899][289]
>
> From General Joubert, Volksrust, To Acting Commandant General, Pretoria. Re: your telegram of today. I agree to the exchange proposed, but am resolved against the exchange of Churchill or of the conceited officer of General White's staff [Walter Adye.][290] captured on 30th October; or of the officers who were last captured in the armoured train beyond Colenso, or of the officers and men who since their capture have expressed themselves improperly against the Government, whether in speech or writing. . . .[291] See also, another copy of the telegram on last page, 309, of Penn State copy.[292]

I have so often since heard suggestions thrown out that Churchill was allowed to go, and that the Boers in any case intended to release him, that I feel it incumbent upon me to state that, in my opinion, nothing was further

[288] Rosslyn reports having the same conversation at a later date: April 18th just before primrose day. Mr. De Souza left me after telling me that Winston Churchill's warrant of release was in their hands the night he escaped, and that they were glad to be rid of him. "He talked so much," he added, "that we all avoided him" I could not restrain a smile; but his escape and Mr. Hellawell's unsuccessful attempt have certainly made matters "difficult" for war correspondents and officers too, since these have been removed to more guarded quarters ever since three of their number escaped about a month ago.

Page 295 of the book and 368 of the scan of Earl of Rosslyn, James St Clair-Erskine, 5th Earl of Rosslyn Twice Captured: A Record of Adventure During the Boer War. United Kingdom.

[289] OBC Vol 1 Part 2 p. 1101.

[290] Adye Walter Major MID LG: 8 February 1901, page: 917. Source: Lieutenant General White. 2 December 1899. Royal Irish Rifles. Adye was one of 954 officers and men to be taken prisoner that day. It is said that the officers of the Royal Irish Rifles were 'so exasperated at the exhibition of the white flag that they set to work and smashed their swords rather than give them up.' Adye was mentioned in despatches and, on being released, served as DAAG at Army HQ from July 1900 until February 1904......He served briefly as Deputy Assistant Inspector of Remounts, Eastern Command, from 1914 until his death in September 1915. Anglo Boer War - Name search

[291] This is probably the information that reached Winston and steeled his resolve to escape.

[292] Sir James Aylmer Lowthorpe Haldane Aylmer Haldane, How We Escaped from Pretoria. Edinburgh: W. Blackwood. 1901 University of Pennsylvania Copy: #11 - How we escaped from Pretoria / by Aylmer Haldane. - Full View | HathiTrust Digital Library.

from their intentions. It requires no intricate process of induction to conclude that such was the case when one recalls the extraordinary hue and cry which followed his departure, and the elaborate arrangements made to prevent a repetition of the occurrence.[293]

An incident happened two days after Churchill's departure which would in any case have led to the discovery of my escape had I succeeded in effecting it. My cousin, Mr. R. B. Haldane,[294] with the assistance of Lord Rothschild,[295] had managed to transmit a draft for £100 to me, foreseeing that it might be of great assistance in furthering any attempt to escape. The commandant would not trust it in other hands, brought it himself, and told me he had directions to give it to me personally and take my receipt.

I tried to cash this welcome addition to my funds, and was told that I might bank it if I liked and draw upon it gradually, but that I could not be permitted to possess at one time so large a sum. Later on, however, during my journey to the border, I found means of converting the draft into a less portable but more profitable form.[296]

I believe that this was the only case in which money reached an officer from outside sources while we were prisoners.

Many drastic changes now took place in the police and other arrangements for the security of the prisoners. By day and night sentries were posted in the adjacent gardens, and their numbers were for some days augmented from nine to seventeen. Roll-calls took place twice daily; beer, newspapers, visiting patients in hospital, walking round the building after 8.30 P.M., and sleeping on the verandah, were all prohibited. For a time we were subjected to many petty annoyances, which displayed to fine advantage the narrow-minded and malicious nature which actuated our warders.

Dr. Gunning did his best to smooth our difficulties, but he ran the risk of drawing suspicion on himself by so doing. Public opinion in Pretoria ran in favour of very stringent measures, and the Government (which endeavoured to deserve the name of a popular one, with the Dutch at least) felt compelled to meet to some extent the wishes of the citizens.

Several detectives and the inspector of police inspected the building and surroundings with a view to prevent a recurrence of the event of the 12th, and the result of their visit showed itself after a few days. The offices at the back of the yard, originally connected with the paling, were isolated, so that the sentries could pass behind and prevent anyone from climbing over; the lower branches of the trees were cut off so that no dark place remained; and the reflectors of the electric lights were adjusted in such a way that no spot was now in shade. I shall refer later to two additional lights which were placed in the streets on the southern and western sides of the building, as they were destined to play a part in a future plan.

The sentries, who now kept their rifles loaded, were placed under the inspector of police, a German; and every night that officer visited them without fail, and saw that they were thoroughly vigilant.

Up to the time of the incident I have just recounted, the Reverend Godfrey a Church of England clergyman, had held a service in the schoolroom on Sundays; but after Churchill's escape he deemed it wiser to forsake us.

[293] See Epilogue where Haldane forwards the Smuts Telegram and the Joubert one to Churchill in 1902.

[294] Richard Burdon Haldane, 1st Viscount Haldane, KT, OM, PC, FRS, FSA, FBA, KC (30 July 1856 – 19 August 1928) was an influential British Liberal and later Labour politician, lawyer and philosopher. He was Secretary of State for War between 1905 and 1912 during which time the "Haldane Reforms" of the British Army were implemented. Raised to the peerage as Viscount Haldane in 1911, he was Lord Chancellor between 1912 and 1915, when he was forced to resign because of false allegations of German sympathies. He later joined the Labour Party and once again served as Lord Chancellor in 1924 in the first Labour administration. Richard Haldane, 1st Viscount Haldane-Wikipedia

[295] Lionel Walter Rothschild, 2nd Baron Rothschild, Baron de Rothschild, FRS (8 February 1868 – 27 August 1937), was a British banker, politician, zoologist and soldier, who was a member of the prominent Rothschild family. As a prominent Zionist leader, he was presented with the famous Balfour Declaration which pledged to a Jewish national home in Palestine. Walter Rothschild, 2nd Baron Rothschild-Wikipedia

[296] PF: Courtesy of Mr. Howard I believe. Who gave him gold Sovereigns in exchange for his paper currency or bank draft.

J. H. Godfrey
To the British Officers, being Prisoners of War.
From Rev. Mr. Godfrey
St. Albans, Pretoria, Z.A.R.
Dec. 15, 1899.
Gentlemen - By the kind courtesy of the Government I have been permitted to hold services for you in connection with the Church of England, which services I have felt it a privilege on my part to conduct. After what has recently occurred—viz., the escape of Mr. Churchill from confinement—I exceedingly regret that, in consideration of my duty to the Government, I must discontinue such regular ministrations, as I desire to maintain the honour due to my position. Of course I shall always be glad to minister to you in any emergency with the special permission of the authorities, who will, with their usual kindness, duly inform me.—With my best wishes, I am, Gentlemen, yours sincerely,
J. H. Godfrey.[297]

I believe he thought that he incurred the danger of being suspected of complicity, and hence of being put over the border. Some of us considered his fears somewhat groundless, seeing that the commandant never left his side during the whole time he was engaged in his ministrations.

Fortunately for us, the Reverend Adrian Hofmeyr, [298] who had been most treacherously taken prisoner on the western border (I think through the instrumentality of Cronje), and in spite of ill-health had been kept for some time in a prison-cell, arrived to join our company. The commandant in charge of us would fain have induced him to play the part of *mouchard*,[299] but he reckoned without his host.

I feel sure that every officer who was a prisoner soon felt that in Mr. Hofmeyr he had the staunchest of friends, and that a most loyal Englishman had joined our throng. Though cut off entirely from wife and family, and allowed to receive no letters from them, suffering no doubt from depression of spirits and ennui, he was ever ready with a kindly word and sympathetic look, which went far to mitigate the trials of our confinement. His able and encouraging discourses will not readily be forgotten by some of us, and as I understand that he may publish them, it is not improbable that they may reach farther than was originally intended. At one time there was a talk of his release, but glad as we should have been to see this come about, many of us secretly hoped that he would not leave us.[300]

About this time the party who had been captured with the guns at Colenso[301] arrived, and from them we gleaned a good deal of news, which, since the newspapers had been forbidden us, we sadly lacked. Dr. Gunning told me in confidence that a battery of Creusot guns of the latest pattern had reached Pretoria *via* Lorenco Marques;[302] and though I afterwards learnt that this important addition to the Boer armament had been smuggled through the custom-house in piano-cases, I have since been told that during the war only one case of that description crossed the border. Stories of the smuggling of contraband of war were very common in Lorenco Marques, and were spread by the Boers themselves with the view of inducing our naval authorities to bring about international complications by searching the merchant ships of other nations which frequent that port.

[297] The letter as reported by Rosslyn. And quoting from his work: "Needless to say, no reply was sent to this epistle. This gentleman, I was also told, did not pray for the Queen! On the 15th February a letter signed by the committee of officers to the State Secretary, Pretoria (owing to the death in jail of Lieut. Tarbutt from enteric fever), "requested the Government to reconsider the position of the prisoners of war who were former residents in the Transvaal, and who were then in prison, and to treat them in a similar way as the British officers were treated, seeing that they were not subjects of the Transvaal, but of the Queen." From: Twice Captured: A Record of Adventure During the Boer War. Earl of Rosslyn, James St Clair-Erskine, 5th Earl of Rosslyn. Wm. Blackwood and Sons, 1900. Page 320 See: Twice Captured - Google Books and: James St Clair-Erskine, 5th Earl of Rosslyn - Wikipedia.

[298] See his account, infra.

[299] An undercover investigator; a police spy, especially in a French-speaking country.

[300] He was released on May 12th, 1900 See Hofmeyr Chapter 38 titled 'Free.'

[301] The Battle of Colenso was fought on December 15th, 1899 and resulted in a British defeat and withdrawal from the field.

[302] Maputo, officially named Lourenço Marques until 1976, is the capital and most populous city of Mozambique. Maputo is situated on a large natural bay on the Indian Ocean, near where the rivers Tembe, Mbuluzi, Matola and Infulene converge. Maputo - Wikipedia

Adrian Hofmeyr.[264] " My little corner in the Staats Model School—8 feet by 6 feet—in a room occupied by eight officers. The table I bought, and the flowers my sister sent me."[265]

For the first few days after Churchill's escape we occupied ourselves by noting the various alterations undertaken for our security, and in thinking out some new plan of operations. One day we made a raid upon some cupboards and broke them open, finding therein three screw-drivers, two triangular files, and a pair of wire-cutters. All of these were destined to be of service in the near future. Tales of escape from prison had always had a peculiar fascination for me, and I was as familiar with the true account of Latude's[303] escape from the Bastille and Jack Sheppard's[304] from the castle-ward in old Newgate, as I was with the fictitious escape of Edmond Dantes[305] from the Chateau d'If. Our next plan was to be on the lines of the last of these.

On first arriving at the school the idea of digging one's way out had presented itself to me, but as there was a prospect of escaping without going below ground, the subterranean project was for the time being laid aside. I personally have always had a horror of being buried alive, and for that reason had avoided suggesting a plan which might involve such a contingency; but Brockie, who had worked in the mines of Johannesburg, had no such fears, and after debating the question we decided to put it in force. The occupants of our room (five in number, since Churchill left) agreed to try to burrow their way to freedom, and decided to start work forthwith. We invited [Frederick Neil] Le Mesurier[306] of the Dublin Fusiliers, my companion in our successful escape, to join us, and being of powerful physique he proved a most indefatigable digger.

On Sunday the 17th December we made a preliminary exploration to ascertain whether there was space under the floor sufficient to contain the earth we should have to remove in our operations. We had reasonable hopes that this would be so, for the Model School is built on a slope, and our room was towards what I may term the downhill end of the building. Before cutting through the floor we examined the planks, which were well laid and

[303] Jean Henri Latude (23 March 1725 – 1 January 1805), often called Danry or Masers de Latude, was a French writer famous for his lengthy confinement in the Bastille, at Vincennes, and for his repeated escapes from those prisons. Jean Henri Latude-Wikipedia

[304] Jack Sheppard (4 March 1702 – 16 November 1724), or "Honest Jack", was a notorious English thief and prison escapee of early 18th-century London. Born into a poor family, he was apprenticed as a carpenter but took to theft and burglary in 1723, with little more than a year of his training to complete. He was arrested and imprisoned five times in 1724 but escaped four times from prison, making him a notorious public figure, and wildly popular with the poorer classes. Jack Sheppard-Wikipedia

[305] Edmond Dantès is a title character and the protagonist of Alexandre Dumas's 1844 adventure novel The Count of Monte Cristo. Edmond Dantès-Wikipedia

[306] Frederick Neil Le Mesurier Captain QSA, QSA known to exist Source: List of QSAs with the clasp Talana from the Talana book Royal Dublin Fusiliers. MID LG: 10 September 1901, page: 5951. Source: Field Marshal Roberts. 4 September 1901. Re: General mentions Anglo Boer War - Name search Page 5927 | Issue 27353, 10 September 1901 | London Gazette | The Gazette.

He died at Ypres in 1915. See: Frederick Neil Le Mesurier Captain 2nd Bn. Royal Dublin Fusiliers Died 25 April 1915 Ypres (Menin Gate) Memorial Panel 44 and 46. Belgium The Commonwealth War Graves Commission | CWGC | CWGC.

PF: He may be related to Gerald John Le Mesurier DFC who was a South African flying ace of World War II, credited with 3 'kills.' After completing his schooling at Diocesan College in Cape Town, he studied at the University of Cape Town. He joined the Special Service Battalion in 1936 but later in 1936 transferred to the South African Military College, joining the South African Air Force Gerald Le Mesurier - Wikipedia.

dovetailed, and to our satisfaction found a trap-door. Removing the screws which held it in its place, we raised it and descended.

As three of us were doomed to occupy this earthy chamber for nearly three weeks, I shall here give a brief description of it. The floor was about two feet and a half above the ground, and the space below, which corresponded with the room above, was divided into five narrow compartments by four transverse stone walls on which the cross-beams that carried the flooring were laid. Each chamber was about eighteen feet in length and three feet and a half in breadth, and there were man-holes in the walls. A certain amount of air came into this damp space through small ventilators under the verandah, but the atmosphere was very close, and one could not see except by candle-light.

We now divided ourselves into three reliefs, established a. system of signals, so that the operators below ground might be quickly warned to pause in their labours should a stranger enter the room above, and on Monday morning commenced work.

It had to be carried out as silently as possible. The scheme was to sink a shaft about twelve feet deep by five in length and three in width, and from the bottom of this to drive a tunnel diagonally under the street on the western side of the building for about one hundred feet. This would bring us under a kitchen-garden, whence we could easily escape after dark, unseen. For digging implements we used our screwdrivers, and succeeded in breaking through some three feet of caked earth as hard as concrete. After this was penetrated the work became easier; but to our dismay, though not surprise, two feet deeper landed us in water. After many ineffectual attempts to get rid of it, by carrying it to the most distant of the five compartments, we found that the inflow, especially after rain, greatly exceeded what we could bale out, and our second plan accordingly fell through. It had been a work of considerable labour, armed as we were with such unsuitable tools, and the disappointment at having to confess ourselves beaten was great. Some lines by Sir Walter Scott[307] kept constantly recurring to me—

"Patience waits the destined day;
Strength can clear the cumber'd way;"

—and the word *impossible* was not to be thought of. The numbers who were now incarcerated in the Model School had increased greatly, and Field-Cornet Malan, before leaving for the front, had started a scheme for locating us in a much more roomy place, which we understood was in course of construction and was approaching completion. We lay on our oars for a time, hoping that the new place might be more favourable to our plans than our present habitation, or that the sentries after a time might grow less vigilant. In the meantime, much could be done which might prove of value in the future. Maps of the country between Pretoria and the border in many directions were made, and for this purpose we used Bartholomew's tourist map of South Africa, which one of the officers when taken prisoner happened to have with him. Brockie gave me some very elementary instruction in the Kaffir language, and I wrote down several sentences which I proposed to pour forth at the proper time, though how I should possibly understand what was said in reply was a matter on which I did not care to ponder.

Christmas-day came, a dreary day for us and one on which many had hoped to be in Pretoria with a victorious army. Two English women employed in the hospital as cooks made us a present of plum-puddings, and the generous Transvaal Government allowed us each a bottle of beer, on payment.

We heard that President Kruger had received several Christmas cards from the Cape Colony and Natal, some of which had caused him considerable wrath.

I had hoped that at this usually festive season our guard might possibly depart from their persistent sobriety, and so give us opportunity to take advantage and escape; but whether it was owing to the lack of the necessary stimulant or that their sense of duty was too strong, no sign that they were in any way affected by the festival could be detected.

The days were now growing more and more monotonous, news was scarce and not encouraging, and beyond reports that we should shortly shift our quarters, and the occasional arrival of a prisoner, nothing occurred to break the unvarying sameness of our fungus-like existence. The possibility of organising a rising and break out

[307] Sir Walter Scott, 1st Baronet FRSE FSA (15 August 1771 – 21 September 1832), was a Scottish historical novelist, poet, playwright, and historian. Many of his works remain classics of both English-language literature and Scottish literature. Famous titles include the novels Ivanhoe, Rob Roy, Waverley, Old Mortality (or The Tale of Old Mortality), The Heart of Mid-Lothian and The Bride of Lammermoor, and the narrative poems The Lady of the Lake and Marmion. Walter Scott-Wikipedia

of our prison had gone no further than the stage of discussion. The difficulty of obtaining reliable outside help and of procuring sufficient arms, and many other points, led to the final abandonment of the project.

All through January Sergeant Major Brockie and I had kept watch night after night, between 7.30 and 8.30, thinking it possible to climb the railings and to slip past a certain sentry when his back was turned. I even went so far as to keep one of my coat-pockets full of pepper, a liberal dose of which I proposed to hurl into his eyes if perchance he looked our way, as I imagined that this substance would probably be more effective than the metaphorical dust.

Fortunately for us, he never gave us the wished-for opportunity, for I am now convinced that some other sentry would have seen us, and that we should not have got far. At the end of January we were allowed once more to buy a daily paper, and to visit, under supervision, our friends in hospital.

The news gleaned from the 'Volkstem,' a Government organ, was as nothing compared to a source from which we obtained the latest and most reliable information. It came about in this wise. We had noticed a man who occasionally passed the Model School, generally accompanied by a St Bernard's dog.[308] From his manner he seemed sympathetic and anxious to communicate with us, and he sometimes muttered some words as he passed.

As time went on he took to signaling to us by the Morse code with his stick. As the police seemed very suspicious of him, he could not send more than a word or two, such as "British victory." The sentries became more suspicious of him, and I believe he was told not to pass the school. For a time, at any rate, he rarely was seen. At length a system of communication was arranged with him through the medium of the hospital, and we found out that we were indebted for our news to Mr. Patterson, one of the principal telegraphists in Pretoria.

Matters were soon on a better footing. Our sympathetic signaler took to visiting a certain house, inhabited by Mr. Cullingworth and his family, and in the afternoon he sent many messages to us from the verandah.

As, however, his duties prevented him from coming daily, and his presence at this house was calculated to arouse suspicion, he instructed two young ladies, who, we imagined, were daughters of our opposite neighbour, in the art of signaling, and they undertook the risky work of transmitting to us the news with which he supplied them.

The police were exceedingly suspicious of the house and of its inhabitants, and did all they could to ascertain what was transpiring between them and the prisoners across the street. But the system on which the messages were sent almost defied detection.

Standing well inside the door, one would signal with a white flag, while the other, seated on the verandah, gave warning when a sentry or passer-by was approaching. By this means we now received, twice daily, the latest news, from the Boer point of view, of what went on at the front; and I believe that we were the recipients of the same telegrams which were laid before his Honour President Kruger. The same news was often served up in the daily papers in a very different guise; sometimes it was omitted altogether.

We unfortunates felt we owed a deep debt of gratitude to those who risked everything to give us news of what was going on in the outer world, a debt which it was impossible for us to repay. I am glad to say, however, that when Lord Roberts reached Pretoria he showed his high appreciation of what had been done for us by calling upon the Cullingworths and later on the officers presented them and Mr. Patterson with mementoes bearing inscriptions indicative of the circumstances under which they came to be given. Shortly before I succeeded in escaping, Messrs. Cullingworth and Patterson were commandeered as undesirable persons and sent to the front. Their last message to us told how they intended to cross over to the lines of the British army at the earliest possible opportunity. They were not fated to succeed, were captured in the act and sent back to Pretoria, and there confined in jail until the British troops released them.

[308] This is Mr. Patterson; who works in the telegraph office, also called "dog-man" by Frankland because he walks his dogs past the school.

It is difficult to be certain of one's dates where no notes or diary have been kept, but I think it was early in February that twelve officers were allowed, on parole, to attend the funeral of an officer of a Colonial corps.[309][310]

On their way back they drove past the President's house and saluted him as he sat smoking on his verandah. The old Dopper[311] statesman at once rose and courteously returned the compliment, raising his tall hat.

President Kruger at his home.[266]

President Kruger on his veranda.[267]

This was the only occasion on which any officer, so far as I am aware, who was a prisoner in Pretoria saw him. In the same month I happened one day to be leaning over the railings of our prison, ruminating on my hard fate and the difficulty of overcoming it, when I saw an unmistakable Englishman, accompanied by another man, of whom I took no particular note. They were passing by on the other side of the street, and their sympathetic looks attracting my notice, I responded.

As the sequel will show, I was unconsciously looking at the very two men who were soon to give Le Mesurier and me invaluable aid in our escape, and, strange to say, they had done as much for Churchill.[312]

The time was now approaching when our last and successful bid for freedom was to take place. On the 12th February I had heard from Sir William Nicholson[313] that, as I expected, there was no possibility of my being exchanged under present circumstances.

309 Lieutenant Charles Tarbutt died on February 13th. Charles A Percy Tarbutt: He died of typhoid fever at Pretoria, February 13th, 1900. He was the eldest son of Percy Tarbutt of the Consolidated Gold Fields of South Africa. Lieutenant Tarbutt was twenty-six years of age, and was educated at University College School (where he was in the football XV.), and at the Bedford Grammar School. When Lieutenant Tarbutt landed in South Africa he was given a commission as Lieutenant in the South African Light Horse. After the battle of Colenso he was reported missing, but it was subsequently discovered that he had been wounded and captured. Source: Donner https://www.angloboerwar.com/name-search.

310 Hofmeyr: And thus it was that every member of the Imperial Light Horse, because most of them had lived at Johannesburg before, were special objects of this hatred. And so poor Lieutenant Tarbutt was sent to the common gaol, to be locked up in a miserable cellar cell fit for a condemned criminal, a thief, or a murderer, but certainly not for a gentleman, though he was a prisoner of war. And this they call civilisation-chivalrous action towards a fallen foe! Weep, guardian angels of Transvaal! For weeks Lieutenant Tarbutt was detained there, and, as could be expected, he fell ill of fever. You see, sanitation is everywhere at a discount here, and especially in the prison—I was there myself, I speak from experience. Imagine a fever patient in a miserable stuffy cell, in summer! But the officials gloated over it. "We'll show him what it is to fight against the Transvaal; we'll pay him out!" Such were the comments. Meanwhile our young friend grew worse, and only when after a fortnight's illness it was proved that he was suffering all the time from a virulent attack of typhoid fever did the Government remove him to a decent hospital. Lieutenant Tarbutt now grew rapidly worse, and, after about twelve days in hospital, died yesterday, February 13th. What a shock the news of his death was to all of us (the hospital was just next door to our prison)! So young—I believe about twenty-seven years—and now cold in death.

311 Dutch Reformed Church members are colloquially referred to by this name.

312 He is referring to Mr. John Howard and Mr. Charles A. Burnham.

313 Sir William Newzam Prior Nicholson (5 February 1872 – 16 May 1949) was a British painter of still-life, landscape and portraits. He also worked as a printmaker in techniques including woodcut, wood-engraving and lithography, as an illustrator, as an author of children's books and as a designer for the theatre. William Nicholson (artist)-Wikipedia

This was no disappointment, but it made me decide that to wait longer was useless. Rumours of our moving to the new building were now of almost daily occurrence, and from a plan which I had seen, it appeared to be a place from which escape would be even more difficult than from our present dwelling.

The scheme upon which we now based our hopes was to induce the English electrician who was in charge of the lighting of the town, and also of the Rontgen ray apparatus[314] in the hospital next door, to extinguish the lights of the school and of that district of Pretoria in which it was situated, for half an hour on some dark night. We inquired of him through Major Adye, who was in hospital, if he would help without divulging anything. The reply came that, being on parole,[315] he regretted he could not.

We heard afterwards that if the lights at the school had been extinguished the searchlights in the forts would have been turned on us at once. This plan having so far failed, we decided to do the extinguishing part of the programme ourselves by cutting the wires which lighted the building and back yard.

Four conspirators were engaged in this scheme. The general idea was to cut the wire on some dark wet night, and, silently crossing the yard with feet encased in noiseless India-rubber shoes, pass between the sentries and climb as quickly as possible over the corrugated iron paling, and so attain the yard behind the hospital. Thence the street could easily be gained; and at the worst, if discovered, we could make a run for it and try to pass the special police, reach the Apies river[316] some two streets off, and, following its course, come eventually to the outskirts of the town.

This project, I need hardly say, was a very risky one, for we had bound ourselves to stop for nothing once the light went out our motto was to be *Vestigia nulla retrorsum*.[317] If seen, we should have to run the gauntlet of two sentries a few yards distant, who would shoot at us without hesitation.

There seemed no other possible way of baffling the guards, and the most dangerous way is generally the safest. We trusted that on the light being suddenly extinguished the sentries would not see well for some seconds, and that their surprise might distract their attention from us.

Brockie had noticed that once weekly on a certain night the sentries, instead of being relieved as usual at eight o'clock, continued their watch till midnight. We therefore decided that, other things being favourable, the preference must be given to this night, for it seemed probable that after so long a tour of duty the sentries might somewhat relax their vigilance.

In a place where we were so crowded, it was extremely difficult to discuss our plans in secret, and indeed privacy was rarely to be found. In discussing our earlier plans, when the occupants of the school were considerably fewer in number, we were able to confer together in an empty room; but now all such questions had to be postponed till night, when we repaired to the gymnasium, at that time generally deserted.

Two possible means of cutting the wires presented themselves—one where they passed into the electrometer, which was fixed on the wall close to the front door of the building; the other by penetrating into the loft and attacking them at the spot where they passed through the roof. The first method was not considered feasible: the wires were at this spot extremely thick, and the place itself sometimes thronged with those going in and out of the doorway. The loft must therefore be invaded. To reach that place it was necessary to remove a heavy ladder from the gymnasium and by its assistance climb through a trap-door in the ceiling of the main passage. By this means we explored it on one occasion, but as a permanent route it was radically bad.

In order to succeed it was necessary to keep our arrangements absolutely secret, and this was impossible unless some other means of reaching the loft could be devised. At length a simple means occurred to me. The roof of the gymnasium, which was of wood, had sloping sides, and by climbing to the top of one of the ladders a trap-door might be cut, and then we should have private access to the wires. I forthwith made a saw out of a table-knife, and the trap-door was cut and cross-battens screwed to it to keep it from falling out. In order to facilitate climbing over the iron paling, we prepared two plank ladders with padded ends: these we painted over with boot blacking to render them inconspicuous.

[314] The X Ray machine. X-ray - Wikipedia

[315] In that he had given his word, as a condition for his freedom, that he would take no part in the conflict or provide aid and comfort to the British.

[316] The Apies River is a river that flows through the city of Pretoria, South Africa. Its source is located just south of the city (south of Erasmus Park) and it flows northward until it drains into the Pienaars River. The word "Apies" is Afrikaans for small monkeys and is a reference to the historical abundance of vervet monkeys on the Apies River banks. Apies River - Wikipedia

[317] "No retreat" or "We never go backward."

Another important matter was the question of food. As we intended making for Mafeking,[318] we expected to be several days on the road, and sufficient provision of a non-bulky nature must be carried. We double-lined our waistcoats, making numerous pockets therein, each of which would hold one or two packets of chocolate.

Should we have to run for it, we should be carrying weight where it would be least felt. These waistcoats, when completed and packed with chocolate, resembled in some degree those cork life-belts to be met with on board ship, and ere long they were to prove themselves veritable life-preservers.

I have mentioned before that two extra electric lights had been placed in the street, and these were, outside our control. On more than one occasion I had made an inspection about 3 A.M., when all the lights in the building were out, and it seemed probable that they would have more effect than we had allowed for. One of them cast a beam into the back-yard past a corner of the building; while the other, situated immediately opposite the front door, sent a stream of light into the passage, if indeed it did not penetrate farther.

Closing the door would have availed nothing, for above it was a plate-glass window. There were lights also in the hospital which were not beneath consideration, but as these were generally extinguished by eight o'clock, we could time our movements accordingly.

We had often remarked that no sooner had we got our plans well matured and every detail carefully thought out, than some new difficulty would arise. The present scheme was no exception. One evening I was sitting on the back verandah when the inspector of police arrived, accompanied by a very objectionable species of sentry; a large white bull-terrier. This animal was allotted a permanent post, to which he was chained day and night, close to where we meant to break out.

He seemed to be a dog of discrimination, for he shewed a violent antipathy to policemen and a most amiable disposition towards us. As the nights were cold, we represented that his health might suffer, and he was shortly provided with a rude kennel, which we hoped would have the effect of inducing sleep, and so reducing his value as a sentry. But a more trying incident was to follow, and at the risk of boring my readers with details I shall record it. One of the rooms of the school had been used as a store-room, and as this was now required for other purposes, the stores were removed into the gymnasium.

We were no longer masters of the situation, for the two officers of mercantile predilections who were in charge of this important department kept their office locked, and were frequently at work with accounts until a late hour. We managed, however, to secure a duplicate key of the gymnasium, which in some degree compensated us. Of course we could have taken the storekeepers into our confidence, but that was undesirable.

Shortly after this inconvenient removal, we discovered that our trap-door had been seen, and that the loft, like the cave of the forty thieves, had been visited by persons unknown. The knowledge that our secret was no more our own, added to the fact that someone, ignorant of our proceedings, suggested to me the very plan which was then so nearly ripe, made us decide to go at the earliest opportunity, for every day's delay presented the possibility of our being forestalled. All was now ready, and a servant had been engaged to cut the wires. He had been introduced to the loft, and instructed what to do on receiving a certain signal.

The afternoon of Friday, 23rd February, was dull and threatening, and there was a possibility that we might have rain at night. The afternoon was passed in some anxiety. Our eyes were frequently directed towards the sky, and, like the prophets of old, we were filled with earnest hopes that a perfect deluge would fall. Provided there were no lightning, all might be well. Before going in to dinner our plank ladders, concealed between two folding-chairs, were carried out and hidden on the verandah, which was to be our rendezvous.

The gymnasium was vacated and locked up for the night at 8.30, and our accomplice was duly consigned to the spot where his wire cutters were to operate. A drizzling rain was falling, and the wind whistled through the trees—the precursor of a stormy night. The sky was covered with inky clouds, and the ground underfoot being damp, would deaden the sound of our footsteps as we crossed the yard. Fortune at last seemed to smile on our venture, for the sentries were one less in number than usual, the absentee being the very one close to whom we should have had to pass. This seemed an omen of success.

At nine o'clock the four conspirators assembled on the verandah, ladder in hand, ready to creep with stealthy steps to the appointed spot. It was an exciting time, the prospect of escape so near and so uncertain. I gave the

[318] At that time, Mafeking was 160 miles to the West and also under siege. It would have been quite difficult to enter. Mahikeng, still commonly known as Mafikeng and previously Mafeking, is the capital city of the North West province of South Africa. Close to South Africa's border with Botswana. Mahikeng - Wikipedia

signal, and the moments seemed like hours. Suddenly the lights in the building and yard went out, and like a flash we made for the gate through the wire-netting; but scarcely had we reached the barrier of wire when the buildings and yard were again illuminated.

Yet even in those fleeting moments of extinction we had noticed that the yard was not quite dark. Back we crept to the verandah, imagining that our accomplice had received a shock which had disabled him, and had failed to cut the wire completely. Several minutes of suspense passed, and the sentries showed unmistakably that they were wondering what had happened. The corporal went his rounds, and seemed to be discussing the eccentricities of the light.

Interior of Gymnasium of the Model School, Pretoria, Showing how Some of the Prisoners Made their Escape[268]

After some time our accomplice came and told us what had happened. He had cut the wire, he said, completely through, receiving a severe shock in so doing, but after a little persuasion agreed to go back and complete the job by cutting the other wire. Unknown to me, one of the party told him not to operate until he again received the signal, and all waited for me to repeat it while I went through minutes of suspense which felt a lifetime. At length, however, our wire-cutter, growing impatient, returned and declined peremptorily to take further part in the proceeding.

After a consultation we decided that the guard being on the *qui vive*[319] it would be madness to cut the other wire that night and think to escape unseen. We therefore deferred the attempt till the next dark night. Next day we insulated the wire-cutters with a silk handkerchief and some India rubber, and obtained the services of another man to cut the wire.

But the next night and the next again were bright and rainless, and unsuitable for an attempt. Monday, the 26th [of] February, came, the day preceding Cronje's[320] defeat[321] and the anniversary of Majuba.[322] The 27th was to be kept in Pretoria as a day of humiliation, and not as usual of exultation.[323] The President deemed that too

[319] A state of heightened vigilance, especially prior to battle.

[320] Pieter Arnoldus "Piet" Cronjé (4 October 1836 – 4 February 1911) was a South African Boer general during the Anglo-Boer Wars of 1880–1881 and 1899–1902. Piet Cronjé-Wikipedia

[321] At the battle of Paardeberg. Battle of Paardeberg-Wikipedia

[322] The Battle of Majuba Hill on 27 February 1881 was the final and decisive battle of the First Boer War that was a resounding victory for the Boers. Battle of Majuba Hill - Wikipedia

[323] On the anniversary of Majuba, Cronjé, with 44 commandants and other officers of all grades, and over 3500 men, surrendered

much arrogance had resulted from the victory at Colenso, with the consequence that the tide of fortune had begun to turn, and that therefore dust and ashes were more appropriate to the 27th than beer and skittles.

About half-past five on Monday afternoon the electric lights in the back-yard, which together with those in the streets adjacent to the school were turned on some time before sunset, suddenly went out. The soldier servants, it was said, were kicking a football in the yard which had struck the wires and caused the mischief.

To us conspirators what had occurred was plain: the wire, which no doubt had been only partially cut through, had fused.[324] I heard afterwards that the English electrical engineer was sent for, but said he could not come till next day. It was said that he suspected that the light had not gone out from natural causes, and that he did not want to spoil sport by repairing it. If this be so, we have a great deal to thank him for. Some workmen came and examined the wire at the point where it crossed over the street into the building, but neither there nor in the school itself did they find the broken place. During dinner a written request was passed round, stating that the commandant desired that we would not walk round the building that evening. On coming out from dinner, a visit to the back-yard showed what an escape we had had on the previous Friday, through the non-extinction of the lights.

The two subsidiary lights in the street illuminated the yard to such a degree that anyone crossing it was distinctly visible. The question now arose, What was to be done? Report said that we were to be moved to new quarters in two days, and from there escape, we believed, was hopeless.

The cut wire and trap-door would be discovered on the morrow, and as the latter was in the quondam[325] gymnasium, now storeroom, the innocent storekeepers would be implicated. We should have to acknowledge that we were the culprits, and this would mean jail and no chance of escape.

The situation was grave. To attempt to make one's way out by the back-yard, in which a bonfire had been lighted, was impossible, the sentries being doubled, and a Kaffir cordon placed outside them. We were at our wits' end. The last hope of regaining freedom seemed gone; yet there is almost always a way out of a difficulty, if only one can find it. Ten o'clock came and still we had not hit on any plan. Le Mesurier, Brockie, and I were seated in our room, our heads together in earnest confabulation.

All of a sudden an idea struck me, and half-jokingly I told them what was in my mind. "Supposing we hid under the floor, do you think we should be found?"

The words were no sooner said than both ejaculated "Of course!" and after a few moments' further conversation our decision was made. While they collected a few necessary articles, and with great difficulty began reopening the trap-door, which had been tightly screwed down, I went round the occupants of the room, all of whom readily consented to preserve the secret of our hiding-place. What we three hoped for was that the Boers would think that we had taken advantage of the darkness, bribed a sentry, and so made off. This was exactly what occurred, and strange to say, despite the precautions taken by the commandant to prevent such an occurrence, he concluded that we had gone. For once the mistrust which one Dutchman has of another stood us in good stead. We had the satisfaction of baffling the whole talent of the Pretoria police.

The entrance to our hiding-place having now been opened, Le Mesurier, Brockie, and I descended to the scene of our former operations. The trap-door was secured, and nothing showed our absence but three empty beds.[326]

unconditionally to Roberts. Sir Redvers Buller's force captured the Boer position at Pieters. This action opened the road to Ladysmith. Boers retired north to Ladysmith.

324 In this case meaning that like a blown fuse, the remaining wire was insufficient to carry the current and had become so hot it had melted or vaporized, severing the circuit.

325 Former.

326 PF: This news reached Winston in Natal and he reported on it to the Morning Post.
British Officers' Escape. Durban, March 5, 12.30 p.m.
Captain Haldane and Lieutenant Brockie, who are reported to have escaped from Pretoria, were the two officers who tried to escape with me, but could not then elude the vigilance of the guards. Lieutenant Le Mesurier, of the Dublin Fusiliers, who is with them, is an officer of great physical strength. All three are men of daring and determination.
Lieutenant Brockie is a colonial, speaks both Kaffir and Dutch fluently, and knows the country.
Their escape from Pretoria, in spite of all the extra precautions taken after my evasion, by the use of electric wires and additional guards, is a marvellous instance of ingenuity, courage, and energy, and there is every hope that they will succeed in the yet harder task of making their way through the enemy's country to the frontier. But the chase will be hot. Source: The Naval Guns (Reports reprinted.) 7 March https://www.britishnewspaperarchive.co.uk/viewer/bl/0000174/19000307/059/0007.

IV. Gone to Ground

"Tout vient à point qui sait attendre."[327]

From the day on which we began the passive part of our escape, as I may term it, in distinction to the active part which followed when we left our underground dwelling, I kept a few notes, and as I brought them from Pretoria, it will be more convenient if I write this chapter somewhat in the form of a diary.

After going below the floor on the night of the 26th February, we prepared to make ourselves as comfortable as the surroundings would permit. We decided to sleep in the compartment which I have described elsewhere, next to the ventilators underneath the verandah, as here the atmosphere was not quite so close as in the other chambers.

We only possessed one blanket apiece, not daring to take more below with us for fear of arousing suspicion, since a person who was believed to have got clear away would hardly take all his bedding with him. The ground made a very hard couch, and this, added to the lack of fresh air and the excitement of the adventure, caused sleeping to be very difficult.

At length, after a night of continual turning over and over, trying to find a softer spot, we saw by the faint light which filtered through the ventilators that another day had come.

This day, we thought, meant for us either discovery and imprisonment or non-discovery and escape. But escape was farther off than we supposed. Ere we shook the dust of Pretoria from off our feet, we had to undergo what it makes one shudder to recall. It was now 5.30 A.M., and stealthy footsteps were audible overhead. It was the commandant going his morning rounds, and counting his charges as they lay asleep. A surprise was in store for him. Suddenly his footsteps ceased. In imagination we pictured what was occurring. He had reached one of the vacant beds; he was looking at it in amazement.

No dummy figure had been deemed necessary this time. His tread was heard again, and twice more it ceased as two more blanks were seen. Then came a sound of voices. He was inquiring from some drowsy mortal where the usual occupants of the vacant beds were. The conversation was of the briefest: what reply he got I do not know, so I cannot give it here; and perhaps it is as well, for the hour was early and the inquirer had no claim to popularity among us. Whatever was its nature, the sounds of his retreating footsteps were soon heard as he left the room. We knew and were satisfied that our absence was at least suspected.

The hours rolled slowly on. At 8 A.M. a roll-call took place. Our absence passed from the region of doubt to certainty. What would happen next? At 9.45 I heard Colonel Bullock's[328] voice bidding all go outside the building. A search was about to be made. An exciting moment was drawing near, yet we felt reasonably confident that our hiding-place would be overlooked. Soon the tramp of many feet, "of armed men the hum," resounds from the tiles of the passage. Nearer and nearer it comes; our hearts begin to beat a little faster.

The door is reached, barred by the arm of no fair brave scion of the house of Douglas.[329] In the throng, a posse of detectives and armed police, with but an inch of deal between them and their prey. We hold our breath as they approach the roof of our humble dwelling.

The room above does not offer many hiding places—no secret panel or priest's hole—nought but four bare walls, ten beds, a cupboard. Soon they leave. All the talent of the Vidocqs[330] and Le Coqs (with apologies for the comparison) of Pretoria has failed to scent our lair. We breathe freely again.

[327] Good things come to those who wait.

[328] Lieutenant General Sir George Mackworth Bullock, KCB, KCMG (15 August 1851 – 28 January 1926) was an officer of the British Army. He served during World War I, rising to the rank of lieutenant general, and was also the one-hundred and eighth civil Governor and military Commander-in-Chief of Bermuda George Bullock (British Army officer) - Wikipedia.
Also: Colonel Bullock MID LG: 26 January 1900, page: 508. Source: General Buller. 16 December 1899. Re: Colenso. Devonshire Regiment. London Gazette January 26th, 1900 p. 508 [During the battle of Colenso on December 15th, 1899] "Colonel Bullock, 2nd Devons, behaved with great gallantry; he did not receive the orders to retire; his party defended themselves and the wounded of the two batteries till nightfall, inflicting considerable loss on the enemy, and it was only when surrounded that he consented to surrender, because the enemy said they would shoot the wounded if he did not." See also his summary at Anglo Boer War - Anglo Boer War

[329] Clan Douglas is an ancient clan or noble house from the Scottish Lowlands. Clan Douglas-Wikipedia

[330] Eugène-François Vidocq (24 July 1775 – 11 May 1857) was a French criminal turned criminalist, whose life story inspired several writers, including Victor Hugo, Edgar Allan Poe and Honoré de Balzac. The former criminal became the founder and first director of the crime-detection Sûreté nationale as well as the head of the first known private detective agency. Vidocq is considered to be the father of

The anxiety of our faithful comrades who remained outside was great, and their relief when they heard we had not been discovered greater still. After leaving our room the crowd of searchers hunted through the other rooms, and finding nothing of us, they repaired to the roof. Here our rude trap-door was immediately seen, and near it, left there by accident, lay my saw. This was pounced on by Dr. Gunning for the Museum, an interesting account of which appeared in the 'Strand Magazine'[331] of April 1900. The break in the electric wire was also noticed. By 10.45 the search was over, and the usual occupants of the building were crowding in.

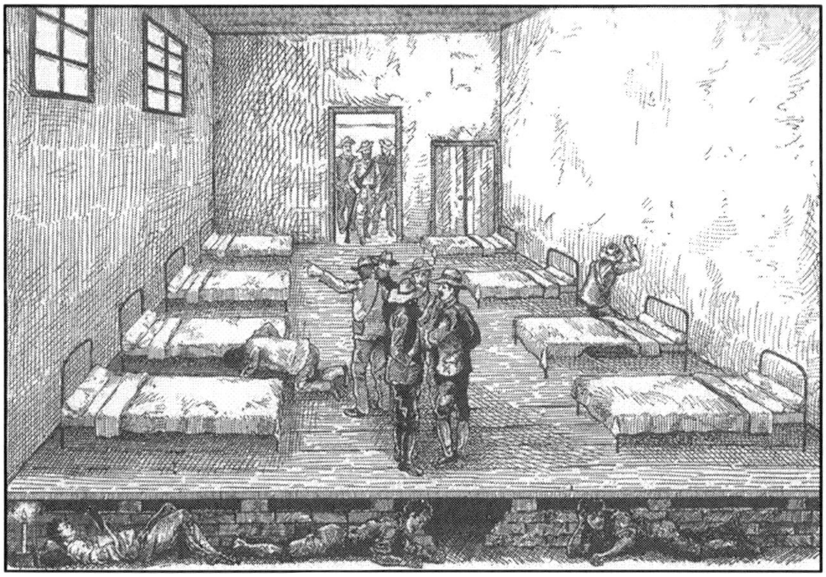

"We hold our breath as they approach the roof of our humble dwelling."[269]

Not satisfied with searching the school, several houses in the neighbourhood also were honoured with a domiciliary visit, and the same precautions to prevent our escape were taken as already described in Churchill's case. The conclusion arrived at by the authorities was that we had made our way out during the darkness of the previous night, and had gone to Mafeking.

Many months later the British Intelligence officer at Johannesburg, in looking through the police telegrams found there when that town was occupied by our troops, came across one which to me was of special interest, and which he was so good as to send me. It had been despatched from Pretoria to the police at Johannesburg as soon as our disappearance was discovered, and the translation is as follows:—

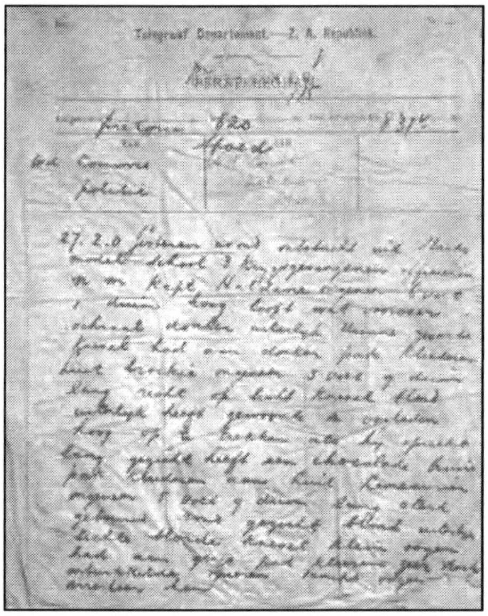

"8.20 A.M., 27th February 1900.

"Yesterday evening three officers, prisoners of war, escaped from the Staats Model School, viz.:—

"Captain Haldane.—About 6 feet 1 inch in height, walks with a slight stoop, thin, complexion and moustache dark, wears a dark-coloured suit.

"Lieutenant Brockie.—About 5 feet 9 inches in height, erect, complexion and moustache fair. Has a habit of raising his eyebrows when talking. Has a long shaped face, and wears a chocolate-brown suit.

"Lieutenant Le Mesurier.—About 5 feet 9 inches in height, strongly built, round face, complexion and moustache fair, small eyes. Wears a grey suit. Muscles powerfully developed, blinks his eyes."

Telegram with description of escaped officers.[270]

modern criminology Eugène François Vidocq-Wikipedia

331 Direct Link to Page 477 "The Boer Tower of London" The Strand Magazine: An Illustrated Monthly : George Newnes (Editor) : Free Download, Borrow, and Streaming : Internet Archive.

I have drawn attention before to the fact that detectives were in the habit of taking notes on our appearance, and that they did so in no perfunctory manner is unquestionable.

Our first day underground seemed endless. Putting our house in order occupied but little of our leisure, and that done the spirit of unrest seized upon us, and till night came we sat listening on the tiptoe of expectation lest word that all were to be moved should come. When darkness fell we found it necessary to close the ventilators as best we could, which procedure, though it made our supply of air somewhat limited, removed the danger of our candles being detected from without.

The day after our supposed flight an account of it appeared in the local paper, the 'Volkstem.' It said that a Kaffir had found the remains of a roast fowl and a hat, some few miles on the Mafeking road, where we were supposed to have bivouacked, and thither the inspector of police and his myrmidons[332] rode in hot haste. I hope someday I may find and reward that imaginative Zulu.

> The following appeared in the 'Volkstem' of the 28th February:—
>
> Escape of British Officers.
>
> On Monday night three of the British officers, who have been for some time confined in the State Model School, succeeded in making their escape. They were at dinner last night. and were missed, we understand, only this morning.
>
> Their names are Captain Haldane of the Gordon Highlanders, Lieut. Le Mesurier of the Dublin Fusiliers, and Lieut. Brockie of the Natal Carabineers. The two former were, we believe, captured in connection with the armoured train incident, at Chieveley, at the same time as Mr. Winston Churchill. and the latter some time before that. The escape was at once notified to the police-office on its being discovered, and every precaution is being taken in order to prevent the fugitives from getting out of the State.
>
> Yesterday afternoon a report was brought into town by a Kaffir to the effect that the three British officers who had escaped during the previous night from the Model School had been seen in the neighbourhood of Koodoespoort. When the news reached Lieut. Du Toit, of the police force, he at once went out with a number of his mounted men and made a diligent search of the neighbourhood in which the men were supposed to have been seen. They succeeded in discovering a spot where the officers had probably spent the night, and also where they had partaken of food, but beyond this nothing was seen of the fugitives. It transpired that one of the Hollander Guards from the town had seen a man who appeared to be acting suspiciously, and upon giving chase to him the man commenced to run, and endeavoured to hide himself in the bushes which are scattered about on the veldt. The pursuer managed to get to within about a hundred yards of his man, when he was prevented from proceeding further by a barbed-wire fence. He stated that the man then made his way into the hill, at the right-hand side of Koodoespoort, but, after a diligent search, no traces were found of him. While conducting the search two Englishmen, connected with one of the local banks, were found amongst the hills. These men had applied to Lieut. Du Toit on the previous day for a permit to leave the town with their bicycles, but had been informed that they must apply to the field-cornet for such permission. This, it appears, they had not done, and as a result they were arrested on suspicion, and conveyed into town by a guard of police, where they were lodged in jail until the matter is investigated.
>
> The escaped officers seemed to have aided their flight by severing the electric wires in the roof of the building, for which purpose they had constructed a saw from a broken table-knife, and cut out a hole in the ceiling sufficiently large to admit the body of a man.
>
> And in the 'Standard Diggers' News': "No further reports have come in regarding the three escaped officers, but important information regarding them has come to the Government.

It was satisfactory to us to know how completely they were off the scent, and how remote from the building their thoughts were. In every room was a trap-door of exactly the same description as that in our own; but the authorities, after the first search, never for a moment wavered in their conclusion that we were outside Pretoria.

[332] A soldier or a subordinate civil officer who executes orders of a superior without protest or pity (sometimes applied to bailiffs, constables, etc.).

In a few days after we were supposed to have escaped, I was in some anxiety lest the guard, annoyed at their dismissal, should put their heads together, and, concluding that it was quite impossible for us to have slipped past them in the dark, demand that the school be searched again.

I heard afterwards that the builder of the Staats Model School was a Scotsman. He must have known of the possible hiding-place we occupied; but if he happened to be in Pretoria he kept his own counsel.

I may here mention in passing that no restrictions were imposed on account of our escape, a matter of some satisfaction to us; but the tide of war was turning, and the victories of Lord Roberts were beginning to be felt Ladysmith and its gallant defenders were relieved,[333] or on the point of relief, and the Boers had given up hopes, so frequently expressed to me earlier by Dr. Gunning, that the townspeople would live to see an army of prisoners led by a general, himself a prisoner, marching through their streets.

The harmonium[334] of which we had been deprived some two months back was now restored, and a piano promised. This day we heard that the move to new quarters was expected to take place in two days, and that Cronje and his men were prisoners.[335]

Strange to say, all inclination to indulge in smoking had left me, and though Brockie consumed a vast quantity of cigarettes, I felt no desire to follow his example. An attempt I made with a pipe failed ignominiously, the tobacco being damp and seeming to draw to itself much of the moisture of our earth-hole.

Our matches, too, gave us great trouble, and if left uncovered for forty-eight hours would neither strike on the box nor anywhere else.

The complement of officers in our room was made up, but the two new occupants never knew that we were hidden below.

Our accomplices above had managed things well. Without our principal ally, Frankland of the Dublin Fusiliers, who was taken prisoner with me, I don't know what we should have done. My only regret was that he was not with us; but we could not have dispensed with his help, for he was clever and reliable and full of expedients. He undertook our commissariat arrangements, and though our wants were not great, he and his assistants found some difficulty in satisfying them. Bread had to be abstracted by them in the view of all at meal times, and as the ration granted by the Boers was small, I fear that in their anxiety for our welfare they did not hesitate to stint themselves.

When we first went below we had to be on "short commons," but as we did not expect to be there long, and as the life was very sedentary, we judged it best to eat sparingly. For a day or two we fed on chocolate, jam, and potted meat, and a little biscuit or bread; but as time went on and we became more exacting in our requirements, we took the officer who presided over the stores into our confidence, and through his agency we lived in plenty. A bottle of cocoa as a night-cap was our greatest luxury. This had to be drunk out of empty tin jampots, for our table necessaries were of the crudest description.

A double knock, twice repeated, was the signal that the trap-door was about to be opened. It was under Frankland's bed, and could only be raised occasionally during the day, and then with great precautions. Unfortunately for us, one of the officers who had taken our place in the room above was of a somewhat slothful disposition, and passed the greater part of the day lying on his bed, which was next to the trap-door. Many a time his proximity to our line of communication made it impossible to send us our daily convoy of supplies, and various artifices had to be tried to lure him to a flank.

The "signalgrams" from the telegraphist and the daily paper reached us daily, and those in our confidence did all they could to make our existence bearable and to encourage hopes. Morning and evening, and sometimes oftener, our post, which took the form of notes pushed through the chinks of the trap-door, arrived. How anxiously we scanned the writing to see if the verb "to move" was present in any of its tenses.

The theories of the other officers as to what had become of us were communicated to us by Frankland. All were very wide of the mark, and some so utterly absurd that they furnished food for laughter, which, however, had to be suppressed. But at times, when to control one's mirth was impossible,—for I had two Irishmen with me, and

[333] On February 28th, 1900.

[334] A pump organ. Pump organ-Wikipedia

[335] On February 27th, 1900, Cronje, with 44 commandants and other officers of all grades, and over 3,500 men, surrendered unconditionally to Roberts.

we felt the humour as well as the misery of the situation,—we would bury our heads in our coats and give way without restraint.

The most trying part of the under-floor life, next to the lack of fresh air, was the constrained position in which we constantly had to remain. To move about we had to creep on hands and knees. I don't know what a "housemaid's knee" may be, but we all developed very delicate knees, after a few days, from creeping on the rough ground.

Our heads, too, suffered a good deal from bumping against the beams and hitting hard walls in the dark. Lack of any kind of exercise, if one is not weak and ill, is more intolerable than might be supposed. In order to preserve a respectable "going-away" suit, we donned, with some repugnance, certain garments which had been discarded and left below the floor on the conclusion of our first attempt to dig our way out. If they had been lying on a dust heap in the open for months they could not have looked or felt more uninviting than they actually did.

Coughing, sneezing, or talking above a whisper was absolutely forbidden, for "love, a cough, and smoke will not remain secret." After a day or two we broke through the cross-wall and got under the next room, and there Le Mesurier and I took up our quarters. We had found that when all three were living in the same compartment the atmosphere at night grew close, and for other reasons the additional space we acquired was very necessary. How we passed our time will not take long to tell. We generally tried to make our night last till 10 A.M., then had a little food; lunch at 1 P.M. over, we read or slept; food again at seven, cocoa, and bed.

Not an exciting way of passing the twenty-four hours, but the only way. Reading and patience[336] filled the greater part of our time, and our friends above had some difficulty in feeding the hunger we expressed for literature. I give my warrant that every sentence was religiously devoured and not a word overlooked. One of the books which somewhat appropriately was sent us, and which I had not read before, was 'For Faith and Freedom.'[337] I felt grateful to the author, for it helped admirably to pass a few of as weary hours as one can well imagine.

Unfortunately we had no means of washing, and to get a bucket of water through the trap-door was too unsafe. We had therefore to join the brigade of the "great unwashed" till better times. The noise of the occupants of the rooms above tramping up and down was very trying, and I suffered much from headaches, aggravated by walking into crossbeams. Such was the damp that our boots, money-belts, and anything of leather, turned green in a night, and burning candles did not help at all in drying the atmosphere.

Six uneventful days passed. Though at first great hopes were held out of a move, they grew less and less as day succeeded day. We thought at first that a week would be about as much of this kind of life as we could endure. Now we hardened ourselves to hold out for a fortnight. We fully recognised that each day that passed would give us a longer start when the time to emerge from the school came; for the officials had no idea that we were in Pretoria, and our having escaped was becoming ancient history. This was almost the only consolation we could draw from the situation. One advantage, however, of our present life was the entire absence of mosquitoes. In the rooms above they had persistently annoyed us; here, with the exception of the homely house-cricket, no variety of the insect class thought us worthy of a visit. Except, too, when some heavy footed monster in the room above tramped loudly and directly overhead, we may be said to have enjoyed comparative peace and quiet, and privacy was always there.

My companions caused me no little alarm, for one was given to talking in his sleep, sometimes with vehemence and gesticulation, and the other snored. At the hour of the commandant's early visit I generally tried to be awake, lest the one should be holding forth, or the other trumpeting.

On Sunday, the 4th March, we heard that Mr. Hofmeyr had taken as the subject of his sermon "Patience," and had otherwise specially remembered us. He did not then know where we were. On the 6th a catastrophe nearly occurred. Our rule was that when we moved for any reason from one compartment to another, the candle, if one were lighted, must invariably be extinguished. The reason for this precaution was that the flooring above us was near, and painted with tar to ward off white ants. On this day I had gone into the adjacent apartment to watch Le Mesurier playing patience, and had left a candle burning. After a time a smell of burning attracted our attention, but we played on and thought no more about it.

[336] Solitaire. Patience (game)-Wikipedia

[337] Sir Walter Besant (14 August 1836 – 9 June 1901) was an English novelist and historian. William Henry Besant was his brother, and another brother, Frank, was the husband of Annie Besant. Walter Besant-Wikipedia For Faith and Freedom. 3 vols. 1889.

At length the smell got so bad that I looked where I had been and where Le Mesurier and I slept, side by side, and found his blanket and a corner of his coat in an advanced stage of smoulder, ready to burst into flames with very little encouragement. With some difficulty we managed to extinguish the burning cloth; but the whole space was now full of smoke, which was flowing out through the ventilators and into the open air in front of the building, not far from where a sentry stood. Hastily closing the ventilators with paper, we prevented them from becoming a source of discovery, and for a long time we sat in the stifling smoke, with smarting eyes, until it gradually became dissipated. No evil resulted from my carelessness, and Le Mesurier accepted the injury to his coat with equanimity.

On the 8th March we heard that no move was likely to take place, and that a few days earlier Le Mesurier and I had been captured on the Natal border. Things were now growing somewhat desperate, and even Mark Tapley[338] would scarce have felt cheerful under similar circumstances. Sometimes we felt a half regret that we had undertaken what was proving itself to be so futile and disagreeable an attempt to gain our freedom. Our friends above urged us to come up and live in the roof, or at least occasionally emerge into the room overhead and get some fresh air; but we set our faces against such proposals. The fewer who know a secret the better. Others would get to know it, and as many of the sentries understood English, they might easily overhear and understand some indiscreet remark as to our whereabouts. We hardened our hearts, and decided to remain where we were until the prisoners were exchanged, the campaign concluded, or ourselves too ill to bear it any longer.

Some twenty persons now knew where we were, among them Mr. Hofmeyr, who supported our resolution to see the matter to an end of some sort. In a state bordering on desperation, I had written to him on the 6th and asked him to obtain an interview with Commandant Opperman, lead him to believe that the officers were about to break out and work on his fears to such an extent that he might be driven at length to recommend their immediate removal to the new habitation. I felt that it was possible that this representation might be successful; for Mr. Hofmeyr being a Dutchman, the Boer commandant found it hard to believe that he could be also a thoroughly loyal Englishman at heart. Some weeks earlier, too, Mr. Hofmeyr had mentioned to me that on first coming to the Model School Opperman had hinted to him that it would be to his advantage to play the spy upon us.

These considerations led me to adopt this course, and that it turned out to be the best under the circumstances the sequel will prove. I think, to show how ably my coadjutor in this affair carried out my suggestion, I cannot do better than quote from his own interesting work, 'The Story of my Captivity in Pretoria'[339] He says:—

"I took our gaoler apart and said to him, very slowly and seriously, 'Remove us as soon as possible.'

" 'Why?' asked he. 'The lights in the new place are not in yet, and we can't get workmen enough.'

" 'Well,' said I, 'hurry u—that's all I can advise you to do.'

" 'But why?' asked he again.

"I looked solemn and grave, and said, 'Do you want to wake one morning and find yourself dead?'

" 'What do you mean?' asked he, as his face visibly paled.

" 'Half a word is enough to him of good understanding,' I answered.

" 'But do you know or hear of anything—any conspiracy, any plot?' and beads of perspiration stood on his forehead.

" 'Certainly not, and if so I would assuredly not tell you; only I judge by the signs of the times—four officers gone, reverses for your troops everywhere—look out—move us at once.'

"Our guardian angel was now thoroughly frightened, and immense was the pressure brought to bear on the authorities and workmen.

"And we, too, were anxious to be away, as our quarters were becoming dangerously crowded, and we were told the new quarters were so comfortable. What otherwise would have taken four or five weeks certainly was now accomplished in about seventeen days."

[338] Mark Tapley is a character in a Charles Dickens novel who is always cheerful (jolly, in his own words). Mark is happy at the inn where he works, which he (Martin the protagonist) decides does not reflect well on him because it shows no strength of character to be happy when one has good fortune. See: Charles Dickens, Martin Chuzzlewit-Wikipedia.

[339] See infra for Adrian Hofmeyr's full account. The Story of My Captivity During the Transvaal War 1899-1900.

Mr. Hofmeyr wisely did not tell me fully the result of the interview, no doubt fearing to raise hopes which might prove false. On the 7th [of March] I wrote again and more strongly, urging him not to spare the commandant, and to tell him that he might rest assured that his life would be the first to be sacrificed.

There had been a discussion among the officers several days earlier with regard to an offer from Mr. Patterson the telegraphist, to assist from outside if they would organize an attempt to break out and seize the town. The project was put to the vote but not carried—rightly, I think. Winston Churchill had been greatly in favour of it when he was with us, but since the men had been removed sixteen miles off to Waterfall,[340] the chances against carrying it out with a reasonable possibility of success were next to none. It was after this plan was finally shelved that I wrote to Mr. Hofmeyr.

The shout of "Letters" which occasionally reached us was very trying: we knew there must be some for us, but to claim them was impossible. They were, however, handed over to the American consul [341], and eventually fell into our hands at Lorenco Marques.

We had now several visitors at night in the shape of rats, who daily grew bolder, running over us as we lay, and at times upsetting tins with alarming noise. Our friend the white bull-terrier still kept watch and ward outside, and would have been of use now could we have borrowed his services.

On the 9th March we definitely heard that there was no chance of moving. Some time prior to this, and about the time we hid below, we had been told that the Committee who were supposed to look after our interests had condemned the new building as unsanitary, and that the State Secretary, Mr. Reitz,[342] concurred in this opinion.

Francis William Reitz.[271]

The Birdcage, British Officer's Prison, Pretoria.[272]

President Kruger, however, considered it good enough for us; and as he had gone to the Free State to rouse the waning courage of the "brave burghers," we hoped that when he returned he would insist on his wishes being complied with. There was no wooden floor to this building, and early in March it was rumoured that one was being laid. We now heard to our dismay that all work had ceased. The only chance left of the officers being removed was the arrival of more prisoners; and there was still room for some twenty more. There was the possibility of the Model School being required as a hospital, should the wounded from Volksrust and Bloemfontein be sent to Pretoria.

[340] This is a different place, close to Pretoria; it is not Waterval Boven.

[341] Possibly Mr. Adelbert S. Hay or Vice-Consul J.G. Coolidge. The American Consul was designated as the 'Protector' of British citizens. See: Adelbert S. Hay - Wikipedia

[342] Francis William Reitz (1844–1934) South African lawyer, politician, statesman, publicist and poet; member of parliament of the Cape Colony; Chief Justice and fifth State President of the Orange Free State; State Secretary of the South African Republic at the time of the Second Boer War; and the first president of the Senate of the Union of South Africa. Francis William Reitz - Wikipedia and Francis William Reitz - Wikisource, the free online library At the advent of the South African War (Second Anglo-Boer War), F.W. Reitz, in his capacity of State Secretary of the South African Republic, published an overview of Anglo-Boer relations in the nineteenth century in Dutch, under the title Eene eeuw van onrecht. The book was an important propaganda instrument against the British. The actual authorship of the book is unclear. The second Dutch edition of the book carried the text 'Op last van den staatssekretaris der Z.A.R., F.W. Reitz' ('By order of the State Secretary of the S.A.R., F.W. Reitz'). J.C. Smuts is indicated as author, but probably only edited the introduction and the end of the book, in collaboration with the historian E.J.P. Jorissen. The rest of the text was probably prepared by J. de Villiers Roos A Century of Wrong - Wikisource, the free online library.

Though things looked black, they had looked so before, and something might "turn up." We heard that officers were trying to dig their way out in the manner which we had attempted in December, but that they had all struck water. Some were so indiscreet as to carry the water from the scene of their operations through the building, and empty it outside in view of the sentries, and the noise they made as they dug was heard outside distinctly. We knew that should the authorities be warned the whole building would be searched, and our discovery would be certain. All this caused us a good deal of worry, which, added to our wearisome life, did not make it more bearable.

March 10.—We decided that, if only for the sake of exercise, we must do something. The only thing was to dig. A reference to the plan of the building will show that the end room on the north side was not far from the hospital. We resolved to get under the floor of this room and drive a shallow tunnel under the pathway, on which the prisoners took their exercise, into the yard of the hospital. A shallow tunnel above water-level, well shored up with timber, would be safe, and we could break out after the eight-o'clock relief had been posted.

When our first plan of digging had failed, we had thought of the plan we were now considering; but as at that time the police were quartered in tents in the back-yard, and in order to get from there had to pass through the gate into the hospital, near where our exit-hole would come, we had not carried it into effect. Now the police tents—probably owing to Mr. Hofmeyr's interview with the commandant—had been removed and pitched in the street south of the Model School, outside the yard, and the hospital-gate was not used after dark, except for the marching in and out of reliefs. The plan was now, therefore, feasible. We had our old tools with us, and arranged with our friends above to collect all the wood they could find—shelves, umbrella-stands, &c.—which would be required for the tunnel. We also ordered a saw and a hatchet, which were supplied; they were supposed to be for use in the kitchen. Some other officers who had discovered that we were still hidden in the school, had hit upon a plan similar to ours. Three of them, passing under the floor of their own room by means of the trap-door in it and creeping under the adjacent rooms, at length penetrated to our lair. We discussed with them how best we could combine and divide the labour so that the work might proceed more quickly; but for some unknown reason, after one day's work, they decided not to continue, and we were left to our own devices.

March 11.—Le Mesurier and I broke through the two-foot cross-wall into the end room after some four hours' work. We had some difficulty in getting through as we had to work silently, so as not to be heard above. We thought that if ever the authorities found that we had been below, they would have their work cut out to restore matters to their original condition.

Emerging into the first compartment of the end room, we crept through the man-holes, till we reached the outer wall. The ventilators here were only as long as the thickness of the wall, there being no verandah. We closed them all up, lest, should we work at night, the light of our candles might shine through. Our scheme was to dig about four feet and a half down, then drive a tunnel about thirty feet long into the hospital-yard. We calculated that if we laboured hard we might complete the work in a fortnight. This day and the next we dug every moment we could, getting down some way in the hard earth, similar to that which we had before encountered.

Our tools had been augmented by a pick, but in using it we made so much noise that we had unwillingly to lay it aside. We found that a bayonet and a steel skewer were the best implements. Some of the officers belonging to our own room offered to take part in our mining operations, and they and Frankland gave us valuable aid. I am again tempted to quote, this time from Mr. Winston Churchill's interesting work, 'Ian Hamilton's March,' in which he includes Frankland's diary of a prisoner.[343] The latter vividly describes the feelings which another officer and he experienced on their first visit to our "catacombs."

> "On going down by the trap-door he remarked what an awful hole it was. Suddenly, in the flickering candle-light, he saw a gaunt, bearded, unwashed face and a half-naked body. At first he could not make out who it was; but when he at last realized it was a brother officer, he said you could have knocked him down with a feather, had it not been that he was already crawling on his stomach. The new shaft is a long way off: when I went down I had to crawl on hands and knees along passages and through man-holes, backwards and forwards in a regular maze of compartments, and indeed, had the candle gone out, one could easily have been lost. Haldane looked very ill, but the others, except for being covered with dirt, seemed well enough."

[343] See infra, where this narrative is reproduced.

On the afternoon of the 12th we heard that the officers in the room above us were complaining that they were disturbed by the noise of digging under their room, and that evening an order was given out that as the police were growing suspicious such work must cease. The conclusion we came to was that it was the noise and not the police that had given birth to the order, and we determined to continue work.

This night I dreamt in a singularly vivid manner that someone came to me and told me not to be anxious, as we should succeed in escaping; and when I woke in the morning the feeling was strong within me that the shadow of coming events had been cast upon me in the night.

March 13.—News of the capture of Bloemfontein came.[344] On the previous evening we had decided to try an experiment to soften the hard earth, and had emptied half a dozen bottles of water into our shaft. On going there this morning we found that it had had excellent results, and the ground was much less refractory. We got on famously, and the exercise did us so much good that we felt different beings. We now had strong hopes of escaping. These had been a good deal crushed by recent events, and now began to reassert themselves. We found that with care we could dig out the hard earth with little more noise than a rat makes in scratching a hole, and quite as fast as before we had watered it, when we had worked regardless of noise.

March 14.—We managed, when the two officers in our room who did not know we were below happened to be out of it, to get down a bucket of water, and perform some much-needed ablutions. Digging in a close atmosphere is not a cleanly operation, and the inside of the lid of a tin biscuit-box showed us what dirty objects we were; but after the use of the contents of the bucket we became recognisable.

The life was beginning to tell on us, and our unshaven faces looked pale and thin. Most of this day was spent in digging; but our hands were bruised and blistered, and we did not make much progress. In the afternoon we got a piece of news which raised hopes in our breasts, but lest they should end in disappointment we tried to stifle them. A certain commandant had made an inspection of the school with a view to ascertaining its suitability for a hospital. We had, of course, no two opinions on that matter.

To us it seemed an ideal building for the purpose, more particularly since it adjoined the Red Cross Hospital. It would do the Boers no harm to occupy a building where we had been living for nearly three weeks under the floor. Our allies were unable to discover what this commandant's opinion was. To have displayed too much interest would have awakened suspicions. At last we had a peg on which to hang the worn-out remnant of our hopes.

March 15.—We continued our mining operations, and by night had far better news than we had dared to hope for. All day we had displayed less vigour in the work: a presentiment that our digging would again be fruitless must have seized us; indeed, we only worked from 11 A.M. till 1 P.M. At 3 o'clock we overheard someone use the word "move," and soon other remarks of a similar tenor caused us to listen intently. At any time it was difficult to hear what was said in the room above, unless it was shouted, and on this occasion we could detect no more encouraging words. At 4.30 P.M. we were in full possession of the excellent news. A note came down, in which we read with feelings indescribable, "We move to-morrow after breakfast. Patience has carried the day: you deserve your luck."

We had been seventeen days and nights below, and had still some twenty-four hours to pass. But the danger of discovery from internal causes was not yet over. Knowing that many officers, who did not know that we were in the building, had been digging, we were strongly of opinion that some of them would hide below to-night, thinking to escape after the move. That this was the case was soon ascertained, and word was sent to me that the same officer who some time previously had unconsciously penetrated our plan of cutting the wires was now encouraging a wholesale descent below the floors.

As no officer had yet escaped from the Model School, I sent word to Colonel Hunt, R.A.,[345] saying that we had been hidden below for seventeen days, and that if any others hid at the eleventh hour, without cutting the

[344] Bloemfontein "Fountain of flowers" was the capital of the Orange Free state. On 13 March 1900, following the Battle of Paardeberg, British forces captured the city and built a concentration camp nearby to house Boer women and children. The National Women's Monument, on the outskirts of the city, pays homage to the 26,370 women and children as well as 1,421 old men (also 14,154 black people, though some sources feel that the records are unsatisfactory, and that this number could be as high as 20,000 who died in these camps in various parts of the country. Bloemfontein - Wikipedia

[345] Mr. H. Hunt Lt Col. Royal Regiment of Artillery. Severely wounded at Colenso. 15 Dec 1899. Source: Natal Field Force Casualty Roll, page 27 line 34 PF: Col. Hunt was probably associated with the artillery unit that was devastated and abandoned during the battle and he remained on the field after the retreat of the army to be captured by the Boers. Anglo Boer War - Name search

electric wires as we had done, and so making some pretext, discovery for all was certain. I begged his assistance, and he came to the rescue in a way which made us feel most grateful to him. I was informed that he saw the senior officer of each room, and told them that there must be no hiding in the building, and that if anyone in their room was at all recalcitrant he would see him, and explain the why and the wherefore. I believe that six officers waited on him, and were told in confidence that we were still in the building. In spite of this assurance we spent an anxious evening, feeling uncertain whether some adventurous spirit would not perhaps defy authority.

As regards our food arrangements for the coming journey, we had between us 7 1/2 lb. of chocolate, half-a-dozen tins of Brand's essence, and four small tins of meat lozenges.

We had also a very little biltong, of which we could have got plenty more, but we were dissuaded from carrying it by Sergeant-Major Brockie, who told us that it was only eatable when cooked. We knew later that it is always eaten dry, and deeply regretted not having brought more. Our allies, as soon as the orders for removal had been received, had sent down large supplies of tinned and potted meats, biscuits, and suchlike; but all these we had reluctantly to leave behind. To carry them would have added to our weight, and a parcel might attract attention in the town.

We had decided to go to Mafeking for the following reasons. Churchill had given an account of his escape in the papers, which, though I now know that, for obvious reasons, the facts were not fully stated, was sufficiently in detail to make the Delagoa route unsafe.

We were supposed to have gone to Mafeking, and as eighteen days had elapsed since our supposed escape, all search would have ceased. Mafeking was the nearest spot where fighting was going on, and we would chance three extra or rather famished mouths being unacceptable to the garrison.[346]

All of us had maps of this route and of several others which we had drawn. In case the main doors were locked, we had provided ourselves with the keys of two rooms which looked out on the back-yard, and in which the windows were broken. After arranging a code of signals for the morrow, the trapdoor was closed and all chinks carefully filled up with putty, and false screws put into the screw-holes. The whole, no doubt, looked most innocent.

It may interest those who play "patience," a game I only learnt for the nonce,[347] and shall probably not play again till I am next in prison, to know that this day, before the news of the move came, I did the "demon" five times out of six, and Le Mesurier did it three times in succession. It was the 15th of March—the Ides of March had come, and the omens were favourable. But as the Scots say, "Them as follows freits,[348] freits follow," and I wanted something more material to go upon than the fall of a card.

I should have mentioned that in addition to food we carried some money and other odds and ends. Cheques up to £10 could be cashed in Pretoria by a confiding agent. After making all arrangements for the following day, Le Mesurier and I made our way under the floor of the dining-room, where we sat for some time enjoying Mr. Hofmeyr's fine voice, accompanied by the harmonium.

I felt certain that it was due entirely to his representations to the commandant that the removal was about to take place, and later I learnt what confirms me in that opinion; for the Boers never used the Model School as a hospital, but left it unoccupied till it was utilised for that purpose by Lord Roberts' army. One of the last things I did that night was to send him [Hofmeyr] a letter containing the grateful thanks of those whose freedom was so near; and Mr. Hofmeyr still, I believe, preserves this memory of a trying time.

Friday, 16 March.—Day at length came, a day that will live long in the memory of my two companions and myself. For the last time we heard the commandant going his rounds, and wondered anxiously if all were present. By 7 A.M. we had not heard of any one having hidden, and all seemed busy packing.

Breakfast at 8, and then the prearranged signal was heard, and we knew that all were present A little later we heard Frankland's voice saying, "All's well; good-bye!"

[346] Mafeking was under siege at this time. They would have had to "break in" to the besieged town. It was towards the west and 160 miles (256 kilometers) away as the crow flies.

[347] The one or single occasion; the present reason or purpose.

[348] A superstitious object or observance; a charm, an omen.

Officers leaving the Model School.[273]

We now moved under the end room, and opening one of the closed-up ventilators, sat looking at what we could see of the prospect outside. The inspector of police, the commandant, and Dr. Gunning were bustling about, looking extremely important, and wagons were arriving for baggage, and, after loading up, moving off. There was no mistake about this being a general move. All was proceeding exactly as we wished. The lady of the £25 fine passed by, looking as sympathetic as ever.

By 10.15 all the officers had left the school, and no doubt were wending their way to the new habitation a mile off. We noticed that they were unaccompanied by a guard, so assumed that they had been granted parole for the occasion. The servants were still busy taking to pieces and loading beds. The din was very pleasant to our ears. A party of native convicts arrived, evidently to strike and remove the tents of the police.

About 12.15 P.M., all the baggage having left, the servants were marched off, and the sentries—their charges having gone—quitted their posts. To make sure of this, I made my way round through the man-holes to the opposite side of the building. We then crept back under our own room, and as we were about to have some food, a number of policemen were heard above us in fits of laughter at the caricatures of President Kruger and others drawn on the wall by Frankland. At that time we did not know the reason of this merriment; but later on, when we emerged into the room above, the whitewashed walls, on which the moon was shining, plainly showed the cause.[349]

[349] Frankland's Description: March 11th—I drew another large picture on my wall, a sequel to the first. It represents Kruger just escaping from Roberts, who with drawn sword appears to be running after him at a good pace. My picture No. 1 is entitled "President Kruger goes to front to exhort his burghers;" No. 2 "But returns on urgent business."

V. On the Veldt

"Unhook the west port and let us gae free."[350]

The moment for our "moonlight flitting"[351] had arrived. All above seemed still and deserted, but lest we might be mistaken we listened intently for some minutes.

Sergeant-Major Brockie,[352] who was deputed to reconnoitre the building, to make sure that no caretaker had been left in charge, now pushed up the trap-door; but so firmly had it been fastened down the previous night, that despite the greatest care he was unable to prevent it from making a loud noise, which echoed in the deathlike stillness. Creeping into the room above, he made his way out through the doorway into the passage, and shortly returned to say that all was clear.

Le Mesurier and I followed, carefully shutting the trap-door after us. We were astonished to find how weak the confinement and cramped position had made us. Le Mesurier's legs gave way and he fell down; and all of us, when we tried to walk, reeled like drunken men. Several minutes passed before we dared leave the room, and it was not till we were some distance from Pretoria that our limbs regained their wonted strength.

Leaving the room which we had lived in and under for four months, and threading our way with care along the passage, which was blocked with forms, we reached the back-door which opens on the yard. It was locked, but the absence of a large pane of glass, which had been broken and completely removed, provided an easy exit.

Strange to say, after my long incarceration I felt almost reluctant to go forth into unknown dangers and difficulties; but such a feeling was of brief duration, and this no time to moralise. We now put on our boots, which we had been carrying, and passing through this opening gained the verandah. Pausing for a moment to see that no one was in the yard, we crossed over to the low buildings on the other side. The moon was full, and the electric lights not being turned off, as we had anticipated they would be, we felt unpleasantly conspicuous. The dog's kennel was in its usual place, but fortunately the occupant was gone.

We had intended to leave the yard by climbing over the iron paling into the next garden; but the moon was far too bright for this, and the windows of a house which looked on to the garden were open and some people were looking out.

We therefore made our way to the railings, near which the police tents had been, and climbing over reached the street on the south side of the school.

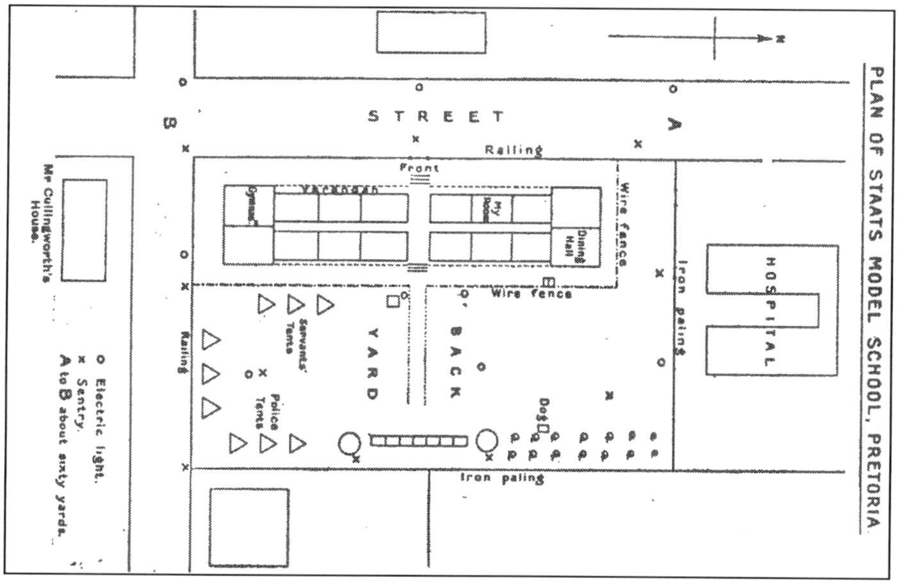

Plan of Staats Model School.[274] Note railings on South side, across the street from Cullingworth's house.

[350] "Bonnie Dundee," is a song about John Graham of Claverhouse, Viscount Dundee, who was known by this nickname. The song has been used as a regimental march by several Scottish regiments in the British army and was adapted by Confederate troops in the American Civil War. "Unhook the west port and let us gae free, For it's up wi' the bonnets o' bonnie Dundee! Then awa' to the hills, to the lea, to the rocks E'er I own a usurper, I'll couch wi' the fox! Then tremble, false Whigs, in the midst o' your glee Ye ha' no seen the last o' my bonnets and me." Bonnie Dundee - Wikisource, the free online library See: John Graham, 1st Viscount Dundee - Wikipedia.

[351] The act of moving from one residence to another; moving house. (Scotland, Northern England).

[352] Brockie, who spoke Dutch and Afrikaans was the logical choice for this task in case there was anyone about.

A couple happened to pass, just after we had got into the street, but they took no notice, of us, and Le Mesurier told me afterwards that a special policeman slowly turned the corner of the Model School at that very moment. Crossing the road, we went up the street which leads towards the fort guarding the southern entrance to Pretoria. Of the town itself we had a small scale plan, which many months before we had removed from a biography of President Kruger, a book belonging to the Staats library.

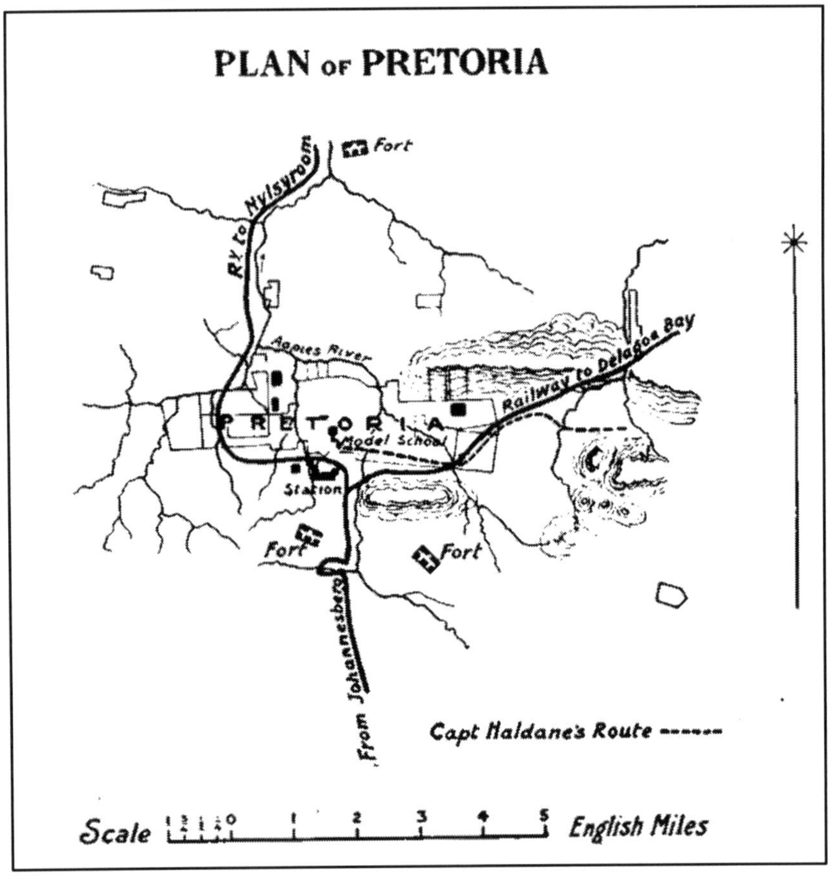

Plan of Pretoria.[275]

I stopped to light my pipe, and Brockie, keeping in the shadow of some trees, donned a white sling in which to place an arm, and this pious fraud, supported by his wearing the Dutch colours round his hat, gave him the appearance of a wounded Boer. We next bore to our left, then to the right, and finally struck a road which, from the plan, we conjectured would cross the Delagoa Bay Railway.

Brockie had warned us both that if we wanted to escape notice we must slouch along with knees bent and backs rounded, as is the manner of the Boer. This to me was somewhat difficult, for the exquisite sense of exhilaration which I was experiencing made me feel much more inclined to run, jump, or indeed do anything but walk soberly along.

So far, although we had passed a few special policemen, our appearance had seemed to attract no particular notice. In the Transvaal, as in many other parts of the world, the country bumpkin, when he honours his county town with a visit, puts off his workaday garments and comes in his Sunday best. Our garb could scarcely be said to come under the latter heading: we looked more like three moonlighters[353] than anything else I can think of.

Fortunately for us, the town had been depleted of many of its inhabitants, most of whom were at the front, and in consequence the streets presented a very deserted appearance; and although we knew that the Dutchman is abominably inquisitive, yet we hoped to evade his notice.

Le Mesurier and Brockie kept to the middle of the road. I followed them on the pathway, a little behind. In this fashion we were tramping along towards the outskirts of the town, with the villas in gardens on our right and left, when we met a special constable. This one, instead of passing us as the rest had done, stopped, and turning

[353] *Moonlighter*: One of a gang who used violence, especially at night, to promote agrarianism in Ireland during the Land War of the late 19th century. (Ireland, historical)

round scanned us with suspicion. I expected every minute that he would ask us in Dutch who we were and whither we were bound.

But Brockie, having noticed that we were running the risk of being challenged, turned half round, exposing to view his quasi-wounded arm. This seemed to partly satisfy the guardian of the peace, for he turned up a side street, stopped to have a good look at us, and ultimately moved off, unaware perhaps that

"Them as is watched out of sight, bide away for many a night,"[354]

which proved to be the case. Perhaps he thought discretion the better part of valour, for we were three desperate men, and he was alone, with no help near.

Whatever may have been his motives, his action was fortunate, for we were prepared to go to any length to avoid a hue and cry. Brockie had vowed that nothing would induce him to be captured a second time, and had armed himself with a sausage-shaped bag of cloth filled and jammed tightly with earth. He was burning to use it on someone, and must have lamented this lost chance.

Pretoria From The East. Photo by Wilson, Aberdeen.[276] Note Church Spire in the middle.

The road along which we were travelling now began to leave the town, and the gates of a level-crossing, betokening the presence of the railway, were plainly to be seen. We halted for a moment to consider whether it would not be wiser to turn aside and avoid a spot where a patrol might possibly be met; but our anxiety to get clear of Pretoria without delay overcame the inclination to make a detour, and pressing on, we found that nothing barred our way. Crossing the railway line, we sat down in some long damp grass to decide whether we should follow road or railway. I had a small medicine-bottle full of whisky, presented by a fellow-prisoner as a parting gift. The occasion was not one to be overlooked, and we proceeded to drink success to our venture; then, having decided to take the railway as our guide, we set off again, and found we were clear of Pretoria and its suburbs.

I inwardly congratulated myself that now no officer or man who wore the Gordon tartan was a prisoner in the hands of the Boers. But we were not yet out of the wood. Straight before us, high up in the eastern sky, shone the moon, dimming the brilliancy of the evening star which followed closely in her wake; to the right the Southern Cross, and low down in the north-east Orion's Belt. With guides like these, to lose ourselves would not be easy.

Reaching the railway, we walked along in single file, halting now and then to listen and make sure that all was clear in front. A coal-train from Balmoral, going westward, passed us, and soon after we had to throw ourselves down in a ditch beside the line to conceal ourselves from a Dutchman going home off bridge-guard, for every bridge and every culvert from Pretoria to Komati Poort had its guard by day and night. These guards (we were afterwards told) had orders to fire on any one walking on the track at night.

When we had gone about three miles, and were covering the ground at a good rate, we almost walked into the arms of a sentry who was sitting on the parapet of a bridge which carried the railway over the road. We came upon him suddenly, as he was sitting partly hidden by a bush, and walked to within a few paces from him; and that he did not see us, unless he was asleep, is a miracle. We dropped like stones beside the rails, lay absolutely still for some moments, then softly crept back down the slope of the embankment in the direction whence we had come. Then along a muddy ditch until we had lost sight of him, climbed a barbed-wire fence beside the

[354] PF: Unable to identify this quotation with basic research in several forums.

railway, which shook and rattled, after which, making a wide detour, and creeping through a mealie-field to avoid a Kaffir kraal, we passed along the face of a hill and found ourselves close to a telegraph wire.

This wire, we discovered later, ran like a Roman road over hill and dale to Komati Poort, and had we followed it we should have made our journey many miles shorter, and escaped the annoyance of losing our way in the dark. All around us dogs in the neighbouring farms were baying loudly at the moon, and this, and occasional shouts from afar, made us hasten on our way. We now followed the wire for a couple of miles, but thinking we were going out of our way, we left it and climbed a rocky ridge not far from Koodoespoort, where Churchill had caught the train when he escaped. Below us we heard the sound of running water. Hastily making our way to it, for we were growing thirsty, we came upon a muddy stream evidently used for irrigating the fields on the hillside, down which it flowed.

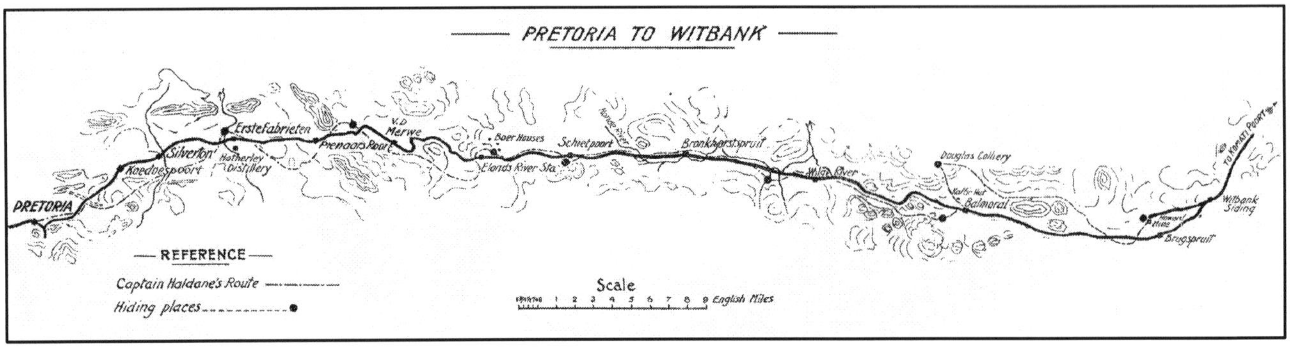

Haldane Escape Route.[277]

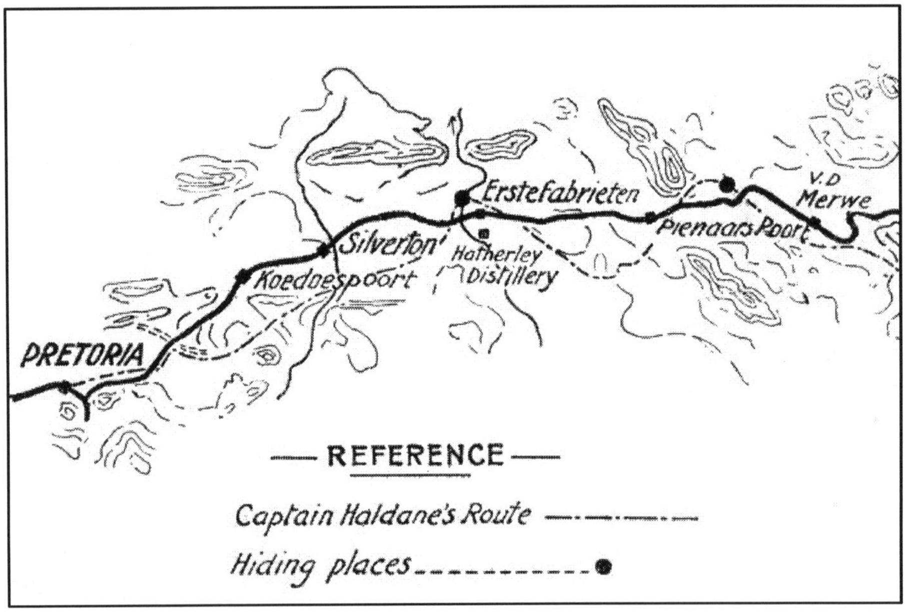

Unfortunately at this juncture Le Mesurier, who had got off the path among some rocks, twisted his ankle. Painful as it was, to walk on, however slowly, was better than to rest, and so allow it to grow stiff. Pluckily recognising this, he kept moving forward, and we made some progress, though slow. Our way led us over the open veldt, where the grass was long and wet. Brockie was growing tired, and insisted that he saw Boers under every tree and bush, and for some time we moved with great caution. We again struck the railway, crossed it, and a black patch in the distance told us we were approaching the first station, thirteen miles from Pretoria, and decided to halt there till next evening. I felt a good deal disappointed at covering so short a distance the first night; but it was no one's fault, and it was as well, perhaps, not to walk too far at first, for we were not in the best of training just then.

Reaching Eerste Fabriken, our farther progress was barred by a ditch some fifteen feet deep, with just beyond it a sluggish muddy river. Hunting up and down the ditch, we at length found a spot where, hidden by the thick bushes, it gradually sloped up to the ground-level. The bottom was damp, but that was soon remedied by covering it over with straw, which we found lying about. We had, however, got into a veritable hotbed of mosquitoes, for they attacked us with the greatest fury and to sleep was impossible.

The arrival of morning and dispersal of these greedy bloodsuckers was a great relief. At 6.30 A.M. the steam-horn of the Hatherly Distillery hard by sounded, bidding the workmen rise for another day of toil. At this establishment, I am told, strong waters of every variety, from Hennessey's three-star brandy to Long John of Ben Nevis fame, can be procured. It is merely a matter of labels, and they are not so hard to copy as bank-notes.

Our hiding-place proved to be admirable, for though it was uncomfortable, it was improbable that anything but a dog would find it, and we were well sheltered from the sun. This was the first day for nearly three weeks that we had seen the sun, and we were not sorry to be protected from his burning rays by the thick foliage. During the day we ate a little chocolate and some meat lozenges; but as regards water we were as badly off as Tantalus,[355] for the river, a few yards distant, could not safely be approached. Some one seemed to be fishing in it close to us—an Englishman we thought, for he whistled, one after another, many tunes that were familiar to us. His *repertoire* was somewhat antiquated, and included amongst others "Oh, what a surprise!" and "After the ball is over." Except for the occasional song of a Kaffir, the noise of cows busily cropping the grass, and the buzzing of many insects, all else was silent. The ubiquitous fly had of course marked us down, and did not spare us his objectionable attentions, and to drive him off by smoking was not considered safe.

Even here, and indeed throughout all our travels, we never raised our voices above a whisper, lest some passerby might hear and discover us. Our determination was if possible neither to see anyone nor to be seen by anyone till we reached the frontier. This would free us from incriminating explanations, if any were demanded.

Time passes slowly when one has nothing to do all day but brush off flies, and we regretted that a pack of cards had not been included among our baggage, for then we might have called the "demon" to our aid. A few trains passed up and down: these and the occasional sound of the distillery horn alone served to break the monotony. At half-past four Sergeant-Major Brockie, armed with Le Mesurier's field-glasses, which he had managed to secrete when taken prisoner, was despatched to do a little scouting, in order to find out whether there was a drift over the river near where we lay hidden. After a short time he returned, bringing the satisfactory news that the ford lay between us and the station, and across it several waggons were outspanned. There had been a heavy fall of rain during the afternoon, but we hoped insufficient to flood the river.

When darkness at length came, we sallied forth from our retreat, crossed the river, then the railway, and reached the road. By the small scale map we possessed, the railway appeared to make a considerable bend to the south-east, and as the road seemed to form the chord of the arc, it would be shorter to follow it. But Transvaal roads are not as other roads, and are as unlike their counter-part in England as a lane in Devonshire is unlike the turnpike to Bath. Looking at a map of the Transvaal, you are tempted to believe that once on the road, so clearly defined on paper, you have only to shut your eyes and go ahead. Try it, and you will find a close resemblance between yourself and the blind led by the blind. The highways in the South African Republic are innocent of metal—Macadam[356] is a name unknown: they consist of nothing but deeply indented wheel-tracks left by the clumsy, ponderous transport of the country—the ox-waggon. As a Dutchman in wet weather leaves the main track where it has become swampy and marks out a line for himself, the natural consequence is that in time the vicinity of a road becomes a maze of tracks, and to find your way in the dark and in an unknown district is nigh impossible.

Tonight we were to experience this. Sergeant-Major Brockie, who posed as a fairly good "pathfinder," took the lead, and we pushed boldly along; for our lame comrade was a little better, and had secured a bough as a walking-stick. We noticed that the track we were following went due south, but thought that doubtless it would soon change its course. A waggon coming towards us on the road, we had to turn off and wait until it had gone by. Not long after we had crossed the railway we came upon a large white wooden gate. Passing through it, we found we were on an avenue, both sides of which were lined with firs,—a most unusual luxury for a Dutchman; for the Boers have done almost nothing for their country in the way of wood-growing, and except for a few trees round a farmstead, you may travel miles without seeing a leaf or finding shade.

[355] Tantalus was a Greek mythological figure, most famous for his punishment. He was made to stand in a pool of water beneath a fruit tree with low branches, with the fruit ever eluding his grasp, and the water always receding before he could take a drink. Tantalus-Wikipedia

[356] Macadam is a type of road construction, pioneered by Scottish engineer John Loudon McAdam around 1820, in which single-sized crushed stone layers of small angular stones are placed in shallow lifts above grade and compacted thoroughly. A binding layer of stone dust (crushed stone from the original material) may form; it may also, after rolling, be covered with a binder (tar) to keep dust and stones together.

A little farther on we noticed the lights of a cottage, which we took for the lodge of some larger building. Passing by as quietly as we could, we soon came to four crossroads, but we were beginning to lose faith in our conductor, and we turned north-east.

Our route now led along what appeared to be a deer-park, for in the moonlight several of these animals could be seen to fly at our approach. We must have diverged considerably, for it seemed a long time ere we again met the railway, and both my companions were footsore and tired. Before us lay a range of hills in which was a *poort*, or gap, through which the railway passed, and on the name-board of the deserted station we read Pienaars Poort.[357] As we drew nearer to the hills we heard the noise of rushing water, and once inside the *poort* we sat down by the river to refresh. The spot looked singularly picturesque, road, railway, river—all were crowded together into the gap, which at its narrowest point was about two hundred yards wide; and on either side the hills, tenanted by chattering baboons, rose steep and bare. The ground ascended before us towards the farther entrance of this defile, and the gradient of the railway was steep. One felt instinctively that this would be a bad place in which to encounter a patrol, for the only way out of it seemed forwards or backwards.

Brand's essence and whisky worked wonders on my tired companions, and, reinvigorated, we started off full of the intention of reaching Elands River[358] station, twenty-nine miles from Pretoria. I led the way along the railway-line. On our left, a stone's-throw distant, flowed the river, its surface shining like a mirror under the rising moon. Before me ran the double rails, looking like bars of silver; and beyond them, on the right, the road. As we trod softly and looked ahead as far as one could distinctly see, some tents appeared on the right of the railway, pitched close under the hill; one of them seemed larger than the rest. I turned to Brockie, who followed me, and asked his opinion. "Only a Kaffir kraal," he replied, so on we pressed. A few paces farther, and the stillness of the night was broken by the angry barking of a dog. Immediate action was imperative. Down we dropped into the long rushy grass which filled the space betwixt the river and the railway. The dog continued barking—he had evidently seen us.

Presently voices became audible, one of them bidding the cur be silent. But he did not cease, and another followed suit. After lying in the grass about twenty minutes, for we did not care to move so long as the dogs remained on the alert, we heard voices coming in our direction, and the barking of the dogs became more distinct. A patrol was on its way to clear up the situation. A whispered conference was held, when it was decided to retreat, and then we dragged ourselves like snakes diagonally back towards the river. Fortunately the rushes had been beaten down by a recent flood, and our tracks were barely visible. Reaching a broad ditch full of water, Le Mesurier, who was following me, came alongside and asked me if I had seen Brockie, who had been following him. Being in front I had not, so we waited a few moments; but seeing nothing of him, and the enemy drawing near, we crossed the obstacle and found ourselves at the edge of the stream. Again we paused, this time for several minutes, and the searchers came in view, following our track.

One of them, whom was walking along almost bent double, we thought was Brockie, but looking through the rushes we quickly found our mistake. At this moment many things rushed through my mind, and more especially one.

Shortly before leaving Pretoria I had been reading again a very ancient friend, 'Tales of a Grandfather,' and now I instinctively remembered how the Bruce,[359] when hunted by John of Lome's bloodhound, had waded down a river and so destroyed the scent.

My mind was quickly made up. The crisis had come: to stay where we were meant probably recapture. I whispered to Le Mesurier to follow me silently and not to splash. Transferring my manuscript map from my pocket to the lining of my hat, the next minute I was in the river, which was out of my depth, and Le Mesurier dropped in beside me.

[357] The Pienaars River is a tributary of the Crocodile River. It originates east of Pretoria, City of Tshwane, Gauteng Province, flowing northwards into Roodeplaat Dam (north of Mamelodi), from where it carries on northward until it passes under the N1 and turns westward. North of Makapanstad the Apies River (or rather a short section of the Tshwane River) joins it from the south, as well as the Plat River from the east. Pienaars River - Wikipedia

[358] Elands River is a river of the Komati River basin in Mpumalanga province, South Africa. Elands River (Mpumalanga) - Wikipedia

[359] Robert I (11 July 1274 – 7 June 1329), popularly known as Robert the Bruce, was King of Scots from 1306 to his death in 1329. Robert was one of the most famous warriors of his generation and eventually led the Kingdom of Scotland during the First War of Scottish Independence against England. He fought successfully during his reign to regain Scotland's place as an independent country and is now revered in Scotland as a national hero. Robert the Bruce-Wikipedia

"Holding on to the reeds, we watch the searchers."[278]

Holding on to the roots of the reeds which lined the bank, we paused and watched the searchers, then carefully pulled ourselves some distance downstream, and paused again. The Boers and their dogs were evidently now at fault, and showed no signs of coming our way, so we continued our downward course, and ultimately swam across and into a ditch on the other side.

We had been a good half-hour in the stream, which seemed to us intensely cold, and our teeth were chattering so that we could scarcely speak. My wrist-watch had stopped; but Le Mesurier's—a Waterbury—was still going, for it had been provided by his care with a waterproof case.

We now crept along the ditch up-stream again, and then turned off towards the hillside, which was dotted with large boulders. Coming round the corner of one of these, we found a tent in front of us, in which a light was burning, and not caring to pass it, we tried to climb up the steep face of the hill. Failing at one point, we found a kind of "chimney," up which we climbed, pulling and pushing each other till the top was gained. A few minutes' rest was necessary, for our clothes were heavy with water and the climb had made us breathless.

Le Mesurier had done wonders with his ankle—the cold water had been most efficacious. Next we walked along the rocky face of the hill, parallel to the direction we had followed below, and gradually descended to the level and struck a path. While on the hillside we had looked down into the valley below, where all seemed quiet and nothing moved. Brockie was irretrievably lost,[360] and it was useless to attempt to find him. He had with him a water-bottle and sufficient food, and knew both the Dutch and the Kaffir language. Following the path, we passed several clumps of bracken, one of which we selected as a suitable hiding-place. To have walked farther in our wet and clinging garments might have been wiser, but we decided that we had had sufficient excitement for one night without trying to add to it.

Carefully avoiding breaking down the fern, we took our way into the centre of the clump, and made ourselves as comfortable as was possible. I took as a prophylactic ten grains of quinine and four of opium, all that was not reduced to pulp by our recent immersion. The dose of opium, I have since been told, was a large if not indeed a lethal one, but it did not make me sleep.

[360] PF: It is also possible that Brockie, knowing the language and country as he did, felt his chances were better if he proceeded alone. He was successful as well, by entirely different means. See infra.

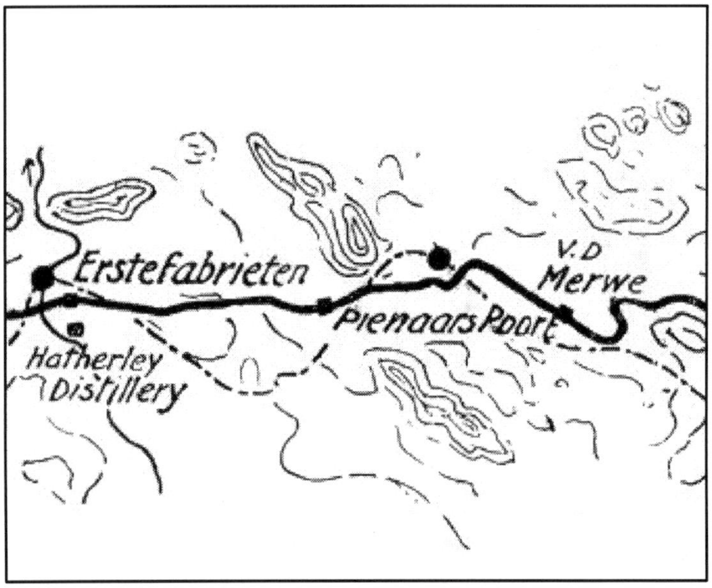

The general area. Haldane Day 2 (13.7 miles, 22 km)

The night was bitterly cold—for where we lay was nearly five thousand feet above the sea-level and, soaked as we were, we shook and shivered. Of course the mosquitoes did not spare us, and we spent the remainder of the night in fighting this useless scourge. When day broke I found I was so stiff and rheumatic that I could not move, and Le Mesurier was not much better. However, when the sun rose and penetrated our wooden bodies we soon found movement possible, and by noon we were dry and ready for any more adventures that might come. We had suffered some loss by that unavoidable dip in the river. Le Mesurier had lost a pocket-book[361] and I the contents of the whisky-bottle, the cork of which had come out when we were crawling along in the grass—by accident, not design. I had also lost a triangular file, a memento of the Model School. Further, our food supply was water-logged, our tobacco spoilt, and likewise our matches. We managed to eat some pulpy chocolate, which was becoming nauseous to us, besides creating a thirst we had no means of quenching. We had trusted to finding plenty of mealies growing but the harvest was just gathered in, and this source of supply was lacking.

Our hiding-place, although overlooked by the hills, was a good one. Only a few Dutchmen and Kaffirs passed by along the paths, and they saw no sign of us. The want of proximity to water was its greatest fault, and we mutually agreed that howsoever uncomfortable a swamp might be as a sleeping-place, it was the only place in which to pass a day under the broiling sun, unless a shady nook were forthcoming. Few trains passed, as it was Sunday. One could not help carrying one's thoughts back to Pretoria, and wondering where our accomplices would think we had reached, and what they would think we were doing.

Only too well we knew, at any hour of the day, how they were passing their time, and how unhappy many of them were. How light-hearted we felt at being free again! All that was necessary for us now was care and caution, and these we had been accustomed to when below the floor, and we were now determined not to leave anything to chance, or what people term luck, or to spare any trouble which might help to ensure our ultimate freedom. In addition to the ordinary risks of being captured or shot—for being in plain clothes we ran the risk of treatment as spies—there were two others which now presented themselves. My fear was lest the trap-door, which we could not shut down very neatly, might have been noticed; and also that Sergeant-Major Brockie should get ahead of us, reach Lorenco Marques, and, forgetting that we were still behind him, talk. That the latter supposition was not devoid of foundation will appear later. On the whole, however, I felt exceedingly confident that our enterprise would prove successful: if we should have the misfortune to be recaptured, at least we would feel satisfied for ever after that we had made a determined bid for freedom.

The hour had now come round when it was time to continue our journey. It was decided to leave the railway and use the moon as a guide till near dawn, when by turning north we could regain the railway with certainty and

361 Rather, more than a year after Le Mesurier lost his pocket-book, it was found at the spot where he dropped it by a corporal of an infantry regiment stationed at Pienaars Poort. This man, ascertaining from the contents to whom it belonged, sent it to the owner, and except that the cover had been eaten by insects, it was in wonderful preservation.

reach Elands River station. The distant roll of thunder and the lightning-flashes gave every sign of a coming storm. The moon was not yet up; but the flashes which from time to time shot forth from the inky clouds made the bare veldt look as bright as day, and we felt that we were very conspicuous. Dropping down into the valley, we crossed at a narrow spot the same river whose waters we had encountered the night before. Soon we came to the railway, which we had decided not to follow, crossed it, and proceeded up a long steep hill.

After we had covered between two or three miles, to my dismay I discovered that I had left my money-belt, containing some £18, and a compass, at our last hiding-place. The risk of returning to look for them was too great, for on quitting our hiding-place we had noticed many lights in the neighbourhood, and the probability of finding the clump of ferns in the pitchy darkness was remote. Besides which, both Le Mesurier and I were even now not entirely devoid of money, and it was doubtful if we should require what we still had.

We therefore pushed on till by the light of the moon we saw in front of us a Kaffir kraal, and found that we were walking past a field of water-melons. The opportunity was too good to be lost. Since we left our hiding-place we had had no water save that which a shower had left in the hoof-marks of some cattle. We now sat down and simply gorged this thirst-quenching pulp. I have often eaten Afghan melons, which the caravans bring on their camels to Peshawar, and these more resembled them than any others that I have tasted. But we had been heard or scented, for the kraal dogs nearby began to bark. I hastily tied up a melon in my handkerchief, and we walked silently away. Treading our way softly through some mealie-fields, we came to a deep donga[362] at the bottom of which was a muddy rushing stream. We followed this for some time, till, thinking that our destination must be near, we crossed over and mounted a hill. Half-way up its side we heard a sound of galloping. The place was very open and the grass burnt short, but after a few moments of suspense nothing worse than a frightened horse appeared.

It seemed as if we were never going to strike the railway; but in this particular locality, where the ground is hilly and the gradients steep, I believe its windings are extraordinary. As with all the railways in South Africa, the contour of the ground was followed, cuttings and embankments being rare. Suddenly the glint of the moonlight reflected from the rails caught our eyes, then the black telegraph poles like spectres, and the wires, through which the chilly breeze was softly moaning. We had reached it at last.

But our last two nights' experiences with railway guards had had their effect, and we foolishly left the line and wandered far from our course. After some miles we saw in front of us a small wood, close to which some waggons were outspanned, and soon after passing them I had what cannot have been anything but a hallucination. Brockie was in my thoughts, and I was wondering where he was and if by any chance we should meet him to-night at Elands River, for we had talked with him of making for that place. On the ground a little to my left I saw what seemed to be a man lying at full length, his elbows resting on the ground, his head between his hands. Telling Le Mesurier what I saw, we went nearer, but the vision of my fancy had gone; yet it was so clear that I can recall it and the spot where it occurred with the utmost distinctness.

Not far from here we came to a ridge of grey rocks, where we sat down. I was tired out, principally from having slept badly when below ground, and from not having slept at all since we left Pretoria, and was prepared to lie down on the veldt and take my chance of discovery. But dawn was near, and before us we saw a signal-post and a little farther off a wayside station. Crossing the railway, we made for a clump of trees which rose prominently before us on the sky-line. No river was to be seen, although the name of the place betokened one: there was nothing in a liquid form visible save a tiny trickling stream. This, however, is a little way of their own that South African rivers have—at one time a roaring torrent, at another a microscopic stream. Hard by a cock was shrilly crowing, betokening the presence of some inhabited dwelling, but, tired as I was, that was of no concern to me.

[362] A usually dry, eroded watercourse running only in times of heavy rain.

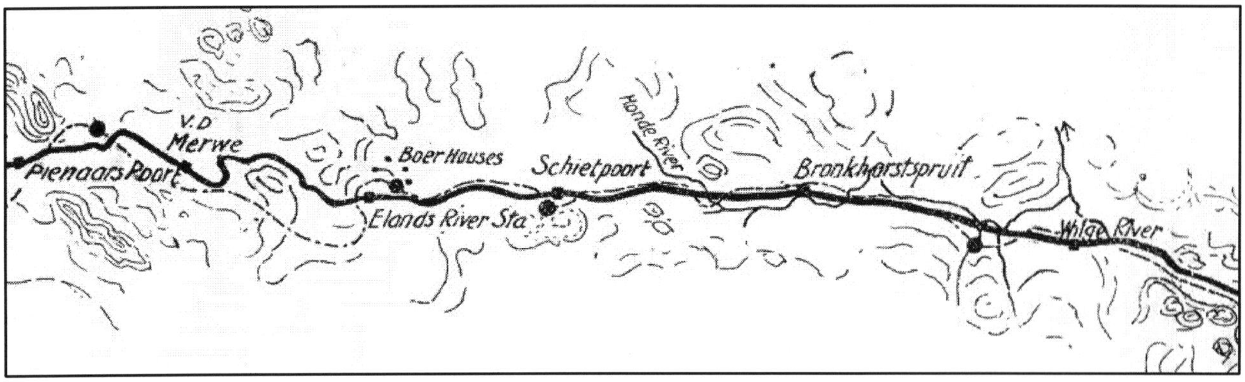

Haldane Days 3-5 (13.6 miles, 22 km) .

Nearing the trees, we walked into a barbed-wire fence, and crossing it, selected a small clump of blue-gum or eucalyptus trees for our lair. These were only saplings, six to eight feet high, interspersed with larger trees of a kind unknown to me. Behind us we had left two alarming tracks in the dew-sodden grass—but the sun would soon rectify that. Throwing ourselves down, we endeavoured to get some sleep before the mosquitoes found us and made repose impossible. We were far from comfortable, for the place was strewn with roots and stones; but my melon made such an excellent pillow that I was soon in the land of Nod, and awoke startled at 7 to hear Le Mesurier whispering, "I think we are discovered!" I could get no more from him just then, so lay thinking what was to be done, and hoping that he was mistaken.

He told me afterwards that someone had passed close by, and, he thought, had seen us. A heavy shower of rain now fell, perhaps hiding our footprints in the grass. Two hours later the sound of advancing footsteps became audible. It was a young Dutchman coming along a path not six feet from where our heads lay on the ground. He stopped when he came opposite to us, and a dog which accompanied him growled. How this animal did not immediately discover us will always remain a mystery to me, but perhaps the same shower which hid our tracks also spoilt the scent.

"It was a young Dutchman."[279]

Visions of bloodthirsty *vrouw* arose in one's mind; for now there were few men on their farms, and it was said on all hands that the women were more than Spartan in their severity. The lad now began shouting loudly what I supposed to be some Kaffir name, and after a brief interval was joined by someone with whom he talked for a space. It seemed to us as if he must have seen us, and, not daring himself to disturb the intruders, had called for help. But it is marvelous how, if one remains absolutely still, one may pass unnoticed, and our imaginations here had led us astray. Nothing happened, they moved away, and the incident only increased our confidence; for we felt that having had two such close shaves, we must eventually escape altogether.

All day long the Kaffirs worked near us, busy cutting wood and drawing water, and the Dutch lad passed more than once again. The Kaffir is more like an animal than a human being in the way he will detect a spoor and traces that would not catch a white man's notice; but we were completely unnoticed that day by black or white.[363]

Of course, by now we knew that we had walked right into the middle of a Boer farm; but I had been too tired the night before to care where I went. The difficulty now was to leave it so quietly at night that even the dogs might not hear us. At 7 o'clock we quitted the gum-tree thicket and made our way through mealie-fields and hedges till we were quite clear of our dangerous surroundings. This took a long time, for each sun-dried mealie-stalk in our path had to be held aside, lest the rustling noise which they would have made if we had attempted to pass quickly through should rouse the watch-dogs. I eagerly scanned each plant as we made our way between them, but not one single corn-spike remained.

It might almost seem as if we had departed from our original plan of trying to catch a night train; but on the three preceding evenings we had wandered far from the railway, and had thought that during our absence from its vicinity the eastward-going train might well have passed. To-night, however, though some doubts had certainly arisen in our minds as to whether the train ran or not, it was our intention to lie in wait for it and board an empty truck as it moved slowly up a gradient. Should we be successful, it would bring us to near Balmoral by daybreak. But we were destined to pass that night and the two next on the veldt. We now made direct for the railway, which was near, and shortly after found a gradient of 1 in 57 for some hundred meters.

Lying down close to it, and shielding our faces from the mosquitoes with our coats, we remained till 2 A.M., but no train passed. It now seemed certain that the night-service no longer ran, and there was nothing left but to go on walking. Our experiences of the previous evening decided us as to our route, and for a couple of hours we followed the line, and then began to look for water and a place to hide in. As there was a ganger's[364] house close to the railway-line, it seemed probable that we should find water of some sort near. In this we proved to be right, for we nearly walked into a circular pit, at the bottom of which were some inches of the precious liquid. Quenching our thirst and filling the bottle, we looked everywhere for some place that would hide us.

There was nothing for it but to lie down by an ant-bear[365] hole upon which we fortunately stumbled. The dwelling of this animal is a hole like a badger's, with a trench about two feet deep, and as broad, leading to the entrance. Here we lay down and prepared, when the sun rose, to be grilled for nine or ten hours, for there was no shade, and we had no means of making it without attracting notice; and our ten-ounce water bottle would not go very far. In two more marches we should reach Balmoral, and once there, we had a very good hope of hiding ourselves in a coal-truck bound for Portuguese territory.

[363] PF: As much as I would like to edit out this sentence I am not privileged to be Captain Haldane's censor.

[364] The dwelling or bunkhouse for the railway workers (gang) to occupy.

[365] An aardvark's den.

VI. From Veldt to Coal-Mine

"We must take the current when it serves, or lose our venture."[366]

We had prepared ourselves for a warm day, but this one far exceeded our anticipations: however, so long as we remained well hidden and knew that at nightfall we should be cool and able to procure water, we had no great cause for complaint. The main road and the telegraph-wire were only a few yards distant, but very few Kaffirs passed our way and not a Dutchman, for except the house which stood close to the railway, no other human habitation was near. As usual, we had got soaked to the waist in going through the long grass; but the hot sun soon remedied such a trifle, and before the day came to an end we would have given something to have been in a condition similar to that in which we had first invaded the ant-bear hole.

I have often since thought that in our wanderings on the veldt we were fortunate to escape a grass fire, an everyday occurrence in these regions. Had it happened that one had come our way,—and the tell-tale smoke we sometimes saw afar off showed how possible that was,—nothing would have remained but to seek safety in flight, a course which would inevitably have led to discovery.

We calculated by the kilometers marked on the white boards at various points along the line that we had only covered thirty-six miles in four days—a poor record—and had still twenty-one to compass before we should reach the neighbourhood of the collieries at Balmoral. Short commons,[367] lack of sleep, and want of water were beginning to have their effect upon us, and we thought we had done well when we had walked ten miles. It was scarcely dark when, our thirst having got beyond endurance, we hurried to the pool which we had enjoyed the night before. Except an army emergency ration, which Le Mesurier had carefully preserved ever since he was captured at Dundee,[368] and a scrap of biltong, we had nothing left to eat.

Seating ourselves by this pool we broke into our last reserve, and seldom have I enjoyed a meal (if such it can be called) so much. This ration is intended to last the soldier for thirty-six hours, and it is divided into two parts, each in a small tin which is enclosed in a larger one. One tin contains what I cannot clearly describe, but it is a species of powder, or pemmican, greyish in colour, caked hard, and tasting slightly of "Bombay duck."[369] The other tin is full of cocoa in a highly concentrated form, with a pleasant acid taste. They are both to be cooked or not, as circumstances will allow. Cooking-pots were not included in our kit, and means to light a fire, had we desired to do so, did not exist. We therefore dined off our delicacy in its raw state, but only ate about one-third of it, for it had to last two hungry men for seventy-two hours, not one soldier for thirty-six. Our supper at an end, we proceeded on our way, and avoiding the ganger's house, from which the sound of laughter could be heard, gained the railway. From constant listening and peering through the darkness, eyes and ears had grown far more sensitive, and the least sound of voices, or dogs barking, or the faint glimmer of a light in a bridge-guard's tent, did not escape our senses.

[366] And we must take the current when it serves, or lose our ventures. William Shakespeare in *Julius Caesar*. (Brutus and Cassius are discussing the final phase of their civil war with the forces of Octavian and Marcus Antonius.)

[367] Food and nourishment.

[368] The Battle of Talana Hill, also known as the Battle of Dundee and Glencoe, took place near the town of Dundee where the garrison was encamped. It was the first major clash of the Second Boer War. A frontal attack by British infantry supported by artillery drove Boers from a hilltop position, but the British suffered heavy casualties in the process, including the mortal wounding of their commanding general Sir William Penn Symons. Battle of Talana Hill-Wikipedia

[369] Bombay duck, (Harpadon nehereus) is a species of lizardfish. The origin of the term "Bombay duck" is uncertain. A popular false etymology claims that the overpowering smell of the dried fish when transported by mail train (the Bombay Daak) led to the phrase "You smell like the Bombay Daak" in the days of the British Raj. However, the phrase is attested as early as 1815, 37 years before the first railroad in Bombay was constructed, making this derivation impossible.
According to local Bangladeshi stories, the term Bombay duck was first coined by Robert Clive, after he tasted a piece during his conquest of Bengal. He is said to have associated the pungent smell with that of the newspapers and mail which would come into the cantonments from Bombay. Bombay duck - Wikipedia

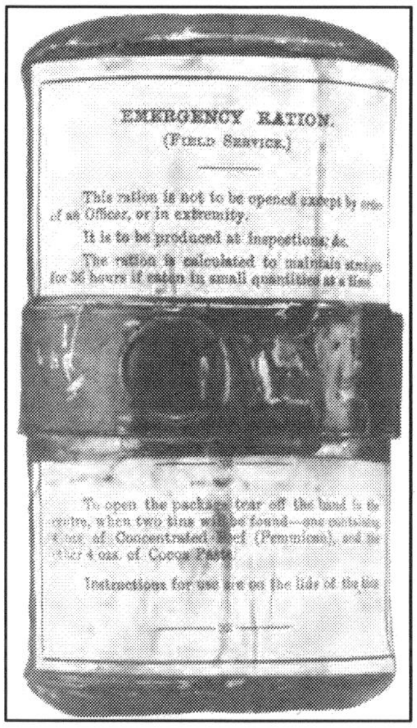

Emergency Ration.[280]

I cannot but think that the use of dogs in war by us British, a dog-loving nation, has been unaccountably overlooked. I believe certain nations—the Germans and French—use them; but we, who have our kennel-clubs, and every encouragement to breed the best and purest of the canine species, do not include these faithful friends among our war-material. In India, where one suffers so much from rifle-thieves, without his dog there would be many a court-martial on the owner of a stolen rifle. There the dogs are chained to the arm-racks, sentries with whose vigilance no human being can compete.

Often in Tirah[370] the wily Afridi[371] would bring his dog with him when bent on disturbing our rest by firing into camp. He knew right well that if the plucky little Ghoorka[372] was engaged in stalking him, his four-footed friend would give him timely warning.

Le Mesurier and I know what a curse were the dogs within whose hearing we came on our way to Lorenco Marques; and I think that if we employed them on outpost duty, and trained them to growl softly when some stranger was approaching, the possibility of surprise would be greatly lessened, and the soldier on his lonely post, where he has to depend on the acuteness of his sight and hearing, would feel the value of this assistant.

The night was wet and windy, which made both hearing and seeing doubly difficult. Keeping along the railway, we passed through a long but shallow cutting, and after going for a few miles saw in front of us a tent in which was a light, and heard the low murmur of voices. As usual, we left the line and made a detour, which brought us to the well-known Bronkhurst Spruit—a fact we ascertained by stealthily creeping along the station platform and reading the name-board. I may be pardoned for recalling to my readers the episode which befell the 94th Regiment under Colonel Anstruther near here, on the 20th December 1880.[373]

[370] The Tirah campaign, often referred to in contemporary British accounts as the Tirah expedition, was an Indian frontier campaign from September 1897 to April 1898. Tirah is a mountainous tract of country near the Kyber Pass. Tirah campaign-Wikipedia

[371] The Afridi tribe had received a subsidy from the government of British India for the safeguarding of the Khyber Pass for sixteen years; in addition to which the government had maintained for this purpose a local regiment entirely composed of Afridis, who were stationed in the pass. Suddenly, however, the tribesmen rose, captured all the posts in the Khyber held by their own countrymen, and attacked the forts on the Samana Range near the city of Peshawar. Afridi-Wikipedia

[372] The Gurkhas or Gorkhas, are soldiers native to South Asia of Nepalese nationality recruited for the British Army, Nepalese Army, Indian Army, and UN peacekeeping forces around the world. Historically, synonymous with "Nepali," which originates from the hill principality Gorkha Kingdom, from which the Kingdom of Nepal expanded. Gurkha - Wikipedia

[373] At the first battle at Bronkhorstspruit on 20 December 1880, Lieutenant-Colonel Philip Anstruther and 120 men of the 94th Foot (Connaught Rangers) were killed or wounded by Boer fire within minutes of the first shots. Boer losses totaled two killed and five wounded. This mainly Irish regiment was marching westward toward Pretoria, led by Lieutenant-Colonel Anstruther, when halted by a Boer commando group. They were halted when they approached a small stream called the Bronkhorstspruit, 38 miles from Pretoria. Its leader, Commandant Frans Joubert, (brother of General Piet Joubert), ordered Anstruther and the column to turn back, stating that the territory was now again a Boer Republic and therefore any further advance by the British would be deemed an act of war. Anstruther refused and ordered that ammunition be distributed. The Boers opened fire and the ambushed British troops were annihilated. In the ensuing engagement, the column lost 56 men dead and 92 wounded. With the majority of his troops dead or wounded, the dying Anstruther ordered surrender. Battle of Bronkhorstspruit - Wikipedia And First Boer War - Wikipedia

It was the opening scene of the war of 1881, which terminated with Amajuba [Majuba].[374] A detachment of that corps was marching to Pretoria, before any declaration of hostilities had been made, when they were ambushed by the Boers and out of 250 persons all but one officer and 120 men fell within a few minutes. The Boers boast of this brutal massacre as a victory and in spite of this, and their abuse of the white flag and use of expanding bullets during the present campaign, there are men who call themselves Englishmen who are ready to extend to them the hand of friendship and of sympathy.

Hurrying past this place of mournful memory, we soon came to the village of the same name. The dogs here seemed strangely restless, and as the moon was shining brightly and the rain had ceased, another wide detour on hands and knees was essential. Then we came to a river, and crossing it we regained the railway. A few miles farther on, with no incident worth recording—except that Le Mesurier had the misfortune to tumble full-length into a marsh,—we began to think it was time to make our nightly search for a hiding-place.

I had been told by a colonial that the grass when at its longest would hide a man from view, and had expected when we left Pretoria to experience no difficulty with regard to hiding-places. But like many tales from South Africa, this one proved untrue. After a good deal of hunting up and down, backwards and forwards, we agreed to lie down in a swamp where the grass and reeds were fairly tall, and near which no human habitation could be seen. It was very wet, probably full of fever-germs; but no matter, we should have the wherewithal at hand with which to fight the sun when he rose.

Next day passed peacefully; we were evidently "far from the madding crowd,"[375] for all day long we did not see a soul except some Kaffirs, who were working on the railway-line some way off. By the side of a tiny stream we sat, paddling our feet in its cooling waters, talking over the adventures we had met, anticipating those which were to come, and longing fervently for anything, no matter what, of an edible nature for the pangs of hunger absolutely declined to be postponed by copious libations of water. Refreshed by a few hours' sleep, we made our way to the railway as twilight turned to night, and, avoiding one or two bridge-guards, reached some running water.

The night was very dark, so much so that two Dutchmen passed close by us while we lay down, and crossing rivers by faith and not by sight is not pleasant work. The river we had reached is called the Wilge river,[376] and being at that season in flood, was broken into several branches. After negotiating four, we thought no more remained; but a girder railway-bridge was dimly visible on our right, a gentle hint that all were not yet passed. The Boer who guarded this bridge must have heard us, for he came out of his tent with a lantern, and, while we lay hidden in the wet grass, listened for some time. Of course at the very moment he appeared the moon began to rise over the crest of a hill in front of us; but as the lantern soon disappeared, we got up and continued our travels.

A few paces brought us to the edge of a most uninviting-looking river—one of those sluggish, stagnant streams such as one meets in Belgium. Neither of us was in the mood for a swim—it was too early in the evening for that; and on trying its depth I found it reached to my waist, and its bottom was composed of soft mud. Along the bank we walked, and were soon rewarded by hearing the sound of running water, and a little farther brought us to a shallow ford: overlooking it on the opposite bank was a cottage, which seemed to be deserted. We crossed over, and found ourselves on a hillside covered with loose stones—an ideal spot for spraining an ankle, and a place which caused my lame companion some distress.

The moon came to our rescue, and we again struck the railway without mishap. We now thought that Balmoral[377] could not be far off, and the hope of reaching there before dawn made us push along quickly—so

[374] The Battle of Majuba Hill on 27 February 1881 was the final and decisive battle of the First Boer War that was a resounding victory for the Boers. Battle of Majuba Hill-Wikipedia

[375] Affected with madness; raging; furious. Also, Far from the Madding Crowd (1874) is Thomas Hardy's fourth novel and his first major literary success. Far from the Madding Crowd-Wikipedia

[376] The Wilge River (Afrikaans: Wilgerivier, meaning "willow river") is a tributary of the Vaal River in central South Africa. This river is important as part of the Tugela-Vaal Water Transfer Scheme where water is transferred from the Tugela River basin to the Vaal River basin. Wilge River-Wikipedia

[377] Balmoral is a town in the Emalahleni Local Municipality in the Mpumalanga province of South Africa. The village was established as a railway station of the Oosterlijn from Pretoria to Maputo in 1894. The village and railway station were erected at the farm Eenzaamheid. For decades the village was nothing more than a railway stop with a primary school for the surrounding farms. During the Second Boer War the British built a concentration camp near the railway station. Balmoral, Mpumalanga-Wikipedia

quickly that we failed to bear in mind the maxim, *festina lente*.[378] In consequence we nearly walked into a tent, and had to hide beside the railway till, after reconnoitering, I ascertained that we had not been seen.

Before us on either side of the railway, and close to it, were farms, easily recognisable by the clumps of trees around the buildings. To follow the railway, which bent sharply to the left, was inadvisable, so we turned aside and made a flank march round the outskirts of a farm. In front was a line of hills which seemed to block the way, but on approaching we found a rocky gap down which a stream of water ran. Following this to its source, we traversed a tract of marshy ground, and after another mile or two rejoined the railway. Suddenly we were aware of an odour familiar to us both, the smell of burning coal. We had not expected to find a colliery so close, nor was there one, for the wind brought from a distance what our nasal organs had at once detected. However, we very soon came upon a siding full of coal trucks. I now made a systematic examination of every truck, to see by the ticket on it where its contents were bound for. It is not the custom to put the full name of a place on these labels, and had I seen L.M., I should have known that it meant Lorenco Marques.

But nearly all these trucks were labelled N.C. or J.K. There was no mistaking the first two letters,—they must stand for Newcastle in Natal; but as for J.K.,[379] it was Greek to us. Standing at one end of the siding was an engine, and close by the trucks a small Kaffir hut, outside which were some embers where the occupants had cooked their evening meal.

We were tempted to try to clear up the situation, and find out if any trucks were bound for our destination; but it was 2 A.M., the Kaffirs were probably asleep, so we returned to the swampy spot which we had passed before. On our way back I found a large-sized mushroom, one-half of which I ate forthwith, while the other served for breakfast next morning. Cutting down some rushes, we spread them in the swamp, and lay down to snatch a few hours' sleep, which the mosquitoes were determined we should not have. The night, as usual, was very cold, and as I had had rheumatism in my knees ever since the night we took to the river, I was right glad when the sun rose.

The day was very warm; but there was more than enough of water, and our hopes were high, for we were determined to secure a truck at one or other of the colliery sidings, and get the rest of our journey over. All day long the noise of the shunting at the siding, where we had been so early in the morning, could be heard, and more than one train seemed to leave it. We hoped that there might be some trucks for Delagoa Bay, where many ships must surely stop to coal.

Towards evening a middle-aged Boer, accompanied by a boy and armed with a sporting rifle, passed us not far off. As luck would have it, a bird, not unlike a curlew,[380] rose from the swamp close to us and flew behind him; but he neither heard nor saw it, and soon they disappeared from view. The weather, which all day had been very sultry, now began to look most threatening. From every quarter black clouds seemed to roll towards that part of the sky which overhung our swamp. There was no mistaking these signs—we were in for a very heavy storm. Half-past five passed and still it did not come, but by six the rain had begun to fall, and a few minutes later we left the swamp drenched to the skin. Amid flashes of lightning which caused the veldt to look bright as day, and the thunder's sullen roar, we made our way to the Kaffir hut at the siding.

Before leaving our hiding-place we had finished every scrap of food we possessed, yet I have no recollection of feeling anxious as to the future. Not even an inch of biltong remained. This succulent and sustaining stuff is always carried by the Boers when travelling or campaigning. It is made from the flesh of the buck or ox, cut from the choicest parts. Strips of this meat are steeped in vinegar and pepper for two hours, then covered with salt, which it sucks in all night. Next day the strips are hung up in a shady place until their exterior is dry, when they are put in the sun, which soon makes them as hard as wood. This is now fit for eating, and as much as is required for a meal can be cut from it as one slices a cucumber.

[378] Make haste slowly.
[379] Possibly Johannesburg.
[380] Like a sandpiper. Curlew-Wikipedia

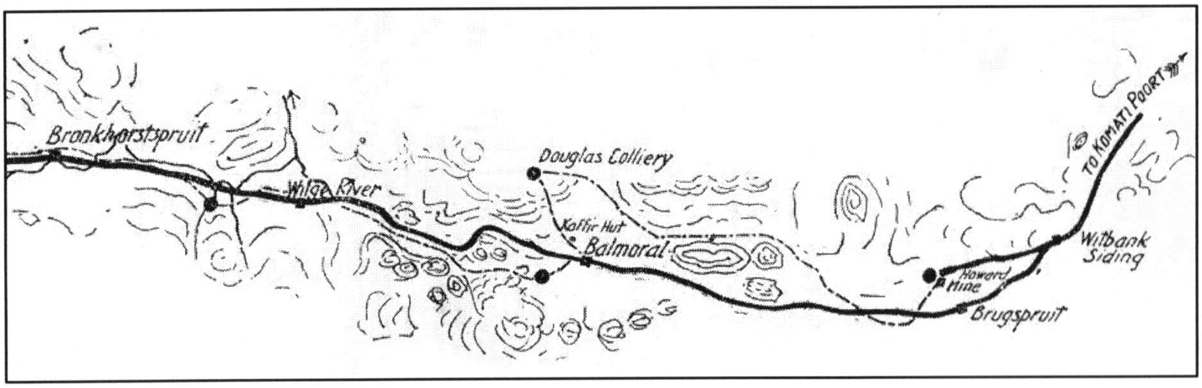

Haldane Days 6-7.

Miserable objects, indeed, were we when we came opposite the coal-siding. Not a stitch of dry clothing on us, starving and weak; but shelter and perhaps food were to be found inside the hut, and Le Mesurier had impressed on me from previous experience that Kaffirs were absolutely to be trusted. He had some years before served in Zululand, and had the utmost confidence that no Kaffir we met would betray us, more especially so now that we were at some distance from Pretoria.

The door was open and a light burned within. We entered, and saluting the occupants with the Kaffir greeting, "*Saca-bona?*" (How do you do?) sat down. There were five thick-lipped, ebony-coloured negroes seated round a caldron, which, turned on its side, displayed its half-eaten contents—thick dry mealie-meal porridge. Without another word we joined the circle, those nearest us drawing aside their mats, for the rain-water was running off us on to the hard mud floor.

We joined the circle.[281]

Indicating our desire to share their meal, the vessel was tilted towards us, and stretching out a hand we drew forth a lump of the coagulated mess. We were both so hungry that the desire for food left us when it was within our grasp; but after a short time the craving returned and we did justice to the simple fare, realising keenly what it was to eat to live.

The Kaffirs exhibited no vestige of a sign of surprise at our sudden inroad, behaving in the same phlegmatic, dignified manner that one meets with in the East. They must have wondered too who we were, for had we been Dutchmen we should have known their language.

Meantime we had tried the few Kaffir words and sentences we knew upon our black friends, one of whom took pride in repeating his very few words of English. The conversation was not long sustained, and having satisfied our hunger we left the hut. It was pouring still, and the prospect of finding other coal-sidings and examining trucks without the aid of a light—for, as I have said before, we had no matches—was not a pleasing one. We therefore entered the hut again, having decided that we would disclose who we were and try to enlist the help of its occupants. It was easily explained that we were Englishmen from Pretoria running away from the Boers and making for Delagoa Bay.

We at once had the sympathy of all, who showed their hatred of the Dutch by signs and gestures, drawing their fingers across their throats, clenching and shaking their fists. So thoroughly did they grasp the situation that the conversation was now conducted in whispers, for outside, not far off, was an engine the driver of which was a Hollander. It was now necessary to try and explain what we wanted them to do. I happened on the previous evening, for no particular reason, to have taken the label off a truck and put it in my pocket; producing it, no word of explanation was required. Two of the Kaffirs left the hut and did not return for some minutes, but when they did they brought with them several other labels. Looking at them, we found that none of them bore the mystic letters L.M. Again they went out. We now gave five sovereigns to the elderly Kaffir who seemed to be the *doyen*[381] of them all, and made him understand that they were his if he could get us a truck for Delagoa Bay. It may be remembered that before we left Pretoria, Sergeant-Major Brockie had given me some slight instruction in the Kaffir tongue, and having some suitable sentences written down, I read them out, much to the curiosity of the listeners, who wanted to see the paper on which they were inscribed. After several more visits had been paid to the coal-trucks, it seemed certain that none was bound for the coast. The old Kaffir then handed back the gold, and giving him a trifle in return for the food, we again left the hut.

Outside everything looked so black and unpromising that we agreed to put our fate entirely in the hands of the five Kaffirs. For the third and last time we entered the hut and asked them point-blank who their *baas*[382] was. Was he an Englishman or a Dutchman? The reply came from all that he was the former and lived not far away, and we thereupon decided to go and ask his help. It might be that we should find that he was a man on parole to take no part in the war, and to remain neutral but it could do no harm to see him and obtain some food and information.

We therefore indicated that we wanted one of them to conduct us to his house whereupon the youngest Kaffir rose, put on an ancient overcoat, and we went forth again into the night. The rain had ceased, and, following our guide, we crossed the siding, feeling that our difficulties were coming to an end. In the distance a burning slack-heap of a coal-mine could be seen against the sable background of the sky. It looked nearer than it was, for almost an hour passed before we got into close proximity to it but the walk seemed to dry our clothes, and we stumbled along in the dark, anxious to know what was in store for us. Passing some sheds, we came to the pit's mouth, and then climbing over some rubbish-heaps, found ourselves facing a row of one-storeyed dwellings. Our cicerone[383] indicated that he had performed his part of the undertaking, and that the manager's habitation lay before us. Knocking on his door, a voice bade us enter. We did so, and found ourselves in the presence of a tall fair man, who later told us that he thought we were Boers. I asked him if he were an Englishman. He replied that he was, but from his accent and appearance he looked more like a Swede.

Eventually he informed us that he was a native of Denmark, [Named Jansen] whereupon I told him my name was of Danish origin, then inquired if he were on parole, as we wanted his help. As he replied in the negative, I added that we wanted to travel to Delagoa Bay concealed in a coal-truck, and asked how far he could help us. He remained lost in thought for several minutes, and then said that it would be difficult; that his mine was sending no coal to the sea-coast, but that three trucks were to be loaded for Lorenco Marques at a small mine close by on the following morning, and that he would try to have the loading postponed till night, so that we could reach our hiding place unseen.

[381] In the English language, the meaning of *doyen* (and the less common *doyenne*) has extended from the French definition to also refer to any senior member of a group, particularly one whose knowledge or abilities exceed those of other members. Doyen-Wikipedia.

[382] An Afrikaans word for boss and Dutch word for supervisor/leader/master, from which the American word *boss* is derived.

[383] *Cicerone* is an old term for a guide who conducts visitors and sightseers to museums, galleries, etc.

Douglas Mine (Colliery) and Jansen's House.[282]

This meant taking another man into our confidence, but as he was a Scotsman there was no objection. He said that he would arrange the matter at once, but the question was where to put us for the night. On my suggesting the coal-mine, he said that there were several Hollanders on duty there, none of whom could be trusted to keep their counsel. Le Mesurier and I then said that the veldt was good enough for us, and that if he could get us some food, we would lie out all night and come to some appointed place next evening. He answered he would bring the storekeeper to us, and, leaving the room, shortly returned with that, to us, most welcome individual.

Mr. Moore, the manager of the Douglas Colliery store, now took complete charge of us. He led us to his house, which stood in a garden quite apart from any others, and soon we were seated at his hospitable table eating as if our lives depended on it. Our first civilised meal for over three weeks consisted of tinned salmon, cocoa, and the usual adjuncts. I think we horrified our host by our indecent rapacity, but it gave him a good idea of what we should require on the morrow. Mr. Moore told us that he was the son of a late Indian general who had served in the 13th Madras Native Infantry,[384] and that his brother, in the Imperial Light Horse, had fought at Elandslaagte side by side with my regiment.

It was a great relief no longer to talk in whispers, and our voices sounded strange to us. Our host occupied himself in considering where to lodge us for the night, and decided to put us in a forage-shed next the store, which was generally occupied by his Kaffir boys, who need not know who we were. The moon was up, and we made our way to the shed. In the next house was an Irishman who was in sympathy with the Boers and had some grievance against Moore, so we were cautioned to keep very quiet. With apologies for putting us in such a place—a palace to what we had been accustomed to of late—Moore left us for the night.

A deep sense of contentment and satisfaction now came over us, partly engendered by the feeling of repletion, and of being able to look forward to a much-wanted night's rest without mosquitoes. A slight arrangement of the forage bundles was necessary, and this made, we were soon fast asleep, despite numbers of mice whose dwellings we had invaded. At a quarter past six next morning we were startled by hearing some one knocking at the small window, which was partly covered by a sack to hide our presence within.

Whoever it was next came round to the door, which was bolted inside. Cautiously opening it a few inches, a black face peered in at me, and the owner thereof having apparently satisfied his curiosity gave a loud guffaw and went off. Shortly after this a man's face, of somewhat forbidding aspect, appeared at the window and shouted something in Dutch. Receiving no answer, he tried us in English, saying, "Hulloo! who told you to doss[385] down there?"

Le Mesurier said by mistake, "Mr. Jansen," which was the name of the Dane who had passed us on to Moore. This seemed to excite the inquirer's ire, probably because this store was Moore's, and saying, "Damn his

384 13th Madras Native Infantry - Wikipedia

385 To sleep in the open or in a derelict building because one is homeless. "I normally have to doss in shop doorways or park benches."

cheek!" he went off. It left with us a most uncomfortable feeling that, even if he were friendly, of which I felt some doubts, not knowing who the strangers were, he might reveal what he had seen to some one better kept in ignorance.

Our fears were dispelled when at 9 o'clock Moore appeared, bringing with him our recent visitor, who turned out to be the butcher of the mine, and lived next door. He told us that his name was Smith, and that he was a Natal-born Englishman, and had been compelled by the Boers, being a burgher, to fight at Elandslaagte and Spion Kop but that he had made a rapid strategical move to the rear at the former fight and escaped the lancers. He said he had avoided pointing his rifle so as to inflict damage on his fellow-countrymen. A door which led from the store into the forage-shed was now opened, and throughout this day a liberal supply of meat and drink was passed in to us. At 10 o'clock we were told that the medical man who had charge of the miners, and happened to be making one of his occasional visits, was coming in to see us.

Dr. [James] Gillespie,[386] one of those fortunate beings whose voice and manner at once inspire confidence, now entered the shed. He told us that, purely on chance of hearing some news at the mine, he had driven over from Brug Spruit on the previous day, and stopping the night had chanced to hear in the morning that the escaped prisoners, of whom every one knew, had at length arrived. The extraordinary chain of circumstances which had brought us to the mine, exactly at the right time, was now made evident to us.

Had not Le Mesurier delayed us owing to his sprained ankle, had not the thunderstorm driven us to the Kaffirs' hut, we should probably never have heard of Dr. Gillespie. It was not his usual day for visiting the mine.

He now told us his plans for getting us safely over the border. To my amazement he said that he and some others had managed it for Churchill, and they would do the same for us. He told us to say nothing to anyone of the fact that they had helped Churchill, and that when it grew dark he would drive us to another mine, where plans for the future would be matured. He added that we might now consider ourselves out of the country, our further movements would be so devoid of risk. Bidding us farewell till evening, and saying that the coal-truck plan was now at an end, he left us.

Another visitor followed, Mr. Haughton, manager of the adjacent mine, which was sending the three trucks of coal to Delagoa Bay. He was an old soldier, and had served in the Zulu war, and was quite disappointed when he heard that our plans were altered and that we were not to travel in one of his trucks. He told us that Dr. Gillespie had an excellent plan for getting us out of the country, and we could entirely rely upon him.

It was here that I happened to read in a paper, to my great surprise and distress, of the death of Sir William Lockhart,[387] Commander-in-Chief in India, on whose staff I had been for three years and a half, and who had allowed me to rejoin my regiment in order to come to South Africa. It had been my very good fortune to have twice served on his staff in the field,[388] and to have realised that magnetic influence which makes some few commanders born leaders of men, and this power he possessed in the highest degree.

During the day a gramophone played numerous topical airs for our amusement, and when it grew dark Moore came and took us again to his house. Here we found, thanks to his thoughtful kindness, a complete change of garments, for our own were fit for nothing but the dust-heap. We also enjoyed the luxury of a bath and were much amused at the reflection of ourselves in a looking-glass.

After we had had supper, a number of Englishmen employed on the mine came and, out of curiosity and interest in our escape, asked to see us. I felt glad that we were not remaining at this mine, for far too many knew our secret; and though each individual would keep it inviolate, still an indiscreet remark might lead to our discovery.

Dr. Gillespie left before us to see his horses inspanned,[389] and shortly afterwards, conducted by Moore, we bade adieu to the occupants of the house, and striking across the veldt went to the appointed place where we were to join the doctor. It had been considered unsafe to meet him at the mine.

386 Dr. James Gilespie (Sandys p. 141).

387 General Sir William Stephen Alexander Lockhart GCB, KCSI (2 September 1841 – 18 March 1900) was a British General in the British Indian Army. William Lockhart (Indian Army officer) - Wikipedia

388 Re: Haldane: Between 1894 and 1895, Haldane was part of the Waziristan Field Force and participated in the Chitral Expedition. He was dispatched to quell the Afridis rebellion in the Tirah campaign for the next two years (1897–1898), was appointed a Companion of the Distinguished Service Order (DSO) on 20 May 1898, and became aide-de-camp to General Sir William Lockhart, Commander-in-Chief in India, later the same year.

389 To bring or force into service. Usually referred to hooking oxen to a cart or horses to a carriage. Contrariwise, *outspan* is to remove

VII. At the Coal-Mine

"A friend in need is a friend indeed."

Following our conductor down the lower slopes of the hill, on the summit of which, strange to say, the coal-mine was situated, the lights from the burning slack-heaps grew less distinct. On reaching the level no signs of the doctor were visible. Matches were lighted, but no responsive flash answered ours. At length Moore returned along the way we had come, and having met his assistant a short distance off, came and told us that the doctor had decided to take another route, one which was better than that he had at first intended to follow.

Retracing our footsteps, we came upon him near the mine. He was seated in a two-wheeled dog-cart, to which a pair of grey horses were harnessed. Climbing up beside him, we took our places on the front seat, and the hood of the trap was raised. We bade farewell to Moore, and, taking with us his earnest good wishes for our freedom, started.

Dogcart.[390]

Cape Cart.[283]

I regret to say that, many months later, when Botha and his men were retiring eastward from Pretoria after Lord Roberts' entry into the capital, some of his force happened to camp at the Douglas mine.

Douglas Mine (miners in coal trollies).[284]

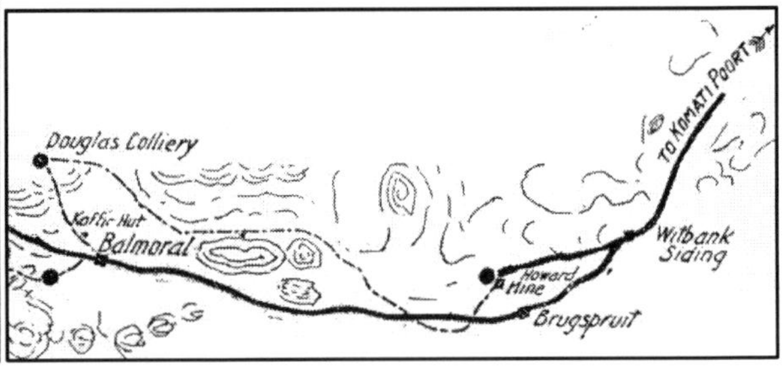

Area around Colliery & Witbank.[285]

As I have mentioned before, the secret of our having been hidden there was known to many, and came to the knowledge of the commando which was passing through. Whereupon some of them without hesitation maliciously destroyed Moore's house and property, and he probably only escaped with his life by happening to be in Johannesburg at the time.

The night was intensely dark, and until the moon rose—for we were driving without lamps—we could barely see a horse's length before us. I have mentioned the difficulty we had on leaving Eerste Fabriken of finding our

them from the yoke or harness.

390 A dogcart (or dog-cart) is a light horse-drawn vehicle, originally designed for sporting shooters, with a box behind the driver's seat to contain one or more retriever dogs. The dog box could be converted to a second seat. Later variants included: A one-horse carriage, usually two-wheeled and high, with two transverse seats set back to back. Dogcart - Wikipedia File:Dogcart (PSF).png - Wikimedia Commons

way by road when walking; now we were to experience a similar difficulty while driving. The doctor knew the road well, but the numerous tracks, of which in parts it consisted, made it difficult to follow the right one; and so dark was the night that we were obliged to keep at a foot's pace almost the whole way.

Many times during that drive of fourteen miles we were on the eve of upsetting; but fortune favoured us, and though we rolled from side to side at the bad places, like a ship broadside on in a heavy swell, we never quite went over. Part of the journey took us close to the railway; and I have heard since that we drove past several Boer dwellings, and in two instances close to their doors. There was a possibility of some inquisitive Dutchman stopping to speak to the doctor, and becoming suspicious when he saw three persons in the cart, for the usual number it carried was two—himself and his servant; but we saw no one, and only at one place where we crossed the railway had I to squeeze down behind the splash-board.

The doctor told us he was taking us to the Transvaal Delagoa Bay Company's colliery. Reaching it, he would hand us over to Mr. Howard, who had arranged for Churchill's escape, and he himself would go home to Brugspruit, four miles off. Even if any of the Hollanders should chance to hear that we had been at the Douglas Colliery for one night, they would have no idea where we had gone afterwards, and, the ground being very dry and hard, the cart would leave but an indifferent spoor.[391] It was half-past one in the morning of Saturday, 24th March, when we drew up near the office of the colliery where we were to be the hospitable manager's guests for several days. All was silent and deserted, except for the presence of the two Kaffir boys on guard outside the office.

Dr. Gillespie got down a short way from the office, going on in advance to see if the coast was clear, and after a few minutes returned and conducted us into a room behind it. Here the occupant of the room [Mr. Adams] and I mutually recognised one another. It was no other than the Englishman whom, as I related before, I had seen, with another, walk past the Staats Model School.

Mr. John Howard.[286] .[287]

Mr. Charles A. Burnham.[288]

That other very soon joined us, and was introduced to us as Mr. Burnham, the manager of the mine store. The first, whose name was Adams,[392] was resident secretary of the mine. They now told me that they had passed the school, wondering how they could communicate with us to help some of us to escape.

Adams left the room in search of the manager, who lived a short distance from the office. When he knocked at the door of Mr. Howard's room a voice said, "Anything wrong?" The reply, "Pumps broken down," was given. Mr. Howard afterwards told us that he instinctively said to himself, "They've come at last." He had made up his mind that we should somehow reach his coal-mine, and had kept Kaffirs out night after night near the railway watching for us. He also had for nights past played "God save the Queen" on his piano, with the windows open, lest we passing that way should hear and crave admittance. Soon the objects of so much solicitude saw before

[391] The track, trail, droppings or scent of an animal.

[392] John R. Adams (Haldane has consistently referred to him as Addams. This spelling has been harmonized with that used by Churchill and Sandys.)

them a tall, spare, clean-shaven man—evidently a man of strong purpose and resolution. The resemblance of the manager to Von Moltke[393] struck me at once.

"The Colliery Offices, where Officers Hid for Many Days."[289] "Room in the colliery offices in which W. Churchill hid for several days, and was fed through the broken window by Mr. Howard."[290]

His arrival was the signal for a decision to be come to regarding our disposal. It appeared that Churchill, or Dr. Bentock (which was the name he assumed while here), had lived down the mine on his first arrival, and afterwards in a room next to the office.

Transvaal and Delagoa Bay Colliery Company's Mine.[291] Head-Gear of a typical deep mine shaft.[292]

This was an inconvenient arrangement, as it involved taking into the secret three English miners,[394] and one day also nearly led to his being discovered;[395] and there was a further difficulty about food. It was decided, therefore, that Mr. Howard, who generally had his meals at the office, should be unwell for the next few days and feed at his own house, and that we should take up our quarters with him. To make the arrangement safe, it would be necessary to take into our confidence two English servant-girls[396] and a Kaffir boy. We now left the room and went to Mr. Howard's house, and, before going to bed, he told them who we were and swore the Kaffir boy to secrecy.

We had not reached the mine till after 2 A.M., and it was now 3 o'clock, and having slept on the hard ground for days without covering of any kind, we longed once more to go to bed like reasonable beings; but before allowing us to do so, Mr. Howard insisted on our having another meal.

[393] Otto von Moltke, was a German-Danish military officer who fought in the Franco-Prussian War in 1870 and, after emigrating to Chile in 1876, in the War of the Pacific in 1879. He was killed in battle in 1881, during this last war. Otto von Moltke-Wikipedia

[394] Daniel Dewsnap, Joe McKenna and Joe McHenry. All three had been very helpful in assisting Churchill to escape and it is unclear why they were not enlisted in assisting Haldane and Le Mesurier.

[395] Churchill was seen by one of the native miners who thought he was a ghost. Subsequently none of the miners would voluntarily work in that section of the mine. (See London to Ladysmith)

[396] See Vol 2 London to Ladysmith. In the Charles Burnham Gold Watch article, WSC includes Ms. Ellen David (received a 'Lady's' watch) and Ms. Ada Blunden (received a 'Lady's' watch).

When morning came we heard with interest how Winston Churchill had found his way to the manager's house. He had left Pretoria, according to plan, the same evening as he had got out of the Model School, and with some difficulty had succeeded in climbing into a coal-truck on the 11.10 P.M. train. Travelling all night till he reached the vicinity of Balmoral, he had left the train, waited till dark, and, following the line, had seen before him the electric lights of the Whitbank Colliery. As this colliery is managed by Frenchmen, and there are very few Englishmen on it, discovery meant capture. Making towards the lights, he was fortunate to lose sight of them in a dip of the ground, and when he got on to the flat again he saw in front of him the glimmer of a slack-heap. He now directed his steps towards this, and eventually arrived at Mr. Howard's house.

As there were at that time seven Boer farms all occupied and within a mile of the mine, he was fortunate not to chance upon one of them. The garden of this house was tidy and looked like an Englishman's, and he knocked at the door. In reply to Mr. Howard, who opened it, he said that he had fallen off a train and lost his way. It soon came out who he was, and he was there and then put down the mine. The name which he used was assumed, so that all who knew the secret at the mine could talk of him more freely.

At this time Mr. Burnham was sending seven trucks of wool to Lorenco Marques, and a hiding-place was made for Churchill in one of them between the bales. Accompanied by Mr. Burnham, who travelled on the same train and bribed the officials to let the trucks go right through, he reached, after some delays, his destination. Suspicion, however, was roused through a member of the firm to whom the wool was consigned noticing some marks of grease on the bales, which seemed to say that someone had occupied the truck.

This person made inquiries from Burnham, who professed ignorance; but he was undoubtedly suspected, for he had been seen walking with a stranger at the time the wool arrived at Lorenco Marques, and a little time after a Dutchman asked him in Pretoria how much he had got for getting Churchill out of the country.

In spite of the risk he ran, however, he was prepared to repeat the experience for our benefit; but it was decided that as a certain amount of wool was still being sent by various persons to Lorenco Marques, and as many goods-trucks were detached for twelve hours at Middelburg, a short distance off, he and Adams should keep a look-out at Whitbank station and note the trucks that passed eastward. Hiding in one of these trucks, we could get out of the country without throwing suspicion on Burnham. If, however, none should stop at Middelburg, Burnham would buy sufficient wool for one truck-load, and send us with it, accompanying it as he had done before.

Mr. Howard set his face against anyone connected with the wool travelling by the same train as it; and in the end—for we were entirely of his opinion—Burnham, for some reason, which made me feel it was connected with the despatch of the wool, was refused a passport. After martial law had been proclaimed in the Transvaal no one could travel without a pass, and this had to be applied for in person from a field-cornet.

The day of our arrival at the coal-mine, Adams and Burnham drove to Whitbank and brought back news of Brockie, to the effect that he had arrived there that morning, and had gone to a store and asked for work. He had been passed on to another store, where he chanced upon an Englishman whom he knew. A passport, the name on which was altered after it had been obtained, had been procured for him as far as Kaapmuiden,[397] forty-six miles from the border. Travelling there by train, he was, on arrival, to go to an Englishman, who would provide him with a Kaffir guide to take him over the border. I will leave him there for the present; but will say in passing that my fears of his getting into Portuguese territory before us were renewed.

Adams had seen some one on the platform of the station, neatly dressed and shaved, who tallied with our description of the missing man [Brockie], and soon after was informed [by a third person] who he was [Brockie], and that he was anxious as to what had happened to us.

[Mr. Adams] Wisely keeping his counsel [not revealing any information about Haldane], he waited till the train passed through, and noticed one truck loaded with wool, which would do well for us if it were detached at Middelburg.

Next day he [Adams] drove there, but no wool-truck had been left. During the time we were at the mine we heard of the death of Commandant General Joubert,[398] and that the officers who were prisoners had sent a

[397] Kaapmuiden (Cape Mouth) is a small farming town situated at the confluence of the Kaap and Crocodile Rivers in Mpumalanga, South Africa. Kaapmuiden-Wikipedia

[398] Petrus Jacobus Joubert (20 January 1831 – 28 March 1900), better known as Piet Joubert, was Commandant-General of the South African Republic from 1880 to 1900. He also served as Vice-President to Paul Kruger from 1881 - 1883. Piet Joubert - Wikipedia

wreath to be placed on the coffin. We were also told that Mr. Cullingworth's house had after our departure been three times searched. The same had happened when Churchill left, for the Boers were determined that he had not left Pretoria, an idea which I of course encouraged.

From Mr. Howard I gathered that he had studied engineering in England—his father having been in the Royal Engineers—had lived in the Free State for many years, and had taken part in several expeditions against neighbouring tribes who had broken out; and as the Boers imagined he was a burgher,—which he was not,—he feared he might have to go to the front. He was on very good terms, however, with the field-cornet of his district, and hoped to avoid this disagreeable contingency. He had contemplated letting the officers in Pretoria know that if any escaped they must come to him, but his plan of sending a message inside a tennis-ball was too risky, with the police living in the back-yard of the school, and he had been obliged to give it up.

I have seldom experienced such care and kindness as he gave us during the few days we were under his roof. It must be remembered that he and the others who were helping us to complete our escape ran the risk of being shot if they were found out, yet they never hesitated to undertake the plan which eventually proved successful.

Both Le Mesurier and I were beginning to feel the effects of our outing, and Mr. Howard did everything he could devise to make us fit and well before we left him. I had lost in weight a stone and a half,[399] and Le Mesurier was still lame and about to have a serious illness.

Except to take a walk with me after dark towards the Whitbank lights, Mr. Howard rarely left the house for long, lest the field cornet or some other dangerous person should chance to call. As for his two guests, they passed the day indoor, blinds down, and ready to conceal themselves should any visitor arrive in the absence of their host. Dr. Gillespie came over and paid us several visits, and promised to put in an appearance the night we were consigned to the truck.

It was explained to us how impossible escape in a coal-truck would have been. If we had selected one in which the coal was stored in bags, we could never have made a hiding place which would have been safe, for the shunting of the trucks would have brought down the sacks of coal on our heads. Also, it was the custom on the railway, in order to prevent theft, to throw chalk all over the topmost layer of bags, and any shifting of them on our part would have been detected at once.

I told them that at the Douglas Colliery the plan had been to build with planks a hole sufficiently large to hold us, which would have been hidden with the sacks. They laughed at the idea, and said we should have been suffocated by the heat and coal dust, and, worse than all, those who had helped us would have been discovered, unless the consignee at Delagoa happened to be in the secret, which in our case he would not have been.

On the return of Adams and Burnham a consultation was held, and it was decided that the latter should buy up sufficient bales of wool to make up a truck-load—sixteen would be required,—telegraph to the same firm at Lorenco Marques, offering them the wool for cash or its equivalent in kind, and say that it must be sold. The anxiety to get rid of the bales might arouse suspicion, but we could not expect to avoid every risk.

Accordingly on Monday, 26th March, a telegram was sent to the firm at Lorenco Marques, and by evening a reply came, accepting the load of wool on satisfactory terms. Burnham knew where he could lay hands on a supply of the necessary article to complete the truck-load, and by the following Wednesday the amount required was collected. The project of our stopping at Komati Poort to see if it were possible to blow up the bridge was discussed, but finally vetoed, since most of the supplies required by the Transvaal and Free State had been already collected in the country, and the quantity now coming in by this route was insignificant.[400]

On Wednesday the truck arrived from Whitbank drawn by one of the colliery locomotives. All day it remained on the line immediately outside the office, and from the manager's house we could see Burnham and Adams busily superintending the special manner in which it had to be loaded for our comfort and concealment.

In order that the reader may better understand the kind of place we were to occupy for sixty-three hours, I will describe it in as few words as I can. The bales of wool, weighing each four hundred pounds, were in size about five feet long by two feet and a half broad and the same in depth. There were in all, I think, sixteen to be loaded into the truck. The truck itself was an empty coal-waggon, about eighteen feet long and seven wide, with sides of some three feet in height.

[399] The stone weight is an English and imperial unit of mass equal to 14 pounds (approximately 6.35 kg). Stone (unit)-Wikipedia

[400] PF: Very intrepid of them to consider this option. They were at several coal mines so I'm sure there was dynamite available.

Bales of Wool.[293] "Mr. [John] Howard, Colliery Manager, who hid W. Churchill and other Escaped Officers."[294]

Three bales of wool were laid end on at one side of the truck and three on the other. Above these were laid three parallel rows of three bales each and as there were only two rows on the floor of the truck, occupying a total breadth of five feet, there remained a kind of tunnel down the centre of the truck. Other bales were placed above those already in position, and when all was done there remained a space for us to sit in at the end of the truck three feet by seven. This space was available because the waggon was eighteen feet long, and three bales endwise occupied only fifteen feet. Le Mesurier and I decided that we would turn our trousers inside out, for we had been told that the floor of the truck, although it had been swept out, would be very dirty, and we afterwards found that, in addition to this, there were numerous splinters of wood.

Although it had been impressed upon us that we were almost as "safe as a house" regarding detection, I had an inward feeling that our adventures were not quite over. Once inside the truck, we knew that no line of retreat was open to us, and after the peace and quiet we had enjoyed at the mine, we did not look forward with unmixed joy to the coming journey.

Mr. Howard made elaborate arrangements for victualling our novel carriage for the journey, since it was quite impossible to say how many days we might have to remain in it. I think we had with us—we had to throwaway a good deal unused—a roast duck, chicken, a tin of smoked beef, butter, jam, and plenty of bread, besides some things I have forgotten. For drink we had nine bottles of cold tea, two of water, and one of whisky. With this we could hold out for a week if need be, but Burnham thought that his efforts to induce the chef (as the local station-master is called) to hook us on to the passenger train had prevailed. Bribery, as I have said before, will do most things with a Transvaal Dutchman; but there are many others besides him who have their price.

That night the doctor [Gillespie] came over to see us and bid us good luck on our journey. He and the others had dined at the manager's house, but we broke up early, as we were to be afoot at 4.30 A.M. The two Kaffir boys who kept guard outside the office had been got rid of on some pretext, for they would have seen us climbing into the truck. Mr. Howard lent me his pocket aneroid[401]—which he has since given to me as a memento of an eventful journey,—and as they do not call out the stations on the Delagoa Bay line, and I had no time-table, by means of this instrument, knowing the height above sea-level of each station, I always knew where we were when we stopped. All arrangements for the morrow being ready, we bade our host good night and went to bed.

[401] A barometer. Pressure measurement-Wikipedia

VIII. Over the Border

"We must be free or die."

Punctually at 4.30 A.M., while it was still dark, we walked to the truck destined to carry us and our fortunes. The tarpaulin which covered it was thrown back, and we began storing the provisions and bottles in the numerous interstices between the woolen bales. This took a little care in arranging, and it was 5 A.M. before Le Mesurier and I climbed into the waggon which was to be our dwelling for the next two and a half days. A parting hand-shake all round being concluded, the tarpaulin was made secure by its numerous ropes to the links fixed to the lower edge of the outside of the truck. One of these ropes was tied in such a manner that we might experience little difficulty in undoing it when the time arrived to get out. As we should find it very warm under this covering, one of the bales had been so arranged as to bulge over the edge of the truck, pushing out the tarpaulin and leaving space for the air to enter by.

We lay on the floor of the truck until ten o'clock when Mr. Howard passed by, and asking us if we were all right, said good-bye again. About this time, as I now know, the field-cornet, Theis Pretorius, rode up. Mr. Howard was standing close to the truck and thought he suspected our presence, and felt in consequence most uneasy. They walked together to the house, and Pretorius at once noticed the tablecloth in the dining-room, and inquired who had been dining there the previous night. He was informed that Dr. Gillespie and some men from Brug Spruit had come over to dine and play cards. This seemed to satisfy him, but Mr. Howard did not feel at ease until he left.

Not long after noon we were taken by a colliery engine to the Whitbank station.[402] One or two trains passed us, and we listened anxiously to hear if we were going to be picked up and taken on by them.

Truck outside the office at Whitbank.[295]

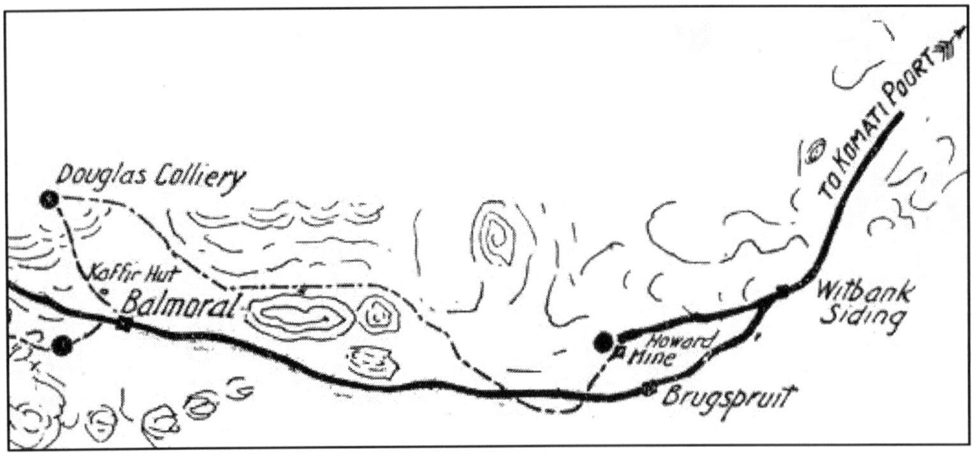

Area around Colliery & Witbank.[296]

[402] Emalahleni, previously named Witbank is a city situated on the Highveld of Mpumalanga, South Africa, within the Emalahleni Local Municipality. The name Witbank is Afrikaans for "white ridge," and is named after a white sandstone outcrop where wagon transport drivers rested. The city is known for its coal-mining in the surrounding region. Witbank-Wikipedia

During the time that our truck was standing in the station, unknown to us within, a somewhat exciting episode was in progress without. Adams and Burnham had gone to Whitbank station in order to see us coupled to the passing train, and what occurred is best told in the words of the former, as related to me.

"When the truck arrived at Whitbank it was halted immediately alongside of an engine which was standing in the station, the driver and stoker of which were both Dutchmen. Thinking that Le Mesurier and you might be lying in the open part of the truck, and might talk or cough, I felt not a little anxious; but to my horror, after a few minutes' suspense, both men got off their engine and talked together, one of them leaning his back against the truck. This was more than I could bear with equanimity, and I strolled over to them and invited them to come with me across the line and have a drink. They declined, stating that they could not leave the engine; and presently the driver took a sheet of paper from his pocket, and flattening it out on the smooth side of the truck, began to write a letter. My anxiety was intense, for the slightest noise would have been heard, and I walked past on the other side and tried by some words which I muttered to give you warning to keep still. While the shunting operations prior to your departure were going on, one of the station employees undid the tarpaulin and looked into the truck."

Le Mesurier and I were in blissful ignorance of what Adams was undergoing, as fortunately we had taken the precaution of hiding ourselves between the wool bales.

At 2.30 a train, evidently from the direction of Pretoria, drew up at the station. Soon after we heard the rumbling sound and clank of an engine coming our way; this was followed by a bump and a jerk forward. We then found ourselves being moved forward, next a halt and a backward push, and we knew that we were on the passenger train.

If we were not detached at Middelburg,[403] we should be at any rate carried to Waterfall Boven,[404] one hundred and thirty miles from the frontier and one hundred and sixty-four from Pretoria. This would be a goodly step towards freedom.

Middelburg was passed, and we were not detached.

I was agreeably surprised that the engine had treated us so politely when it took us up, and not charged us as if to test the buffer-springs. My only previous experience of goods-train travelling had been when escorting ammunition from Dublin to Belfast in 1888. Then, though I travelled in a first-class carriage attached to the train, I was most unmercifully dealt with, more than once during the journey being shot off the seat on which I was lying. The driver of our engine on the present occasion was evidently of a more gentle nature, and treated us, who were lying on the wooden floor with nothing to deaden the shock, more kindly.

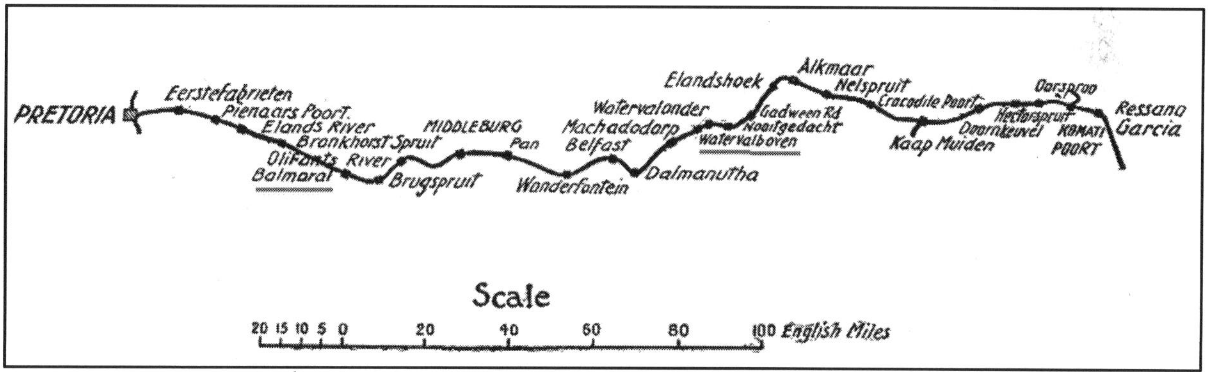

Journey on December 19th. 70 miles, 114 kilometers, from Balmoral/Whitbank to Waterval Boven.

The journey was singularly uninteresting, for until I cut a hole in the tarpaulin the next day, we could see nothing of the country we were passing through. At every thirteen miles or so there was a station, and at every

403 Middelburg is a large farming and industrial town in the South African province of Mpumalanga. Middelburg, Mpumalanga-Wikipedia

404 Waterval Boven (officially known as Emgwenya) is a small town situated on the edge of the Escarpment on the banks of the Elands River above the 75m Elands Falls on the railway line from Pretoria to Maputo in Mpumalanga, South Africa. Hence the name, which is Dutch for "above the waterfall."

It is the sister town of Waterval Onder which is at the base of the Escarpment below the waterfall. Both settlements were established in 1895 because of the building of the Pretoria-Delagoa Bay railway line, built by the Netherlands-South African Railway Company (NZASM). Waterval Boven-Wikipedia station we stopped. Occasionally we heard a few words of English, but far more often it was Dutch. I made it a rule on approaching each station or halting-place to retire into the tunnel, lest anyone should lift up the tarpaulin and look in. This was quite unnecessary, but it was best to be on the safe side.

Now we were puffing and panting up a steep gradient, and again, shutting off steam, we glided smoothly down. During the afternoon's journey we passed the highest point on the line, called Belfast.[405] Here the line attains to an elevation of six thousand four hundred and fifty feet above the sea-level, and in the twenty-seven miles which follow drops nearly two thousand feet to Waterfall Boven. At this station the train stops for the night, the passengers remaining in their carriages or stopping at the diminutive hotel which I believe exists here. We remained in our carriage, and were kept company by a few mosquitoes.

We passed a fairly uncomfortable night, and started again at 6.20 A.M. After leaving the station the line drops at a gradient so steep that the engine here used is provided with a cog-wheel,[406] as on the Righi, Monte Generoso,[407] and other European mountain railways. I believe the line passes through some very fine scenery; but if so it was lost upon us, for we could not see outside our dingy truck.

Cog Wheel Railway.[297]

Waterval-Boven, circa 1901. Note steep incline at top left.[298]

At one point we passed through a tunnel—the only one, I think, on any Transvaal railway. Before reaching Komati Poort,[408] a station on the western side of the Komati river,[409] we thought it advisable to part with nearly all our stores. At this place, which is the last station in the Transvaal, we knew the truck would be searched, if it were to be searched at all. We therefore pitched bottles, tins, &c., out on the line, and reserved only a very small amount, in case we were unable to leave the truck as soon as we expected. We fully thought that, having got so close to the border, we should be in Portuguese territory this day; and on reaching the frontier station it was a bitter disappointment to find ourselves detached and pushed into a siding.

Before reaching the station we had retired into our tunnel, and I, being the longest and thinnest, had entered it first, feet leading. By squeezing myself a good deal and keeping on my side, I could get nearly to the other end of the truck; and Le Mesurier, who followed me, had his head, which was to be covered by a coat in the event of a search, about five feet from the entrance to the hole. We had retired into this uncomfortable region, as I have said, about 2.20 P.M., and when we found ourselves detached we heard the chattering of several Kaffirs just outside the truck. It was very warm and exceedingly uncomfortable, and I was glad to hear our train move off about three o'clock, as our fate would be decided soon one way or the other.

[405] Belfast is a small town in Mpumalanga Province, South Africa. It is situated in the eMakhazeni Local Municipality in the Nkangala District Municipality. Belfast, Mpumalanga-Wikipedia

[406] Rack railway-Wikipedia

[407] The Monte Generoso Railway or Ferrovia Monte Generoso (MG) is a mountain railway line in the Italian speaking canton of Ticino, in south-east Switzerland. The line runs from Capolago, on Lake Lugano, to a terminus near the summit of Monte Generoso. Monte Generoso railway-Wikipedia

[408] Komatipoort is a town situated at the confluence of the Crocodile and Komati Rivers in Mpumalanga province, South Africa. The town is 8 km from the Crocodile Bridge Gate into the Kruger Park, and just 5 km from the Mozambique border and 65 km from the Eswatini border. Komatipoort-Wikipedia

[409] Komati River-Wikipedia

The idea possessed me that the Kaffirs had been placed where they were to guard the truck until such time as it could be searched. Indeed I felt that it was known that there were stowaways within. I therefore prepared for the worst, and drew from my pocket the bag containing a hundred sovereigns[410] which Mr. Howard had kindly given me for my cheque, intending therewith to bribe whoever came. I also tore out of my diary several pages which contained compromising matter and kept them in my hand.

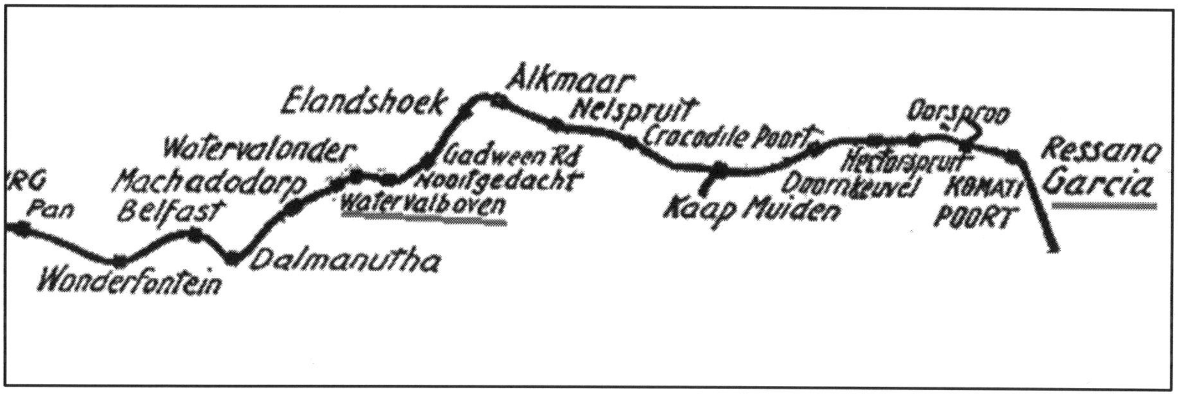

107 miles, 173 km, from Waterval Boven to Ressano Garcia.

A few minutes of suspense passed, and then we heard a gruff voice giving some order to the Kaffirs. The next moment a chill shot through me, and my thoughts returned to Pretoria. We heard the rattle of the links as the ropes of the tarpaulin which were tied to them at our end of the truck were unloosened. The moments that followed defy description: so many thoughts were crowded into one's mind, thoughts of recapture and ruin to all hopes of seeing more of the campaign.

Soon the tarpaulin was lifted up and thrown back over the top of the truck.

What happened neither of us occupants will ever know, for we dared not move to look but the daylight from above and from the end of the tunnel flooded in upon us, and we felt that discovery was unavoidable. The search must have been most perfunctory, though in our excited imaginations it seemed to last an age. Then the tarpaulin was returned to its place, the ropes made fast, and the Kaffirs resumed their chattering. The thought that crossed my mind was that we had been seen, but that the searcher, remarking to himself that there were two of us had deemed it wiser to go for assistance.

In this stifling hole—and there was very little air where I was—we lay till five o'clock. I could bear it no longer, and my limbs were becoming cramped so, catching hold of my companion's foot, I intimated my desire to get into a more open space. We moved forward into the empty part of the truck, and concluded that we had had a marvelous escape from recapture. Shortly afterwards we found the cause of the Kaffir colloquy. There was a water-pipe to one side of the truck, and they were washing when we were first put into the siding. We found that we had to keep absolute silence till after dark, as the path to this stand-pipe was soft, and no footstep could be heard on it. On looking out through the hole I had cut in the tarpaulin, I saw that we were in the station, three lines distant from the platform, and with trucks on both sides of us. About five o'clock several Boers with their women folk came on to the platform and sat there till dark, while a dog ran here and there about the lines.

The next day was Saturday, and we hoped greatly that we might not be left on the wrong side of the river, for on the Sunday following no train might be running. In our expectation of reaching the Portuguese territory on Friday, we had parted with more of our food and drink than was wise.

When morning came a certain amount of shunting seemed to be going on in the station, and for a long time no engine came our way. At length at 9.30 A.M. we felt the shaking caused by an engine moving on the same line of rails as our truck. In considerable anxiety we waited. Then came a very welcome bump, and we were hauled forward and then backed into the station.

At 9.40 the train steamed off, and a minute later we were thundering over the fine bridge which spans the Komati river.[411] We grasped each other's hands, and with difficulty suppressed a shout of exultation. Five

[410] The sovereign is a British gold coin with a nominal value of one pound sterling (£1).

[411] Near the town of Komatipoort, it is joined by the Crocodile River. Crocodile River (Mpumalanga) - Wikipedia The Crocodile tributary rises, as the Elands River, in the Bergendal (1,961 m) near the upper waters of the Komati, and flows eastwards across the highveld, Komati River - Wikipedia.

minutes later, on looking through the tarpaulin, I saw what I had been told to look out for, a white conical pillar. This was the boundary between the Transvaal and Portuguese territory.

Crossing the Komati River.[299]

Komati Poort Railway Bridge. [300]

We were free! Of three of us taken in the Chieveley armoured train, two had escaped. Now that what one had hoped and struggled to attain for so long had at length come, it was difficult to realise it, difficult to believe that one's Anabasis[412] was at an end. A minute more and our goods train stopped at the back of Ressano Garcia[413] station, the first in Portuguese territory.

As we did not want to be seen getting out lest the owner of the wool might thereby become suspected, we decided to remain in the truck till dark. We had finished our water, and it was very warm, and for eight or nine hours we should get none. Many a time during that day I thought of Byron's lines-

"Ring for a valet, bid him quickly bring
Some hock[414] and soda-water, then you'll know
A pleasure worthy Xerxes, the great king."

One would have given a good deal for an iced drink of that description.

It was a long weary day. At 3:15 the Transvaal train rattled into the station, evidently carrying some refugees, for the strains of the National Anthem reached our ears. At half-past six Le Mesurier leant as far out of the truck as he could and with difficulty undid one of the tarpaulin ropes. I handed out to him all that was left of our food and followed him.

While we were engaged in doing this no fewer than six Kaffirs passed the truck, and he had to hide underneath it. Keeping well beyond the platform, we crossed the main line, and after throwing away the remains of our provender, went to a Kaffir kraal which was quite near. Here we found out that the hotel which is in the station was kept by two Englishmen, and we at once proceeded thither. Seeing one sitting outside on the platform, I asked him if he were an Englishman. Receiving a reply in the affirmative, I informed him who we were.

Champagne was clearly indicated, and after suitably toasting the occasion, we sat down to supper, which was continued to a pretty late hour. Before going to bed we remembered that we had left the tarpaulin rope untied. Returning to the truck we made all fast, and I removed as a memento one of the two tickets describing the inanimate portion of its contents.

[412] Greek, a going up, an ascent, also a military march up-country.

[413] Ressano Garcia is a small town in the Maputo Province, Mozambique. The town is adjacent to South Africa. Ressano Garcia-Wikipedia

[414] A Rhenish wine, of a light yellow color, either sparkling or still, from the Hochheim region.

N.Z.A.S.M. Model 106.

WITBANK

Van

Naar Lourenco Marques

Afzender Burnham

Geadresseerde

Inhoud 17 balen wol

Datum 29 Trein

Wagen N°.

Truck Label. “I took the label off the truck as a memento.”[301] Note: “Afzender: Burnham.” Had they been discovered Charles Burnham would have been the first person the police would have contacted.

There were no Transvaal agents here, we were told, and no one to bother us; but it would be best to hide the fact of our being escaped prisoners until we reached Delagoa Bay.

IX. Men Must Endure

"Their going hence even as their coming hither."[415]

I would not have exchanged my feelings on waking on Sunday, 1st of April, for a very great deal. Had we really escaped? Were we not under some deception, born of this day—All Fools'-day—the New Year of the British army? No! Here we could behave like ordinary mortals; there was no longer need to whisper, to emulate the habits of some wild animals, crouching in their dens by day and only coming forth at night.

Strange, that the difference of being on one side or the other of a stone pillar should have brought this about!

But one could not lie all day indulging in this entrancing reverie—other things had to be thought of. Our landlord provided us with a guide and some clean garments, and despatched us to the river.

When lying all night in the truck at Komati Poort, I had been tempted to leave it, and, swimming the river, settle at once the question of freedom. Fortunately I had not done so, for we were told that it was infested with crocodiles, which, when they got the chance, carried off any unfortunate Kaffir who came their way. It was not a pleasant-looking river—muddy, rocky, and fast-flowing—and we did not stray far from the edge.

The village of Ressano Garcia—if it can be so called, for there are only half-a-dozen houses—is a very quiet, unhealthy spot, and, except for one or two people who boarded at the hotel, we saw no one till the up-train from Lorenco Marques arrived about noon. There were many inquiries as to who we were, and one man on his way to the Transvaal, who spoke to me, asked me if I were not an escaped prisoner?

Nearly six months later, after we had taken Lydenberg in the North-east Transvaal, I was engaged in buying stores for the officers' mess, of my regiment in that town. While crossing the market square I came across a man whose face I knew, and accosting him I asked him if he remembered to have ever seen me before. He expressed ignorance; but when I reminded him that he had spoken to me at Ressano Garcia, thinking I was an escaped prisoner, and that I had a beard then, he recollected the circumstance at once. He told me that his reason for suspecting that I was one of the escaped officers arose from his hearing people at Lorenco Marques talking about our expected arrival there from the Transvaal.

Before we left Ressano Garcia we heard that several of the men had tried to escape from Waterfall, [Waterval POW Camp] and that one had actually reached Komati Poort and had been passed by the custom-house officer there. As he was getting into the train to continue his journey a field-cornet noticed that he was wearing "ammunition boots," called him back, and asked him how he had been employed at Johannesburg, whence he said he had come. On replying that he had worked on the Village mine, he was asked who was the manager, and his failure to give a satisfactory reply resulted in his recapture when on the very verge of freedom.

At three o'clock our train arrived. And shortly afterwards we left, travelling at last in comfort. Our carriage, which had a corridor, was fairly crowded. and soon after the train was in motion a party of Dutchmen in another compartment burst forth into their national Volkslied.[416] This was more than British flesh and blood could stand. We immediately rose and speedily drowned their voices with our own. Scarcely had we got through the opening words when Uitlander[417] after Uitlander, their faces beaming at hearing this familiar tune, crowded into our compartment and into the passage, and joined us. The Dutchmen wisely ceased, and for some minutes I thought it would end in their leaving the train by an exit other than the door. After our vocal efforts had come to a close with the usual cheers, our neighbours began "La Marseillaise," and we constrained ourselves to permit them this safety-valve, although its words, if referring to the British, were far from inoffensive.

[415] A line in Shakespeare's King Lear: "Men must endure their going hence, even as their coming hither: Ripeness is all." It speaks to being ready for inevitable death, but also, as children's book author Maurice Sendak explains, savoring your life and making the most of every moment until the end.

[416] *Volkslied* (literally: folk song), is a genre of popular songs in German which are traditionally sung. Volkslied-Wikipedia

[417] *Uitlander*, Afrikaans for "foreigner" (lit. "outlander"), was a foreign (mainly British) migrant worker during the Witwatersrand Gold Rush in the independent Transvaal Republic following the discovery of gold in 1886. The limited rights granted this group in the independent Boer Republics was one of the contributing factors behind the Second Boer War. Uitlander-Wikipedia

Transvaal National Anthem

Dost know that folk of the heroes' might, and yet so long cast down?
It gives its blood and weal outright, great liberty to crown.
Come Burghers, set your banners streaming,
Our sufferings are past; proclaim your glorious heroes gleaming:
We're free-born folk at last!
We're free-born folk, we're free-born folk, we're free-born folk at last!

Dost know that land of dwellers rare
And yet so wondrous bright,
Where Nature's marvels past compare
Are lavishly in sight?
Transvaalers, loud your joy be singing,
There where our brave men stand,
Where festive volleys loud are ringing,
There is our Motherland!
That glorious land,
That glorious land,
There is, that is our Motherland!

Dost know that State still but a child
In the world's comity,
But yet by mighty Britain mild
Erstwhile declared free?
Transvaalers, noble was your fighting,
Though held in scorn so great;
But God ,Who's given us our righting,
Be praised for our free State!
Praise be to God,
Praise be to God,
Praise be to God for land and State! [418]

South African Republic (Transvaal) Flag.[302] Vierkleur, (Top to Bottom) Red, White and Blue with Vertical Green Stripe.

[418] Heroes of the Boer War. Review of Reviews, 1903. See Also: National Anthem of the Transvaal - Wikipedia.

April 1st, 1900; 50 miles, 80 km, from Ressano Garcia to Lourenco Marques. Google Image © 2023 Image Landsat/Copernicus.

The evening was approaching as we neared the end of our journey, and from the windows we could see the calm blue sea which runs far inland. Like Xenophon[419] and his Greeks of old, we gazed upon it with delight, and realised with what heartfelt thanks they had broken into their cry, "Thalassa! Thalassa!"[420]

Soon the station was reached, crowded with those who had come to see more refugees arrive. Passing quickly through them, for we were troubled with no baggage, we drove to the Cardoza Hotel, picturesquely situated on the high ground overlooking the bay.[421]

Lorenco Marques from the east.[303]

[419] Xenophon of Athens (430– 354 BC) was an Athenian-born military leader, philosopher, and historian. At the age of 30, Xenophon was elected a commander of one of the biggest Greek mercenary armies, the Ten Thousand that marched on and came close to capturing Babylon in 401 BC. Xenophon-Wikipedia

[420] Greek primordial goddess of the sea.

[421] Frankland: *April 3rd*—Hurrah! the papers this evening report the safe arrival of Haldane, Le Mesurier and Brockie at Lourenço Marques, having travelled through Swaziland.

Fort on the Left. "Delagoa Bay has been described as the 'Key of South Africa,' but we allowed it to slip through our hands."[304]

We had intended to keep our secret; but (how I did not know) every one seemed to be already in possession of it, and one of the British consul's assistants had met us at the station. That evening I gave Mr. Douglas, the courteous correspondent of the 'Times,' a brief account of our adventures, omitting everything about our having hidden under the floor of the Model School. My reason for doing so was that Le Mesurier, Brockie, and I had, in the interests of those who had helped us and fed us and remained behind, bound ourselves to say nothing about that part of our story until the end of the war.

Later in the evening I received a request from Mr. Fritz Pincus, Reuter's correspondent (who with his colleague, Mr. Minzesheimer, an American gentleman, treated us most hospitably), that I would kindly call on him. I did so, and gave him a brief account similar to that I had given to the other correspondent. After I had finished he remarked to me, "How long did you remain hidden under the floor? for you have told me nothing about that." I was dumfounded. Here was a man in Lorenco Marques in possession of what we were sworn to keep secret. Questioning him, I found out how it had reached his ears, and indeed those of most of the inhabitants of Lorenco Marques, including Mr. Carnegie Ross, the British consul. He did not then tell me who was his informant, but only that it was some one who had left Pretoria after we had hidden. I felt sure that no one had left who knew me except three British officers released on parole, and they had been ignorant that we were hidden, and even had they not been, they were to be trusted.

Here was a puzzle that I could not solve. Next morning I gave the 'Times' correspondent that portion of my tale which I had omitted the previous evening. Concealment was no longer of any use, as Lorenco Marques swarms with Transvaal agents, who doubtless long ago had reported to their Government.

One thing Le Mesurier and I felt thankful for, which was, that considering it must have been known in the Transvaal that we had not escaped on the 26th of February, but only after the prisoners were removed, on the 16th of March, we were remarkably fortunate to have escaped at all. After making some inquiries, I succeeded in tracing the matter to its source.

It seemed that Sergeant-Major Brockie, on reaching Kaapmuiden, had found that the person to whom he was consigned had just died. He got employment at the station bar at that place, and, strangely enough, his and our descriptions were posted up there in the view of all. A few days after, a clergyman, whom I met, saw him and got part of his story from him; but, unknown to us, a certain person, not an officer, had been released on parole from Pretoria.

Here was the offender, though not primarily responsible. This person had seen Brockie at the Kaapmuiden bar, who told him our adventures from beginning to end. On reaching Lorenco Marques he had indiscreetly parted with what he had been told, and so the facts of our escape got abroad. We felt greatly annoyed on account of the officers who had helped us at the Model School, and I have no doubt that until they read this they will naturally think we betrayed them.

One day while we were walking through the streets of the town, an English-speaking man came up to me and almost without preliminaries said, "That was a cute dodge of Patterson's, sending you news across the street at Pretoria." I looked blankly at him, and he appeared a good deal taken aback; and on my pointing out to him that I was not in the habit of talking to utter strangers in the public street, he replied that he had something of an important nature to communicate to me. I gave him my address and told him an hour at which he could call, but, as I expected, he did not appear, and I have since thought that he was a Boer spy, and was merely trying to "pump" me. We had just missed a steamer to Durban, so we had to wait for the Konig[422] on Friday, April 6. During the days we remained in Portuguese territory we spent a great deal of time on board the British cruisers, five of which were in harbour, and from their officers we received a very warm welcome.

On board the Forte[423] Le Mesurier found an old schoolfellow, and I discovered that her commander, Captain C. H. Dundas, and I were Scotch cousins.

HMS *Forte.*[305]

SS Königin Luise.[306]

Post Escape Pictures

Captain Haldane and Lieut. Le Mesurier on reaching Lorenco Marques.[307] And: Captain Haldane and Lieut. Le Mesurier the day after their arrival at Lorenco Marques.[308]

[422] SS Königin Luise ("Queen Louise") was a Barbarossa-class ocean liner built in 1896 by Vulcan Shipbuilding Corp. of Stettin, Germany, for the North German Lloyd line of Bremen. She served on the company's Australian, Far East, and North Atlantic routes for nearly two decades. SS Königin Luise (1896) - Wikipedia

[423] HMS *Forte* was an Astraea class cruiser of the Royal Navy launched on 9 December 1893. She was constructed under the Naval Defence Act of 1889 along with several other Astraea class cruisers. HMS Forte (1893)-Wikipedia

CAPT. HALDANE, D.S.O. LIEUT. LE MESURIER.

Sketch of both men.[424]

On Friday night we left for Durban, there being on board Sir Ellis Ashmead-Bartlett,[425] full of his Swaziland[426] trip, and Sergeant-Major Brockie, who had arrived the previous evening.

The last I had heard of him [Brockie] was that he was employed at Kaapmuiden[427]. There he had heard that a general exodus of British subjects was about to take place from the Transvaal, and his friends at Whitbank had procured for him a passport as far as Komati Poort. Travelling there with a crowd of refugees, he had bribed one of the guard at the frontier bridge to let him cross over to get a drink at Ressano Garcia, as no liquor, under martial law, was permitted to be sold in the Transvaal. As soon as he was beyond the river he made for the Portuguese frontier station, and, rejoining the train there, arrived in safety at Lorenco Marques.

Le Mesurier and I parted at Pietermaritzburg—he to go round to the Free State with General Hunter's Division,[428] I to rejoin my regiment at Ladysmith.[429] The trying experience we had gone through gave him enteric fever,[430] and after being invalided to England and remaining there for some time, he returned to South Africa only to be sent home again shortly afterwards with malarial fever. Brockie fared better, but he too was obliged to go to hospital. I hear that he has recently been given a commission, and is now recruiting-officer in Johannesburg.

As regards myself, during our tramp to the coal-mines I must have imbibed some fever-germs, for I have had two attacks of malarial fever. We have both obtained our desire—to get back to the front to duty; and as I write these words, and the end of the campaign seems near, I cannot help regretting deeply the wasted months of captivity, and try to console myself with the well-known couplet—

"Be cheerful; wipe thine eyes; some falls are means the happier to arise."[431]

[424] Haldane and Le Mesurier as depicted in the Exmouth Journal, 7th April 1900 Per Neville Chamberlain: Anglo Boer War - Armored Train Ambush 11-15-99 Chuchill & Haldane - Page 15 - Boer War Forum.

[425] Sir Ellis Ashmead-Bartlett (24 August 1849 – 18 January 1902) was an American-born British Conservative politician who sat in the House of Commons from 1880 to 1902 Ellis Ashmead-Bartlett (politician)-Wikipedia

[426] Eswatini and formerly and still commonly known in English as Swaziland is a landlocked country in Southern Africa. It is bordered by Mozambique to its northeast and South Africa to its north, west, and south.. Eswatini-Wikipedia

[427] Kaapmuiden (Cape Mouth) is a small farming town situated at the confluence of the Kaap and Crocodile Rivers in Mpumalanga, South Africa. Kaapmuiden - Wikipedia.

[428] Sir Archibald Hunter. Archibald Hunter - Wikipedia

[429] Gordon Highlanders Gordon Highlanders - Wikipedia

[430] Typhoid fever-Wikipedia

[431] From Shakespeare: *Cymbeline*. As soldiers can. Be cheerful; wipe thine eyes Some falls are means the happier to arise.

Durban from the Bluff.[309]

Durban, St. Pauls Wharf in Durban Harbour.[310]

The saga relating to both Winston's and Alymer's successful escapes would continue to generate ink and contention for decades afterwards. See infra for Aylmer's 1931 letter to Lord Knutsford.

At the time, Winston, with his usual generosity and thoughfulness both congratulated Aylmer, offered help, and gave advice to his esteemed friend and fighting comrade.

Congratulations Telegram.

April 9th, 1900 [432] Norval's Pont

My dear Haldane,

My heartiest congratulations on your wonderful exploit which will mark you as a man of daring, endurance and resource among all soldiers. I am delighted to think you are safe. I feared they had murdered you in the veldt.

I hear that at the court of Inquiry on [regarding the events of] 15 Nov [Colonel] Long [433] said that he had told you not to go beyond Frere and in consequence they may ask you for explanations. If you want any evidence, write to me. I do not think however that they will make any trouble.

I should advise you to send a copy of your report to Nicholson[434] as well as to Buller. I am not in favour here and it would help me as well as your own case.

I will talk to Nicholson about the fight and explain matters. The incident has become rather famous in the story of the war.

Write to me at Bloemfontein.

Yours very sincerely

Winston S. Churchill

Colonel Long at the Mount Nelson Hotel, Cape Town.[311]

[432] OBC Vol 1 Part 2 p. 1167-1168.

[433] C. J. Long: Colonel Long (Royal Horse Artillery) is a gunner of much distinguished service, and it is impossible not to admire the gallantry that led him into the crucial mistake made in the handling of the guns at Colenso on December 15. Upon that unfortunate occasion Colonel Long was dangerously wounded. It was intended that he should come into action covered by the 6th Brigade, but the 14th and 66th Batteries were led into the open, where they were subjected to a fire in which it was impossible for men to live. Anglo Boer War - Name search

[434] Sir William Nicholson was director of transport at Roberts' headquarters.

Epilogue

Frederick Le Mesurier[435] will continue to serve in the 2nd Battalion of the Royal Dublin Fusiliers where he rose to the rank of Captain. During the First World War, he died in 1915 at the second battle of Ypres.

In Memory Of Captain Frederick Neil Le Mesurier.[436] Sketch of Le Mesurier.[312]
2nd Bn., Royal Dublin Fusiliers who died on 25 April 1915
Remembered with Honour - Ypres (Menin Gate) Memorial - Panel 44 and 46.

As noted above, Aylmer Haldane will serve in the military until his retirement in 1925.

Alfred Adam Brockie was killed in a mining accident, by a fall of rock in one of the Johannesburg mines[437] sometime before 1904. His widow, Annie, wrote to Churchill in 1902 to ask for assistance and he sent her a courteous reply and a draft for £10.[438]

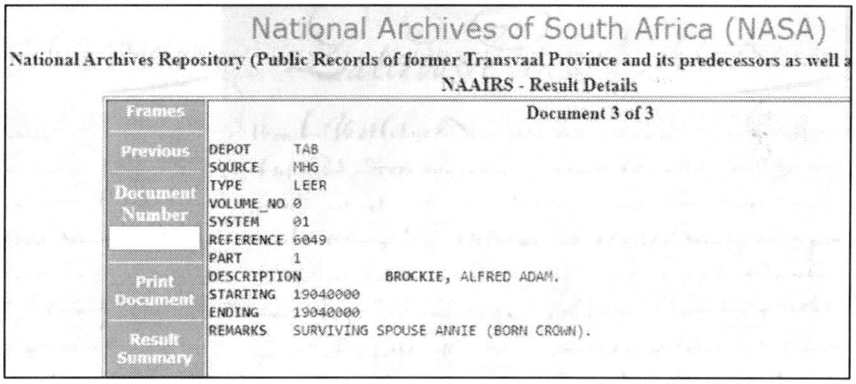
National Archives of South Africa (NASA)

National Archives Repository (Public Records of former Transvaal Province and its predecessors as well a

NAAIRS - Result Details

Frames | Previous | Document Number | Print Document | Result Summary

Document 3 of 3

DEPOT	TAB
SOURCE	MHG
TYPE	LEER
VOLUME_NO	0
SYSTEM	01
REFERENCE	6049
PART	1
DESCRIPTION	BROCKIE, ALFRED ADAM.
STARTING	19040000
ENDING	19040000
REMARKS	SURVIVING SPOUSE ANNIE (BORN CROWN).

Adam Brockie.[439] Certificate Search.[440]

[435] Unable to obtain a biography of Frederick Neil Le Mesurier that was published by the Bushey Museum. https://busheymuseum.org/

[436] Frederick Neil Le Mesurier Captain QSA (4) QSA known to exist Source: List of QSAs with the clasp Talana from the Talana book Royal Dublin Fusiliers. MID LG: 10 September 1901, page: 5951. Source: Field Marshal Roberts. 4 September 1901. Re: General mentions Anglo Boer War - Name search Page 5927 | Issue 27353, 10 September 1901 | London Gazette | The Gazette.
He died at Ypres in 1915. See: Frederick Neil Le Mesurier Captain 2nd Bn. Royal Dublin Fusiliers Died 25 April 1915 Ypres (Menin Gate) Memorial Panel 44 and 46. Belgium The Commonwealth War Graves Commission | CWGC | CWGC PF: He may be related to Gerald John Le Mesurier DFC was a South African flying ace of World War II, credited with 3 'kills.' See Gerald Le Mesurier - Wikipedia

[437] See Haldane letter to Viscount Knutsford infra.

[438] Sandys p. 111.

Adrian Hofmeyr's Account of Brockie's Adventure.

Let me tell what happened to [Brockie] the latter before I close this chapter [Chapter 34] [441]—the story of the former [Haldane and Le Mesurier] has been told in several papers already. After losing his companions Brockie made his way out alone to the Portuguese border. The last station on the Transvaal side is Komati Poort, the first on the Portuguese side is Ressano Garcia, and in between is the bridge over the Komati River. The two stations are little more than a mile apart, the river being the boundary between the two countries. Brockie eventually arrived at the Komati Poort Station, looking for all the world like a dilapidated tramp. He felt convinced that only bounce and the proverbial tramp's cheek could now help him out of as tight a corner as one could well get into. He had a fund of both, fortunately, and so marches up boldly to the manager of the refreshment room, and playing the *role* of tramp in the uttermost stages of destitution begs to be taken on as barman.

The manager has compassion on him, and soon our hero is installed behind the bar, where he does the honours worthily. Inside the bar-room there is a Government notice on the wall—it is a description of the three fugitive officers, and an offer of reward for anyone recapturing them. There are men reading this notice and discussing it. Brockie joins the group and launches forth, as a good and loyal *burgher* of the State, in fierce invective against the English in general and the escaped officers in particular! He tells all and sundry what he would do in case he caught them, and promises drinks all round when he takes the reward. Little did the others surmise how near "the reward" they were standing themselves, and how easily they could have earned it.

For a few days Brockie played his part of barman to perfection, selling wines and beer, the sale of spirits being prohibited in the Transvaal during the war. Then, all of a sudden, the manager noticed that his model barman was seized with ague, a precursor of fever—and Komati Poort is a frightful fever hole. What now? With chattering teeth Brockie himself makes a suggestion. "Give me some money," said he, "and I'll go over to the Portuguese side to buy whisky." "But the bridge is guarded," said the manager. "I'll square the sentry," answered Brockie. And off he crawls didn't he have the ague terribly! Moaning and miserable he approached the sentry and begs to be allowed over the bridge, just to get a little whisky for himself and his manager, who was seized with ague even worse than himself. He begged so piteously and promised the sentry such a grand share for himself, that with an injunction to return at once he was allowed to cross over.

Brockie, be it known, had timed his ague and expedition very nicely—just when he knew that a train on the other side was ready to start for Lorenco Marques. And so over the bridge he crawled, and over the intervening ground, and when the sentry was out of sight, he ran, fresh and strong and healthy, till he reached the station. Never was man so quickly relieved of ague—it was a marvelous and quick cure. It must have affected Brockie's mind, however, for instead of returning at once and bringing the thirsty, longing sentry the whisky he had promised him, this absent-minded beggar of a fugitive took a seat in the train, and was whirled away to Delagoa Bay! How could he be so forgetful!

They say that oft in the silent midnight watches a sentry's impatient steps can be heard on the Komati bridge, and in between the curses deep and heavy, over the waters of the river ever floats the sad refrain of the song that sentry sings: "Oh, where and oh, where has my whisky-bottle gone?" And the crocodiles crawl out on the banks and each one wipes a tear from the other's eye. Thus Brockie escaped.

[439] Adam Brockie is listed in four places at Anglo Boer War - Name search My supposition is that they all refer to the same individual. He may have been promoted because Haldane states that he is tasked as a recruiting officer in Johannesburg after his escape.

Brockie Adam No. 489 Sergeant QSA (5) Source: List of QSAs with the clasp Elandslaagte Imperial Light Horse

Brockie A. Lieutenant QSA (5) Source: List of QSAs with the clasp Elandslaagte, Johannesburg Mounted Rifles

Brockie A. Lieutenant QSA (5) CC Eland DoL Tr SA01. QSA (5). Officially renamed. Sotheby October 1987 Provisional list of recipients Source: Ladysmith Siege Account and Medal Roll Johannesburg Mounted Rifles

Brockie Adam Corporal BSACM Mashonaland 1897 (0). Source: BSACM rolls British South Africa Police Roll

The South African National Archives website shows evidence that Brockie died in 1904 as there is a deceased estate file for him. He died in the Transvaal. http://www.national.archsrch.gov.za/sm300cv/smws/sm30ddf0?20210926024800646 84B63&DN=00000003 Image source: Vera Gallony.

[440] http://www.national.archsrch.gov.za/sm300cv/smws/sm30ddf0?20231109182436 85D79463&DN=00000002

[441] Adrian Hofmeyr The Story of My Captivity During the Transvaal War 1899-1900, Publisher Edward Arnold 37, Bedford Street, Strand, London, 1900 Printed by: Unwin Brothers, The Gresham Press, Woking And London The Story of My Captivity During the Transvaal War 1899-1900 - Google Books.

Haldane Forwards Telegram

October 8th 1902.[442] Intelligence Division, Winchester House

St James's Square

My dear Churchill,

I enclose two telegrams which will be of some interest to you, which I came across in going through a pile of papers recently received from S. Africa. I have had them translated.

They bear out what I have always maintained and stated in my book,[443] that there was reason to doubt as to the sincerity of the Boers in their expressed intention to release you.

The one from Smuts shows that at any rate he had heard no word of an intention to release you, and after reading them it is improbable that anyone would still maintain that you were purposely allowed to escape. If at any time you refer to them, it will not be advisable to say that they came into your hands through the Intelligence Division, as I have thought it unnecessary to ask permission to send them to you, and I don't want to get into trouble for doing so!

Very sincerely yours

A. Haldane

Smuts' Question

Jan Christian Smuts to State Attorney's Office, Pretoria

16 December 1899.[444] General Burger's Headquarters, Ladysmith

What truth is there in the rumour that Churchill has escaped but has been caught again?[445]

J. C. Smuts

Second Telegram

[December 12th, [sic 10th] 1899][446]

From General Joubert, Volksrust, To Acting Commandant General, Pretoria. Re: your telegram of today. I agree to the exchange proposed, but am resolved against the exchange of Churchill or of the conceited officer of General White's staff [Walter Adye.][447] captured on 30th October; or of the officers who were last captured in the armoured train beyond Colenso, or of the officers and men who since their capture have expressed themselves improperly against the Government, whether in speech or writing. . . .

Compiler's/Editor's Note: The following image gave me the first indication that this telegram existed. This partial page was scanned as the last page of the Penn State copy of Aylmer Haldane's 'How We Escaped From Pretoria.' It is dated December 10th, two days before Churchill escaped. The comments penned in the margins throughout the book were evidently written by someone who knew the subject intimately.

[442] OBC Vol 1 Part 2 p. 1119.

[443] How we Escaped From Pretoria, originally published in Blackwoods Magazine, and first published in book form in 1900.

[444] South African Government Archives and OBC Vol 1 Part 2 p. 1091.

[445] Aylmer Haldane will later send a copy of this telegram to Winston as confirmation that the Boers had no intention of releasing him. See infra.

[446] "On the 13th July, 1903 a copy of a Boer telegram, which came into the hands of the British after Pretoria was occupied by Lord Roberts in 1900, was sent me by an officer, on the staff of the military historian of the war, whose duty it was to examine captured Boer correspondence. (Aylmer Haldane Papers) OBC Vol 1 Part 2 p. 1101.

[447] Adye Walter Major MID LG: 8 February 1901, page: 917. Source: Lieutenant General White. 2 December 1899. Royal Irish Rifles. Adye was one of 954 officers and men to be taken prisoner that day. It is said that the officers of the Royal Irish Rifles were 'so exasperated at the exhibition of the white flag that they set to work and smashed their swords rather than give them up.' Adye was mentioned in despatches and, on being released, served as DAAG at Army HQ from July 1900 until February 1904...He served briefly as Deputy Assistant Inspector of Remounts, Eastern Command, from 1914 until his death in September 1915. Anglo Boer War - Name search

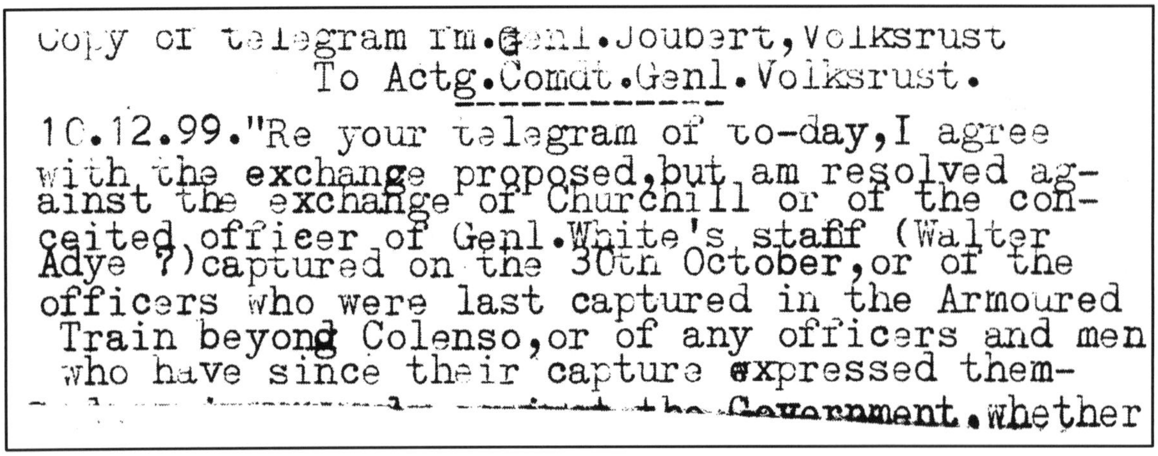

Copy of telegram fm.Genl.Joubert,Volksrust
To Actg.Comdt.Genl.Volksrust.

10.12.99."Re your telegram of to-day,I agree with the exchange proposed,but am resolved against the exchange of Churchill or of the conceited officer of Genl.White's staff (Walter Adye ?)captured on the 30th October,or of the officers who were last captured in the Armoured Train beyond Colenso,or of any officers and men who have since their capture expressed them- [illegible] the Government.whether

Telegram on last page, 309, of Penn State copy.[448]

PF: Note this list of prisoners from the 'Gram'[313] Adye, W is noted as released under R.C.S. 3,1900. [Red Cross Service?] He is noted being in the hospital.[449] He had given his parole and was not able to help Haldane & Co. to help cut the electricity or otherwise help with their escape.

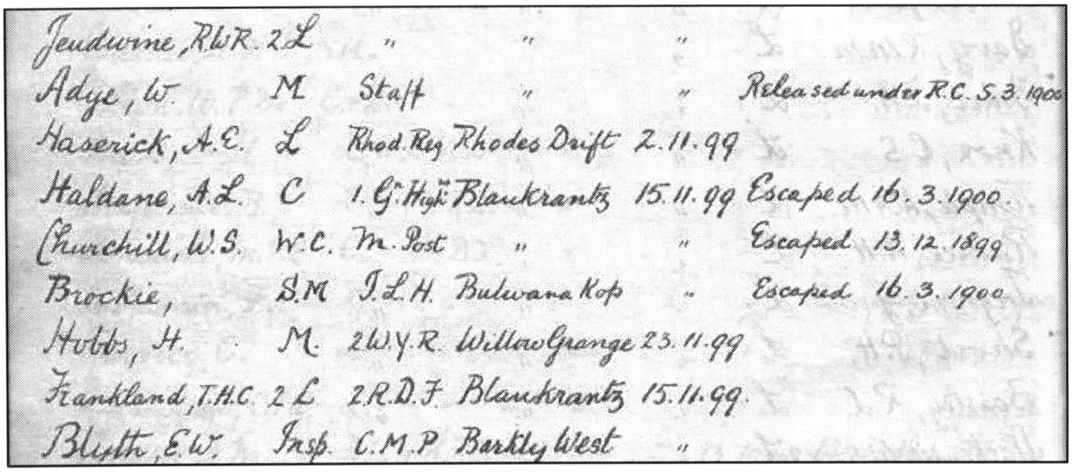

Jeudwine, R.W.R.	2L	"	"	"	
Adye, W.	M	Staff	"	"	Released under R.C.S. 3. 1900
Haserick, A.E.	L	Rhod. Reg	Rhodes Drift	2.11.99	
Haldane, A.L.	C	1. G. High.	Blaukrantz	15.11.99	Escaped 16.3.1900.
Churchill, W.S.	W.C.	M. Post	"	"	Escaped 13.12.1899
Brockie,	S.M	I.L.H.	Bulwana Kop	"	Escaped 16.3.1900
Hobbs, H.	M.	2 W.Y.R.	Willow Grange	23.11.99	
Frankland, T.H.C.	2L	2 R.D.F.	Blaukrantz	15.11.99.	
Blyth, E.W.	Insp.	C.M.P.	Barkly West	"	

List of prisoners and their particulars from the 'Gram.'[314]

Haldane quote from, supra: "I have so often since heard suggestions thrown out that Churchill was allowed to go, and that the Boers in any case intended to release him, that I feel it incumbent upon me to state that, in my opinion, nothing was further from their intentions. It requires no intricate process of induction to conclude that such was the case when one recalls the extraordinary hue and cry which followed his departure, and the elaborate arrangements made to prevent a repetition of the occurrence."

[448] Sir James Aylmer Lowthorpe Haldane Aylmer Haldane, How We Escaped from Pretoria. Edinburgh: W. Blackwood. 1901
University of Pennsylvania Copy: #11 - How we escaped from Pretoria / by Aylmer Haldane. - Full View | HathiTrust Digital Library.

[449] Haldane, supra. The scheme upon which we now based our hopes was to induce the English electrician who was in charge of the lighting of the town, and also of the Rontgen ray apparatus in the hospital next door, to extinguish the lights of the school and of that district of Pretoria in which it was situated, for half an hour on some dark night. We inquired of him through Major Adye, who was in hospital, if he would help without divulging anything. The reply came that, being on parole, he regretted he could not.

Rosslyn Twice Captured

"Mr. De Souza left me after telling me that Winston Churchill's warrant of release was in their hands the night he escaped, and that they were glad to be rid of him. "He talked so much," he added, "that we all avoided him" I could not restrain a smile; but his escape and Mr. Hellawell's[450] unsuccessful attempt have certainly made matters "difficult" for war correspondents and officers too, since these have been removed to more guarded quarters ever since three of their number escaped about a month ago." [451]

James Erskine, 5th Earl Rosslyn.[452] April 18th [1900] just before primrose day.[453] [454]

Blackwoods & Continuing Saga

This offhand comment by Rosslyn, contained within a 90,000 word book created a storm of controversy that lasted over a decade.

> "Talking of escapes, the story of Winston Churchill's was already known to me through his own description, and the officers were most interested to hear about it, though, I am afraid, he was not a persona grata[455] with his fellow prisoners, and, as far as I could ascertain, did not play quite fairly with the others who concocted the plot, but, according to them, followed the principle of sauve qui peut[456] rather than "shoulder to shoulder." [457]

[450] Rosslyn mentions: on page 292 of the book and 365 of the PDF: Mr. Hellawell, another Daily Mail' correspondent, is coming here from Waterval, we shall probably start a better mess a trois. Or: Possibly, A. Hellawell Source: QSA and KSA medal rolls. Western Province Mounted Rifles.

[451] Earl of Rosslyn, James St Clair-Erskine, 5th Earl of Rosslyn Twice Captured: A Record of Adventure During the Boer War. United Kingdom, William Blackwood and Sons, 1900. Page 295-296 of the book and 368 of the scan. See: Twice Captured - Google Books and: James St Clair-Erskine, 5th Earl of Rosslyn - Wikipedia File:James Francis Harry St Clair-Erskine, 5th Earl Rosslyn,.jpg - Wikimedia Commons.

[452] James St Clair-Erskine, 5th Earl of Rosslyn: File:James Francis Harry St Clair-Erskine, 5th Earl Rosslyn,.jpg - Wikimedia Commons.

[453] Primrose Day marked the anniversary of the death of the British statesman and prime minister Benjamin Disraeli, 1st Earl of Beaconsfield, on 19 April 1881. The day was marked each year into the 1920s, with arrangements of primroses left at Disraeli's tomb at St Michael and All Angels Church, Hughenden and his statue in Parliament Square and many supporters wearing primroses as buttonholes, garlands, and hat decorations. Primrose Day - Wikipedia

[454] The Primrose League was formed in 1883 to continue the legacy of Disraeli by spreading and popularising traditional Conservative ideals, including the British monarchy, the British Empire, the Anglican Church, and free enterprise. Promoted by Lord Randolph Churchill, who advocated Tory Democracy, the League grew rapidly, with over half a million members by 1887, a million by 1901 and over 2 million members by 1910, roughly equally women and men. The Primrose League faded with universal suffrage after the First World War when the Conservative Party pursued direct mass party membership, and was formally wound up in 2004.

[455] A person who is welcome or acceptable.

[456] Every man for himself - more literally "save yourself if you can"; most literally "may he save himself, whoever can."

[457] Page 335 of the book and page 412 of the PDF scan of Rosslyn's 'Twice Captured.'

Letter to Haldane

October 5th, 1900.[458] 105 Mount Street

Winston sends an 'extract' from Rosslyn's book containing the quote above and possibly others.

He finds the aspersions made about him so offensive that he placed the matter in the hands of his attorney Sir George Lewis.

In a letter of ~600 words he goes on to explain:

Exterior of 105 Mount Street, London.[315]

Entrance 105 Mount Street, London.[316]

My dear Haldane,

I have wanted for a long time to write to you upon this subject but I am so pressed and hunted and have but little time.[459]

He then goes on to describe his part in the escape plan and aver that Haldane was in favour of waiting 'night after night' for the best opportunity. Haldane being his superior officer as a Captain in the Gordon Highlanders, had every right to direct his Lieutenant enlisted in the affair as to the order of operations, but I digress.

Winston, opines that he was in favour of making a desperate venture of it at any opportunity.

So far this is all true to form.

One thing divulged is that prior to the attempt on December 11th the conspirators had enjoyed a bottle of burgundy before their abortive attempt was to begin.

After the attempt on the 11th was not successful; Winston brings Brockie into the picture alleging a conversation with him in which Winston asserts he could have surmounted the wall undetected on the 11th.

Brockie replied "why the hell didn't you?"

To which I answered, 'where should I have been if I had got over and you had found it impossible to follow' and he said to me distinctly 'if you will get over, we will follow all right, don't be afraid of that!' [460]

Turning to the 12th, he alleges factually that when he returned to the latrine a second time that he had no inkling that the plan had been abandoned for the evening. He had told Haldane to "Follow Presently" as per Haldane's own diary.[461] [462]

[458] Please see the original for the full text: OBC Vol 1 Part 2 p. 1106-1108.

[459] During the Khaki Election of 1900 he won his seat in Oldham early on during the election held between 26 September and 24 October 1900. He will then tour other constituencies on behalf of Conservative candidates to speak at rallies that had not yet voted.

[460] Note by Haldane: To have done so would have been impossible while we were together in the latrine. That, we both agreed upon. What Churchill meant by saying 'he could have got away' is incomprehensible. See OBC cite supra.

[461] [December 12th] .... We walked back to the verandah where Brockie was sitting. **He jeered and said that we were afraid and that he could go any night**. [Haldane] said to him 'Then go and see for yourself.' He crossed over to the latrine and W.C., who I suppose thought that Brockie might go and leave us, said to me "I am going across, **follow me presently**." This clearly showed that he fully understood that he and I were to go together. Aylmer Haldane Papers pp. 124-5 and OBC Vol 1 Part 2 p. 1087-1088

[462] In 1935 Haldane mis-remembers his diary entry stating:

New information disclosed was that it was a man named Southey (R.I.F.) [463] who spoke to Winston from inside the fence saying that "all was up" and that Haldane had attempted to go over but had been seen by the sentry and stopped.

Here again, just a few months after the events Winston relates the next moments when Haldane came to the fence and Southey had suggested that Winston climb back in since it was pretty hopeless for him to continue without "compass, map, opium or silver money" and not being able to speak any of the native languages.

Winston did not think that was possible alleging in the letter that the wall was nearly 12 feet high from outside. Haldane apparently made a note on the letter that it was "7 ft. 6 in. at most."

A third correspondent says that he measured the fence in 1963 and that it was only 6 feet tall.

L.C. B. Howard [John Howard's Son] to Randolph Spencer Churchill

31 May 1963.[464] 10 Coronation Buildings, Germiston Transvaal

First of all, on reading through your Father's narrative in the Reader's Digest [September 1940][465]............. [He makes several observations, including] ....

And lastly, your Father mentions the corrugated iron fence at the Staats model School in Pretoria, which he scaled in his escape from there as being about ten feet in height; it was only 6 ft. & is there to this very day. I went & had another look at it, on the afternoon of our gathering of the 20th instant.[466]

The front of the school in 1988.[317]

He [Brockie] had crossed the yard and entered the latrine when Churchill said to me, possibly thinking that Brockie might go without us, 'I am going over again, **don't follow immediately**', thus proving once more that he understood that we were to go in company. Not grasping what he had in his mind I waited while he pursued the same direction as Brockie had taken. As Churchill drew close to the latrine I could see Brockie emerge from it and both stop for an instant and apparently talk to each other. Then the sergeant came back to the verandah when he angrily said, 'That d—d fool Churchill wanted to stop and talk within earshot of the sentry. I told him that it was useless to try to escape then.' Meanwhile Churchill had disappeared into the latrine, and as neither Brockie nor I for a moment imagined that he would attempt to go while I was not with him, we went into the dining-room, **intending after a hurried meal to try again**. Churchill did not come into dinner, and after about ten minutes I went to our room and found that he was not there, nor had he been there, for I looked under his pillow - the place where he hid his soft hat, an article not easily acquired by a prisoner and therefore carefully preserved - but there was no sign of it. [Odd, Churchill would have had Adrian's hat with him during the first try that evening.] I began to wonder what had become of him, when almost immediately someone arrived with the news that he was on the other side of the enclosure. This report was shortly after confirmed by an officer [Southey] who had spoken to him through the open ironwork of the round house over which he had climbed. OBC p. 1103.

463 Lieutenant Charles Elliot Southey, Royal Irish Fusiliers.

464 OBC Vol 1 Part 2 p. 1132.

465 PF:Unable to locate actual issue or article: The Reader's Digest Index for 1940: Vol 37 Index: https://archive.org/details/sim_readers-digest_1940_37_index Second Index: https://babel.hathitrust.org/cgi/pt?id=ien.35556000799700&seq=10.

466 Per RSC: On 20 May 1963 a bronze plaque commemorating the history of the school. and recording WSC's escape was unveiled at the former State Model School. The school, since converted into a library, was declared a monument in 1962. See: Staats Model School - Wikipedia Retrieved Nov. 7, 2024.

Regardless, the fence was taller on the outside and there was no ledge to perch upon before going over.

The letter concludes with a recollection that Haldane offered to toss his compass over the wall but Winston declined. Haldane bade [him] goodbye and he began his journey.

Winston goes on to aver that Brockie was the hope of the venture with his colonial knowledge and reasserts that he was told by Brockie to go over first.

> "I was particularly instructed by B. to get over first that night. I don't think he would attempt to deny that, though you know what these Colonials are and I must have waited in the garden upwards of an hour and twenty minutes at what seemed to me the peril of my life."[467]

Winston says that Rosslyn's remarks were copied into newspapers and placarded about before his election but fortunately to no avail.

He askes Haldane to write a letter saying that Winston acted in good faith.

Haldane indicates that the rest of the letter is of 'no consequence' but I want to bring up one quote from it.

> I hope the Buller command incident has been satisfactorily settled, Nicholson told me that my evidence was perfectly sufficient to vindicate your character. Please do the same for me. Our roads have run curiously together on several occasions. Perhaps they may do so again. I count on you.

This refers to the military enquiry held regarding the command decisions made during the armoured train incident. See the April 9th, 1900 letter of congratulations on Haldane's escape supra.[468]

Haldane's Response

This letter threw a quite unexpected light on the matter, and I replied that I laid no blame on him for departing when he did. The actual words of my letter I cannot recall, but on other occasions when his political opponents have tried to induce me to throw mud at him I have invariably replied in the same term as on this occasion.[469]

That Brockie should have taken upon himself to direct Churchill to escape when alone and not in company with me, which as I have previously stated was a definite arrangement, was indeed a surprise to me, more especially in view of what the sergeant had said of Churchill immediately after his escape.

It is difficult to believe that Brockie would have consented to that arrangement being altered without at once informing me, more especially as we occupied the same room, about this time, constantly in one another's company. Nor can I forget his astonishment when it came to our knowledge that Churchill had given us the slip.

Back in London again and amongst my friends and relations and having myself succeeded in escaping, the matter did not seem one of particular importance and it never entered my head to write to Brockie for an explanation. I probably felt pretty sure as to the nature of his reply and decided to let the matter rest where it was. Moreover any possibility of clearing up the disputed point was removed by the fact that the sergeant not long after the conclusion of the war lost his life by a fall of rock in one of the Johannesburg mines.

I should mention that on the 30th April, 1901, I saw Churchill for the first time since the eventful night of the 12th December. All I have regarding what passed between us on this occasion is, according to my diary 'he said that he had no idea that we were to get over the wall together.' I can well remember how skillfully he presented his point of view, which was very different from my own.

As I had no means of refuting what he stated, and it was evident that even if Brockie did so Churchill would flatly deny the imputation, I decided that the less said the better, which is the policy I have consistently followed in the matter. It has sometimes amused me, when Churchill and I have met, that he never can keep off the subject of his escape, though it and my own, so far as I am concerned, are such matters of ancient history that I never give them a thought.

[467] Note by Haldane: much exaggerated.

[468] OBC Vol 1 Part 2 p. 1167-1168.

[469] OBC Vol 1 Part 2 p. 1108.

Prince of Wales Letter

Winston also brings this matter to the attention of the Prince of Wales.[470]

Winston as Candidate for Oldham, 1899.[318]

Portrait of King Edward VII.[471]

In this letter of ~ 125 words dated October 17th, 1900,[472] he encloses extracts of Rosslyn's book to his Royal Highness and expresses that, he can find no words to convey his "indignation at such a cruel and wicked falsehood"(s); that are presumably contained in the extracts.

There are several passages that are causing offence. The remarks about Winston's escape and some comments made in Rosslyn's last chapter titled 'Idle Reflections' regarding "mad flight"[473] and "desertion of guns," by a regiment.

Winston also goes on to thank the Prince for the kind remarks that Edward made to his mother Jennie regarding his candidature in a private letter.

Winston had previously written to Edward while a prisoner in the Staats Model School, on November 30th 1899[474] which was coincidentally Winston's 25th birthday. It is ~2,500 words and describes quite eloquently the circumstances of his capture, the conduct of the participants, the weapons employed against them and his observations about the Boers. Upon reading this a second time it occurred to me that this correspondence could be or probably was among the documents which Winston gave to Francis Reitz the Secretary of State and which Deneys refers to in Commando. If anyone in the SAR had the ability to send uncensored correspondence through diplomatic channels to the Prince of Wales it would have been Francis Reitz.

[470] Edward VII (Albert Edward; 9 November 1841 – 6 May 1910) was King of the United Kingdom and the British Dominions, and Emperor of India, from 22 January 1901 until his death in 1910. Edward VII - Wikipedia

[471] Given to National Portrait Gallery by Terence Pepper: File:King-Edward-VII (cropped) (b).jpg - Wikimedia Commons.

[472] Royal Archives and OBC Vol 1 Part 2 p. 1116.

[473] Rosslyn's last chapter 'Idle Reflections p. 441 of book p. 528 of PDF; to end. 'Aspersions on Crack Regiments' p. 449/536.

[474] See OBC Vol 1 Part 2 p. 1080-1082.

Deneys Reitz Excerpt

Deneys Reitz[475] - Excerpt from 'Commando.'[476]

I was in Pretoria for only two days. Before returning to Natal, I walked with my father[477] from his office in the morning and we touched at the State School where a number of captured British officers were confined, one of whom had asked for an interview.

We passed through the sentries into a large classroom where he was playing games with his fellow prisoners. His name was Winston Churchill, a son of Lord Randolph Churchill, of whom I had often heard. He said he was not a combatant but a war correspondent, and asked to be released on that account. My father, however, replied that he was carrying a Mauser pistol when taken, and so must remain where he was.

Winston Churchill said that all war correspondents in the Sudan had carried weapons for self-protection, and the comparison annoyed my father, who told him that the Boers were not in the habit of killing non-combatants.

In the end the young man asked my father to take with him some articles which he had written for a newspaper in England and if there was nothing wrong with them to send them on via Delagoa Bay. My father read portions of the articles to us at home that evening, and said that Churchill was a clever young man, in which he was not far wrong, for soon after the prisoner climbed over a wall and escaped out of the Transvaal - how, I never heard.

Francis William Reitz, 1890.[319]

Deneys Reitz during War.[320]

[475] Deneys Reitz (1882—1944), son of Francis William Reitz, was a Boer warrior who fought in the Second Boer War for the South African Republic against the British Empire. After a period of exile in French Madagascar he returned to South Africa, where he became a lawyer and founded a major South African law firm. In the First World War he fought for the Union of South Africa against the German Empire, and then was an officer in the British Army, commanding several battalions. In later life he was a politician. Deneys Reitz was educated at Grey College, Bloemfontein. While in exile in Madagascar, he wrote about his experience of the Second Boer War (1899–1902). When it was eventually edited and published in 1929 as Commando: A Boer Journal of the Boer War, it still had the freshness and detail of an account written soon after the war. The account is unique in that he was present at virtually every major event of the war. Deneys Reitz - Wikipedia

[476] Deneys Reitz. Commando: A Boer Journal of The Boer War. London Faber & Faber, 1929 (Chapter 6) Commando A Boer Journal Of The Boer War: Reitz Deneys : Free Download, Borrow, and Streaming : Internet Archive And:

[477] Francis William Reitz, Jr. Francis William Reitz - Wikipedia Francis William Reitz, Jr. (5 October 1844 – 27 March 1934) was a South African lawyer, politician, statesman, publicist, and poet who was a member of parliament of the Cape Colony, Chief Justice and fifth State President of the Orange Free State, State Secretary of the South African Republic at the time of the Second Boer War, and the first president of the Senate of the Union of South Africa.

Rosslyn Quarrel

Daily Malta Chronicle and Garrison Gazette - Wednesday 07 November 1900.[478]

Winston Churchill

Mr. Winston Churchill M.P., has been very much in evidence this week. The quarrel over Lord Rosslyn's book, is generally considered at an end.[479] It is impossible to accept Lord Rosslyn's gossip—for such it really at best is,—about the "mad flight"[480] and "desertion of guns,"[481] by a composite regiment of British cavalry.[10th Hussars] All that kind of washing family linen—public, people agree ought to be ever avoided.

Mr. Churchill gave an interesting lecture on the South African War, at St. James's Hall,[482] and presided over by Lord Wolseley,[483] where his mother, accompanied by ducal relatives, were present. He has much of his father's blunt and impetuous manner.

He eulogized the conduct of Sir Redvers Buller, and testified to the gallantry, patient perseverance, and bravery of the British soldier: Lord Wolseley saw in the officer, the highest type of humanity and warrior. The lecture was interesting only too long—two hours.

St. James Hall.[321]

[478] Daily Malta Chronicle and Garrison Gazette - Wednesday 07 November 1900 p. 4 https://www.britishnewspaperarchive.co.uk/viewer/bl/0004654/19001107/020/0004.

[479] Earl of Rosslyn, James St Clair-Erskine, 5th Earl of Rosslyn Twice Captured: A Record of Adventure During the Boer War. United Kingdom, William Blackwood and Sons, 1900. See: Twice Captured - Google Books.
Also: Catalog Record: Twice captured: a record of adventure during... | HathiTrust Digital Library.
And: James St Clair-Erskine, 5th Earl of Rosslyn - Wikipedia.

[480] Rosslyn's last chapter 'Idle Reflections p. 441 of book p. 528 of PDF; to end. 'Aspersions on Crack Regiments' p. 449/536.

[481] This quote is not found I the book in these words. Other words are used.

[482] St. James's Hall was a concert hall in London that opened on 25 March 1858. It was situated between the Quadrant in Regent Street and Piccadilly, and Vine Street and George Court. There was a frontage on Regent Street, and another in Piccadilly. Taking the orchestra into account, the main hall had seating for slightly over 2,000 persons. St James's Hall - Wikipedia

[483] Field Marshal Garnet Joseph Wolseley, 1st Viscount Wolseley, (4 June 1833 – 25 March 1913), was an Anglo-Irish officer in the British Army. He became one of the most influential and admired British generals after a series of successes in Canada, West Africa and Egypt, followed by a central role in modernizing the British Army in promoting efficiency.
He was appointed by the new Conservative government to succeed the Duke of Cambridge as Commander-in-Chief of the Forces on 1 November 1895. After holding the appointment for over five years, he handed over the command-in-chief to his fellow field marshal, Earl Roberts, on 3 January 1901. Garnet Wolseley, 1st Viscount Wolseley - Wikipedia

Twice Bankrupt Reported

Two War Yarns.

Dr. Conan Doyle. On British. Prisoners. And Boer Gaolers.[484]

Presiding over the gathering thronged St James's Hall to hear Mr. Winston Churchill's second [UK] lecture on Monday evening. [October 30th, 1900][485] Dr. Conan Doyle made an excellent speech. packed full of illustrative yarns.

The lecturer himself figured to far greater advantage on the occasion of his initial discourse. He spoke from notes only, and many of his happiest allusions were obviously impromptu.

The savage attack on Lord Rosalyn with its cruel reference to the latter's "Twice Captured," - and suggestive proposal that the noble lord should furnish a companion volume. called "Twice Bankrupt" was wisely left out. [It seems that the reviewer has attended both the first and second lectures.]

It is dubious. indeed, whether Mr. Churchill may not, ultimately have to pay up for originally indulging in this gamesome sally. Lord Rosalyn is a bit. of a peacock, but no coward; and should he come face to face with Winston, M.P., there will undoubtedly be "wigs on the green."[486]

Dr. Conan Doyle, in the course of his remarks, said he never had the pleasure of meeting Mr. Churchill, and perhaps, under all the circumstances, it was well, for the peoples he had met had been usually unfortunate. (Laughter.)[487]

When he was at Pretoria. he remembered be had pointed out, to him, as an object of interest, the exact spot, over which Mr. Churchill bed climbed when he escaped from the State Model School, and he saw, the historic tree that threw its black shadow to screen him.

It was evident that the lot of the Boer commander who had been in charge of the school must have been a particularly unhappy one; for if one looked under a bed one's attention was directed to "another trap-door." And in wall or roof others were shown; the whole place had indeed been made into a sort of rabbit warren by British officers, in their attempts to escape. (Laughter and cheers.)

He did not know what had become of the Boer commander, but be had A shrewd suspicion that his experiences might have rendered him a fit candidate for a lunatic asylum. [Either Field Cornet Malan[488] or R.W.L. Opperman.][489 490]

[484] Lyttelton Times-Christchurch, New Zealand Thursday 27 December 1900 p. 3 Col 5-6 https://www.britishnewspaperarchive.co.uk/viewer/bl/0003346/19001227/024/0003.

[485] OBC p. 1209 Woolsey letter agreeing to preside at the lecture.

[486] PF: It dawned on me what this might mean. If two 'gentlemen' were going to engage in physical combat they would probably take their expensive and carefully maintained wigs off beforehand, or alternatively, these items would be one of the first items lost during the physical engagement.

[487] Doyle, Arthur Conan. The Great Boer War. Vol 1 1900. Vol 1 ISBN:9781535224338, 1535224339 Volume 1 concludes with Spion Kop. Tauchnitz, edition The Great Boer War - Google Books.

Second Edition: https://archive.org/details/greatboerwar00doylgoog.

Doyle, Arthur Conan. The Great Boer War: Seventh Impression 1900 (September) London, Smith, Elder & Co. 595 pages This edition concludes with "Halt at Pretoria" The great Boer war : Doyle, Arthur Conan, Sir, 1859-1930 : Free Download, Borrow, and Streaming : Internet Archive.

Doyle, Arthur Conan. The Great Boer War 17th Edition, 1902 with the two volumes are available as one document from Project Gutenberg: The Great Boer War by Arthur Conan Doyle - Free Ebook (gutenberg.org).

Doyle, Arthur Conan, Sir, 1859-1930 The war in South Africa, its cause & conduct; 1902 McClure, Phillips & Co. (163 Pages discussing causes, negotiations, concentration camps and farm burning.)

The war in South Africa, its cause & conduct; : Doyle, Arthur Conan, Sir, 1859-1930 : Free Download, Borrow, and Streaming : Internet Archive And: https://archive.org/details/warinsouthafric00doylgoog/mode/2up.

Sir Arthur Conan Doyle had served as a volunteer doctor in the Langman Field Hospital at Bloemfontein between March and June 1900. In his widely distributed and translated pamphlet 'The War in South Africa: Its Cause and Conduct' he justified both the reasonings behind the war and handling of the conflict itself. He also pointed out that over 14,000 British soldiers had died of disease during the conflict (as opposed to 8,000 killed in combat) and at the height of epidemics he was seeing 50–60 British soldiers dying each day in a single ill-equipped and overwhelmed military hospital. Quote from: Second Boer War concentration camps - Wikipedia

[488] Field Cornet Malan was a grandchild of President Kruger. (Sandys p. 82) This was not exceptional because Kruger had 17 children and over 100 grandchildren. He was captured in August of 1901 and exiled to Ceylon. (Haldane) Over 5,000 Boers were interned in Ceylon (Sri Lanka) during the Boer War. The Ceylon camps were the second set of overseas POW camps after those in St. Helena had filled. http://libcom.org/library/pow-camps-ceylon-during-boer-war

When he was in Pretoria lie went out to Waterval, where three thousand "Tommies" had been kept, and he obtained some information as to what had occurred there from a communicative Boer farmer's wife. She said, "Yes, yes. they were terrible men, they gave a great deal trouble." Under pressure of persistent questioning as to what they did to give trouble., she said at times they sang "God Save the Queen." and supplemented this at others by "Soldiers of the Queen." And, she continued,

"They were very mischievous: they made a **khaki man**." Asked why this was an evidence of mischief, she added that there was a part, of the camp where no one was allowed to go after nightfall. They put their khaki man there, and when, after ordering it to off, the sentry challenged it, and eventually fired, the whole 3,000 prisoners got up and laughed and shouted at him. (Laughter.)

One point Mr. Churchill had not spoken much of, and that was the extreme good humour of the soldiers engaged in the campaign, though they were often surrounded by the most sordid and difficult circumstances. Wherever he himself went he always heard the same story. One which had reached him was, perhaps, apocryphal but it would bear repeating.

The story was told of a Tommy who had to go up a hill at Colenso. and he had to dispose of a Boer who was behind a rock. Tommy was shot by the Boer, but before he fell he bayonetted his antagonist.

When they recovered consciousness. Tommy asked the Boer if he would have a drink out of his flask, an offer which the Boer was very glad to accept. In return. the Boer made Tommy a present of some of his biltong. By-and-by the ambulances of the respective parties came up; Tommy was taken to the British ambulance, and the Boer by the Transvaal contingent, and as they passed, Tommy turned round to his whilom[491] companion and said, "Well, good-bye mate; what a blessing it is we met each other!" (Laughter.)

Blackwoods to Lewis

Wm. Blackwood & Sons to Messrs. Lewis & Lewis

November 1st, 1900.[492] 45 George Street, Edinburgh

Dear Sirs,

[Re:] Twice Captured

Adverting[493] to our correspondence we have now finally considered our position here, and the representations made by you and while we emphatically repudiate your assertion that the paragraph in question suggests dishonourable conduct or cowardice on the part of Mr. Winston Churchill, we have decided to delete it in future issues of the book.

We need hardly say that no copy of the old issue has been given out since we received your letter of the 20th ult.[494]

We further beg to state very frankly that while repudiating your suggestion, we are sorry, and would apologise for any personal annoyance felt by Mr. Churchill owing to the paragraph in question appearing in the book.

We are, Dear Sirs Yours faithfully

Wm Blackwood & Sons

489 "Commandant [R.W. L.] Opperman, who was directly responsible for our safe custody, was in times of peace a Landdrost or Justice. He was too fat to go and fight, but he was an honest and patriotic Boer, who would have gladly taken an active part in the war. He firmly believed that the Republics would win, and when, as sometimes happened, bad news reached Pretoria, Opperman looked a picture of misery, and would come to us and speak of his resolve to shoot his wife and children and perish in the defence of the capital." (WSC London to Ladysmith.)

490 Opperman Notice Mr. Churchill's Gaoler [Laffan's Telegram.] Spearman's Camp, Feb. 3, 8.55 P.M. Among the Boers who were killed last week was one named Opperman. He had been gaoler in charge of the British prisoners at Pretoria, but was dismissed from his post after Mr. Winston Churchill escaped. Spirit of Buller's Force: https://www.britishnewspaperarchive.co.uk/viewer/bl/0000174/19000205/037/0005.

491 Whilom: From Middle English whilom (adverb) at one time, formerly, once; once upon a time; at times, sometimes.

492 OBC Vol 1 Part 2 p. 1116.

493 To turn attention to, to take notice of (something).

494 Abbreviation of Ultimo: in or during the previous month. In this case, October.

1912 Litigation

The controversy was resumed in 1912 when Winston again initiated a libel action against William Blackwoods & Sons Magazine. This action is in regard to a satirical poem titled, A Lost Letter of Ancient Rome, that was published in the February issue.

In order to prepare for the litigation and bring his counsel up to speed on the matter Churchill wrote a wrote a 4,000 word memorandum regarding the circumstances surrounding his escape.[495] In it he brought up new facts and elaborated on some aspects that inform the overall narrative.

This undated document is typed on Admiralty Paper,[496] and must have been written sometime in prior to the case being heard on May 20th, 1912. In this hearing the issue was not contested and the publisher apologized for the libelous statement in the ~2,000 word poem.

In his memo, Winston recalls how he was privy to the planning of the attempt from the beginning but was primarily interested in obtaining his release as a civilian known to be an accredited press correspondent. After that opportunity evaporated he turned to Haldane's plan.

The most relevant fact recited is that Winston repeats virtually word for word what Aylmer wrote in his diary contemporaneously with the event a decade earlier on December 12th, 1899. To wit, that Brockie goaded both Haldane and Winston. Here as reported by Winston:

> "Brockie then came up and asked us why we had not got over. Haldane explained the difficulty of the sentry's position, and Brockie said: 'You're afraid'. Haldane replied: 'You can go and see for yourself'. Brockie then went across the yard, got into the roundhouse, and remained there some time."

As noted above, this was very unwise of Brockie; to question Winston's courage and determination. If anyone in the world was going to show themselves courageous, I believe a 25 year old Winston Churchill would be very close to the top of the list.

After a decent interval, Winston returned to the outbuilding, meeting Brockie on the way out, having used up a reasonable but not excessive amount of time in the 'Office' 'Roundhouse', latrine, waiting for his opportunity. They passed each other in the yard. Winston, in this memorandum says they did not speak for fear of being overheard by guards but Brockie is adamant that they did exchange words. I choose to believe Brockie in this case. I think some words passed between them and I'm sure that Winston was on a mission to find a way out and show he was not afraid of risk.

This quote is crossed out in the typed copy of the memo but I feel summarizes the situation accurately it also adds the detail of turning to light a pipe of tobacco which is different from talking to another guard as earlier and is also repeated in his later accounts as a cigarette.

> I had come to the conclusion that we should waste the whole night in hesitations unless the matter were clinched once and for all; and as the sentry turned to light his pipe, I jumped on to the ledge of the wall and in a few seconds had dropped into the garden safely on the other side.

Compare to London to Ladysmith.

> Through a chink I watched the sentries. For half an hour they remained stolid and obstructive. Then all of a sudden one turned and walked up to his comrade and they began to talk. Their backs were turned. Now or never!
>
> I darted out of my hiding-place and ran to the wall, seized the top with my hands and drew myself up. Twice I let myself down again in sickly hesitation, and then with a third resolve scrambled up. The top was flat.[497]

[495] See: OBC Vol 1 Part 2 p. 1094-1099.

[496] In October 1911, Churchill was appointed as First Lord of the Admiralty and served until May of 1915 when the Asquith Coalition ministry was assembled.

[497] "My waistcoat got entangled with the ornamental metal work on the top. I had to pause for an appreciable moment to extricate myself. In this posture I had one parting glimpse of the sentries still talking with their backs turned fifteen yards away. One of them was lighting his cigarette, and I remember the glow on the inside of his hands as a distinct impression which my mind recorded." Strand Pt. I.

And in Strand Magazine, later in 1924.[498]

> On the 12th, in the evening after my two friends had made an attempt but had not found the moment propitious, I strolled across the quadrangle and secreted myself in the circular office. Through an aperture in the metal casing of which it was built I watch the sentries. For some time they remained stolid and obstructive. Then all of a sudden one turned. and walked up to his comrade and they began to talk. Their backs were turned. Now or never!

Scaling the interior Prison Wall. [322] [323]   Mr. Churchill at the top of the wall. Note circular "Office."[324]

> Lying on it I had one parting glimpse of the sentries, still talking, still with their backs turned; but, I repeat, fifteen yards away. Then I lowered myself silently down into the adjoining garden and crouched among the shrubs. I was free. The first step had been taken, and it was irrevocable. [499]

One further fact recited in this memorandum is that because of Winston's issues with his right shoulder that Haldane was to give him a leg up and over the wall and then follow.

> It is quite true that both on the 11th and on the 12th, Haldane and I went into the roundhouse together and waited for an opportunity to get over, and that we had arranged that he should give me a 'leg-up', as I was thought to be the least active of the party, but the sole object in view was that we should all three get over the wall the best we could, and wait for each other in the garden beyond.[500]

[498] Churchill's full account of his escape appeared in the Strand magazine in 1923-24. The Strand Magazine - Wikipedia The relevant issues are available from the: HathiTrust Digital Library | Millions of books online Catalog Record: The Strand Magazine | HathiTrust Digital Library.

My Escape from The Boers-Part I. The Rt. Hon. Winston Churchill; The Strand Magazine Vol 66 July to December 1923 Pages 537- 547 (Compilation P. 557- 567) Illustrations by Ernest Prater and from Photographs. #5 - The Strand Magazine. v.66 1923 Jul-Dec. - Full View | HathiTrust Digital Library.

Boers, My Escape from The-Part II. The Rt. Hon. Winston Churchill; The Strand Magazine Vol 67 January to June 1924 Pages 14-23 (Compilation P. 24-33) Illustrations by Ernest Prater and from Photographs. #5 - The Strand Magazine. v.67 1924 Jan-Jun. - Full View | HathiTrust Digital Library.

Also Published in full in Cosmopolitan Magazine in January, 1924: Cosmopolitan Magazine, Vol 76 Issue 1 "My Escape from the Boers" Page 30-34 and 161-166 of the PDF Cosmopolitan 1924-01: Vol 76 Iss 1 : Free Download, Borrow, and Streaming : Internet Archive.

It is very similar to the account he published in his autobiography 'My Early Life a Roving Commission', London: Thornton Butterworth, September 1931 [fifth impression] My Early Life. A Roving Commission. (fadedpage.com).

[499] No wonder Haldane was annoyed. Our protagonist is taking a bit of literary license and I'm sure it did not sit well with, by now, in 1912, General Haldane.

[500] Haldane's observation p 1102: "I had noticed that Churchill did not take part in the various games, such as fives, rounders and skipping, which had been started by the prisoners in order to keep themselves fit mentally and physically. This led me to believe that his agility might be at fault; and, moreover, as the height of the roof of the latrine was about seven feet, he might not be able to reach it without assistance or pull himself up on to it on account of his [right]shoulder, which he had dislocated while playing polo in India and which was in consequence weak." Winston was left handed so this was not ordinarily a problem day to day.

Winston's use of 'self help' to manage this feat on his own is except for the timing according to plan. He was to be the first over the wall one way or another.

He then goes on to talk about obtaining the attention of Southey, inside the compound; "who had come to the roundhouse for a private purpose"[501] and telling him to fetch Haldane and Brockie pronto.

The memo continues along familiar lines that:

Haldane did try and the guard saw him and told him to get down. It was not seen as a serious effort and no further action was taken by the guards.

Per Haldane's later account:[502]

> Some time passed when word came that the sentry, who up till now had stuck to his post, had moved, and as it was clear that if Brockie and I were to escape we must at all costs do so without losing a moment, I at once crossed to the latrine with three other officers.
>
> The sentry who, as it happened, was standing close to the building but was invisible in the darkness, saw me as I pulled myself on to the metal roof, for the moon had just begun to rise and shone on my face. He put his rifle to his shoulder a few feet from me and pointing it at my head called out 'Go back you fool.' At this moment the relief marched up, and as I dropped back, the corporal in charge, to whom the sentry had spoken, rushed to the entrance of the latrine and, seizing the officer nearest him marched him off to the Commandant's office.
>
> This officer having explained that he had no thought of escaping was released; and except that the sentries seemed to be more alert than usual, making it impossible to evade them, no extra precautions were taken. That this was so was no doubt because it was not yet suspected that anyone had managed to get away.

Haldane did confer with Winston through the fence and inform him that he and Brockie would be unable to follow.

> A little later an officer came to me [Southey] and said that he had been whispering through the grille to Churchill, whereupon I went to the same place and told the latter that I could not get away.
>
> He replied 'Tell Brockie to come.' I well remember feeling annoyed at this utter disregard of myself and replied that it was now impossible for either Brockie or me to escape.
>
> As he then said he would go without us I offered to toss my compass and some chocolate over the roof to him. These he declined, fearing, as I afterwards knew, lest the noise of their falling in the garden where he was might draw the attention of the sentry. He then said that he was off, but I have no recollection of wishing him 'bon voyage!'

The last thing that Winston wanted to do was to go alone. He had no map, no compass, and did not speak the language. It was agreed that it was pretty hopeless starting out this way.

Winston concludes with several strong factual assertions regarding what ultimately come down to the fortunes of war.

> Further, the fact that I had got over, in no way prejudiced his chances of escape. I made no noise, did not disturb or alarm the sentry or lead him in any way to suspect that anything was going on…
>
> Haldane had his chance just as good and clear as if I had never come into the project of escape at all. He had the bad luck to be seen and intercepted by the sentry, either because he made a noise, or just through bad luck.
>
> My action in getting over had nothing whatever to do with his being prevented from coming. He had a perfectly fair run at it; and indeed it was a great advantage to him not to have to think about me or help me over at the same time.
>
> His own chance of getting over would have been much reduced if we had both been scrambling over together. His chance was in no way affected by my action, except that he had to go that night and could not put it off to a later occasion.[503]

[501] How does he come up with these apt phrases on demand?

[502] OBC p. 1104.

And to this it must be remembered that we had already tried the previous night, and frankly shrunk from the risk, and that I had made it quite clear to him and to Brockie in the afternoon that I was resolved at all costs to be out of the place that night, and we had all agreed that we would go and would not be put off by the danger when it came to the point.[504]

Brockie's disgust when he found we had not succeeded in our first attempt on the night of the 12th, and his insulting remark to Haldane,[505] is a clear proof of the decision we had taken to make the attempt even at a very great risk.

Hmm. Courage, decisiveness, and willingness to take risk. Gosh, who is speaking? Also, if he had been caught in the attempt! Just factually, Haldane and Brockie would have really let him have it. He would have been subject to the derision of the whole prison. Not only that, he would have been taken to a real prison in Pretoria and locked up tight until the city fell to the British in June of 1900.

My conscience is absolutely clear on the whole episode; I acted with perfect comradeship and honour the whole way through. I would not consent to escape on the 15th November on the engine while he and others were left behind. I gave up a certain means of escape, with honour and credit, from a scene of disaster, in order to return and fight it out with those who were left behind.........

There was no more agreement, or bargain, or stipulation, as to who should go first, or how we should go, than there is among a dozen people in the hunting field who are waiting to take their turn at an awkward gap.[506]

When I had got over, I waited for an hour and a half at imminent risk of recapture, for the others to come, and the fact that they were not able to come seemed to me and seemed to them to deprive me of all reasonable chance of escape.

Haldane will remain a true and stalwart friend to his brother officer while still maintaining his private opinion that Winston did not follow orders in conforming to Captain Haldane's plan. How could it be any other way? [507]

Back at the State Model School Prison in 1899, Haldane is catching grief from all sides.

I must admit that I was surprised and disgusted to find myself left in the lurch, for Churchill had walked off with my carefully thought-out plan or what he knew of it, and had simply taken the bread out of my mouth. Brockie, I need hardly say, was furious, and the escaped correspondent of the Morning Post came in for a full dose of opprobrious epithets of which Brockie had a liberal command. It requires no effort of memory at this distance of time to recall his sneering allusions to 'Your trusted friend - a nice kind of gentleman!' and so on who contrary to his advice I had allowed to join in our plot.

Nor was he alone in his abuse of the escaped war correspondent, for many, indeed most, of our fellow-prisoners, some of whom had pressed me to let them share in the enterprise and who now began to realise the far greater difficulties of evasion in the future, joined in the chorus of vituperation that arose and continued for some days. These would include Capel-Cure[508] and Lonsdale, [509] See Haldane's diary.

[503] He says the same thing three times but each one containing facts and insight that elaborate the theme.

[504] Here he is justifying his spontaneous change in the plan of when to go and under whose direction.

[505] Interesting that in this memo he does aver that the remark was made to him personally. I beg to differ.

[506] This is a great analogy but a little overreaching. He'd jumped the gun. So did Haldane by trying again too soon. Why didn't he tell Winston to leave and they would try in an hour. But remember the moon was going to rise at 7:45 that evening.

[507] This could be a four-act play. The language, drama and structure is already extant. 1899, 1900, 1912, 1948. Either that or a film that focuses solely on the time frame from the rainy morning of the disaster when he leaves Leo Amery behind to his boarding the Induma for the journey to Durban. Maybe include the evening before when Leo and Winston convince Col. Long to not evacuate Frere and the artillery is backed off of the train. Leo's quote of Winston saying; "I did that...er, we did that."

[508] Captain Alfred Capel-Cure, Royal Artillery, and Captain Malcolm Percy Eustace Lonsdale, Royal Dublin Fusiliers.
Capel Cure, Herbert, Captain, was born 23 October 1859, at Eaton Place, London. He was considered in the first ranks of polo players in India between 1883 and 1894. He captained the 2nd Gloucestershire Regimental Polo Team in India, when the regiment won the Inter-Regimental Cup; played at Lucknow in 1892, and again in 1893. He was also captain of the Gloucester team which played in the Final and lost by one point (a disputed one), and again in the Baroda Tournament which was won by the regiment. Burma 1885-7 Burma 1887-9 (Capt. Gloucester Regt), QSA (1) Natal (Major DSO Gloucester Regt). Source: DSO recipients (VC and DSO Book) Anglo Boer War - Name search.

[509] Captain Malcolm Percy Eustace Lonsdale, Royal Dublin Fusiliers. QSA, Clasp of Talana Hill Anglo Boer War - Name search.

I said little, swallowing my chagrin as best I could, and on the principle that:

'Things without remedy Should be without regard: what's done is done.' [510]

Haldane continues to adopt a mature and expansive view of the possibilities.[511]

............after reflection, I decided to keep an open mind regarding Churchill's escape. In my letters home I avoided reference to the subject, and in only two - for all letters to my own people were preserved - can I find any allusion to Churchill's escape. I had come to the conclusion that, in all probability, Churchill on that occasion had acted on a sudden impulse, greatly excited as he was at the time, and I had buried all the resentment which I admit I felt soon after the event.

General Stonewall Haldane

While going to the Admiralty to meet Winston might have been intimidating for some, Aylmer could hold his ground as a General Officer and also on the strength of his cousin, Richard Burdon Haldane[512] who was then serving as Secretary of State for War, down the street or across town at the farthest.

April 25th, 1912

I was then Brigadier-General, General Staff of the Eastern Command and had just been selected to command the 10th Infantry Brigade at Shorncliffe.[513] On that date I received a letter from the private secretary to the First Lord of the Admiralty, in which he said that Mr. Churchill would like to see me at 6 p.m. that day. I had good reason to suspect what the subject of conversation between us would be, for on the 21st Mr. William Blackwood, head of the well-known firm which had published the account of my escape, had mentioned to me that a libel action was pending against him for a statement in 'Maga'[514] that Churchill had broken his parole when a prisoner in 1899.

I said 'but there was no parole for us, and nothing would have made me give mine.' On this he replied that I must know perfectly well what the magazine meant, for if Churchill had not broken his parole he had not at any rate played fair with his fellow-prisoners.

I said that he knew that I had never accused Churchill and I would not consent, were I asked, to give evidence against him.

On the 25th I duly went to the Admiralty at the appointed time, and for nearly an hour Churchill harangued me on the subject of his escape. He recalled what he had previously said and written, and when I asked him what he wanted, replied that he wished me to go into the witness box and state that he had behaved quite fairly and properly.

I thereupon said that, believing him to be a man of honour, I would be prepared to go into the witness box and depose that, having heard his statement which, however, differed from my own view, I would accept his word in the matter and that he had escaped with no idea that he was not behaving fairly to Brockie and myself.

He would not, however, hear of such a qualified statement, and, as I held my ground, he assumed a browbeating air and said that if I would not give him my support he would state in Court that Brockie and I had funked going, on which he had clinched the matter, and if he had not given us a lead we would never have gone.

[510] Lady Macbeth says, "Things without all remedy should be without regard: what's done is done" to Macbeth at the beginning of Act 3 Scene 2 when they are talking about the continual plague of thought and fear that killing King Duncan has brought to them.

[511] OBC p. 1105.

[512] Aylmer Haldane is a cousin of Richard Burdon Haldane.

Richard Burdon Haldane, 1st Viscount Haldane, KT, OM, PC, FRS, FSA, FBA (30 July 1856 – 19 August 1928) was a British lawyer and philosopher and an influential Liberal and later Labour politician. He was Secretary of State for War between 1905 and 1912 during which time the "Haldane Reforms" of the British Army were implemented. As an intellectual he was fascinated with German thought. That led to his role in seeking detente with Germany in 1912 in the Haldane Mission. The mission was a failure and tensions with Berlin forced London to work more closely with Paris. Richard Haldane, 1st Viscount Haldane - Wikipedia PF: He also befriended Oscar Wilde and helped to ameliorate the conditions of his penal servitude.

[513] OBC p. 1109-1110.

[514] This is shorthand for Blackwood's Magazine.

He also accused me of spreading tales about the escape reflecting on him. He added that he would call Frankland - captured with me on the armoured Train - as a witness to corroborate what he had said.

I thereupon said that he could state what he liked, as I was indifferent in the matter, but that the truth was that at Pretoria I was thinking of three individuals and he of one man only, himself, and that I knew that unless the chances were favourable three of us could not get away together, which was proved by what had actually occurred.

I added that he could send me his proposed statement, and what he wanted to say, and that I would then inform him what I would be prepared to say.

I must admit that I particularly wished to avoid anything of the nature of an open rupture with him, as no good could possibly come to me from making him a declared enemy. Nevertheless, rather than state anything contrary to what I felt to be true, I was prepared to take that risk, no matter what the consequences might be.

On the 27th April, having taken over the command at Shomcliffe,[515] I heard from Churchill, who said that in place of writing out his statement, his Council, Mr. F. E. Smith would like to have a talk with me. I was advised to reply to the effect that I should be obliged in the first instance if he (Mr. Churchill) would send me a copy of what he wished me to say.

Nothing further occurred till the 2nd May when a second letter arrived from Churchill repeating his desire that I should meet his council. As I was obliged to go to town for a farewell dinner, I saw my solicitor, and, on his advice, the following letter was despatched to Churchill, on the 3rd May.

> 'I am in receipt of your letter of the 1st. When I was with you I understood that you were going to send me a copy of your statement, and I should like that sent, together with a copy of the statement which you say Frankland has furnished, and that which I am asked to make, and I will then write again.'

On the 16th May, I received a letter dated the previous day from Churchill's Council Mr. F. E. Smith, a copy of which I made and handed the original to my solicitor. Whether this letter was marked 'private' or not I cannot say, though I imagine from the reply I was advised to send that it was so marked. In any case I reproduce it here as I consider I have every right to do.

He lays stress on the danger which, once out of the enclosure, he ran in waiting for us, and the almost hopeless prospect of his being able to reach safety alone. This may be useful matter wherewith to gloss over the fundamental cause of his forlorn situation and throw dust in the eyes of his readers, but it carries no weight with those of his fellow prisoners who had not been kept in the dark regarding the intended escape.

The situation in which he found himself was entirely of his own creating, and for that reason he saw himself later on forced to endeavour to regulate it, by untruthfully affirming that Sergeant Brockie, and not himself, was responsible for it.

He next says 'Nevertheless I was determined that nothing should stop my taking the plunge next day.'

Here we seem to have alighted on the truth, but it is an unfortunate admission for the writer. In point of fact, and as Churchill's letter of the 5th October, 1900 clearly shows, it was on the 12th and not the 11th December that he was told by Brockie, assuming that he was to go independently of his companions. He had evidently determined to go with or without us, and as I have already said, it was easy for one to do so, but not for three or more.

[515] Shorncliffe Army Camp is a British Army installation near Cheriton in Kent, established in 1794. The camp, described as "the birthplace of the modern British Army" Shorncliffe Army Camp - Wikipedia

F. W. Smith Letter May 15th (Wednesday)

F. E. Smith[516] to Brigadier-General Aylmer Haldane

May 15th 1912.[517] 70 Eccleston Square

I think that Churchill has told you that I am advising him in the action which he is bringing against Blackwood's Magazine. It is possible that the hearing of the case may be expedited and it may even be dealt with before the Whitsuntide Vacation.[518]

I understand from Churchill that you very naturally desired to be informed.

(a) of the tenor of the proposed statement,

(b) of the statement you were invited to make. I write with the object of giving you this information. It is proposed that Churchill's statement should consist of only three or four sentences.

He will say:

(a) 'I never gave any parole to the Military Authorities at Pretoria and consequently never broke my parole,

(b) In making my individual escape I did not break any agreement, expressed or implied, with those of my companions who concocted with me a plan of escape. On the contrary I waited for two hours outside the wall of the prison, at the greatest personal risk, in the hope that one or more of them might share my good fortune. In fact I did not think that a single prisoner could possibly make good his escape alive.'

Your evidence would be taken at even greater brevity; and as the other side are apologizing it will not of course be subject to cross-examination.

It is suggested, if you are willing, that your statement should take this form –

'I was one of those who concocted a plan of escape. The others were Mr. Churchill and an officer, who is now dead – Lieut. Brockie. The actual circumstances under which Mr. Churchill escaped are of course well known to me. Those circumstances fully justified Mr. Churchill in leaving the prison as he did, and his escape gives no grounds for any reflection whatever on his honour.'

I am advising Churchill in this matter as a friend rather than professionally otherwise I should approach you through a solicitor and not personally. I should be very grateful if you would let me know your views at 70 Eccleston Square, S.W., and if you would like to see me upon the matter I will make an appointment with you tomorrow. The most convenient hour for me would be between 5.45 and 6.45.

Frederick Edwin Smith Photograph by Elliott and Fry.[325]

516 Earl of Birkenhead was a title in the Peerage of the United Kingdom. It was created in 1922 for the noted lawyer and Conservative politician F. E. Smith, 1st Viscount Birkenhead. He was Solicitor-General in 1915, Attorney-General from 1915 to 1919, and Lord High Chancellor of Great Britain from 1919 to 1922. Earl of Birkenhead - Wikipedia

517 OBC Vol 1 Part 2 p. 1112.

518 The Sunday of the feast of Pentecost, seven weeks after Easter.

Haldane's reaction is: "On receiving this letter I at once wrote to my solicitor, forwarding it and pointing out that I was not prepared to say what was required without a considerable qualification. Next day I had a telephone message from him, which necessitated a visit to London, and the dispatch, that evening from Folkestone, to Mr. F. E. Smith of the following letter."

Haldane May 17th Letter (Friday)

Brigadier-General Aylmer Haldane to F. E. Smith Radnor Club 17 May 1912.[519] Folkestone

Dear Mr. Smith,

I am in receipt of your letter of the 15th instant. Churchill had told me that you were advising him, and that you had suggested that I should give evidence in an action he was bringing.

I see by the papers, however, that the case is fixed for Monday next[520] when it is impossible for me to be in London. As it is undefended I do not suppose that my absence will make any difference to either Churchill or you.

I quite understand from what you say why you approached me personally and not through a solicitor.

Yours faithfully A. Haldane

Resume 1912 Narrative

On my return shortly after 8 o'clock, to the hotel where I was living, I found that Churchill had telephoned there and desired that I would call him up at the Admiralty as soon as I returned.[521]

Soon after I sat down to dinner a telegram came from him saying: it is 'of serious importance for you to send me your statement. Will see you up to 4 p.m. tomorrow (Saturday).'

I replied laconically 'Have written to Smith'.

On Monday the 19th [sic 20th] the case came on and a full account of it may be found in The Times law report of the 20th.[522]

I will say nothing regarding it except that the statement of Council as reported, did not contain the first sentence of para (b), namely that Churchill, in making his escape did not break any agreement, expressed or implied. He and his Counsel were quite prepared that I should make that statement, but it seems from the law report that they were not ready to do so themselves.

During all the years since the war in South Africa, I rarely chanced to meet my friend Captain Le Mesurier,[523] for he had left the army and obtained employment under the Colonial Office. On the 27th December 1912 we met again and I told him of the case against Blackwood. On that date I made the following entry in my diary..............Le Mesurier said that there can be no doubt that, after talking frequently at Pretoria to Brockie on the subject, the latter never said to Churchill on the afternoon of the escape or at any time that Churchill should try to escape alone and not with me, as agreed.

Le Mesurier thought the idea was fantastic and, that Brockie being dead, had been invented by Churchill to provide a plausible excuse for going off alone and leaving his pals in the lurch. .... Brockie made no secret of his rage at the way we had been let in by Winston Churchill.[524].............

Le Mesurier was killed in the Ypres Salient in 1915 a few days before he was to take up the appointment of Assistant Provost Marshal, which I had managed to obtain for him as I had no vacancy for ADC on my staff.

Captain Frankland, his great friend, was my Staff Captain for the first three months of the Great War and I know what he thought about the whole business of the Pretoria escape. He was a gallant officer and after leaving my staff was killed at the landing at Gallipoli.

[519] OBC Vol 1 Part 2 p. 1113.

[520] Is this May 20th or May 27th.

[521] This is a continuation of the narrative that begins on page 1099.

[522] See Blackwoods Libel Action section infra with the newspaper coverage and images. There was Sunday the 19th coverage of Friday's scheduling hearing. The case was heard on Monday the 20th.
So far as I can tell the 19th was a Sunday and Monday was the 20th. Was this a leap year for the UK? I need to get the Times Law Report at all costs. Times Digital Archive, 1785–2006 (Gale Cengage) | The English Historical Review | Oxford Academic.
And: Primary Source Collections Online Trial | Gale.

[523] Frederick Neil Le Mesurier (1875-1915), Royal Dublin Fusiliers.

[524] Perhaps because Brockie had in fact said something of the sort to Winston and needed to cover his mistake.

I mentioned in the preface to this amplified account of my adventures certain works that had appeared of late years bearing on the escape of Churchill. That of '**A woman of no importance'** by an anonymous writer - whose name is known to me - is so inaccurate that I take no notice of it.[525]

As regards what has been written by Mr. John Buchan[526] (now Lord Tweedsmuir), who has repute as a writer, I have one remark to offer.[527]

He credits Churchill with the 'audacious plan' of scaling the wall of the enclosure, while relegating to me - as a cautious Scot no doubt! - that of tunneling my way out of prison. I perhaps have seen Churchill in a situation of greater danger than have others, and can affirm with confidence that he possesses one at least of the attributes of his great ancestor. Besides this he can be splendidly audacious at times and, sometimes, at the wrong time.

Mr. Buchan is, however, in error in crediting him, in the account he gives, with being the author of the 'audacious plan', as a perusal of the previous pages, would, I think show.

Personally I have a horror of the possibility of being buried alive, and I only undertook tunneling operations at Pretoria when there seemed to be no other means of gaining my freedom.

I must allow myself a few words regarding articles by Churchill in the Strand Magazine of December 1923 and January 1924, and the book which he has since published regarding his life up to the time of his marriage.[528] From what he says he evidently knew little regarding the plan of escape prepared by Brockie and myself with the help of Major Adye;[529] a plan which, according to his tale, was vague and incomplete.

Far from that being the case, had he, on the night of the 12th December, carried out the arrangements which had been thought out with care, and were for the general advantage of himself and his companions, he would have found that most of the difficulties he encountered had been foreseen and, as far as possible, provided for.

Had Churchill only possessed the moral courage to admit that, in the excitement of the moment, he saw a chance of escape and could not resist the temptation to take advantage of it, not realizing that it would compromise the escape of his companions, all would have been well.

A frank admission of that nature, made early after his escape, would have gone far to disarm much criticism as might have followed. I myself would have been the first publicly to do anything in my power to silence the tongues of any who might rail at him or taunt him with 'not having played. the game'.

Indeed it will be remembered how I have stated that any resentment which I felt regarding his action on the 12th December had quickly passed away, and since those days I have never borne the least animosity towards him for what occurred.

But it was not to be, and the false step, once taken, made the difficulty of retraction, if ever contemplated, a thousand times more difficult, until, as time went on, it became impossible; for what would have been overlooked in the spontaneous admission of an impetuous youth of twenty-five, would have been condemned in the maturer man.

As I come to the conclusion of an effort to compile a strictly accurate and, I hope, unbiased account of the occurrences at Pretoria in which Churchill and I were involved, I can honestly say that I feel nothing but regret that a man of such splendid ability and brilliant parts, whom I once regarded as my friend, but whom, ·since the events recorded, I have never felt I could really trust, should have allowed himself to commit an action followed by prevarication, which I firmly believe in his innermost soul he has bitterly regretted ever since.

Aylmer Haldane

525 What is he referring to? It might be a parody that was published. The Play of the same name, by Oscar Wilde was performed in 1893.

526 John Buchan, 1st Baron Tweedsmuir (26 August 1875 – 11 February 1940) was a Scottish novelist, historian, and Unionist politician who served as Governor General of Canada. He was a popular author of numerous biographies, travel books and novels chiefly about the Empire. John Buchan - Wikipedia

527 A book of escapes and hurried journeys by John Buchan, John, Houghton 1923. From Pretoria to the Sea. Pages 93- 117. https://archive.org/details/bookofescapeshur00buch/mode/2up

Interesting quote from the end of the story: "Brockie had also escaped, but all three paid for some time the penalty of their wild adventure with malaria, and in the case of Le Mesurier with enteric. In a few weeks, however, they were back on duty at the front. Captain Haldane, as we have seen, was to rise to be one of the most successful British generals in the Great War. Brockie was killed by a mining accident a few years after the escape. Le Mesurier fell at the Second Battle of Ypres, and Frankland, who had assisted them to escape, died in a reconnaissance at the Dardanelles."

528 My Early Life, published in 1930.

529 Walter Adye (1858-1915), Royal Irish Rifles, Deputy Assistant Adjutant General of the Army in Natal. Retired as Colonel; CB 1909.

Haldane 1931 Letter to Lord Knutsford

In 1931 (April 22nd) now, Sir James Aylmer Lowthorpe Haldane, GCMG, KCB, DSO is 68 years old and a retired General. He is writing a history, probably a regimental history, and Viscount Knutsford [530] has made some suggestions that Haldane agrees to incorporate in the work. In his reply, Haldane takes the opportunity to talk about their mutual connections and reviews the escape saga and his relationship with Churchill. By this time General Haldane is an accomplished genealogist and writer with three other books to his credit.[531] [532]

Viscount Knutsford.[326]

Haldane reminds Knutsford that they had met previously at Buckingham Palace on 19 May 1922 in the Equerries' room when, after his return from Iraq, He was "commanded to an audience and given a G.C.W.G."[533] He also recalls that his brigade was once on maneuveres and camped at Knutsford's home Kneesworth Hall.[534] General Haldane then describes his disappointment in being taken prisoner and the efforts he made to escape.

He also notes that he served eight months after his escape which would have been in December of 1900. When Haldane later saw Churchill in London, it was brought up that Adam Brockie had said that if he and Haldane could get away they would do so. It is implied that Churchill took this as license to act without the others when his opportunity arose. Haldane, not wishing to harm Churchill, he accepted the explanation and dropped the matter. Haldane does not think highly of the account that Churchill published in 'My Early Life' calling parts of the escape narrative a 'fiction'. He also recounts how Churchill tried to enlist Haldane to 'whitewash' Churchill during the libel case in 1912.[535] He concludes nobly; declining to make any more of the matter and remarking with a backhanded compliment;

> "I have not much faith in him, I admire his undoubted courage, physical and moral, and we can't afford in these days of sloppy statesmen to ignore that quality."

This is in 1931, near the nadir of Winston's career, how honorable and prescient of Aylmer to take this tack. Ten years later Aylmer will live to see Winston's best statesmanship and more, until 1950. In closing he mentions his distant relationship to Knutsford and that he had enjoyed reading Knutsford's book.[536] He also mentions 'this interesting book' "The Escaping Club" [537]

[530] Henry Thurstan Holland, 1st Viscount Knutsford, GCMG, PC (3 August 1825 – 29 January 1914), known as Sir Henry Holland, Bt, from 1873 to 1888 and as The Lord Knutsford from 1888 to 1895, was a British Conservative politician, best known for serving as Secretary of State for the Colonies from 1887 to 1892. Henry Holland, 1st Viscount Knutsford - Wikipedia

[531] See catalog record for: James Aylmer Lowthorpe Haldane Catalog Advanced Search Results | HathiTrust Digital Library.

[532] Summary Information: Repository: Churchill Archives Centre Title: HALDANE, Aylmer, General Sir, 1931-04-22 Reference Code: GBR/0014/WCHL 2/3 Date: 1931-04-22 Scope and Contents: Typed copy (with some omissions) to Lord Knutsford re WSC's escape. Finding aid date: 2004-07-26 14:19:10.983000+00:00. This previously unpublished work must be paraphrased per the UK's 2039 Rule.

[533] Order of St Michael and St George - Wikipedia

[534] Kneesworth Hall, also known as Kneesworth House or Kneesworth House Hospital, is an Edwardian mansion in England, located in Bassingbourn cum Kneesworth, 22.3 miles (35.9 km) south-west of Cambridge in South Cambridgeshire. It was the aristocratic seat of the Nightingale Family and Sydney Holland, 2nd Viscount of Knutsford. Kneesworth Hall - Wikipedia

[535] Part of the litigation involved the escape. Churchill, then First Lord of the Admiralty, called Haldane into his office and attempted to get him to agree to a set of facts regarding the escape. Haldane was not willing to go to bat for Churchill, and stood his ground.

[536] Probably: In black & white, by Sydney Holland, viscount Knutsford. London, E. Arnold & co., 1926. It appears to be an autobiography. #9 - In black & white, by Sydney Holland, viscount Knutsford. - Full View | HathiTrust Digital Library

[537] The Escaping Club, Alfred J. Evans New York, The James A. McCann company [c1922].

1948 Book note.

Randolph Churchill provides the best and most insightful concluding observations to this saga when he relates in his father's biography:[538]

> In 1908 Frankland sent Winston and Clementine a wedding present of Maxwell's[539] two volume Life of Wellington, bound in red leather with the Churchill arms emblazoned on them in gold with the family motto 'Fiel Pero desdichado' (Faithful but unfortunate). A note inserted read: "From J. H. C. Frankland in Memory of November 15th, 1899."
>
> Haldane also kept in touch with his old comrade in arms. When his [Haldane's] autobiographical work, 'A Soldier's Saga' was published in 1948 he sent a copy to Churchill with the inscription: "With profound admiration of an old ally."

Conclusion

Aylmer Haldane was an excellent judge of character and man who, as a general officer, was accustomed to thinking through fully the threads of possibilities of any given situation. He also had full command of his own experiences. The phrase, 'I own my own experiences' was explicit in his character as befits a British officer of his accomplishments and rank.[540] Even when pressed by the First Lord of the Admiralty, he stuck to his guns.

Like several plays by Shakespeare, it all works out well in the end. As Haldane so aptly quotes:

"Be cheerful; wipe thine eyes; some falls are means the happier to arise."[541]

In 1920 when he was assigned to conduct the campaign in Iraq he had several meetings with Winston who was then serving as Secretary of State for War. Regarding the of relief of the city of Samawah, Churchill sent a telegram which began:—

"During these difficult months your patience and steadfastness have been of great value, and I congratulate you upon the distinct improvement in the situation which has been effected by you." [542]

Coronation of Emir (Prince) Faisal I as King of Iraq on 23 of August 1921.[543]

[538] Winston Churchill – Youth 1874-1900 Hughton Mifflin Co, Boston, 1966. p. 486.

[539] William Hamilton Maxwell (30 June 1792 in Newry, County Down, Ireland – 29 December 1850 in Musselburgh, Scotland) was an Irish novelist, historian and clergyman. William Hamilton Maxwell - Wikipedia

[540] He was also truly a British General. His Air Chief of Staff resigned rather than carry out his orders for aerial bombing in Iraq. Expecting to be cashiered when he returned. No action was taken against him.

[541] From Shakespeare: *Cymbeline*. As soldiers can. Be cheerful; wipe thine eyes Some falls are means the happier to arise.

[542] The insurrection in Mesopotamia, 1920, by Sir Aylmer L. Haldane. Edinburgh and London, W. Blackwood and sons, 1922. p. 229-230.

Portrait of Aylmer made in March of 1917.[327]

Photograph taken around the same time.[328]

[543] Photo Post Card of the Coronation of Emir (Prince) Faisal I as King of Iraq on 23 of August 1921. Emir Faisal has already served as King of Syria from March to July. The event was conducted, at six A.M., in the court-yard of the Ottoman built Serai, his temporary headquarters upon arriving in Baghdad. File:Coronation of Prince Faisal as King of Iraq, 1921.jpg - Wikimedia Commons.

Blackwoods Libel Action

Introduction

This is a fascinating tale about the court proceedings themselves. Churchill's 'Character' in the poem, A Lost Letter of Ancient Rome, was purported to be 'Clodius'. He was mentioned five times and alluded to in many other passages. The poem was a scathing parody of the Asquinth administration of the time. Churchill pocketed all the slings and arrows aimed at him and chose his ground carefully to include only that which was libelous. The allegation that he broke his parole. It was a lead pipe cinch he would win and also admirable in the way he framed the action; to ignore the extraneous, and also not to seek any damages other than his costs.

What I find interesting is that the actual libel had nothing to do with his 'going over the wall' on his own. This is an unnecessary detail that Churchill has added to the case. While there are references in the poem to his crossing the floor and shifting his positions around; those are all public record and were made according to his principles and his desire to serve according to his 'lights'.

Haldane's testimony would have been superfluous to the point of irrelevance on the going over the wall issue. It only brought attention to a topic that had not been brought up. Just FYI, the 'Wall' is now 'nine feet high' for this telling of the tale.

Reynolds's Newspaper

Reynolds's Newspaper Sunday 19 May 1912

Mr. Churchill's Action.

Case Against "Blackwood's Magazine" Mentioned.[329]

Mr. Winston is the plaintiff in a libel action against the editor and proprietors of "Blackwood's Magazine," which was mentioned before Mr. Justice Darling[544] in the Bench Division on Thursday.

When the matter came before the Court Mr. F. E. Smith, representing the First Lied of the Admiralty, said this was a case already set down for trial, but was some little distance out of his Lordship's list. It was a case in which Mr. Winston Churchill brought proceedings for alleged libel against the editor and proprietors of "Blackwood's Magazine."

He (counsel) understood that his learned friend, Mr. Douglas Hogg. had been advising the defendants, and it desired to set the down on Monday morning next. [the 21st] The matter would take very little time to dispose of, not more than half an hour. Mr. Winston Churchill desired to make a statement. and Mr. Hogg, so Mr. Smith understood, would also have something to say on behalf of the defendants.

Mr. Hogg was in court, but without his wig and gown. He was, therefore, technically invisible and unable to address the Judge to tell him that he agreed with Mr. Smith's application. Mr. Smith laughingly indicated his learned friend's predicament.

Mr. Hogg: I am afraid I cannot be seen.

The Judge (to Mr. Smith): You can see him, but I can't. (Laughter.) assented to the matter being heard on Monday, and his Lordship accordingly fixed the hearing for that day.

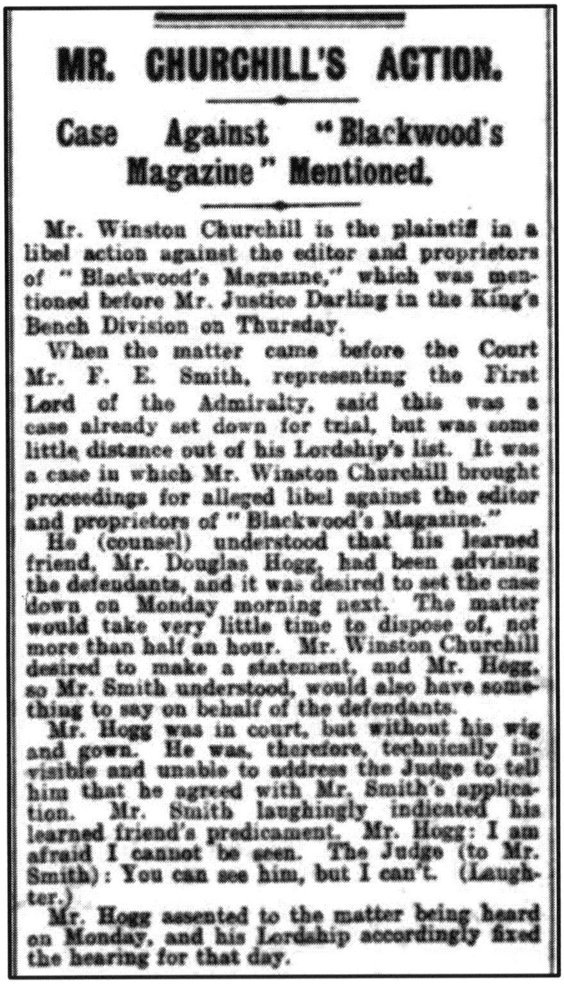

MR. CHURCHILL'S ACTION.

Case Against "Blackwood's Magazine" Mentioned.

Mr. Winston Churchill is the plaintiff in a libel action against the editor and proprietors of "Blackwood's Magazine," which was mentioned before Mr. Justice Darling in the King's Bench Division on Thursday.

When the matter came before the Court Mr. F. E. Smith, representing the First Lord of the Admiralty, said this was a case already set down for trial, but was some little distance out of his Lordship's list. It was a case in which Mr. Winston Churchill brought proceedings for alleged libel against the editor and proprietors of "Blackwood's Magazine."

He (counsel) understood that his learned friend, Mr. Douglas Hogg, had been advising the defendants, and it was desired to set the case down on Monday morning next. The matter would take very little time to dispose of, not more than half an hour. Mr. Winston Churchill desired to make a statement, and Mr. Hogg, so Mr. Smith understood, would also have something to say on behalf of the defendants.

Mr. Hogg was in court, but without his wig and gown. He was, therefore, technically invisible and unable to address the Judge to tell him that he agreed with Mr. Smith's application. Mr. Smith laughingly indicated his learned friend's predicament. Mr. Hogg: I am afraid I cannot be seen. The Judge (to Mr. Smith): You can see him, but I can't. (Laughter.)

Mr. Hogg assented to the matter being heard on Monday, and his Lordship accordingly fixed the hearing for that day.

[544] Charles John Darling, 1st Baron Darling, PC (6 December 1849 – 29 May 1936) was an English lawyer, politician and High Court judge. Charles Darling, 1st Baron Darling - Wikipedia

Daily Mirror

Daily Mirror - Tuesday 21 May 1912.[545]

First Lord's Honour Mr. Churchill's Libel Action on Pretoria Escape Story.

Apology Accepted.

FIRST LORD'S HONOUR

Mr. Churchill's Libel Action on Pretoria Escape Story.

APOLOGY ACCEPTED.

MR. WINSTON CHURCHILL LIBELLED IN A POEM

The First Lord of the Admiralty is seen in court listening to the statement of his counsel, Mr. F. E. Smith, K.C., in the action reported above.—(D.M.P.)

More fond of warlike words than blows.
When captured by our Asian foes,
How cleverly he homeward stole,
And broke his prison, and parole.
Behold him just from school released,
Playing at soldiers in the East;
From discipline he seeks relief
In mutiny against his chief.

These lines, which appeared in the issue of "Blackwood's Magazine" for February last, led Mr. Winston Churchill to bring an action for libel yesterday against Messrs. Blackwood, the publishers. The verses were published under the title of "A Lost Letter of Ancient Rome," and Mr. F. E. Smith, K.C., who appeared as leading counsel for Mr. Churchill, said their object was to satirize prominent members of the present Government. The Ministers referred to were in no case mentioned by name, but under transparent disguises. The verses, which it was contended related to Mr. Churchill, spoke of him as "Clodius," and cast a reflection on his honour as a soldier during the South African War. What happened in Mr. justice Darling's Court will, it is expected, effectually put a stop to any further circulation of the foolish story that Mr. Winston Churchill broke his parole when he made his escape from Pretoria in December, 1899.

Never Gave Parole.

Mr. Churchill, in the witness-box, said that for four years he held a commission in the 4th Hussars, and for many years past he had held a commission in the "Queen's Own" 4th Oxfordshire Hussars.

And, said Mr. Smith, you are the First Lord of the Admiralty to-day?—Yes.

Mr. Smith: Did you ever give parole to the military forces in South Africa when you were captured at Pretoria?—I was neither asked for parole nor did I give it.

Mrs. Winston Churchill, attired in navy blue, was present to listen to the vindication of her husband's honour.

Mr. F. E. Smith told the true story of the escape from Pretoria, "a daring, successful and honourable exploit," including the scaling of a nine feet high wall with a sentry standing close by.

Continuing, Mr. Smith said that Mr. Churchill effected his escape under circumstances that, in view of the nature of the country and the distances, made it an extremely courageous and almost hopeless undertaking.

Utterly Unfounded.

He said this in order to show that the libel contained in the verses he had read converted an enterprise in itself both daring and successful and honourable into the foundation of an atrocious charge, and the offence was not diminished by the fact that the lines were written with undoubted literary ability.

His Lordship : Not too much, I thought. (Laughter.)

Mr. Douglas Hogg (for the defendants) offered a full and frank apology. The story was utterly unfounded. It would never have appeared in "Blackwood's" if more care had been taken, and if those who control that journal had known that reference was being made to Mr. Churchill. The latter had behaved very generously in being willing to accept an apology and an indemnity for costs. A complete apology had appeared in the magazine. Counsel now repeated it in open court.

Mr. Justice Darling allowed the record to be withdrawn, and in doing so pointed out that any repetition of the libel by malicious persons might result in a trial for criminal libel.

[545] Daily Mirror - Tuesday 21 May 1912. p. 5 col 3 https://www.britishnewspaperarchive.co.uk/viewer/bl/0000560/19120521/053/0005.

Westminster Gazette Monday 20 May 1912

Mr. Winston Churchill

Libel Suit Against "Blackwood's Magazine." [546]

Offending Verses.

Result of Cabinet Minister's Action:

There was a large attendance in Mr. Justice Darling's Court in the King's Bench Division this morning in anticipation of proceedings in connexion with the libel action by Mr. Winston Leonard Spencer Churchill, M.P., the First Lord of the Admiralty, against Messrs. William Blackwood and Sons, of Paternoster-row, printers and publishers of "Blackwood's Magazine." The libel was published in February last.

It will be recalled that on Thursday, Mr. F. E. Smith, K.C., applied to Mr. Justice Darling to expedite the hearing of the case, and his lordship agreed to put it in the list to-day, as Mr. Churchill desired to make a statement on oath. Counsel for Mr. Winston Churchill were Mr. E. Smith, K.C., M.P., Mr. R. F. Colam, and Mr. Harold Smith, M.P.. and the defendants were represented by Mr. Douglas Hogg.[547]

Punctually at eleven o'clock Mr. F. E. Smith walked into court, followed by Mr. Winston Churchill, who was accompanied by Mrs. Churchill and a lady friend[548]. Mrs. Churchill was wearing a smart blue costume and a small white straw hat trimmed with a white feather. Mr. and Mrs. Churchill were shown to a seat at the solicitors' table, in front of Mr. F. E. Smith.

Mr. and Mrs. Churchill picture entering Court.[549]

Immediately his lordship had taken his seat Mr. F. E. Smith said this was an action for which his lordship had permitted to he put into the list to-day, brought by Mr. Winston Churchill against the proprietors and publishers Blackwood's Magazine." In the issue dated February 1912 the magazine published some verses tinder the title of "A Lost Letter of Ancient Rome," the object of which was to satirise prominent members of the present Administration. The Ministers referred to were in no case mentioned by name, but under very transparent disguises.

546 Westminster Gazette - Monday 20 May 1912 p. 9 Col 1-3. https://www.britishnewspaperarchive.co.uk/viewer/bl/0002947/19120520/062/0009 & https://www.britishnewspaperarchive.co.uk/viewer/bl/0002947/19120520/063/0009

547 Son of the later Mr. Quintin Hogg, the philanthropist millionaire. (From Daily News & Leader Article, May 21st.).

548 In one article referred to as Mrs. Martell. But I am unable to re-locate the exact article.

549 Daily News (London) - Tuesday 21 May 1912 p. 12. https://www.britishnewspaperarchive.co.uk/viewer/bl/0003212/19120521/318/0012

The Offending Verses.

The verses relating to Mr. Winston Churchill (said counsel) spoke of him under the title of "Clodius," and included the following:

"More fond of warlike words than blows,
When captured by our Asian foes;
How cleverly he homeward stole
And broke his prison and parole.
Behold him just from school released,
Playing at soldiers in the East;
From discipline he seeks relief
In mutiny against his chief."

The substance of the statement complained of, continued Mr. Smith, was that Mr. Churchill broke his parole as a prisoner of war at Pretoria many years ago during the South African war. Mr. Churchill for four Years held a commission in the 4th Hussars, and subsequently was a. major in the Queen's Own Oxfordshire Hussars, and, was well known, he was now First Lord of the Admiralty.

No charge could be more injurious, said Mr. Smith, to a soldier as the head of a great naval department than that he dishonourably broke an obligation that every soldier held sacred. Nobody in public life expected to be free from criticism—indeed, he might add that nobody was more accustomed to it than Mr. Churchill. (Laughter.) Certainly nobody was less than Mr. Churchill likely to object to censure of a political or party character, or to hostile criticism.

Statements of Ill-disposed Persons.

For many years Mr. Churchill had become: mare that statements of the character now complained of were being circulated reflecting on his reputation as a soldier and a man of honour; that had been made by ill-disposed and usually ill-informed persons. These statements had been made under circumstances that made it difficult to bring the responsibility home to persons of position and substance. They had taken various forms, but the commonest was the statement; to which the defendants had given currency in their widely read periodical. The circumstances it desirable that certain details should he in regard to Mr. Churchill and the Pretoria Incidents. The destruction of the armoured train Natal was in November 1899. Mr. Churchill captured on November 5 and taken to Pretoria. Although a civilian, he was held as a prisoner of war by the Boers on the express ground of the damage he had done to their operations, which had involved a large number of soldiers and wounded being carried back to safety to the British lines.

Mr. Churchill (continued Mr. Smith) was confined in certain schools at Pretoria with fifty or sixty other commissioned officers who had fallen into the enemy's hands in the early part the war reached Pretoria, and was held as a prisoner of war without any parole being afforded him, and on terms exactly the same as the other prisoners of war. This lasted until December 12, when Mr. Churchill effected his escape by climbing over the wall of the courtyard of the place he was confined at a point in close proximity to a sentry who was on guard. His two companions attempted to follow him later, but were by the sentry and compelled to desist. Mr. Churchill waited in the vicinity for two hours for his companions, till it became evident that some accident had prevented their departure. The wall that Mr. Churchill scaled was 9 ft. high, said counsel. The idea of returning would in any case have been extremely hazardous, and eventually Churchill escaped under circumstances which were well known.

His Lordship: There was no obligation for him go back?

Mr. Smith (smiling): In any event it would been absolutely impossible.

A Courageous. Undertaking

Continuing, Mr. Smith said that Mr. Churchill effected his escape under circumstances that, in view of the nature of the country and the distances, made it an extremely courageous and almost hopeless undertaking. He said this in order to show that the libel contained in the verses he had read converted an enterprise in itself both daring and successful and honourable into the foundation of an atrocious charge, and the offence was not diminished by the fact that the lines were written with undoubted literary ability.

His lordship: Not too much, I thought. (laughter.)

Mr. Smith—That's a matter of judgement. (laughter)

His lordship—What was the line—"Broke his parole"?

Mr. Smith.—"Broke his prison and parole."

I would not enter into a literary disquisition.

His lordship.—No, no (laughter)[550]

Mr. Churchill was not unwilling to recognise, continued Mr. Smith, that the editor of "Blackwood's Magazine" himself was wholly unconnected with the statements contained in the libel, and Mr. Churchill accepted the editor's statement that had he known he would never have permitted the inclusion of the verses in the magazine. As soon as the editor's attention was called to the fact that the libel had appeared he published a complete apology, which stated that Mr. Winston Churchill had pointed out to them that certain passages in the poem constituted a libel on his personal honour, and they desired to say that it was never the intention of the editor or of the author that such a libel should appear, and they published an unqualified apology.

Mr. Douglas Hogg, said Mr. Smith, who had advised the defendants in the matter, had made a communication to him, and desired to make a statement on behalf of his clients. Mr. Churchill had never desired to use attacks of that kind as a means of making money. It was obvious, on the other hand, that the liability to pay damages might be the only method of deterring malicious persons from making statements of that kind in the future. This did not, however, appear to be one of those cases.

The defendants here had given a complete indemnity in respect of costs, and were prepared to make a public apology that day before their lordships. He (counsel) therefore took that opportunity of saying that whenever and wherever those falsehoods or any variation of them were repeated, Mr. Churchill promised to take steps to punish anybody who, whether by ignorance or malice, aided their circulation.

Mr. Churchill then entered the witness-box and took the oath,[551]

He said for four years he had held a commission in the 4th Hussars, and for many years past he had held a commission in the "Queen's Own" 4th Oxfordshire Hussars.

And. said Mr. Smith, you are the First Lord of the Admiralty to-day? —Yes.

Mr. Smith: Did you ever give parole to the military forces in South Africa when you were I captured at Pretoria?—I was neither asked for parole nor did I give it.

Mr. Smith.—Were the facts contained in my statements state by me with accuracy? —With perfect accuracy.

Mr. Hogg: I have no question to ask.

Defendants' Unqualified Regret.

Mr. Douglas Hogg said he had advised the defendants in the matter, and he desired to make a statement on behalf of the defendants to his lordship and to add a few words to those Mr. Smith had said. As his friend had stated, the editor and proprietors of "Blackwood's Magazine" were totally unaware, until they received a communication from Mr. Churchill's solicitor, that the poem which they had published in the February number contained any personal imputation on Mr. Churchill. It had never been their desire to make any personal attack upon a member of the Government. and least of all to make an attack of that character which everyone would recognise must have caused Mr. Churchill great pain and annoyance.

His Lordship: Do you say this on behalf of the proprietors and editor?

Mr. Hogg: Yes. The magazine is edited by Messrs. Blackwood.

Did you say, went on his lordship, that the editor did not read the poem?

Mr. Hogg: Apparently he did not go through it with the care he ought to have taken, or we should not have been here to-day.

[550] London Evening Standard - Tuesday 21 May 1912 p. 5 col 1-2
https://www.britishnewspaperarchive.co.uk/viewer/bl/0000183/19120521/098/0005.

[551] London Evening Standard - Tuesday 21 May 1912 p. 5 col 1-2
https://www.britishnewspaperarchive.co.uk/viewer/bl/0000183/19120521/098/0005.

Mr. Hogg, continuing, said the proprietors at once took steps not only to write to Mr. Churchill's solicitors, but to put in a prominent part of the magazine an apology which his learned friend had referred to, and call attention to it in the contents page of the magazine.

His Lordship: I gathered it was uncommon for people to pay attention to poetry, but I did not know the editor did not even read it. (Laughter.)

Mr. Hogg: Perhaps he did not get so far; my clients recognise that a charge of this kind having been made Mr. Churchill was perfectly entitled in saying that he could not rest satisfied merely with an apology printed in the paper, and the defendants would repeat in open court the apology which had already been printed in their paper, they also would express their unqualified regret that the magazine should have given currency to such a charge, which they recognised was absolutely unfounded and which should never have been brought.

Continuing, Mr. Hogg said they recognised that Mr. Churchill had behaved handsomely and generously in resting satisfied with this public apology, and not asking for a substantial sum of money as well. The defendants were glad that the apology should be published in their magazine, and they could only express their hope that indirectly that publication might be the means of enabling Mr. Churchill once and for all finally to dispose of a charge which defendants now said was unjust and entirely unjustified.

Mr. F. E. Smith said he thought a convenient course would be that the record in the case should be withdrawn on terms endorsed on counsels' briefs.

Mr. Justice Darling' s Warning.

His Lordship: I will permit the adoption of that course, and I hope it is thoroughly understood that these proceedings in this Court dispose once and for ever of any kind of rumour such as gained currency by this poem in "Blackwood's Magazine." If there had been the slightest reason for any such imputation as was there made it would have been very easy on this occasion to substantiate it.

Mr. Churchill was in the witness-box, he might have been cross examined, and witnesses might have been called. If there were any shadow of foundation for the allegation against him on this occasion it could have been proved. The fact that currency was given to it by such a firm as Blackwood's—long known, with an honourable connexion with literature—made it absolutely necessary, in my opinion, for Mr. Churchill to take these proceedings. .

As Mr. Smith indicated, if anybody should hint at such conduct on the part of Mr. Churchill—and I hope it has been finally disposed of to-day—they have fair warning that he will bring them before the Court, and should they come up before any Court of justice I cannot see the slightest doubt that very heavy damages would be awarded against them.

It is not necessary in such a case as this for people to prove any particular damages. Juries in cases of attacks on private character give exemplary and what are called vindictive damages —I think probably the word originally was "vindicative"—(laughter)—and I mention this because a person of high position does not quite understand the difference I noticed the other day. (Laugher.)

The jury give such damages as are called "vindictive" where conduct such as this has been brought before the Court, and one need not prove any particular damage. I may also state this: Publication of libels is not only a civil wrong, it is a criminal offence, and that is particularly true where they are published concerning people holding the sort of position in the State which Mr. Churchill holds—people, therefore, whose character is of importance not only to themselves. but to the public at large and to the State, and it is just as well that they should know that repetition of this kind of charge may involve them in proceedings before judges of this Court sitting in a slightly different capacity. The record may be withdrawn

Mr. F. E. Smith: There will be payment out of court of the sum mentioned in paragraph one of the defence?

His Lordship: Yes.

Off to Malta & Gibraltar

Daily Mirror - Wednesday 22 May 1912.[552]

MR. ASQUITH'S VISIT TO THE MEDITERRANEAN.

DEPARTURE WITH THE FIRST LORD.

The Prime Minister and Mr. Churchill left London yesterday morning for Genoa, where they will embark in the Admiralty yacht Enchantress for an official inspection of the naval establishments at Malta and Gibraltar.

The First Lord was accompanied by Rear-Admiral David Beatty, C.B., Mr. J. E. Masterton Smith, and Mr. E. H. Marsh, and will be joined on the way by Vice-Admiral Prince Louis of Battenberg, Second Sea Lord. Mrs. Churchill, Miss Violet Asquith, and Miss Hosier also travelled with the party. At Victoria Station to see them off were Mrs. Asquith, Mrs. Cornwallis West, Mr. Edward Montagu (Under-Secretary for India), Sir Francis Hopwood, Sir W. Graham Greene, and Admiral Ernest Troubridge, Chief of the War Staff.

The party travelled in a Pullman car attached to the 11 o'clock boat train express to Dover.

THE DAILY MIRROR

MR. WINSTON CHURCHILL'S NEW STRAW HAT.

Mr. Asquith and Mr. Winston Churchill, accompanied by Mrs. Winston Churchill, Miss Violet Asquith and Miss Hozier (Mrs. Churchill's sister), left Victoria yesterday for Genoa, where they embark on the Enchantress for Malta. Mr. Churchill is making a tour of inspection of the Mediterranean defences. He is seen in the photograph at Calais, wearing a new straw hat, which comes down to his ears.—("Daily Mirror" photograph.)

Departure

Evening Mail - Wednesday 22 May 1912.[553]

Mr. Asquith's Visit to the Mediterranean.

Departure with the First Lord.

The Prime Minister and Mr. Churchill left London yesterday morning for Genoa, where they will embark in the Admiralty yacht Enchantress[554] for an official inspection of the naval establishments at Malta and Gibraltar. The First Lord was accompanied by Rear- Admiral David Beatty, C.R. Mr. J. E. Masterton Smith. and Mr. E. Marsh, and will be joined on the way by Vice-Admiral Prince Louis of Battembarg, Second Sea Lord. Mrs. Churchill, Miss Violet Asquith. and Miss Hosier also travelled with the party.

At Victoria Station to see them off were Mrs. Asquith, Mrs. Cornwallis West, Mr. Edward Montagu (Under-Secretary for India). Sir Francis Hopwood, Sir W. Graham Greene, and Admiral Ernest Troubridge, Chief of the War Staff. The party travelled in a Pullman car attached to the 11 o'clock boat train express to Dover.

[552] Daily Mirror - Wednesday 22 May 1912. p. 4 https://www.britishnewspaperarchive.co.uk/viewer/bl/0000560/19120522/016/0004.

[553] Evening Mail - Wednesday 22 May 1912 p. 5 https://www.britishnewspaperarchive.co.uk/viewer/bl/0003187/19120522/091/0005..

[554] HMS Enchantress - Wikipedia No full article, just this reference: HMS Enchantress (1903) was an Admiralty yacht launched in 1903, sold for breaking up in 1935.

A Lost Letter of Ancient Rome

To the Editor of 'Blackwood's Magazine.[330]

Sir,— A collection of ancient MSS. recently came-it matters not how-into my possession. On examining them, I found that they were letters written by eminent people in the last days of the Roman Republic and dealt with different aspects of the contemporary political situation. Among them was a letter from Cicero to Atticus, which, rendered into the plain manner of Swift's metrical Epistles,[555] might, as it seemed to me, be of interest even today to such readers of 'Maga'[556] as are familiar with the history of that period. I accordingly submit it to your judgment.—Yours respectfully,

A Student.

M. Tullius Cicero[557] To T. Pomponius Atticus.[558]

While, happy man, afar from home,
Through Athens' porticoes you roam,
And hunt all day in secret nooks
For lurking debtors and old books,
His hours, dear Atticus, your friend
Must in the chattering Senate spend
On talk and trifles without end;
Tied to the Forum, watch the tricks
Of tribunician politics
Or waste a world of art and patience
On fruitless party combinations.
"Why"—with your serious face you ask—
Engage then in a hopeless task?
And, as an Optimate, why not
With Cato's[559] self cast in your lot?
Some time you needs must intervene;
And his the cause to which you lean."
Well, first by circumstance I'm tied:
I started on the other side.

[555] Probably: "Ode to the Athenian Society," Swift's first publication, printed in The Athenian Mercury in the supplement of Feb 14, 1691. Archived 13 May 2023 at the Wayback Machine See: Jonathan Swift - Wikipedia.

[556] PF: Shorthand or slang for Blackwood's Magazine. See Haldane's reference to it which I now see as referring to this poem.

[557] Marcus Tullius Cicero; (3 January 106 BC – 7 December 43 BC) was a Roman statesman, lawyer, scholar, philosopher, writer and Academic skeptic, who tried to uphold optimate principles during the political crises that led to the establishment of the Roman Empire. [Optimates are seen as supporters of the continued authority of the senate. The populares are seen as supporting popular assemblies, generally in opposition to the senate.] Cicero - Wikipedia

[558] Titus Pomponius Atticus (November 110 BC – 31 March 32 BC; later named Quintus Caecilius Pomponianus Atticus) was a Roman editor, banker, and patron of letters, best known for his correspondence and close friendship with prominent Roman statesman Marcus Tullius Cicero. Atticus was from a wealthy Roman family of the equestrian class (lower aristocratic non-ruling class) A close friend since childhood, Cicero dedicated his treatise, Laelius de Amicitia (Latin for 'Laelius on Friendship'), to Atticus. Their correspondence, often written in subtle code to disguise their political observations, is preserved in Epistulae ad Atticum (Letters to Atticus). Titus Pomponius Atticus - Wikipedia

[559] Marcus Porcius Cato Uticensis ("of Utica") (95 BC – April 46 BC), also known as Cato the Younger, was an influential conservative Roman senator during the late Republic. His conservative principles were focused on the preservation of what he saw as old Roman values in decline. A noted orator and a follower of Stoicism, his scrupulous honesty and professed respect for tradition gave him a political following which he mobilised against powerful generals of his day, including Julius Caesar and Pompey. Cato the Younger - Wikipedia

Marius,[560] my grand old countryman,
Who these mad practices began,[561]
Led me, and with so strong a will
I find I'm half his follower still
Then there's this Consulship of mine:
I can't forget friend Catiline.
No decent Party would deny
I saved my country; who but I?
And yet these Nobles" grudging spirit
Will but allow me half the merit.
Besides-it is an ancient trouble! -
In civil conflicts I see double;
And if the Parties can't agree
To things that might, but will not, be,
I stand aloof, inert and wise,
To view the warfare I despise,
Because the fools won't-compromise.

You know how stiffly our good Knights
Fought for their old judicial rights,
Which Sulla[562] did so late retrench,
And for his friends swept clean the Bench:[563]
I laboured long with voice and pen
To reinstate the moneyed men,
And from the Senate gained thus much;
But when I'd brought the two in touch,
Our Cato must of course proscribe
Each juror who'd received a bribe;
So, when defaulters were dismissed,
I found myself with half a list.
Contractors, too, he tied, like Fate,
To their strict bargains with the State.
In short, the Knights believed their class
Insulted-and myself, an Ass![564]

Well! since it quite beyond my art is
To reconcile in gross the Parties,
Let us through all our "Statesmen" run,
As individuals, one by one:
Judge you my possible allies,
And when you have adjudged, advise.

[560] Gaius Marius (c. 157 BC – 13 January 86 BC) was a Roman general and statesman. Victor of the Cimbric and Jugurthine wars, he held the office of consul an unprecedented seven times. Gaius Marius - Wikipedia

[561] Cicero and Maris were natives of Arpinum, and Cicero began his political career in connection with the Marian or Democratic party.

[562] Lucius Cornelius Sulla Felix (138–78 BC), commonly known as Sulla, was a Roman general and statesman. He won the first major civil war in Roman history and became the first man of the Republic to seize power through force. Sulla - Wikipedia

[563] Sulla deprived the Knights, who represented the moneyed middle-class, of the Judicial powers which they had acquired under the Gracchi, and he restored these to the Senate.

[564] See Cicero to Atticus i. 17. 8; ii. 1. 8.

Pompey the Great![565] Our natural Leader?
Pompey's the Party's Special Pleader.
None, with so fair a show of right,
Can make the Party black look white;
None then so fit, it may be said,
To be the Party's Figure-Head.
Let Pompey be but Head in name,
To Pompey it is all the same
Which cause he pleads; he'll choose for Figure
The side that makes him *look* the bigger!
Not long ago I hoped that he
Would for the Empire work with me.
Full praise I deemed he'd, sure, assign
To all I'd done with Catiline.
Would you believe? the other day,
When, in the House, there was a way
The fitting compliments to pay,
Praising my energy and skill,
The shabby fellow sat quite still:
He scarce alluded to the thing;
And all the official buttering,
All the superlative degrees,
He left to-Crassus, if you please.[566]

Pompey, when angling for the Knights,
Would talk forever of their "Rights."
His words were full of milk and honey;
None need be scared about their money:
But when he found himself the neighbour
Of Clodius[567] [568]and his Guilds of Labour,[569]
His "nay" was changed at once to "yea":
The bully Tribune had his way.

Though Pompey would by Law become
Lord of the Liberties of Rome,
He means not, in the execution,

[565] Gnaeus Pompeius Magnus (29 September 106 BC – 28 September 48 BC), known in English as Pompey, was a general and statesman of the Roman Republic. He played a significant role in the transformation of Rome from republic to empire. Early in his career, he was a partisan and protégé of the Roman general and dictator Sulla; later, he became the political ally, and finally the enemy, of Julius Caesar. Pompey - Wikipedia

[566] Cicero's vanity and his annoyance with Pompey for not "buttering" his Consulship sufficiently are amusingly illustrated in his Letter to Atticus, i.14. 3. He did not like Crassus.

[567] During the Late Republic, the spelling Clodius is most prominently associated with Publius Clodius Pulcher, a popularis politician who gave up his patrician status through an order in order to qualify for the office of tribune of the plebs. Clodius positioned himself as a champion of the urban plebs, supporting free grain for the poor and the right of association in guilds Clodius - Wikipedia

[568] Publius Clodius Pulcher (c. 92[1] – 18 January 52 BC) was a Roman politician and demagogue. A noted opponent of Cicero, he was responsible during his plebeian tribunate in 58 BC for a massive expansion of the Roman grain dole as well as Cicero's exile from the city. Leader of one of the political mobs in the 50s, his political tactics – combining connections throughout the oligarchy with mass support from the poor plebs – made him a central player in the politics of the era. Publius Clodius Pulcher - Wikipedia

[569] Clodius organised in his own behalf the Colleges or Guilds of Trades, and with their help maintained a reign of terror in the streets.

To violate the Constitution.[570]
To Despotism he fain would rise,
But still by decent party cries;
So, if through him the country fall,
Its end is constitutional.
Used as the "moderate men's" decoy,
Freedom he'll legally destroy,
Slow, timid, calculating tool,
Mad Clodius' butt, and Caesar's fool

Clodius by nature is no slug
Like Pompey: dagger, dungeon, drug,
All urge him on his wild career:
He craves excitement, far or near.
Behold him, just from school released,
Playing at soldiers in the East!
From discipline he seeks relief
In mutiny against his Chief.[571]
More fond of warlike words than blows,
When captured by our Asian foes,[572]
How cleverly he homeward stole,
And broke his prison and parole!
At home a thousand apish tricks
He plays on Rome's dull politics.
Through the whole party round he'll run;
All sides he joins, and sticks to none;
To-day helps Catiline, for sport;
To-morrow drags him into Court;[573]
And-Combination sage and pleasant! -
Is Caesar's jackal, just at present.

Scarcely had Clodius ceased to be a
Male votary of Bona Dea,
When, client of the injured spouse,[574]
Though sprung himself of noble House,
He flings abroad a thousand bribes
To woo the Jurymen and Tribes,
Attacks My Lords with shout and paean,
And begs to be enrolled "plebeian."[575]
Tribune elect, he's now aflame

[570] Plutarch says that Cato undertook to vote for Pompey as sole consul partly because any government was preferable to anarchy, and partly because Pompey promised, with the protection of the City, a Constitutional Government. Pompey, however, "held it but one step from anarchy to absolute Power. "-*Life of Gato.*

[571] Plutarch, *Lucullus,* 34.

[572] Dion, XXXV. 15.

[573] Plutarch, *Cicero*, 29.

[574] Clodius was suspected of an intrigue with Caesar's wife, Pompeia.

[575] Patricians could not be elected to the office of Tribune. Clodius when seeking election to that office, had therefore to be "adopted" into a plebeian House.

With patriot rage: I see his game.
I've marked for his revenge, since I
Exposed his perjured "Alibi"[576]
Nor will his fury cease until
He's either driv'n me by a Bill[577]
To exile, or with kicks and ouffs
Dispatched me by his banded Roughs.

Agrarian Flavius[578] finds his job,
Like Clodius, with the flattered Mob;
The Treasury's spendthrift, clever fellow!
None with more brazen lungs can bellow
Against the noble pirate band
Who from the People steal the land.
Then, - when he's raised the party storm,
He, he alone, can plan "Reform."
Pretty, in a tight situation,
To see him front a Deputation!
So mild, so plausible, so glib;
Flavius, you'd think, can never fib.
Yet, though in specious phrases skilled,
He scarce deceives the Clodian Guild.
"Take heed!"- I overheard one say -
"If we to Flavius grant his way,
We're done! He will as surely gull us
As Pompey's self or rich Lucullus."
Unless I'm much mistaken, I,
He's tuned the party key too high:
Splendid in theory, but in act
I doubt he'll find his fiddle cracked.

But why, you say, why bid me hear
Of Flavius and the like "small deer,"
And do not rather ask my mind
Of Brutus, Cato, and that kind?
Well, yes! In Brutus I record
A champion equal to my sword;
Praised, envied, blamed, abused, admired
Of course (but most when he retired),
None could match Brutus in debate;
His style was worthy of the State,
The Patriot Party's helm and shield -

[576] Clodius, who was prosecuted for intruding himself into the mysteries of Bona Dea, at which only women were allowed to be present, set up an "Alibi" as his defence. Cicero, called as a witness against him, swore that he had seen him in Rome at the time.

[577] Clodius passed a Bill through the Comitia Tributa, B.C.. 58, providing that "Any one who had put citizens to death without trial should be outlawed." This of course had reference to the execution of Lentulus during the Catilinarian conspiracy, and caused the exile of Cicero.

[578] Lucius Flavius, Tribune of the People, author of the Agrarian Law referred to by Cicero in his Letter to Atticus, i. 19.4. The purchase money was to be provided by revenue flowing into the Treasury from foreign sources.

But Brutus now has left the field.
He strove, good "honourable man,"
To be a Stoic partizan.
Through supersubtlety of wit
To lead the multitude unfit,
His mind sagacious, sound, eclectic,
Turned policy to dialectic;
And often, at a Party Meeting,
Athirst for phrases high and heating,
I've known him pause to puzzle out
The whole Philosophy of Doubt.
Scorning the frenzy and the fuss
Strained from the dregs of Romulus,[579]
He into exile goes with Plato,
Disdainful as his Uncle, Cato.

For Cato - Does he hope to gain
A Party triumph by Disdain?
He penetrates the Tribunes' shams,
But can he rule, with Epigrams?
If in his cause the Knights invest,
They will exact high Interest.
And if, on strictly moral ground,
Their claim seems bad - why not compound?[580]
Should Cato their demands deny,
His only choice will be - to die;
And if he please his Stoic whim,
And dies, - why, who will die with him?
Those valiant "Die-Hards," who to-day
Encircle him in brave array,
He'll find their principles have veered,
Whene'er their Mullets grow a beard[581]

Look round the world in short, I see
No soul with whom I can agree;
For always, in another's line
There's something contrary to mine;
I cannot make the two combine.
Hence, all things said, it seems not wise
To act, but still to criticise.
Far from the world of State I live,
In wisdom "deep contemplative";
Sigh for my Country, and expand
My lonely furrow in the sand.

[579] 'In Faece Romuli' Cicero to Atticus, ii. 1. 8.

[580] Cato, by his straightforward honesty, spoiled Cicero's diplomatic policy of brining about a union between the Senate and the Knights.

[581] "Cum . . . nostri priniciples digito se ecelum putent attingere, si mulli barbati in piscinis sint, qui ad manum accedant, " &c. – Cicero to Atticus, ii.1.7.

Apology by Blackwood's

April 1912 Edition.[582]

A Lost Letter of Ancient Rome.
An Apology.

In the February number there appeared a poem, "A Lost Letter of Ancient Rome." Mr. Winston Churchill has pointed out to us that certain lines in this poem constitute a libel upon his personal honour. We desire now to state that such was not our intention, nor the intention of the writer of the poem, and to offer to Mr. Churchill our unqualified apology for the statement, and to contradict the impression which the lines apparently convey, that Mr. Churchill, when in South Africa, broke his parole.

The Editor.

BLACKWOOD'S MAGAZINE.

No. MCLVIII. APRIL 1912. Vol. CXCI.

OUR LOST MAGNA CARTA.

THE TRADE DISPUTES ACT AND REVOLUTION.

"Principiis obsta. Sero medicina paratur
Quum mala per longas convaluere moras."

The country is all too late beginning to realise the perils and the possibilities of Syndicalism. A general strike in the coal industry, following within a few months the disastrous railwaymen's strike of 1911, is stirring the feelings and thoughts of all classes in all civilised communities. Public opinion is not easily aroused in this country. "Le bal cherche le bon joueur" is a maxim equally true of politics and of tennis; but unfortunately the power of intelligently anticipating the trend of political events is a weapon rarely to be found in the intellectual armoury of the normal Englishman: the nation is often forewarned, she is seldom forearmed; and so it almost invariably happens that when the blow falls, England, all unprepared, is left to "muddle through somehow" as best she can. In this crisis, however, the danger is too imminent to be overlooked, too near at hand to be disregarded. It is obvious that a temporary suspension of the railway or mining systems spells commercial disaster, that a prolonged suspension must result in the starvation of the people. Why are the strikers "holding the nation to ransom"? Can nothing be done to stay the progress of the industrial Juggernaut? Is it not possible to establish a permanent settlement, or is Syndicalism destined to develop into Revolution? Such questions as these are freely asked, and no satisfactory

VOL. CXCI.—NO. MCLVIII. 2 H

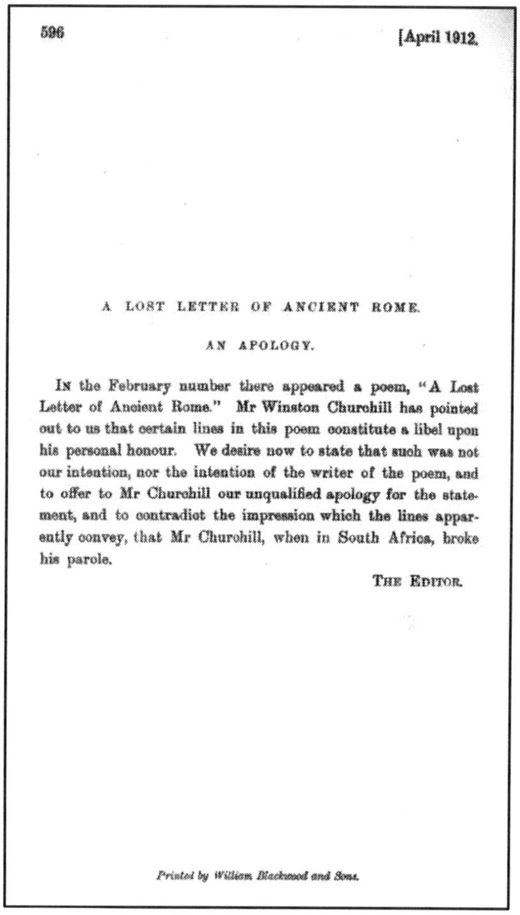

596 [April 1912.

A LOST LETTER OF ANCIENT ROME.

AN APOLOGY.

In the February number there appeared a poem, "A Lost Letter of Ancient Rome." Mr Winston Churchill has pointed out to us that certain lines in this poem constitute a libel upon his personal honour. We desire now to state that such was not our intention, nor the intention of the writer of the poem, and to offer to Mr Churchill our unqualified apology for the statement, and to contradict the impression which the lines apparently convey, that Mr Churchill, when in South Africa, broke his parole.

The Editor.

Printed by William Blackwood and Sons.

582 Blackwood's Magazine, April 1912-04: Vol 191 Iss 1158 Page 596 (Last Page of 156 page issue) https://archive.org/details/sim_blackwoods-magazine_1912-04_191_1158.

Truth Review of Libel Law

Truth - Wednesday 29 May 1912.[583]

Scrutator.

Penal Damages for Libel.

The action of Mr. Winston Churchill against Blackwood's Magazine is another illustration of the muddle into which the law of libel has been allowed to drift. For a long time past Mr. Churchill has been pursued by his political opponents with insulting references to a story about his having broken his parole when he escaped from captivity at Pretoria. He has very properly taken legal proceedings against one or two parties for the purpose of clearing himself of this imputation. ....

It is the confusion of the provinces of civil and criminal jurisprudence. In every one of the cases referred to above —and in a dozen others which might be culled from the law reports of the last two or three years—it is manifest that the principle on which the damages were assessed by the jury was, not to compensate the plaintiff for his sufferings and loss, which were a negligible quantity, but to punish the defendants. And on the same principle the Court of Appeal has declined, whenever there has been an appeal, to pronounce the damages excessive or contrary to reason. ....

The punishment of offences is essentially a criminal matter. It is quite true that in certain specific cases an injury may be at law either a criminal offence or a "tort" or civil wrong. Libel is one of these cases. But it is utterly wrong in principle that the two elements of crime arid tort should be confused in one legal proceeding, and there is no precedent for its being done in the case of any other injury than a libel. ....

A man may bring a civil action for assault, but a jury would not be allowed to award several thousand pounds damages for a trifling assault on the ground that crimes of violence must be severely punished. To punish them is the business of magistrates and judges of assize. ....

[583] Truth - Wednesday 29 May 1912 p.14.
https://www.britishnewspaperarchive.co.uk/viewer/bl/0002961/19120529/037/0014
Page 15 https://www.britishnewspaperarchive.co.uk/viewer/BL/0002961/19120529/039/0015.
Page 16 https://www.britishnewspaperarchive.co.uk/viewer/BL/0002961/19120529/041/0016.

Haldane Appendix

Other Train Ambush Accounts

The following apply equally to Haldane's exploits as well as Churchill's. For space considerations in the series volumes they have been included in this work.

Natal Witness Account

The following account of the attack on the armoured train was published in the Natal Witness, on November 17th, 1899.[584]

Thursday 9:40 a.m. Estcourt

... Capt. Wylie,[585] who is doing well, describes Mr. Winston Churchill's conduct in the most enthusiastic terms as that of as brave a man as could be found. It was on Mr. Churchill's initiative that Capt. Wylie and a number of his men worked to get the trucks blocking the line out of the way.

After that task had been accomplished, amid a hail of missiles, and the engine and tender were able to move forward, Mr. Churchill again directed the engine-driver to steam back to the rear trucks, which were still on the line, with the object of hooking them on and so saving most of the men. Unfortunately, he found the leading truck had been injured, probably by shells, as it could not be coupled on, and Mr. Churchill then directed the driver to proceed slowly while he got as many wounded as possible on the tender.

It was while he was co-operating with Mr. Churchill in this work that Capt. Wylie was shot through the hip. Before collapsing Capt. Wylie succeeded in getting his head behind a large boulder, and there he lay on his face, helpless. While in that position a shell struck the boulder and smashed it to pieces, without further injury to the wounded officer.

When the engine came past again, he was placed on the tender. Previous to being shot down Capt. Wylie had seen that all his men were on the move in retreat. Mr. Churchill accompanied the wounded to Frere Station, and then returned to the scene of the fight to attend to the wounded left on the field. All firing had then ceased, so that Mr. Churchill, like those missing, must be a prisoner.

All is quiet so far this morning... .

Benett Burleigh's Account

The following account of the attack on the armoured train was published by Mr. Benett Burleigh who was the correspondent for the Daily Telegraph and stationed in Estcourt.[586]

He also went out on one of the train reconnaissance missions as shown in the photograph below.

The event of the week has been the attack upon the armoured train. On Wednesday morning it was ordered out at 5 a.m. to proceed carefully from point to point to Frere. Under Captain Haldane, of the Gordon Highlanders, who for the time was attached to the Dublin Fusiliers, seventy-two non-commissioned officers and men of that battalion, with Lieutenant Frankland, forty-five non-commissioned officers and men of the Durban Light Infantry, under Captain Wylie and Lieutenant Alexander,[587] and five bluejackets from her Majesty's ship Tartar, under a petty officer, manned the train. Besides these were the engine hands and seven platelayers to repair damages.

[584] Recited in OB p. 450-451.

[585] James Wylie No. 28 Lieutenant Colonel Source: Nominal roll in WO127 Durban Light Infantry. Anglo Boer War - Name search "Sixty men of the Durban Light Infantry formed part of the personnel of the armoured train which at this time patrolled daily from Estcourt to Colenso. On 15th November a rail was removed or twisted, and the train was attacked; 2 men of the Durban regiment were killed, Captain J. Wylie and 15 were wounded, and 19 were taken prisoners, of whom 8 were wounded, 1 mortally. A squadron of Carbineers and one of Imperial Light Horse came out to the help of the armoured train. These reinforcements drove back the enemy, killing 3." See: Durban Light Infantry article @ Anglo Boer War - Anglo Boer War.

[586] Benett Burleigh The Natal Campaign Chapman Hall Ltd. 1900 Pages 75-83.

[587] Unable to identify this individual. Too many possibilities without rank or a first name. Anglo Boer War - Name search

Mr. Winston Churchill, of the Morning Post, also rode upon the train. Correspondents were wont to proceed in rotation upon the trips, but that morning others whose turn it was, either too sleepy or indifferent, did not embark. A Scotch or Glasgow telegraphist in the Natal Government employ[588] also accompanied the troops, so as to wire back any information either from the stations or by hitching on to the telegraph lines en-route.

Bennet Burleigh in 1898.[589]

The train stopped for a few minutes at Ennersdale, the first station, and about seven miles north. From there a message was sent back that the line was clear, and, receiving authority, the train proceeded onward to Frere, the next station. There another brief stoppage was made, and the police patrol, being interrogated, stated that they had seen no Boers, although it was known that a few of the enemy had occupied a farm but four miles off the previous night. A wire was sent in to headquarters from Frere that all was clear, and, without waiting for any reply, the train ran on to Chieveley. On the way several natives tried to warn the train back.

At Chieveley Station about 300 Boers were seen upon both sides of the line, but some distance away, riding furiously towards Frere, as if to intercept the return of the train. It consisted of, in front, an open flat truck, upon which was a 7-pounder manned by the bluejackets. Next it was an armoured truck with some of the Dublins. Behind that was the engine and tender, and again came two armoured trucks and an open flat truck with railway plant. In the rear armoured trucks were a few of the Dublins, the Durbans, and the railway staff.

Three miles nearer Frere, rounding a curve, and under the slope of a hill, the Boers poured volleys at the train, and at the same moment a repeating 3-pounder Maxim- Nordenfeldt cannon began blazing at them. Two similar guns also opened fire on the devoted train. Suddenly, in a moment, and without warning, as the train was steaming rapidly back—it is a single line—a point was reached where the Boers had removed the fish-plates and propped up one side of the lines with stones. Instantly the flat rear truck—then in front, as the engine was backing—jumped the metals, followed by the next two trucks.

They ran, bumping, for a little distance ; then the flat truck and the armoured truck next turned over, throwing all the occupants into the field. The flat truck turned over, wheels uppermost, whilst the top-heavy armoured carriage almost turned a complete somersault. Luckily the engine and tender and the other trucks kept the metals. A platelayer was killed outright, being pinned under the waggon; but most of the others miraculously escaped with a shaking and slight bruises. Then it was that the Boers began firing shells and bullets, faster than ever, at those struggling upon the ground, striving to free themselves from the wreckage.

With magnificent intrepidity, although being shot at from three sides, the Dublins and volunteers began returning the enemy's fire, taking what little shelter they could alongside the line. Mr. Winston Churchill and Lieutenant Frankland clambered out of the truck next the gun, and, proceeding to the derailed waggons, called for volunteers to assist in clearing the line. A score of willing hands responded, Captain Wylie and others of the Durban Light Infantry assisting. Amid a hail of bullets and bursting 3-pounders the flat truck was tilted over,

[588] Mr. R. T. McArthur (See his account infra.).

[589] Bennet Graham Burley (1840 – June 17, 1914) was a Scottish-born pirate, Confederate spy and journalist. Later in life, he changed his surname to Burleigh and became a celebrated war correspondent for London's The Daily Telegraph. Bennet Burleigh - Wikipedia File:Bennet Burleigh 1898.jpg - Wikipedia The Natal campaign : Burleigh, Bennet, d. 1914 : Free Download, Borrow, and Streaming : Internet Archive

and a few men released. Then they tried to push away the armoured truck, but it was too heavy for their united efforts. Churchill and the others encouraged the men, rallying them again and again.

Meanwhile the Dublins, Durbans, and a few others were pouring volleys into the almost unseen Boers hidden behind the rocks, about 1,000 yards away. Independent firing was also being kept up at them, and the bluejackets, bravely commanded by their petty officer—who was the incarnation of coolness—got their 7-pounder into action. They sent two, if not three, well-aimed shells at the Boers, several hundred of whom lined the hills. But just then a shot from the enemy's 3-pounder or field gun hit the small naval 7-pounder, knocked gun and carriage on to the veldt, and wounded several of the seamen. But the men were not a whit beaten.

Sergeant E. Basset,[590] of the Dublins, standing up, shouted his orders to the men, giving them the direction and ranges in the coolest manner. Nearly everybody had clambered out of the armoured trucks, which were being pelted with the enemy's shells. The "character" of the Dublins, Private Kavanagh—that day one of the stretcher-bearers—chaffed and encouraged his comrades, telling them the Boer shells could hit nothing. He it was who at Dundee, after the long day's battle, being asked if he was hungry and did not wish for something to eat, said, "No. How can I with my mouth full?" "Full!" said his officer—"what do you mean?" "Why, my heart's been in it all day, sir!" replied Kavanagh, with a grin. And so the "hard case" of his battalion shouted and joked, walked about amid a tempest of bullets, and stirred the gallant, glorious Dublins to shoot well and thrue (*sic*).

The trucks could not be man-handled, so engine- driver Wegner [Wagner] and his mates uncoupled the waggons that had been in front, ran them down the line a little way, and commenced to butt and smash a road through the overturned waggons. By dint of pulling back with chains and smashing forward, the engine and tender won past the wreckage. Then they tried to return and pick up the other two waggons, the armoured and flat truck, but they were baulked by one of the others having swung round across the line, and finding the coupling- chains smashed. Captain Wylie,[591] who had been shot through the leg early in the action, got Sergeant Tod,[592] of the Durbans, to place some stones to protect his head as he lay upon the ground. Scarcely had Tod done so and turned to help the other wounded, when a shell knocked Captain Wylie's little parapet away, and Tod was himself thrown up by a kick from a piece of the projectile which hit him severely.

All the wounded, with the help and direction of Churchill, were placed upon the tender. Meanwhile the engine was struck in several places by shells. The feed- injector was broken. A shell passed through a corner of the tender, killing one of the wounded men and maiming others as they lay prone upon the coal-sacks, whilst another missile struck the smoke- box, and missed penetrating the boiler by the merest shave, but it started a leak or two. Wegner, the driver, was hit and stunned by a piece of shell. Then, there being no more to be done, the engine and tender, with its freight of fully a score wounded and well—there were a dozen injured—steamed on to Ennersdale and then to Frere to summon assistance.

Those left behind spread out along the line were still blazing away at the Boers, fighting as a forlorn hope, whilst the train drew off. Several of the Durbans and Dublins followed it a little way on foot, shooting at the enemy as they slowly retired. At Frere it was found that the station hands and police had gone and taken the telegraph instruments with them. Those McArthur took out with him had been broken in the action, and, there being no other way out of it, with their mangled freight of dead and wounded, engine and tender steamed back into Estcourt. Mr. Churchill, how- ever, after handing in his revolver and field glasses, said he was going back to the scene to assist the wounded and stand by the men. The last seen of him was as he trudged alone away down into the arena of battle, where the shot and shell were still screaming, splintering rock and ploughing the ground. A few of the luckier fugitives passed him on the way, but failed to turn him back from his purpose. A small body of mounted volunteers, who rode out from Estcourt and Frere, got upon the right flank of the Boers, and inflicted some loss upon the enemy before they were compelled to retire before superior numbers.

[590] Unable to identify this individual at Anglo Boer War - Name search.

[591] James Wylie No. 28 Lieutenant Colonel Source: Nominal roll in WO127 Durban Light Infantry. Anglo Boer War - Name search "Sixty men of the Durban Light Infantry formed part of the personnel of the armoured train which at this time patrolled daily from Estcourt to Colenso. On 15th November a rail was removed or twisted, and the train was attacked; 2 men of the Durban regiment were killed, Captain J. Wyllie and 15 were wounded, and 19 were taken prisoners, of whom 8 were wounded, 1 mortally. A squadron of Carbineers and one of Imperial Light Horse came out to the help of the armoured train. These reinforcements drove back the enemy, killing 3." See: Durban Light Infantry article @ Anglo Boer War - Anglo Boer War.

[592] James Tod OR Sergeant Source: Nominal roll in WO127 Durban Light Infantry.

The news of the disaster to the train caused a shock to the little garrison, all of whom were anxious to be led out to their comrades' rescue. But it could not be; larger interests held the commander's hands, no doubt, and the firing which we had all heard went on for some little time longer, and then ceased. Subsequently an ambulance train was sent out, under Surgeon-Captain Briscoe, to bring in the wounded. The Boers met him, but declined to give any information or return a man until General Joubert had been communicated with. The officer was told to return next day, when an answer would be given him. He did so, taking as before medicines and stretchers with him; but the answer was that the wounded and prisoners would be sent to Pretoria. Incidentally, Surgeon-Captain Briscoe learned that, according to the Boers, three of our men had been killed and twelve wounded.

That does not include those killed and wounded brought in by the engine and tender. The exact number missing are—Dublin Fusiliers, two officers (Captain Haldane and Lieutenant Falkland) and forty-five non-commissioned officers and men; Durban Light Infantry, twenty-four non-commissioned officers and men missing.

During the afternoon and evening: several of the men straggled back, including a platelayer or two. There are four platelayers missing, and Mr. Winston Churchill, about whom the Boers would give no information beyond saying that the list and report of the affair would appear in the Pretoria newspapers.

In periods of great excitement small matters become lost to the recollection. Of those I interrogated very few agreed as to various details of the disaster.

Robert Taylor McArthur's Account

"The telegraphist, Mr. R. T. McArthur,[593] furnished, perhaps, the clearest and most succinct account." (To Benett Burleigh) He said:—

"We left Estcourt at 5.30 am and ran on to Ennersdale, and reported by wire, 'All clear'. Shortly before reaching Frere we met and spoke to some Natal Police, who had been bivouacking upon the kopjes.

Natal Native Police.[331]

African Scouts.[332]

They told us that the Boers had all gone back the previous night. Then we went to Frere, where we wired the General, 'all well' and without waiting for a reply ran on to Chieveley. Shortly before entering that station we saw 50 Boers going west at a canter with some waggons, as we thought. We waited a few minutes at Chieveley, and then started back. About 3 miles out, or 2 miles north of Frere, we noticed several hundred Boers about 800 yards off, on the west side. Then we saw more on the east of the line. They began firing at us, first with rifles and then with Maxim-Nordenfeldt cannon. One of their shots made a big dent in the rear armoured truck, but did not enter. Their guns were behind the kopje: at least two repeating cannon, which shot a steady flame and a heavier piece, that threw shrapnel at us. But a few yards on, our rear, then the front truck ran off the line,

[593] Robert Taylor McArthur was born in Scotland in 1877 and came to Natal in 1897 to take up a position as a telegraphist based at Glencoe Post Office (about 5 miles from Dundee). After the battle of Talana, McArthur joined the stream of refugees going to Ladysmith. He offered his services to the Military and was one of the two telegraphists on the Armoured train which was ambushed by the Boers on 15th November 1899. Benett Burleigh The Natal Campaign Chapman Hall Ltd. 1900 Pages 75-83 Credit to Mike Rowan See: Anglo Boer War - The telegraphist on the Armoured Train - Boer War Forum.

shaking and jolting terribly, and ours, the next or armoured one, followed suit, and soon all three left the rails. The next thing I knew we were all upset, and strangely, only one man was killed. We all scrambled out. Sitting down, the soldiers began firing volleys at the Boers, who responded by peppering us with more shot and shell. In about 5 minutes, Mr. Churchill came from what was the front of the train, took charge, and asked for volunteers to shift the trucks. About 15 men helped to do so, but the waggons were too heavy to move. Then the engine managed to smash through, breaking them up, and getting knocked about in doing so. All this was done under heavy fire. We tried to couple up the trucks that had been in front, but could not, the line being blocked. Then, picking up all the wounded we could see, we started for Frere to get assistance. My instruments were smashed, and we found that those at Frere had been carried off by the police for safe-keeping. From there we pushed ahead to Ennersdale, whence I wired to the General, giving him a few details. We had several men shot down whilst we were putting the wounded on the tender, although our troops did their best to cover the operation by firing volleys. Several of the shells struck the telegraph wires and poles, cutting and knocking down the line…."

Natal Witness 2

Natal Witness, November 17th, 1899.[594]

The railway men who accompanied the armoured train this morning ask me to convey to you their admiration of the coolness and pluck displayed by Mr. Winston Churchill, the war correspondent who accompanied the train, and to whose efforts, backed up by those of the driver, Wagner, is due the fact that the armoured train and tender were brought successfully out, after being hampered by the derailed trucks in front, and that it became possible to bring the wounded in here.

The whole of our men are loud in their praises of Mr. Churchill, who, I regret to say, has been taken prisoner. I respectfully ask you to convey their admiration to a brave man.

Charles Wagner (Driver) and Alexander Stewart.[595]

The Albert Medal.[333]

[594] Recited in OB p. 451-452.

[595] Image source: Alexander M. Stewart In 1910 Churchill would obtain the Albert Medal for Lifesaving for each of them when he was Home Secretary. Sandys p. 67 and WSC My Early Life. Ch. XIX Note 4. Albert Medal for Lifesaving - Wikipedia

Kroonstad Telegram

Official report of the armoured train engagement as received and released in Bloemfontein on 16 Nov 1899, reporting the events and the capture of "Windstone Spencer Churchill."

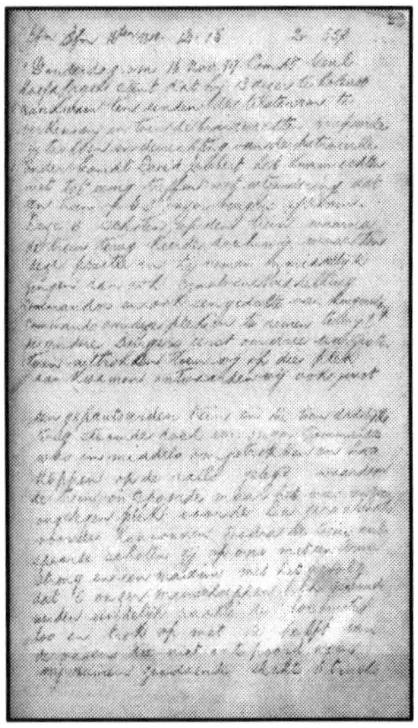

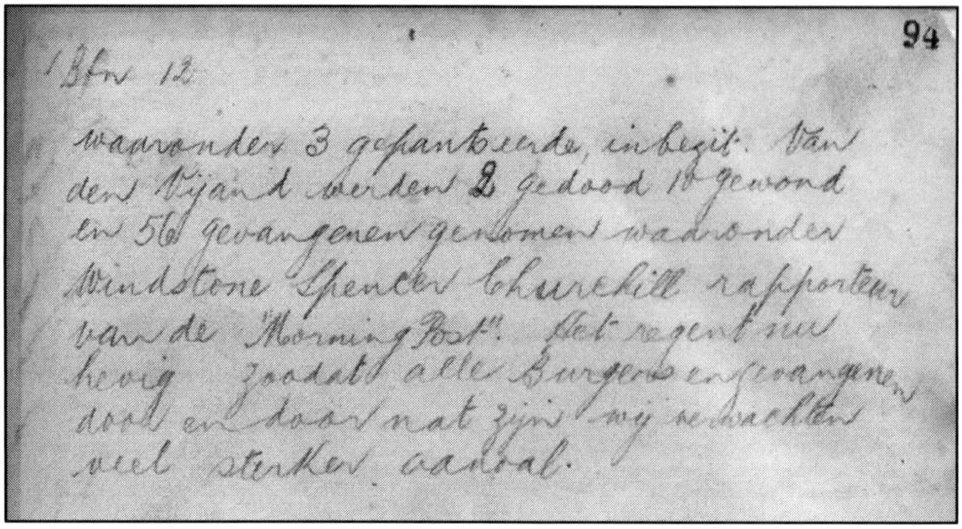

94

Bfw 12

waaronder 3 gepantserde, inbegrip. Van den Vijand werden 2 gedood 10 gewond en 56 gevangenen genomen waaronder Windstone Spencer Churchill rapporteur van de "Morning Post". Het regent nu hevig zoodat alle Burgers en gevangenen door en door nat zijn wij verwachten veel sterker aanval.

Original Telegram citing Dutch 'Windstone Spencer Churchill' of the "Morning Post."[596]

Bloemfontein 16den Nov. 12.15, Ontv. 2.55 p.m. [Issued November 16, 1899][597]

Commandant General, Head Laager, Ladysmith, wires that he arrived at Colenso on the 13th inst., in order to acquaint himself with the positions, and saw the British advance guard. They advanced towards the patrol of Commandant David Joubert. Nothing happened, excepting that a train advanced on three of our burghers. These three fired on the train, after which it returned, but we wished to occupy this position.

The Ermelo and Middelburg Commandos went at once, together with a portion of the Krugersdorp Commando to take up the position, as the other burghers only left at 3 o'clock.

When we arrived at the spot we discovered an armoured train, which steamed back immediately, but one of our commandos had crossed over and placed stones on the line, which derailed her. The place was a very unsuitable one for our burghers, as also for our cannon. As soon as the train was derailed they began to fire on us with an Armstrong and a Maxim, with the result that five of our burghers and some horses were slightly wounded. At last the locomotive was freed, and proceeded with half the trucks that were not derailed. Thus we captured five trucks, three of which were armoured.

Two of the enemy were killed, 10 wounded, and 56 prisoners taken, amongst whom is Mr. Winston Spencer Churchill—reporter for the "Morning Post."

It is raining heavily, so that all the burghers and prisoners are wet through. We are expecting a stronger onslaught.

[596]Two pages (92 & 94) of the Kroonstad Telegraph Office ledger book for 20 Oct 1899 to 4 May 1900 (the day before the town surrendered to Lord Roberts) Neville Constantine: Anglo Boer War - Armored Train Ambush 11-15-99 Chuchill & Haldane - Page 20 - Boer War Forum There is no page 93. Every other page was a "copy" page, which was forwarded to an undisclosed authority (presumably so that the news could be posted for local inhabitants to read). It is possible that these bulletins were printed. The Transvaal version of these news reports was printed as a book at the time: "Boer Version of the Transvaal War" (Vryheid 1901). Unfortunately it is quite different to the OFS text.

[597] Translation of the Kroonstad telegram, as printed in "War Telegrams. Officially Issued by the Late O.F.S. Government up to the Occupation of Bloemfontein", Curling's Standard Printing Works, Bloemfontein, 1900. Anglo Boer War - Armored Train Ambush 11-15-99 Chuchill & Haldane - Page 21 - Boer War Forum

Reuters Interview 11/25

South Wales Daily News, November 25th 1899. [598]

The Armoured Train Fight.

Interview With Mr. Churchill.

Boers Commend British Pluck.

(Press Association War Special.)

Pretoria, November 23rd, via Lorenzo Marques, Friday. Through the courtesy of the authorities, in company with the Under-Secretary for Foreign Affairs (Mr. Grobler) and the Secretary to the War Department (Mr. Desouze), [De Souza] I was able to visit Mr. Winston Churchill, who is now confined in the State School with the British officers who are prisoners here. Mr. Churchill beyond a slight bullet wound in his right hand was well, though naturally chafing at his enforced idleness.

Being a prisoner of war for the time being, he is not allowed to communicate with his journal (the Morning Post) direct, but with the government's consent I offered Reuter's Agency as a channel to convey his experiences to the outside world.

From his references to the incident of his capture, exactly a month after leaving England, I gathered that when the armoured engine with the wounded escaped back to Estcourt on November 15th the Boers galloped from the hills among the retreating infantry and captured 58 of them, including Capt. Haldane, Lieut. Frankland, and Mr. Churchill. The Boers treated the prisoners with much kindness, praising the gallant defence of the train, and expressing surprise that the engine should have been saved under artillery fire from an apparently hopeless wreck.

The prisoners were then marched in pouring rain to Colenso, and on the following morning they were taken to the Boer camp before Ladysmith and thence to the railway at Modder Spruit, thus arriving at Pretoria on the 18th. In the course of the journey a great number of burghers crowded to see the prisoners, while others offered cigarettes in courteous fashion.

With regard to the British casualties incurred in defence of the train and in saving the engine, Mr. Churchill states that the list is still incomplete, but it is shown that the Boers buried three, sent 10 who were severely wounded to Ladysmith, and kept those who were less severely wounded in their field hospitals. Fifteen more wounded were reported to have been carried off with the engine and seven of those brought here as prisoners are lightly wounded. These together make 38 wounded out of a total of 120.

In reply to a question as to his general impressions, Mr. Churchill he had had many discussions with the Boers with regard to the rights of war, and had been much impressed to find so large a number speaking English. The Boers, he gathered, mostly regretted the war, saying it had arisen through misrepresentations, but he found the spirit of the burghers in the field, despite their many privations, most determined.

He concluded that there was no chance of an early peace, and feared the struggle would be bloody and protracted.

598 South Wales Daily News, dated 25th November 1899; p.5: Article Credit: Moranthorse1: Anglo Boer War - Armored Train Ambush 11-15-99 Chuchill & Haldane - Page 20 - Boer War Forum.

Thomas Walden Account

Thomas Walden [599] [600] Letter to Lady Randolph.[601]

17 November 1899 Horseshoe Hotel, Maritzburg, Natal.

My Lady,

I am sorry to say Mr. Churchill is a prisoner, but I am almost certain he is not wounded. I came down to Maritzburg yesterday to bring all his kit until Mr. Winston gets free. I am going back to Estcourt this afternoon. I have joined the Imperial Light Horse on Colonel Long's[602] advice, although I asked him first if I could join as I wanted to be near Mr. Winston as soon as he is free. So Colonel Long said he would arrange for me to leave the regiment as soon as Mr. C. wanted me. I came down in the armoured train with the driver, [Wagner] who is wounded in the head with a shell. He told me all about Mr. Winston. He says there is not a braver gentleman in the Army.

The driver was one of the first wounded, and he said to Mr. Winston: "I am finished." So Mr. Winston said to him: "Buck up a bit, I will stick to you," and he threw off his revolver and field-glasses and helped the driver pick 20 wounded up and put them on the tender of the engine. Every officer in Estcourt thinks Mr. C. and the engine-driver will get the V.C. It took them two hours to clear the road. They had to knock the iron trucks off the road by running into them with the engine.

The shells had knocked them all to pieces. How the engine escaped being blown up I don't know; it was a total wreck. The engine, with Mr. C. on it, got back to Frere station safe, and then Mr. C. would get off and go back to look after Captain Haldane. Mr. C. left his field glasses and revolver on the engine, and the driver says he had lost his hat.

It was a frightful morning too. It had been raining for about twenty-four hours, and it rained in torrents all day, so he must have been very wet; he had a good mackintosh on though. The driver says he was as cool as anything and worked like a nigger,[603] and how he escaped he doesn't know, as about fifty shells hit the engine. Everyone in Maritzburg is talking about Mr. Churchill. I am pleased to say he was in the best of health when I last saw him, as I believe he is now.

I remain, your obedient servant

T. Walden

[599] Thomas Walden was Winston's Batman/Valet. A batman or an orderly is a soldier or airman assigned to a commissioned officer as a personal servant. Before the advent of motorized transport, an officer's batman was also in charge of the officer's "bat-horse" that carried the pack saddle with his officer's kit during a campaign. Batman (military) - Wikipedia

[600] South African Light Horse - details of John S Churchill and Thomas Walden's entitlement to the South African Medal and Clasps, [3 copies]. Personal: South Africa: Correspondence and Papers. https://www.churchillarchive.com/explore/page?id=CHAR%201%2F30%2F34-36#image=0

[601] Published in the Morning Post on December 12th, 1899 on pages 2-3 also reproduced in OB p. 452-453.

[602] Colonel C. J. Long In October, 1899, Colonel Long was appointed to command the Royal Artillery upon the staff of the Natal Field Force with Sir George White, but, on his arrival in Natal, took charge of the lines of communication at Estcourt. He was engaged in the operations in Southern Natal, and served with Sir Redvers Buller's column in the advance to the Tugela. Colonel Long is a gunner of much distinguished service, and it is impossible not to admire the gallantry that led him into the crucial mistake made in the handling of the guns at Colenso on December 15. Upon that unfortunate occasion Colonel Long was dangerously wounded. It was intended that he should come into action covered by the 6th Brigade, but the 14th and 66th Batteries were led into the open, where they were subjected to a fire in which it was impossible for men to live. The story of the gallantry displayed in the effort to recover the guns is well known. Anglo Boer War - Name search

[603] This is regrettable but as in all cases of first person accounts, I must decline to be a censor. Oftimes it was meant in admiration of prowess and accomplishment, not pejoratively.

Times History

The Times history of the war in South Africa account.[604]

They [Armoured Trains] had already been used with success in Bechuanaland, and were destined to play an important part in the later stages of the war. But for reconnoitering purposes, except on perfectly flat country, little reliance can be put in them. Between Estcourt and Colenso the line ran like a switchback, up and down a number of narrow valleys, with hardly a point from which an unbroken view of 800 yards on either side could be obtained. On the other hand, even when it was hidden in cuttings the puffing of its approach could be heard for miles. The particular train in question was of the most primitive character, consisting simply of open trucks boxed in with loopholed walls of thick boiler plate to a height of seven feet from the floor of the trucks. It would be hard to devise a better target than a soldier laboriously climbing in or out of that death-trap.

General Joubert's expedition had started trekking from Colenso before dawn that same morning. An advanced party under Adjutant B. Van der Merwe[605] had nearly reached Frere when the ill-fated train was heard puffing along in the distance. The Boers at once concealed themselves. After a halt [of the train] at Frere to consult with a picket of Natal Police, Captain Haldane, finding all quiet, pushed on to Chieveley.

No sooner had the train passed on than the Boers made their preparations. Large numbers of them had hurried up, and, knowing what was in the wind, had taken care to keep the kopjes near the line between themselves and the unsuspecting train. A long steep slope a mile and a half from Frere was chosen for the execution of their plans. Towards the bottom of the slope the line curved sharply towards Frere, a guide rail along the curve being placed there to prevent trains running off the rails. The Boers filled up the space between the guide rail and the rail next to it with stones to help the train to run off if it was at all hurried—and they meant to hurry it.

Stones on the track.[606] Note Guard Rail installed to keep train on track around the curve.[607]

On both sides of the line a broad glacis-like slope extended 100 to 800 yards up to higher ground, and on the edges of this the Boers posted three guns and a "pom-pom," and spread themselves out. When all was ready they lay down in the pouring rain, philosophically smoked their pipes, and waited for the train to return. Meanwhile the train puffed on to Chieveley. From the station parties of Boers—as a matter of fact the rearguard of the main body, which was already laagering near the scene of the intended surprise—were seen some little way off moving southwards. After reporting by telegraph to Colonel Long, Haldane steamed back. As the train entered the mouth of the trap prepared for it the Boers opened fire with their guns and Mausers. As they had expected the engine driver turned on full steam and dashed down the incline. At the corner the leading trucks jumped off the line, two of them capsizing completely, and the third remaining jammed across the track in front of the engine, which was in the middle of the train.

The sailors at once replied to the enemy's fire with their 7-pounder from the rear truck, but before many seconds the little gun was knocked clean off its carriage. While the Dublin Fusiliers and Durban Light Infantry scrambled out and pluckily tried, under Haldane's direction, to keep down the enemy's fire, Mr. Winston

[604] The Times history of the war in South Africa, 1899-1902; by Amery, L. S. (Leopold Stennett), 1873-1955; Childers, Erskine, 1870-1922; Tallboy, G. P; Williams, Basil, 1867-1950 Sampson Low. Marston and Company, Ltd. Times History Vol 2 https://archive.org/details/timeshistoryofwa02amer Extract from Volume 2 p. 306-308. See: Leo Amery - Wikipedia.

[605] Many Van De Merwe(s) listed at AngloBoerWar.com but no B.

[606] Reported to be the stones used to throw the train off the tracks. Credit: Neville Constantine, Neville C. at Angloboerwar.com.

[607] See Leo Amery's account in Times History p. 306-308 of Vol. 2. Where he mentions this feature of the track.

Churchill, who was present as correspondent of the Morning Post, led a brave little party of volunteers in an effort to clear the line in front of the engine. For over an hour they toiled away, shells and bullets playing furiously upon them. At last things were so far clear that the engine could ram its way past the corner of the overturned truck. The engine and tender got through, but the heavy trucks behind stuck at the impediment, and the couplings parted. Further attempts to move the trucks were fruitless. The engine had been struck several times, and the boiler might be burst by a shell any moment. The only thing to be done was to get the engine away. The wounded were lifted on board, and the engine steamed slowly towards Frere. Haldane made his men run alongside, hoping, if possible, to get back to Frere. But the fire was too hot. To save itself the engine had to increase its pace, and the men fell behind, Haldane and Churchill staying to rally them. A few minutes more and all was over. The Boers galloped down and took some 70 prisoners, including Haldane, Churchill, and Lieutenant Frankland of the Dublins. There were in all five killed and 45 wounded. Twenty wounded and 70 unwounded escaped with the engine, or struggled back during the day. The Boers lost two killed and five wounded. A handful of mounted troops who had arrived post haste from Estcourt were too late to save the situation, but kept up a skirmishing fight for some time against superior numbers.

Lieutenant Wichmann's Account

Sidelights on Artillery in the South African War.

By Lieut. von Wichmann.

(Translated from the "Militär Wochenblatt).

BY LIEUTENANT K. B. FERGUSON, R.G,A.

Lieutenant Friedrich Wilhelm von Wichmann's Account of the Armored Train Ambush, 11/15/1899. [608] [609]

Chapter: III. – Ammunition and Armoured Trains.

As soon as the war broke out, the English began to use armoured trains on their railway lines. They built great hopes upon them, and at first the Boers themselves treated them with profound respect. But when the latter found what short work artillery could make of them, armoured trains fell very low in their estimation, and came to be looked upon as most harmless engines of war.

They consisted, as a rule, of an engine and four or five carriages; the engine was in the middle of the train. The carriages were armed with plates about the height of a man, and some about ½ an inch thick, with very often two tiers of loopholes; they were open on top. These wagons were manned by infantry, but sometimes they also carried small guns, either behind shields or unprotected on the wagons. In the latter form, however, they were very soon given up, because the detachments, when exposed to our fire, preferred to disappear quietly from their dangerous positions.

The engine was also armour-plated. In most cases they were also hung with rope mantlets, and the same thing was done later on with the wagons, outside the armour-plates. The armoured trains were always painted khaki. The engine was placed in the middle of the train so that in case of dynamite being placed on the line, or other obstructions, it would not be injured, and could steam away at once with the rest of the train.

[608] Lieutenant Friedrich Wilhelm von Wichmann Reports, Precis, Translations, Etc. Vol. XXXI, No. 5 (August 1904). Published with the "Proceedings" of the Royal Artillery Institution, Vol. XXXI., 1904-5 Thanks to Neville Constantine for assembling this material. See: Anglo Boer War - Armored Train Ambush 11-15-99 Chuchill & Haldane - Page 22 - Boer War Forum.
None of the other accounts mention the initial engagement that caused the train to accelerate towards the stones laid on the track. They conflate them into one event of continuous firing on the train.

[609] Lieutenant Wichmann is listed in the German Volunteers Article (ID3243): F. W. Von Wichmann Volunteer from Germany serving with the Boer forces. See German Volunteers: 863 Records https://www.angloboerwar.com/?option=com_content&view=article&id=3243 Source: Anglo Boer War Museum database, August 2016.

A protected armoured train.[334] "Hairy Mary" Locomotive at siege of Ladysmith.[335]

At first, armoured trains were mainly used for reconnaissance, but in this capacity they were not very successful. We immediately captured them wherever they appeared. The English then began to use them more cautiously, and it was not until the second phase of the war that they once more assumed an important role; even then, it was only in the protection of railway communication. By that time they were very much assisted by the fact that we no longer possessed any artillery. Dynamite took its place; and for the time, the best that the English could do was to strengthen and guard their lines so that no one could come near to damage them.

The Boers used to think at first that these trains were so strongly armoured that ordinary fragments of shell, or smaller projectiles, would never be able to pierce them; they consequently constructed special steel points to screw on to the projectiles of field guns. But it was very soon found that these measures were quite unnecessary.

Lieutenant Friedrich Wilhelm von Wichmann.[336] One of the "special steel points." This example was found at the site of the Armoured Train Disaster, so is probably from one of the rounds fired by von Wichmann.[337] (with thanks to Dougie McMaster / MC Heunis)

During the advance upon Estcourt, in Natal, while in command of a battery, I happened to come in contact with an armoured train. One used to come out early every morning from Estcourt and run up to Chieveley Station, from which a view of the country could be obtained for about eight miles all round.

In the night of the 15th-16th November, 1899, we set out with a force of some 2,000 men, crossed the railway by Chieveley Station, and waited, under cover of some small hills, until the train came by. I then galloped out with my battery, and took up a position where the train, on its return journey, coming out of a cutting, would pass within 1,100 yards of me. As soon as the enemy saw us, they came back at full speed. Our field guns and pom-poms opened fire, but without any apparent results; a hail of rifle bullets showered around us, and the train was gone. But about a mile further on, our people had put stones on the rails, and the two front wagons were thrown off. I promptly limbered up, and took up another position at a range of 1,390 yards, and opened fire at once. They replied with what turned out afterwards to be an old mountain gun, a prehistoric muzzle-loader; it fired three rounds, and was then silenced. In the meantime the English infantry had opened a brisk fire upon me. I fired the first few rounds with our steel-pointed shell, but as the train was gradually abandoned, I went on to ordinary common shell and shrapnel. The English made tremendous exertions, and managed to push the two wagons aside so as to leave the way free for the others; but the engine was badly crippled by bullets, and only had just enough strength to drag itself along. The remaining wagons stayed where they were. After a time the Boers advanced to the attack They were met at first by sharp rifle fire, but very soon the white flag went up. I immediately rode down to them, and came upon a scene of sad havoc. The plates had been pierced like paper; all kinds of fragments, and even pom-pom shell, had gone through them. On a later occasion I even saw instances of Mauser bullets having pierced armoured trains. Our steel points had therefore been quite superfluous. Thirty or forty men were killed and wounded, and we captured seventy, without the loss of a single man on our side.

The same sort of thing happened to nearly all the earlier armoured trains; later on, they were mostly blown up by dynamite. In theory, and against an enemy who has no artillery and does not know how to use dynamite, they may be very good; but in practice, against an enemy armed with modern weapons, they are an indifferent success.

Held by the Enemy

Extracts from the Journal of Lieutenant Thomas Hugh Colville Frankland, Royal Dublin Fusiliers, lately prisoner of war at Pretoria.

"Lieutenant Frankland was captured by the Boers when the armoured train was ambushed at Chieveley, in Natal, on the 15th of November, 1899. He was carried as a prisoner to Pretoria, where he arrived on the 19th of November, and where he remained until the 5th of June, 1900, when Pretoria fell and the greater part of the prisoners were set free by their victorious comrades." [610]

Thomas Hugh Colville Frankland.[611]

Frankland's sketch of the armoured train disaster, with his signature.[612]

November

November 19th—To wake up and find oneself enclosed in the space of a few acres for an indefinite period is scarcely pleasant; however, one cannot always be miserable. The monotony will, I have no doubt, become very trying, but for the first few days I have a good deal to do. The State Model School, which has been turned into a prison for the officers, is a building of rectangular shape. A long corridor runs through the centre, and on both sides of this are the rooms, where the officers sleep. They are supplied with a spring bed and two blankets apiece, while the whole place is lighted by electricity. At one end is the dining-room and gymnasium.

In front is the road, from which the building is separated by iron railings. Behind there is a sort of back garden where the police and soldier servants live in tents, and where the kitchen and the bath-room are situated. This piece of ground is surrounded on three sides by a six-foot fence of corrugated iron, and the whole place is watched by a cordon of armed police, about fifteen being on duty always.

The Government here generously supplies the officers with bread and water, half a pound of bully beef a day, and groceries. We have a small piece of ground and a gymnasium for exercise. As there are, alas! about fifty

[610] Originally Included in Winston Spencer Churchill's book; Ian Hamilton's March, 1900 The Copp, Clark Company, Ltd. Toronto.

[611] Photo Credit: Thomas Hugh Colville Frankland (unknown-1915) - Find A Grave Memorial Memorial ID 56009464 Photo added by Nancy Wright: Member Profile: Nancy Wright - Find A Grave.

[612] A sketch by Mr. T. H. C. Frankland, an officer in the train, forwarded by Mr. Churchill from the States School Prison at Pretoria to the Editor of The Graphic with the comment: "The enclosed sketch seems to me to be a very accurate presentation of what actually occurred." By permission of the "Daily Graphic," From: Winston Spencer Churchill by A. MacCallum Scott, Methuen, London 1905.

officers here, we have formed a sort of mess, and for the sum of three shillings a day we improve our scanty allowance of food.

They have supplied us with a suit of clothes each, but mine was much too big for me. I began to write my diary this evening, and had a long talk with Garvice[613] in my regiment, who told me how he had been captured. Dinner 7:30; bed, and sleep.

November 20th—It looks as if the rest of my diary for several months would contain each day the words, "the same as usual." I have only been here forty-eight hours, but the monotony has already begun to show itself. Not the monotony only, but the want of freedom, the want of news, the knowledge that the rest of the war will be carried out without my share in its victories, when, had it not been for some unhappy fate, I might yet have seen many an action—all these combine to oppress and irritate my mind. I tried to make a sketch of the armoured train, but it was not a success, and I must begin again to-morrow. The very length of empty time in front of me makes me quite patient.

November 21st—It is getting extremely hot. The lack of open space to walk in makes me feel lazy, and one gets quite tired after going a few times around the building. What one most looks forward to are the meals, and these are not very satisfying. But of course I am still suffering from the appetite of freedom, and I have no doubt that a month or so of this sort of life will make me feel less ravenous. I wrote some of my diary, and commenced another sketch of the armoured train, which I hope to be able to send to the "Graphic." Churchill has written asking to be released, but he does not expect any result. The mosquitoes here are very troublesome, and I have been constantly bitten.

November 23rd—The mail was supposed to go to-day, so I found occupation in a few letters. It is still very sultry. I succeeded in getting through a good deal of my diary, and, after writing nearly all day, played a game of rounders in the evening. This last occupation appears to cause much annoyance to the police, who frequently get hit by the ball. Another game here is fives, which we play with a tennis ball in the gymnasium. There seems to be some news about, but we can get nothing out of these people. By these people I mean Malan[614]—a spiteful, objectionable animal—who ought to be at the front, were he not a coward; [R. L. W.] Opperman,[615] a

[613] PF: I am confident enough in the identity of this individual to include his sketch biography here, despite the fact of his captivity not being mentioned in the summary.
Chudleigh Garvice, Lieutenant, Royal Dublin Fusiliers, 2nd Battalion. He was born 12 January 1875, at Cookham, Bucks, son of Charles and Elizabeth Garvice. He was educated at Blundell's School, Tiverton, and was gazetted to the Royal Dublin Fusiliers 9 December 1896, and was promoted to Lieutenant, 30 June 1898. He served in the South African War from the commencement of hostilities until peace was signed, employed with Mounted Infantry, and was present at operations in Natal in 1899, including the action at Talana; operations in the Transvaal, east of Pretoria, July to 22 November 1900; operations in Orange River Colony, May to July 1900; operations in the Transvaal 30 November 1900 to January 1902. He was mentioned in Despatches [London Gazette, 10 September 1901]; received the Queen's Medal with three clasps, the King's Medal with two clasps, and was created a Companion of the Distinguished Service Order [London Gazette, 27 September 1901]: "Chudleigh Garvice, Lieutenant, Royal Dublin Fusiliers. In recognition of services during the operations in South Africa." He was promoted to Captain 28 May 1902. In 1903 he served in the Aden Hinterland operations, and in the Sudan (Egyptian Army) from 1904 to 1912, in command of the Arab Battalion in various patrols, etc. (Sudan Medal, 1910); retired from the Royal Dublin Fusiliers 9 April 1913, and became Major, Reserve of Officers, 31 August 1914. In the Great War he served on the Western Front of Egypt in 1915-16, and later on the Staff in Egypt. Major Garvice was mentioned in the Despatches of General Maxwell, dated 1 March, 1916, and of General Allenby in the London Gazette, dated 15 January 1918. He was awarded the OBE in 1919, and had the 1914-15 Star, the General Service Medal, the Victory Medal, the 4th Class Osmanieh, 3rd Class Order of the Nile, 1st Class Order of the Redeemer (Greece). He married, 31 January 1920, at Alexandria, Egypt, Isabel, daughter of Andrew Ormiston. He died 3 March, 1921. His father was the famous novelist, of whom his friend, Mr. Douglas Sladen, writes as follows: "The late Charles Garvice, father of this distinguished soldier, was the most popular novelist of his time, and a great figure in the literary world. He was Chairman of the Authors' Club, one of the most prominent members of the Garrick, and Chairman of the Committee of the new After Dinner Club, of which he was co-founder with Mrs. Baillie-Reynolds and Mr. Douglas Sladen. He was the life and soul of the clubs to which he belonged, for, in addition to his sociability and wit in the smoking-room, he was an admirable, after-dinner speaker, and when it was necessary for the club to assert itself on any occasion, he was always willing to take the responsibility, although it might make enemies. He was fearless in the performance of a duty." QSA (3) Tal OFS Trans (Lt DSO R Dub Fus). Holditch 1985 £285. Source: DSO recipients (VC and DSO Book) Also: Mounted Infantry. MID LG: 10 September 1901, page: 5953. Source: Field Marshal Roberts. 4 September 1901. Re: General mentions.

[614] Field Cornet Malan was a grandchild of President Kruger. (Sandys p. 82) This was not exceptional because Kruger had 17 children and over 100 grandchildren. He was captured in August of 1901 and exiled to Ceylon. (Haldane) Over 5,000 Boers were interned in Ceylon (Sri Lanka) during the Boer War. The Ceylon camps were the second set of overseas POW camps after those in St. Helena had filled. http://libcom.org/library/pow-camps-ceylon-during-boer-war

[615] "Commandant Opperman, who was directly responsible for our safe custody, was in times of peace a Landdrost or Justice. He was too fat to go and fight, but he was an honest and patriotic Boer, who would have gladly taken an active part in the war. He firmly believed that the Republics would win, and when, as sometimes happened, bad news reached Pretoria, Opperman looked a picture of

slightly more agreeable person, of large dimensions, and Dr. Gunning,[616] a much more amiable fellow. It seems absurd that they do not allow us to buy papers. What harm could we do with them?

Some of the restrictions are so childish, and tend to make life here so sickening, that I am sure if curses could harm the Transvaal Government it would not be long-lived.

This morning Churchill was visited by De Souza, the Secretary of War, by the Under-Secretary for Foreign Affairs, and others, and there followed a very animated discussion about the causes and the justice of the war. It was a drawn game, and they all talked at once at the end, especially Churchill. I am afraid for his sake he is not likely to be exchanged or released. The Boers have got to hear of the part he played in the armoured train episode.

Louis de Souza.[338]

Louis de Sousa and his spouse Marie.[339]

November 24th—There is some news abroad to-day. The Free Staters have been attacked at Belmont by the British, probably under Buller, but the result is uncertain. Of course the Boers report a victory on their side, but one gets quite accustomed to their "victories." Dundee was a [British] victory, likewise Elandslaagte. I am getting on slowly with my diary, and manage to make it occupy a great deal of time.

November 25th—Evidently we have won a victory at Belmont; its results are immediately apparent here. They have suddenly become much more lenient and complacent. We are actually allowed newspapers, and the President is considering the question of beer. The papers admit that the British drove the Free Staters from their position at Belmont, but with great loss, while that of the Boers is practically nil. Rumours say that General Joubert is cut off between Estcourt and Mooi River; how I hope it is true!

November 26th—The Rev. Mr. Hofmeyr is a prisoner here, and held service this morning, when he delivered a most eloquent address. There is a harmonium[617] in one of the rooms, and Mr. Hofmeyr, who sings very well, gives us some very good music. He knows a lot of old English songs, which are pleasant to hear, although they rather suggest the Psalm beginning "By the waters of Babylon." Hofmeyr, though a Dutchman, is an ardent supporter of the Imperial cause, and he has in consequence been very cruelly treated by the Boers before he came here.

misery, and would come to us and speak of his resolve to shoot his wife and children and perish in the defence of the capital." (WSC London to Ladysmith.)

[616] Dr. J. W. B. Gunning is one of the administrators of the State Model School Dr. J. W. B. Gunning, per Brian Davis, Churchills in Africa p. 201 See also: The Strand Magazine, No. 112, April 1900; University of Michigan and Google Books for an article on his Staats Museum of which he is the director.

[617] A pump organ.

Adrian Hofmeyr.[340]

Mason & Hamlin Organ.[341]

It is quite touching to see how the Boers try to hide their defeat. All the accounts are cooked, but even De Souza acknowledges that if things go on as at present the war will soon be over. There have been several days' fighting south of Kimberley, and Buller is advancing steadily. On the Natal side Joubert passed Estcourt, and reached Mooi River, where he was attacked by the new division and defeated. In retiring he was attacked by part of the Estcourt garrison, result unknown. He will probably retire on Colenso.

November 27th—Not much news to-day. According to the "Volksstem" British lost fifteen hundred at Belmont, and the Boers nine killed and forty wounded. However, they can't deny that the Free Staters were licked, and De Souza admits that Kimberley will probably be relieved shortly. Moreover, Khama[618] is said to have risen. This has disturbed them all exceedingly, and Opperman is highly indignant.

November 30th—I find nothing to record here except the scraps of news one gets in the newspapers, all else is monotonous—appalling monotony. In the evening one feels it most, and sometimes I don't think I can endure it for another month. All sorts of absurd rumours are spread about here by that intelligent paper the "Volksstem." The latest is that four British regiments have refused to fight, being in sympathy with the Republican cause. I wonder whether Buller will desert to the Boer side? The fact remains that the papers give no news whilst there must be plenty, and this looks as if the untold news must be bad for them. We hear that General Forestier-Walker[619] has been killed, and that Lord Methuen is seriously wounded. This morning the rumour runs that our troops have occupied Colenso. The regiment is sure to be there. How I wish I were with it!

December

December 4th—No real news, but various and contradictory rumours. The Boers have begun to acknowledge their losses, and the paper have long lists of killed and wounded. Major ——, of the West Yorks, arrived to-day, having been captured near Estcourt. From him I learned that all was well there. A few days ago three battalions—West Yorks, Borderers and Second Queen's—went out and attacked the Boers. Apparently the engagement was indecisive, and the losses on either side not very great. The rumour goes that Buller is in Natal, and not in the Free State after all. Of course he is advancing to the relief of Ladysmith. We all think that his plan will be to hold the Boers in front of Colenso while he takes a large force around by the flank. The Boers have retired beyond the river, and have blown up the Tugela railway bridge. On the other side, Lord Methuen's

[618] Khama III (1837?–1923), referred to by missionaries as Khama the Good also called Khama the Great, was the Kgosi (meaning *king*) of the Bangwato people. Khama III - Wikipedia

[619] PF: This does not appear to be correct. See: General Sir Frederick William Edward Forestier-Walker, GCMG, KCB (17 April 1844 – 30 August 1910) was a British senior military officer and Governor of Gibraltar. In 1899 he again returned to Africa, becoming GOC Cape Colony and acting as lieutenant general in command of Lines of Communication, South Africa Field Force, 1899–1901. He was thus responsible for disembarkation of troops and military stores and sending them to the front. In a despatch dated 31 March 1900, the Commander-in-Chief in South Africa, Lord Roberts, wrote how Forestier-Walker carried out his duties "with credit to himself and with advantage to the public service." He was promoted to Knight Grand Cross of the Order of St Michael and St George (GCMG) in November 1900 for his services in South Africa. Following the end of the war, Forestier-Walker was promoted to the rank of general on 6 July 1902. Frederick Forestier-Walker - Wikipedia See Also: Anglo Boer War - Name search.

Division is having severe fighting; he has defeated the Boers at Modder River, and the relief of Kimberley is imminent.

The papers do not publish much news themselves, but occasionally publish some of the English cuttings with sarcastic editorial comments. In the Dutch version of the "Volksstem" they slate[620] the Free Staters unmercifully for having run away at Modder River.

Oh, that we might be exchanged. Joubert has wired via Buller to England advocating such a step.

December 15th— "Tempus fugit,"[621] and it has not been quite so dull as usual. First, and most important of all, Churchill has escaped. Whether he has made it good or not is still uncertain; but he has now been gone two days, and I have great hopes. Besides the excitement there has been a very amusing side to the affair. Of course Churchill was the very last person who ought to have gone. He was always talking and arguing with the officials, and was therefore well known, and, indeed, scarcely a day passed without Dr. Gunning or Mr. de Souza inquiring for him. His plans for escape were primitive; but, being still in prison, I must not write anything about this part of the affair. Let it suffice that Churchill got away without any trace left behind. Next morning, as it chanced, it was the day for the barber to come and shave him, and having only just woke up I put the barber off rather feebly by saying that Churchill had gone to the bath-room, and would not need shaving. What should the detective who accompanied the barber do but wait outside the bath-room, and, finding no Churchill, began to suspect. Gunning then came upon the scene, closely followed by Opperman, both asking and seeking anxiously for their captive. Their distress at finding him gone was really pathetic. They immediately put on all kinds of restrictions. No papers, calling rolls, not allowing anyone into the yard outside the building after 8 P.M., and stopping all beer. I am reminded of the fable "Le Corbeau et le Renard,"[622] which ends, "Le Corbeau ... jura, mais un peu tard, qu'on ne l'y prendroit plus."[623] Curiously enough, the day after Churchill had escaped an order is said to have come from General Joubert for his release. However, I have no doubt but that this was all made up to excuse themselves for not being able to catch him. I do hope he gets away.

Our spirits are constantly on the rise and fall. At one time we are about to be exchanged, at another nothing has been heard of it; at one time there is a brilliant British success, greatly modified, of course, by the enlightened "Volksstem" editor, at another a crushing British defeat, with all the Generals and thousands of soldiers killed and wounded.

Yesterday we heard of the splendid achievement of the British troops in Ladysmith in smashing up the 84-pounder at Lombard's Kop, several Howitzers and a Maxim.[624]

Then came the defeat of General Gatacre at Stormburg, and the capture of 600 prisoners, and on the top of this the victory which the Boers claim at Magersfontein.

All this is very terrible. I think I feel almost as miserable as I did the night I was captured. Are the British troops ever going to drive the Boers back? Will they ever come and take Pretoria? or will they, on the other hand, be driven back, and the people at home get sick of the war, like in 81, and—no, impossible—and yet who will dare predict? It is too awful to hear all these shocking reports, and to be able to do nothing oneself. One always imagines on these occasions one's presence at the scene of fighting absolutely indispensable if there is to be a victory. However, these miserable days cannot last for ever. Perhaps they are even now at an end. De Souza, with a faltering voice, has confessed that Buller is advancing at last in great force. He must win.

December 19th—Worse than ever. Buller has attacked in full strength at Colenso and has been defeated with a loss of ten guns and many hundred men. This is too awful—I could have cried. The hand of fate seems to be raised against us. The only thing to do is to wait patiently till the next disaster. The Stormburg prisoners have arrived, the Colenso prisoners are expected to-morrow. Everybody is cursing the Generals; but they always think they could do better themselves. I hear that Hart's Brigade, with our regiment in it, were caught in quarter

620 From Middle English, "to split, break, or cut up."

621 "meanwhile, the irreplaceable time escapes," literally "but it flees meanwhile: irretrievable time flees."

622 The Fox and the Crow (Aesop) - Wikipedia In the fable a crow has found a piece of cheese and retired to a branch to eat it. A fox, wanting it for himself, flatters the crow, calling it beautiful and wondering whether its voice is as sweet to match. When it lets out a caw, the cheese falls and is devoured by the fox.

623 The Raven ... swore, but a little late, that he would no longer be taken there.

624 December 8th, 1899. British sortie from Ladysmith, Lombard's Kop being carried. Creswicke Chronology Source.

column at close range. They must have suffered terribly. Never mind; Methuen has relieved Kimberley. The officials all deny it, but it must be true.[625]

December 23rd—No more news. The authorities are getting more and more silly and disagreeable; all kinds of babyish restrictions are invented to annoy us. Churchill has got to Delagoa Bay, and has wired his safe arrival to De Souza. Hurrah![626]

I have not dared until now, when all is a failure, to set down in this book any account of the one occupation that has prevented us from going mad with disappointment in these sad times. About the middle of the month Haldane devised a plan of making a tunnel from under our room across the road. The five fellows in our dormitory and Le Mesurier, who shifted his abode for the purpose, began about ten days ago. First, we thought of cutting a hole in the floor, but, on looking round, we suddenly found a trap-door already made. Beneath the floor there is a curious place. The rafters are supported by stone walls, so that underneath there is a series of compartments about twenty-four feet by four, with access from one to another by means of man-holes in each wall. We commenced digging in the compartment next to the one under the trap-door. The ground at first was very hard, but with chisels and implements taken from the gymnasium, we managed to get down four feet of the shaft in about four days. It was a queer sight to see two half-naked figures digging away by candle light, for we used to work in reliefs of two—one to dig and the other to cast away the earth in boxes or jugs. Suddenly, one day, we broke through the hard crust and came to some soft clay soil. We were delighted at this, and expected to get through it in no time; but, alas! with the soft earth came water, and without pumps, bale as we would, we could not get rid of it. Every morning the shaft was completely bilged; so, having dug down six feet, our plan was brought to an end, and we had to screw up our trap-door again in bitter disappointment. The officers of the Gloucester Regiment are digging too, but they are sure to find the same difficulties.

Christmas Day, 1899.—I can scarcely realise that it is Christmas, the day I have hitherto spent at home with family and friends. I can see the rooms decorated with holly and "Merry Christmas" cut in white paper and pasted on red Turkish twill hanging over the doorway. A Merry Christmas! What irony! The time, of course, was bound to come when the circle at home would be broken; but little did I dream where or under what unhappy circumstances. A Merry Christmas! to a prisoner—not when his countrymen, victorious and full of enthusiasm, are marching rapidly to his release, but when the armies of his country, beaten back, lie far away; when, helpless himself, despair seizes his heart; when reverses grow into disasters and the might of the dear old land in which he trusted seems to have weakened and died. A Happy Christmas! with the New Year black, uncertain, and unknown. Of course we drank the health of the Queen at dinner—in lime-juice. Twas all we had; but we meant it none the less.

December 30th—They say there were only 1,200 casualties at Colenso; but we have just heard that —— and —— of our regiment have been killed. O, God! it seems too awful. To hear of all one's friends crippled or dead; all the best are picked off, and here are we tied up quite safely with our beastly skins unhurt, and not likely to run into the slightest danger while our comrades are losing their lives. We must win this war.

January

January 1st—I have had many arguments as to whether this is the commencement of a new century or not, and after much reasoning I have decided that as it is the year 1900, or the nineteen hundredth year, it is the last of the nineteenth century and not the beginning of the twentieth. Whatever it may be, this is a hateful place to spend the beginning of anything in. The "Volksstem" printed a list of casualties to-day, and I see that our regiment lost forty-two killed at Colenso. What must the numbers of the wounded have been? {Here follows a list of wounded officers.}[627] Sergeant Gage[628] was killed, and they say he was one of the first to cross the waggon bridge. This looks as if the regiment had stormed the bridge, which is much better than being mown

[625] The siege of Kimberley would not be relieved until February 15th, 1900. Siege of Kimberley - Wikipedia

[626] In the wire Churchill made sure to mention that the guards at the prison were not involved with his departure since they had come under fierce suspicion of having been bribed or colluded in some manner.

[627] The list of names was not included in the original publication.

[628] Gage J C V 3467 Cr Sergeant 2nd Btn Royal Dublin Fusiliers, Died of wounds at Colenso. 20 Dec 1899. Source: Natal Field Force Casualty Roll, page 186 line 4 Also: Source: In Memoriam by S Watt Anglo Boer War - Name search.

down in quarter column. All these losses are terrible, but I believe that Colenso is only a reconnaissance in force. What must a battle be like?

Passage of Infantry over a bridge of Wagons.[342]

The last week has been, if possible, more dreary than usual. One of the fellows in our room has made himself very obnoxious lately, and has had to be sat upon severely. I have never met such an ungentlemanlike creature. It is all the more unpleasant in a place like this, where we are so closely packed. There are rumours of fighting near Colesburg, probably by General French. The Boers say the action is indecisive, which means a victory for us.

January 7th—Nothing of importance has occurred lately. There has been a bit of a fight with Opperman, who tried to take away from Boscher, [629] the local grocer, his contract for the supply of our mess, on the ground that Boscher had helped Churchill to escape: Result a complete victory for us and the reinstatement of Boscher. More Zarps, as the policemen who guard us are called, and poor little Gunning has been commandeered. He prepares himself to go. His reason is peculiar. Should his children, in after years, ask him if he fought for the freedom of the State, he would like to be able to say "Yes." However, if he goes I hope he will find a large rock to get behind and so come back safely.

This afternoon a most alarming rumour was started by somebody, namely: that Ladysmith had fallen. Though I did not actually believe it, we are always having such frightful disasters that I felt very uncomfortable. Later, however, we learned that all was well.

January 10th—Ladysmith has not fallen. The news of the defeat of the Boers on the Platrand has been confirmed, and, in spite of their lies, we know their losses were heavy. [630] At Colesburg there was a night attack, and a half battalion of the Suffolks got much knocked about. [see infra] Two of their officers came in as prisoners yesterday; they say the Boers have received large reinforcements at Colesburg. There is a rumour that Dr. Leyds[631] has been arrested in Germany for trying to enlist German Reservists. A British force is said to be at Douglas, west of Kimberley. They made a night attack and captured some stores and ammunition. The

[629] This is Mr. Bosscher.

[630] The Battle of Waggon Hill; January 5th and 6th, 1900. The Boers around Ladysmith were also growing weak from lack of forage. With little action, many fighters took unauthorised leave or brought their families into the siege encampments. Eventually, with the Tugela in flood, preventing Buller from giving any support, some younger leaders persuaded Joubert to order a storming attempt on the night of 5 January 1900, before another relief attempt could be made. The British line south of Ladysmith ran along a ridge known as the Platrand. Under Ian Hamilton, they had constructed a line of forts, sangars and entrenchments on the reverse slope of the Platrand, of which the Boers were unaware. In the early hours of 6 January 1900, Boer storming parties under General C.J. de Villiers began climbing Wagon Hill and Caesar's Camp. They were spotted and engaged by British working parties who were emplacing some guns. The Boers captured the edge of both features, but could not advance further. British counter-attacks also failed. At noon, de Villiers made another attack on Wagon Hill. Some exhausted defenders panicked and fled, but Hamilton led reserves to the spot and recaptured some empty gun pits. Late in the afternoon, a terrific rainstorm broke, and the Boers withdrew under cover of it. The British suffered 175 killed and 249 wounded. 52 dead Boers were left in the British positions, but their total casualties were not recorded. Siege of Ladysmith - Wikipedia See Also: Platrand Monument. Johannesburg Light Horse Regiment - Wikipedia

[631] Willem Johannes Leyds (1 May 1859 – 14 May 1940) was a Dutch lawyer and statesman who served as state attorney and state secretary of the South African Republic. From 1898 to 1902, during the crucial period of the Second Boer War, he was the Republic's special envoy and minister plenipotentiary in Brussels, accredited to several European states. Willem Johannes Leyds - Wikipedia

Transvaalers in their excitement succeeded in firing into the Free Staters, shooting, among others, Opperman's nephew. We offered our sympathies, but after all it is one the less. This evening we received a most excellent rumour that the Boers had lost 900 men near Colenso. I hope it is true, and that the Tugela has, therefore, been crossed. This will be a step towards the relief of Ladysmith. At Colesburg the Boers are in a critical position. Things seem to be looking up a bit. I wish that we could get just a little truth. These rumours torture and deceive.

January 14th—All kinds of startling rumours have been about to-day: The British fighting in overwhelming numbers around Ladysmith; Buller surprised and taken prisoner at Pieters Station. Boers in a tight corner at Colesburg. What can one believe? All men are liars—in Africa! Life is getting very unbearable. I am sure we shall be a lot of lunatics when we are set free.

January 29th—How we clamour for news, and how our spirits rise and fall as the rumours are favourable or bad. The other day the prisoners arrived from the Spion Kop fight. The result of the attack on Spion Kop is not known. We took the hill, but, for some reason, the rumour goes that we have left it again and re-crossed the river. Can this be another lie? We hear that the regiment did not cross the waggon bridge, but tried to swim the river at Colenso last month. Very few got over. Hensley was killed the other day at Spion Kop. One can scarcely realise these losses, and I don't think we shall until we join the mess and see the sad gaps among familiar faces.

Capt. C. A. Hensley.[632]

Also thought to be Captain Hensley pictured on the train at Estcourt.

February

February 5th—We have been getting a fair share of good news lately, or, at least, good rumours. The relief of Kimberley is an established fact.[633]

Colesburg is on its last legs, though news of its surrender to French needs confirmation. There is fighting at the Tugela, concerning which the latest bulletin is "British have taken a position—Vaal Krantz." Nor is this all, other factors are at work besides the British Army. There is considerable dissension between the Transvaalers and the Free Staters. The former complain that they are always put in the fore front of the battle, while the latter rejoin that not only are they invariably sent to the more exposed kopjes, but that while they are aiding the Transvaalers to fight in Natal they are receiving no help in the defence of the Free State.

632 Captain Charles Albert Hensley. He died January 20th 1900, of wounds received in the operations on the Upper Tugela, near Venter's Spruit. He was the elder son of Albert W Hensley of Halifax, Nova Scotia, was born in Prince Edward Island September 1865, and educated at the Collegiate School, Windsor, S, and the Royal Military College, Kingston, Canada. Being an enthusiastic and successful sportsman, he was known in his regiment as "the young Shikari." He entered the army September 1885, being promoted Captain July 1895. Having served in Egypt and India, he proceeded to Natal with his battalion September 1899, and was present at the battle of Talana and in the retirement on Ladysmith. Just before the siege began, he was sent with part of his battalion to garrison Colenso, and was present at the battle of that name and the fighting on the Tugela, being twice wounded. Captain Hensley, whose family was well known in Canada, married eighteen months before his death the daughter of H Wylde Brown of Pietermaritzburg. He was mentioned in despatches, LG February 8th 1901, for his services. Source: Donner See Also: Vol. 6 No. 6 of SAMHS South African Military History Society - Journal - LETTERS OF CAPT C.A. HENSLEY TO HIS FATHER (samilitaryhistory.org).

633 The siege will not be lifted until February 15th. See cite supra. On this date per Creswicke: General French reached and relieved Kimberley, captured Boer laager and supplies, and forced the enemy to withdraw. The Boers evacuated Majersfontein and Spyfontein, retreating to Koodoosrand Drift.

February 12th—It would take too long, even when time is nothing but a curse, to record all the items of news we have lately received. So many startling rumours have been confirmed and denied that I long to know what is the real truth, but in the Capital of this doomed country—in the very metropolis of lies and liars—we shall never learn the truth until our friends come to bring it with them.

I have just finished reading "Esmonde,"[634] which I enjoyed very much. One advantage of my forced sojourn in this country is that I may improve my education. Indeed, reading occupies the greater part of our time, though I myself cannot fix my attention on a book for very long under these miserable circumstances. The State Library has a fair selection of books, and by paying a small subscription the prisoners are allowed to take out books therefrom. The only forbidden fruits are the books of South Africa; for these volumes, recording the evil wrought by the British race on this chosen people, are carefully stowed away for fear of the English trying to destroy the histories of their crimes.

This morning an officer of the South African Light Horse was buried.[635] To all intents and purposes he was murdered by the Transvaal Government. Although he had typhoid fever he was thrown into prison, and not until the authorities were pretty certain he would die was he sent to the hospital. Ten officers on parole went as pall-bearers and we all subscribed for a very pretty wreath.

Patience[636] is played as a game here largely by ancient Colonels and Majors, and practised by us all with indifferent success as a cruel necessity.

February 17th—Good news at last! Kimberley has been relieved! Boers are retiring in all directions. Lord Roberts, with the British Army, has entered the Free State. Warrenton has been occupied, there is great consternation in Pretoria. Opperman is furious. Perhaps the tide has begun to turn.

To explain how we get news: Brockie, a Sergeant-Major in the Imperial Light Horse, knows a Zarp here and gets a certain amount of news from him, which is not, however, very trustworthy. When we first came here an Englishman named Patterson, employed in the Government telegraph office, used to pass by the railings and whisper the news. He only used to come when there was good news to tell, and generally ended with the words, Hurrah, hurrah! Since he was always accompanied on these occasions by a large St. Bernard, we called him the Dogman. Lately he has elaborated and improved his system of giving us news and has begun to signal with a flag from the passage of Mr. Cullingworth's house opposite. Either he or one of the Misses Cullingworth stands some way back in the passage so as not to be visible to the Zarps and sends messages, which are read by Captain Burrows[637] from the gymnasium window. As he is in the telegraph office and sees all that passes, the Dogman sends very truthful information.

February 18th—More good news this morning. Cronje is lost, strayed or stolen. The Boers have been driven back at Dordrecht.[638]

The British Army is within forty miles of Bloemfontein. Buller has taken the Tugela position.[639] All this needs no comment. "*Quo plus-—eo plus——*." I meant to quote a Latin phrase—the only one I ever knew—but I cannot risk the tenses and moods of the verbs. It means, however, the more we have the more we want.[640] We live, as it were, from news to news. Two officers arrived from Colesburg this morning. They say Colesburg has never been quite surrounded, only hemmed on three sides. General French began to withdraw his Cavalry about three weeks ago, sending away detachments every night until only an Infantry Brigade was left to sit in front of

[634] Unable to identify this work or author. Rapporto di due cause mosse dal Rndo padre Bartolomeo Esmonde della Compagn... - Google Books inauthor:Esmonde - Google Search

[635] This is Lt. Charles A. Percy Tarbutt, Lieutenant: He died of typhoid fever at Pretoria, February 13th, 1900. He was the eldest son of Percy Tarbutt of the Consolidated Gold Fields of South Africa. Lieutenant Tarbutt was twenty six years of age, and was educated at University College School (where he was in the football XV), and at the Bedford Grammar School. When Lieutenant Tarbutt landed in South Africa he was given a commission as Lieutenant in the South African Light Horse. After the battle of Colenso he was reported missing, but it was subsequently discovered that he had been wounded and captured. Source: Donner. Anglo Boer War - Name search

[636] Solitaire. Patience (game)-Wikipedia

[637] Possibly: Arnold Robinson Burrowes. Captain, (Princess Victoria's) 1st Battalion Royal Irish Fusiliers QSA (4) QSA known to exist Source: List of QSAs with the clasp Talana from the Talana book Anglo Boer War - Name search.

[638] 2/17/1900 Cronje's force overtaken and surrounded at Paardeberg. General Brabant engaged the enemy near Dordrecht.

[639] 2/16/1900 Capture of Cingolo Hill by Sir Redvers Buller's force. Flight of Cronje's force and occupation of Majersfontein by the Guards. General Kelly-Kenny, in pursuit of Cronje retiring east with 10,000 men on Bloemfontein, captured 78 waggons with stores, 2 waggons with Mauser rifles, and 8 waggons with shell belonging to Cronje's column.

[640] *Quo plus habent eo plus cupiunt.* (The more they have, the more they want.)

Colesburg, occupying exactly the same extent of front as before. The Boers never spotted this, so that French and his Cavalry succeeded in joining the Free State column, and the Infantry Brigade, by making a great show of their forces, was able to keep up the ruse until the other day, when it was decided to retire. Everything went well with the retirement except for two companies of the Wiltshire who were cut off and captured after a gallant fight. I suppose all Governments lie to a certain extent about their defeats, but this Boer one takes the cake.

February 19th—I have caught the patience disease. I spent most of the day at this interesting game, but found by 7 P.M. I was rather sick of it. Le Mesurier told me to-day that Haldane, Brockie, Grimshaw and he had thought of a plan of escape. The idea was to put out the electric light in the house and in the yard by cutting the wire as it entered the building in the roof above the entrance. The sudden extinguishing of the lights on a dark night would enable them to creep to the back wall and climb over unobserved by the Zarps, whose eyes would not have become accustomed to the sudden darkness. They had made small ladders, by means of which they could climb over the corrugated iron more easily and with less noise. Once outside, they were going to trek for Mafeking, which is only about one hundred and eighty miles off. They had meant to go to-night, but, though it was wet, there was too much lightning.

February 21st—More good news both from Stormburg [area] and the Tugela. Our friend Opperman is getting excessively polite. I think one can best describe him as a greasy, unwashed bully, oily physically and morally, cruel to anyone in his power, cringing to those he fears.

February 22nd—We hear that Cronje is completely surrounded. De Wet tried to break the encircling cordon, but was defeated with great loss. Buller has taken the Boschkop and all the British troops have crossed the Tugela.

A very amusing article appeared in one of the papers the other day, in which Napoleon was termed "the Botha of the early 10's." Botha the Napoleon of these days is presumption, but Napoleon, the Botha of the early 10's! I cannot help pitying the editor of the "Volksstem," as he is only allowed to publish good news, and must really be at his wit's-end to know what to put in now.

Haldane and the others had arranged to go to-night, but unfortunately the sentry was walking about the place which had been chosen for getting over, so that the escape was prevented.

February 24th—Haldane and Co. have tried again. This time they were determined to go. Clough,[641] the servant, was sent up via the gymnasium on to the roof to cut the wire. I gave the signal by going into the room under the main switch and asking for a map. The light went down temporarily but came up again almost immediately. We were much alarmed lest Clough should have got a shock, but he came down all right, surprised that the lights had not gone out. Of course the escape was off.

February 25th—We were all sure that Clough had not cut the wires at all last night. He had received a slight shock and then left it, so it was arranged that Cullen[642] should try. However, the position of the sentry again prevented any attempt.

February 26th—Best,[643] of the Inniskilling Fusiliers, arrived to-day from the Tugela. He said that all were well down there, though the fighting had been very severe, and that the troops were beyond Pieters. Cronje had no food and must surrender shortly.

This evening the lights went out without any mistake. Opperman was greatly alarmed, and the electrician could not find out what was up. They all believed a football must have hit the wire outside and put the light out. Probably Clough had partially severed the wires, and the football had completed the damage. Now, however, the wire being broken before it was quite dark, the advantage of surprise would be lost. It was, moreover, a bright night, and we noticed that the light in the streets shone on the wall where we had meant to climb over it.

The sentries were doubled, so we finally gave up the plan and tried to think of another. We are told that they will remove us to a new place on the 1st of March, and, perhaps, this will give us a better chance.

When I went into my room at about 9.30 I found that Le Mesurier, Haldane, and Brockie were having a discussion. As we were to move in two days to the new prison they argued "why not go to earth now." The

[641] Unable to identify this individual. Too many possibilities without rank or a first name. Anglo Boer War - Name search

[642] Unable to identify this individual. Too many possibilities without rank or a first name. Anglo Boer War - Name search

[643] Possibly T.A.D. Best, but the date of release does not match. 2nd Lieutenant 1st Btn. Missing - released at Natal. 23 Feb 1900. Source: Natal Field Force Casualty Roll, page 98 line 16. Royal Inniskilling Fusiliers.

authorities would think they had escaped under cover of the light going out and would, if anything, hasten the removal of the prisoners, leaving these three under the floor to depart in peace when opportunity offered.

February 27th—This morning Opperman came into our room as usual to count the number of prisoners in bed, and on seeing three beds empty he fairly staggered with astonishment. I was looking at him with one eye and chuckled to myself at his dismay. He went and asked Brett[644] if he knew anything about it. Brett asked innocently, "About what?" Then I pretended to wake up and ask Opperman what the hell he meant by disturbing us at this hour. He left the room in a fury, but presently returned with Gunning and later with Du Toit, the Chief of the Police, who examined everything *à la* Sherlock Holmes, and expressed, with a smile, his confidence in the recapture of the flown birds. After breakfast the whole house was cleared and searched. The rooms, the cupboards, the roof—everywhere except under the floor. Then they brought in a dark lantern, and I really thought they had discovered the fugitives at last, but Sherlock Holmes never thought of the floor; his reasoning did not carry him there.

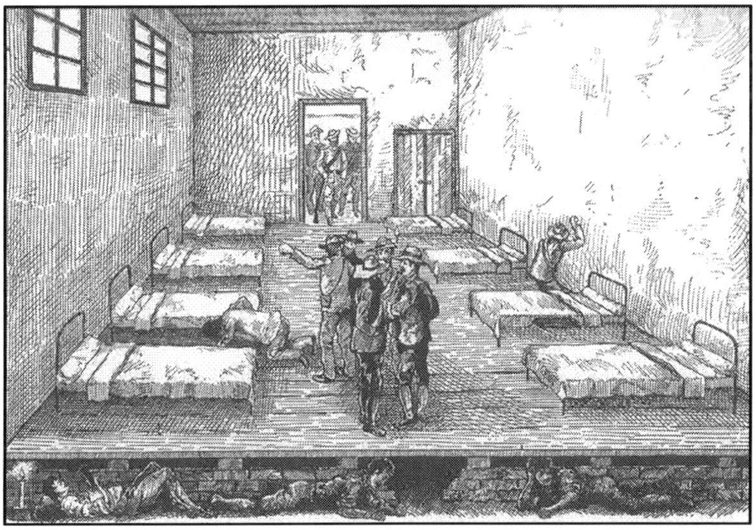

Hiding under the floor. "We hold our breath as they approach the roof of our humble dwelling."[343]

He found Haldane's saw made out of a table knife, and connecting this with the hole in the roof of the gymnasium, and the wires cut, he was sure they had gone away in the darkness. The rest, such is their mutual trust of one another in this country, were quite sure somebody had been bribed. The theories of the other officers in the prison are diverting. The discussions as to how the escaped had got out and where they had gone were full of imagination, but quite off the mark. In the afternoon Opperman and Sherlock Holmes came in with a hat and said the prisoners had been seen going over the hills towards Mafeking and had dropped the hat in question. By nightfall they had been tracked to Koodoosburg, about thirty miles out; and, indeed, the remains of their midday meal had been found. O wise detectives! This evening the Dogman[Patterson] went into Cullingworth's house in a great state of excitement and lit a candle at the verandah—a sign of good news, and on Majuba day too!

February 28th—We received the good news which the Dogman's excitement last night portended. Cronje has surrendered. This was received through the British Consul at Delagoa Bay. Buller has also driven back the Boers, and Botha wired: "No use; Burghers here won't face British." In the afternoon we received the following

[644] Charles Arthur Hugh Brett, Captain Suffolk Regiment, 1st Battalion Wounded and prisoner. Colesberg, 6 January 1900. Source: South African Field Force Casualty Roll. He was born at Muttra, India, 28 March, 1865, son of Lieutenant Colonel Arthur Brett, APD (formerly 2nd Dragoon Guards, Queen's Bays), and Georgina, his wife, daughter of Hugh Hannay, Paymaster, RN. He entered the Army as Lieutenant in the Suffolk Regiment 9 May 1885. He served in the Hazara Expedition in 1888, and received the Medal with clasp; was promoted Captain 15 June, 1894. From January 1895 to January 1899, he was Adjutant of the 1st Battalion of the Suffolk Regiment, and Adjutant, 4th Battalion, 15 November 1900 to 14 November 1905. He again saw active service in the South African War, taking part in the operations south of the Orange River, 1899-1900, including the action at Colesberg when he was severely wounded. Later he saw service in the Transvaal, east of Pretoria, and in the Orange River Colony from July to October 1900. He was mentioned in Despatches 10 September 1901, and received the Queen's Medal with three clasps. For his gallantry at Colesberg, in 1900, when, though shot through the lungs, he kept up command of his company actually charging the Boer trenches in this state until he fell exhausted from loss of blood. He was promoted Major, 24 February 1906. When European hostilities commenced he left the Curragh, August 1914, in command of the 2nd Battalion, to go to France, and fell in action at Le Gateau 26 August 1914, whilst leading his men. Major Brett married, in 1909, Enid Geraldine, daughter of Lieutenant Colonel H H St George, Senior Ordnance Officer, Scottish District, and they had one daughter, Ione Moncrieff St George, born 10 October 1909. Source: DSO recipients (VC and DSO Book).

wire: "Cronje's surrender unconditional. Boers retreating on the Biggarsburg," and in the evening we heard that the British were entering Ladysmith.

Three more officers replaced the three escaped in my room. We did not let them know about those underground, but I managed to send food, news, and water down as usual, also some hot cocoa at night.

March

March 1st—Ladysmith is relieved. Joubert wires: "On Lancers coming out of Ladysmith my mounted men retired leaving waggons and stores behind them." This afternoon the Cullingworths signaled over: "No more news, furthest telegraph station Elandslaagte." Kruger has gone to the front to exhort his burghers with texts. He was preceded by a telegram which was sent to all laagers. It is too long and too profane for me to copy out. Nothing but texts and psalms, showing that they are bound to win "though the enemy compass them about," as the Almighty is their own exclusive and peculiar property. The "Volksstem" says: "There seems to be some foundation for the rumour that Cronje has surrendered, but the report that Ladysmith has been relieved is quite untrue, our burghers are still fighting bravely south of that town. Should, however, Ladysmith be relieved, the war will only enter upon a new phase. We will then have to defend our borders against the greedy grasp of an unholy race. Now will the British see what fighting with the Boers really is. Now will the war begin in earnest."[645]

(Sherlock Holmes & Co. are completely off the track and all is well below.)

March 2nd—There are no signs of our moving into our new prison. This is very disconcerting as our friends cannot stay below much longer without getting ill. The Zarps' tents have been moved into the road. Opperman says because the yard was damp, but I fancy they are afraid of an attack on the Zarps. With the dumbbells in the gymnasium it might be possible to overpower them. The day was wet and dreary; I wrote letters, Mr. Hofmeyr prayed for the escaped. I have had to divulge the secret to No. 12 room, owing to one of them unfortunately seeing the trap-door open. They were very nice about it, and will do nothing to compromise the chances of success.

March 6th—Our signals this morning informed us that the President [Kruger] had wired to Lord Salisbury,[646] "Is it not time bloodshed ceased? Will send peace proposals." These people have got some nerve. First they declare war against an Empire, and then they expect that when they have had enough they can demand a cessation of hostilities.[647] There are no signs of moving.

March 7th—The Ides of March [sic[648]], but I don't expect Kruger will be murdered in the forum of Pretoria. Those below are still all right, though their condition is not enviable.

March 8th—The following telegrams were received to-day by our signaler-in-chief Burrows: (1) Fighting with De Wet; (2) Occupation of Bloemfontein on the 6th.[649] I busied myself in drawing a picture of Kruger going to the front to exhort his burghers, on the wall my room. There seems no chance of moving. Opperman says they have not even put down the floor in our new abode. Haldane wants to try to make them move. He thought that if Grimshaw vanished too it might alarm the authorities, and make them anxious to move us to a more secure place, but I feel sure—and Grimshaw agrees with me—this would only lead to the discovery of everything.

[645] 2/23/1900 The Boers' position at Grobler's Kloof attacked.

2/26/1900 Finding the passage of the river near Colenso commanded by strong entrenchments, Sir Redvers Buller sent his guns and baggage back to the south side of the Tugela, and found a new crossing.

2/26/1900 Colesberg and Rensberg, having been evacuated by the Boers, were occupied by General Clements, while Jamestown was occupied by General Brabant.

2/27/1900 (on anniversary of Majuba, 1881) Cronje, with 44 commandants and other officers of all grades, and over 3500 men, surrendered unconditionally to Roberts at Paardeberg. Sir Redvers Buller's force captured the Boer position at Pieters. This action opened the road to Ladysmith. Boers retired north to Ladysmith.

2/28/1900 Relief of Ladysmith after 120 day's investment.

[646] He is the Prime Minister of the UK at this time. Robert Gascoyne-Cecil, 3rd Marquess of Salisbury - Wikipedia

[647] 3/11/1900 Presidents Kruger and Steyn received reply from the Prime Minister refusing to entertain their overtures for peace.

[648] The Ides of March is the 74th day in the Roman calendar, corresponding to 15 March. Ides of March - Wikipedia

[649] 3/13/1900 Roberts occupied Bloemfontein. His despatch ran:—"The British flag now flies over the Presidency vacated last evening by Mr. Steyn, late President of the Orange Free State. The inhabitants gave the troops a cordial welcome."

March 11th—I drew another large picture on my wall, a sequel to the first. It represents Kruger just escaping from Lord Roberts, who with drawn sword appears to be running after him at a good pace. My picture No. 1 is entitled "President Kruger goes to front to exhort his burghers;" No. 2 "But returns on urgent business."

Similar Cartoon

No record of Thomas Frankland's cartoon exists, but Clive M. Dixon sketched a similar cartoon about an event that was observed outside of the siege of Ladysmith.

Cartoons showing a Boer Commandant inspecting a position and then having his horse bolt when the gun is fired.[650] [344]

March 12th—The man who came for grocery orders reported this morning that Bloemfontein had fallen, but our signal was that the British were within seven miles of the Free State capital. Opperman saw my portraits of Kruger this morning; I am afraid he did not appreciate them as he should have done. However, I told him that with a pail of whitewash and a brush he might obliterate them if he chose. (N.B.—Such is the procrastinating nature of these Boer-Hollander people that Opperman never had the pictures removed, and this with other things had, I believe, a good deal to do with his own eventual removal.)

No. 12 decided to have nothing more to do with the digging plan. We have therefore arranged that Grimshaw, Garvice, and I shall take part in the operation. Garvice has not been informed of Le Mesurier's whereabouts, but has decided to dig. The Colonials in No. 20 room are also digging, but theirs is to be a deep tunnel and I doubt if they can master the water question.

March 13th—Tragedy. The Dogman [Patterson] and Cullingworth have been commandeered as undesirables, but intend, I fancy, to escape to the British lines. We signaled to him, "Good-bye, eternal gratitude, God bless you!" The Dogman replied, "British twenty miles north of Bloemfontein; Good-bye; speedy release; will return with Bobs."[General Roberts] [651]

We started our shaft under the big room No. 16. Apparently we made a good deal of noise, for the old Colonels were very much alarmed and threatened to stop all digging, though they did not know who the culprits were. Opperman came into the room when mining was in full swing below, and it was all the occupants could do to hustle him outside, drowning the noise of the pick by stamping. We were rather distressed and decided to wait a few days. Garvice was very much startled when he saw Le Mesurier.

[650] Major Clive MacDonnell Dixon (10 February 1870 in Middlesbrough – 5 November 1914 in Ypres) was an English illustrator and soldier, best known for the charming images in his book The Leaguer of Ladysmith, created during the four-month Siege of Ladysmith in South Africa. This material also appeared in the Ladysmith Lyre at the time of the siege. Clive MacDonnell Dixon - Wikipedia And: The leaguer of Ladysmith, by Captain C.M. Dixon. London, Eyre and Spottiswoode [1900] Catalog Record: The leaguer of Ladysmith | HathiTrust Digital Library.

[651] He is referring to Lord Roberts who is commanding those forces in the Orange Free State.

Commando mustering (Commandeering) at Bloemfontein.[345]

He describes his feelings vividly. On going down by the trap-door he remarked what an awful hole it was. Suddenly, in the flickering candle-light he saw a gaunt, bearded, unwashed face, and a half-naked body. At first he could not make out what it was, but when he at last realised it was a brother officer he said you could have knocked him down with a feather had it not been that he was already crawling on his stomach.

The new shaft is a long way off; when I went down I had to crawl on hands and knees along passages and through man-holes, backwards and forwards in a regular maze of compartments, and, indeed, had the candle gone out one could easily have been lost. Haldane looked very ill, but the others, except for being covered with dirt, seemed well enough.

March 14th—Grimshaw went down this evening to hold a confab. They have managed to dig without making a noise by wetting the earth. Grimshaw and I made the trap-door into one piece by securing the planks together and also made it so as to batten down from underneath. I sent them down jugs of water during the day to wash in.

March 15th—All went as usual this morning. Grimshaw descended and did a little digging. In the afternoon Opperman brought the news that we were to be moved to-morrow! Most of the officers were very annoyed, but Grimshaw and I sent the information below with gladness. Well, there was no time to be lost. Food enough to last them a week, all the bottles filled with water, and everything that could possibly be of any use to the cave-men was sent down. We heard, however, and not to our surprise, that others were thinking of going into their respective holes so as to escape after we had moved. As this could have had no other effect than to cause the discovery of all, we were determined if possible to stop it. We told Colonel Hunt,[652] and he managed to persuade all concerned to abandon their schemes.

This settled, we set to work, after final good-byes and handshakings, to putty up the cracks between the boards of the trap-door, which had already been fastened down from underneath. This we succeeded in doing to perfection, and after covering the place well with dust, the trap-door could scarcely have been located by anyone; certainly not by those who did not know of its existence.

March 16th—The Staats Model School at an early hour was more than usually busy. We were all packing up such belongings as we had. I rolled everything in my mattress and rugs, and secured with rope. Then the gates were opened and all baggage was moved out on the road ready to be packed on the trolleys provided for the occasion. To be outside those gates was to breathe fresh air; to pass those barriers which had so long defied our efforts and our wits was like going out into another world. I went back into my room, and by prearranged taps on the floor Grimshaw signaled that all was well. I then sang "For Auld Lang Syne" as a parting farewell.

The Government had generously provided cabs for the convenience of the officers (who afterwards found they had to pay), and at about 10 A.M. the first cabs rolled off amid the friendly farewells of many neighbours. The

[652] Mr. H. Hunt Lt Col. Royal Regiment of Artillery. Severely wounded at Colenso. 15 Dec 1899. Source: Natal Field Force Casualty Roll, page 27 line 34. Anglo Boer War - Name search

long column of vehicles was escorted by a motley guard, consisting of very old men and tiny boys armed with Sniders[653] and sporting guns of ancient pattern.

Officers leaving the Model School.[346]

We soon passed out of the town and, crossing a small river, began to crawl up a steep hill. The roads outside of Pretoria appear very much neglected, but, of course, the money that should have been devoted to general improvements was all spent in secret service or in preparations for the war. We soon arrived at our destination. The building stands halfway up the side of a hill, and is probably a much healthier place than the Model School. Besides, the view is really pretty. To the north, indeed, it is limited by the tops of two hills. Southward lies Pretoria, a collection of large Government buildings and of small villas amid masses of trees, nestling beneath a high range of hills, along the crest of which rise the famous forts. The view on the west is merely a vast plain which reaches to the horizon, and a large hill obliterates any view to the east.

The place itself consists merely of a long white shanty with a fairly large compound enclosed by formidable barbed-wire entanglements. Outside are Opperman's house and the Zarps' tents. There are electric lights all round the enclosure, making escape a matter of considerable difficulty. Inside, the place looked more like a cattle-shed than anything else.

Pretoria from the East.[347]

A long galvanised-iron building, divided into a servants' compartment and kitchen, eating rooms, sleeping room, and four small bath-rooms. The sleeping-hall is eighty-five by thirty yards long and accommodates 120 officers, our beds being, roughly, a yard apart. There is no flooring. The drains consist of open ditches, while the sanitary arrangements are enough to disgust any civilised being. A strong protest was at once sent in to the authorities, but I doubt that it will have any effect.

[653] The British .577 Snider–Enfield was a breech-loading rifle. The American Jacob Snider invented this firearm action, and the Snider–Enfield was one of the most widely used of the Snider varieties. The British Army adopted it in 1866 and used it until 1874 when the Martini–Henry rifle began to supersede it. Snider–Enfield - Wikipedia

The Birdcage, British Officer's Prison, Pretoria.[348]

Inside Camp.[349]

Inside the Waterval prison.[350] Note the extensive electric light poles.

How the British prisoners were guarded at Waterval.[351] Note the high fence in the background and light pole.

The Bathing Tank for the Privates at Waterval.[352]

British Prisoners at Waterval. [353]

March 18th—The greatest disadvantage of this place over the Staats Model School is that we can get no news.

March 22nd—Gunning gave us a small baboon the other day, which was very fierce at first, but has tamed wonderfully. There are many different kinds of curious insects here, not curious for this country, of course, but which I have never seen before. The "Praying Mantis" or "Kaffir God" is one of the queerest. The whole place seems to be a large ants' nest, and we have often witnessed great fights between the different kinds. Snakes also abound. A night-adder was killed the other day. It was about thirteen or fourteen inches long and very poisonous, so Gunning says.

We hear Gunning and Opperman are going to the front to-morrow. I am very sorry for the former, though the departure of the latter is a great advantage.

March 23rd—The Zarps and Opperman departed for the front this morning. Their place was taken by a new guard selected from the Hollander Corps. The Commandant is a pleasant fellow and a great improvement on Opperman. [Westernant] [654]

March 25th—We had service as usual this morning. This evening an attempt to escape was going to be made by Ansell[655] and Co., but it never came off. There has been no news of Haldane and the others, so I suppose they are well away by now. This evening the new Commandant had roll-call. We call him "Pyjamas," because he wears a suit of clothes for all the world like a pair of pyjamas. His real name is Westernant.

March 30th—There has not been anything very important to record for some days. On Tuesday an attempt to escape was made by Best. While one sentry was gossiping with another he crept under the barbed wire. As luck would have it, when Best had got half way through, the sentry finished his *tête-à-tête* and returned to his post. At first he thought Best was a dog and called out *footsack*,[656] but seeing he was a human being, merely told him to go back. He might have shot him with some excuse, so Best was lucky in striking a kind-hearted man.

On Wednesday Joubert died.[657] In respect to him we sent a wreath. I don't think this will have any effect on the war, as (and the papers say as much) his moderate attitude in the recent crisis had taken away much of his popularity.

There was great excitement this evening caused by an attempted escape. The electric wires had been tampered with, and at about 10.30, by some device, Horne, a colonial, who is also an electrician, made the current travel on a shorter circuit, thus blowing out the main fuse and extinguishing all the lights round the building. Hardly had this happened than two shots were fired in quick succession, and then another. The escape failed, but all got back into the building unwounded. Apparently the lights had gone down, then up for a second, then finally out.

During the momentary flash Hockley,[658] [one] of the escapees, had been seen and fired at. However, "All's well that ends well," though some say that two bullets went through the dining-room. Sentries were doubled for the night and patrols sent out.

April

April 3rd—Hurrah! the papers this evening report the safe arrival of Haldane, Le Mesurier and Brockie at Lourenço Marques, having travelled through Swaziland. We were so glad to hear this news. Alas! We also hear that sixteen officers arrive tomorrow, and that seven guns were captured with them.

The Cullingworth girls came up this evening and signaled with a handkerchief that Mafeking had been relieved.[659] I hope it is true. We all admire the pluck of those girls. We have already collected a large subscription to get them and the Dogman handsome presents.

[654] Unable to identify this individual.

[655] Possibly: George Kirkpatrick Ansell, Captain & Brevet Major QSA (3). Prisoner, Arundel, 20 Feb 00. Source: QSA medal rolls 6th (Inniskilling) Dragoons. Also: MID LG: 17 January 1902, page: 374. Source: General Kitchener. 8 November 1901. Re: General mentions.

[656] Note in Frankland's Original, it would be Dutch or Afrikaans for "Be off."

[657] 3/27/1900 Death of General Joubert.

[658] Probably: W. Hockley, Lieutenant, Prisoner. Schoeman's Kop, 23 February 1900 Attached FID Source: South African Field Force Casualty Roll, Montmorency's Scouts.

[659] Not until 5/17/1900 Colonel Mahon, at the head of the relief force, entered Mafeking.

There was a large swarm of locusts yesterday. So thick was the cloud that it quite obliterated the view of the distant hills. They continued passing over nearly all day.

April 5th—The prisoners arrived this morning. They mostly belong to U Battalion, R.H.A.; [Royal Horse Artillery] some to the M.I. [Mounted Infantry] and Cavalry. I have not quite gathered the circumstances of their capture, but they seem to have been caught in a trap, owing to the want of the ordinary precautions. The convoy and one battery were practically held up without firing a shot, but the other battery got away. When marched off they heard that another British force was pursuing so that the guns may be recaptured.[660]

They bring very little news; apparently they have heard nothing about the relief of Mafeking, though Warren[661] was on his way thereto. Roberts has been delayed in his advance for the want of horses, but as this has been remedied the forward movement should begin shortly.

Had the horses not been so done after Abrams Kraal, they say De Wet would have been caught and the war over. Such is the fashion of war. If so-and-so had happened—always "if"!

April 6th—How the fortunes of war vary! We seem to be going through a series of small disasters. To-day the papers have the report of a "Brilliant Boer Victory, thirty-six miles south-east of Bloemfontein;[662] 450 prisoners!!!" The only hope is that the account is not "official." But we must be ready for the worst. The leading article says: "Within a few days Roberts will be forced to evacuate the Free State. *His retreat from Bloemfontein will be like Napoleon's retreat from Moscow*."

April 11th—The prisoners reported captured some time ago have not arrived yet. They always seem to be "expected to arrive somewhere," but apparently have not yet been actually seen by anybody. On Saturday their capture was reported officially. On Thursday English wires said that 300 Royal Irish were surrounded. To-day they say the prisoners are expected at Pretoria to-morrow! Well, we shall see.

The last few days we have had many good rumours about the capture of Boers and British victories. To-day the papers say that Lord Methuen is advancing on Boshof (he must be there by now), and that Colonel de Villebois[663] has been killed. He apparently and his men (100, so they say—probably 500) were all killed, wounded, or taken prisoners. A distinguished ex-French officer and his foreign legion is a good bag.

The next piece of information is, quoting from Boer paragraphs or head lines, "Fifteen hundred English in a corner;" "Brabant's Horse in a trap." Then, again, "There is every hope of their surrender." So much for this. But on the Dutch side we read that all telegraphic communication with Ladybrand and the south has been cut, so I rather fancy the Boers have over-reached themselves for once.

The Boers have attacked our camps at Elandslaagte, and because, when they shelled, our camp tents were struck, they report that the British fled. I wonder if Le Mesurier was in this show.

In all these fights, as usual, the Boers "By the grace of God had (about) one man killed and four wounded." This is heavy; generally it is one horse and three mules. "The enemy," of course, "must have lost heavily." So the paragraphs run on. Many are the funny expressions. "One brave burgher succumbed to the explosion of a bomb." "One of our guns *in firing* damaged its sight and one of its wheels!" They always end up with "Our burghers are full of courage, and determined to withstand the enemy to the last."

[660] 4/4/1900: Capture of British troops by the Boers near Reddersburg.

[661] General Sir Charles Warren, GCMG, KCB, FRS (7 February 1840 – 21 January 1927) was an officer in the British Royal Engineers. He was one of the earliest European archaeologists of the Biblical Holy Land, and particularly of the Temple Mount. Much of his military service was spent in British South Africa. Previously he was police chief, the head of the London Metropolitan Police, from 1886 to 1888 during the Jack the Ripper murders. His command in combat during the Second Boer War was criticised, but he achieved considerable success during his long life in his military and civil posts. Charles Warren - Wikipedia

[662] Sanna's Post (a.k.a. Korn Spruit) was an engagement fought during the Second Boer War on March 30th And 31st, 1900. In all, the British suffered 155 men killed or wounded. 428 men, seven field artillery pieces and 117 wagons were captured. The Boer force suffered three killed and five wounded. But even more serious than the losses in the action was the loss of Bloemfontein's water supplies. This greatly aggravated an epidemic of enteric fever dysentery and cholera among the occupying British army, which eventually caused 2,000 deaths. Sanna's Post - Wikipedia

[663] George Henri Anne-Marie Victor count de Villebois-Mareuil or by his shortened name George de Villebois-Mareuil (22 March 1847, in Montaigu, Brittany, France – 6 April 1900, in Boshof, OFS George de Villebois-Mareuil - Wikipedia.

Various officials came up the day before yesterday to inquire into the causes of the protest we had sent in, signed by all the officers here. They promised that everything would be seen to; but they are all—well they are Boer officials, and I doubt if our lot is to be in any way improved.

The weather is getting much colder now, though the sun is still hot by day. A few stray shots whistled over the building to-day, probably "accidentally on purpose." I hope they do not begin sniping regularly.

Rosslyn Twice Captured

James Erskine, 5th Earl Rosslyn[664] April 18th just before primrose day.[665]

"Mr. De Souza left me after telling me that Winston Churchill's warrant of release was in their hands the night he escaped, and that they were glad to be rid of him. "He talked so much," he added, "that we all avoided him" I could not restrain a smile; but his escape and Mr. Hellawell's[666] unsuccessful attempt have certainly made matters "difficult" for war correspondents and officers too, since these have been removed to more guarded quarters ever since three of their number escaped about a month ago."

"Talking of escapes, the story of Winston Churchill's was already known to me through his own description, and the officers were most interested to hear about it, though, I am afraid, he was not a persona grata with his fellow prisoners, and, as far as I could ascertain, did not play quite fairly with the others who concocted the plot, but, according to them, followed the principle of sauve qui peut[667] rather than "shoulder to shoulder."[668 669]

May

May 8th—We have had an immense amount of news lately. Roberts has begun his big advance. Brandfort is in our hands, also Winburg. The force advancing via Boshof has reached Hoopstad, while the British have crossed the Vaal at Fourteen Streams. De Wet has not been heard of for a considerable time. So much is acknowledged in the papers. Rumours say that we are behind Kroonstadt!! That De Wet, Steyn, and 8,000 Boers have been taken!![670] The English in the town think we shall be released by the 24th of May. A panic seems to have seized the Boers, and excited meetings have been held. Kruger summoned the Volksraad on Sunday, and addressed them in stirring words, which, while acknowledging the serious nature of the situation, exhorted the burghers to continue the struggle trusting in the Lord. General Schalk Burger,[671] while addressing the townspeople, said that

[664] Earl of Rosslyn, James St Clair-Erskine, 5th Earl of Rosslyn Twice Captured: A Record of Adventure During the Boer War. United Kingdom, William Blackwood and Sons, 1900. Page 295-296 See: Twice Captured - Google Books And: James St Clair-Erskine, 5th Earl of Rosslyn - Wikipedia File:James Francis Harry St Clair-Erskine, 5th Earl Rosslyn,.jpg - Wikimedia Commons.

[665] Page 335 Twice Captured.

[666] Possibly, A. Hellawell Source: QSA and KSA medal rolls. Western Province Mounted Rifles.

[667] Every man for himself - more literally "save yourself if you can"; most literally "may he save himself, whoever can."

[668] Page 335 Twice Captured.

[669] Rosslyn will pay dearly for publishing this observation. Churchill and others will force a retraction.

[670] Saturday, April 28, 1900 Fighting near Thabanchu Mountain.
Tuesday, May 1, 1900 General Hamilton captured Houtnek.
Saturday, May 5, 1900 British occupation of Brandfort. Roberts's advances towards the Vet River.
Sunday, May 6, 1900 The Vet River (Tributary of the Vaal River) passed and Smaldeel occupied.
Monday, May 7, 1900 General Hunter occupied Fourteen Streams.

[671] Schalk Willem Burger (6 September 1852 – 5 December 1918) was a South African military leader, lawyer, politician, and statesman

a stand might yet be made, if not, the independence of the Republic was at an end. The Church of Pretoria has addressed petitions for peace to the Churches of Great Britain and of Europe and America. They pray that this unholy bloodshed may cease. Kruger says "Continue the struggle to the end." Is it for England or for Kruger to give in?

We have started a newspaper; it is progressing. We call it the "Gram," because at the Staats Model School all our news came in under the popular names of signal-gram (when news was signaled), Kaffir-gram (when brought through the Kaffir),[672] Brockiegram (when [Adam] Brockie succeeded in getting information from the Zarps), and so forth. Rosslyn is editor; Major Sturges[673] sub-editor. White, R.A., Wake, 5th Fusiliers, and I, are the artists. The paper has been all written out by Rosslyn, and is now being hectographed.[674] We hope to bring out seventy good copies of the first number.

Rosslyn at his desk. Note initials THCF for Thomas Hugh Colville Frankland.[675]

who was acting president of the South African Republic from 1900 to 1902, whilst Paul Kruger was in exile. Schalk Willem Burger - Wikipedia

[672] During the Dutch and English Colonial Period the word Kaffir was used to describe monotheistic peoples (Nguni ethnic groups in particular) of South Africa, who were not of a Christian or of an Islamic religious background, without derogatory connotations, until the early twentieth century. It appears in many historical accounts by anthropologists, missionaries and other observers, as well as in academic writings. See Also: Kaffraria and British Kaffraria (Queen Adelaide's Province) were historical regions along the coast north or Port Elizabeth. See: File:Eastern Frontier, Cape of Good Hope, ca 1835.png - Wikimedia Commons Subsequently it became a racial slur. Kaffir (racial term) - Wikipedia See also the British military unit Kaffrarian rifles which were white and black units which were still organized as late as 1947. East London, 1 March 1947. King George VI inspecting 'Kaffrarian Rifles' guard of honour. - Atom site for DRISA And: Kaffrarian Rifles AKA: Buffalo Volunteer Rifles Buffalo Volunteer Rifles - Wikipedia Also: a historical region in Afghanistan and the setting of most of Rudyard Kipling's famous 1888 novella "The Man Who Would Be King." Kafiristan - Wikipedia And: an Arabic and Islamic term which, in the Islamic tradition, refers to a person who disbelieves in God as per Islam. Kafir - Wikipedia PF Note: I have no authority to change the wording used by original authors. There are instances where a pejorative or admiring modifier is used. While this is racist in and of itself, to identify someone by their complexion, in this narrative, identifiers such as Dutch, Boer, English, Uitlander, Indian, and many others are used as means of identification sometimes with pejorative or admiring modifiers.

[673] W. E. Sturges Major Prisoner. Stormberg, 10 December 1899 Northumberland Fusiliers 2nd Battalion. Source: South African Field Force Casualty Roll.

[674] The hectograph, gelatin duplicator or jellygraph is a printing process that involves transfer of an original, prepared with special inks, to a pan of gelatin or a gelatin pad pulled tight on a metal frame The special aniline dyes for making the master image came in the form of ink or in pens, pencils, carbon paper and even typewriter ribbon. Hectograph pencils and pens are sometimes still available. Hectograph - Wikipedia

[675] "The Gram" A social Magazine Founded by British Prisoners of War in Pretoria, Facsimile Edition Eyre & Spottiswoode, Her Majesty's Printers, London #1 - The Gram ; a social magazine founded by British prisoners ... (no.1-3). - Full View | HathiTrust Digital Library Scanned from the University of Minnesota Copy. Image: #60 - The Gram ; a social magazine founded by British prisoners ... (no.1-3). - Full View | HathiTrust Digital Library Cover Image: #61 - The Gram ; a social magazine founded by British prisoners ... (no.1-3). - Full View | HathiTrust Digital Library.

No. 0 "Cheapside."[676]

May 13th—Though two or three prisoners have arrived lately, we can get no particular details of the news. There is no doubt that a general advance has been begun, but what point our troops have reached is uncertain. Also, it is still a question whether De Wet is captured or not. This morning the most serious rumour came in, to the effect that Mafeking had fallen, but I can scarcely believe it.[677]

Yesterday Mr. Hofmeyr received the welcome order to pack up his things and go. He seemed very affected at saying good-bye and nearly broke down. We all liked him very much, and bade him a hearty farewell, cheering him as he left the enclosure, and singing "He's a jolly good fellow." We shall miss him as well as his services.

Our paper came out yesterday and was a great success. We hope to bring out a new one on the Queen's Birthday, though it is an awful labour.

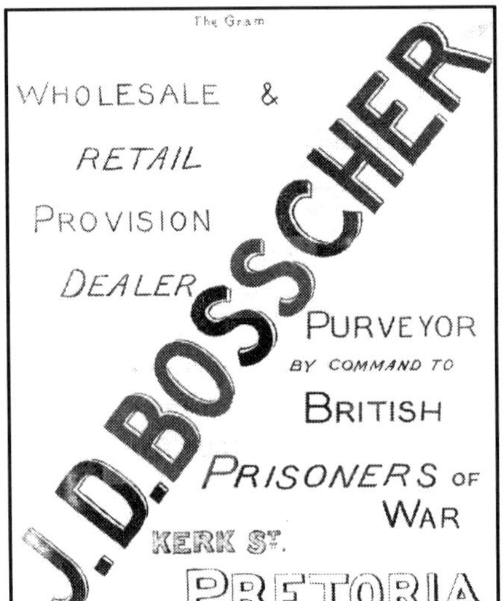

Cover page and an advertisement Mr. Bosscher took out in the prisoners' newsletter 'The Gram.' [678] [354]

[676] Believed to be Rosslyn's own sketch of his quarters. p. 330 of Twice Captured and p. 407 of the scan.

[677] Tuesday, May 8, 1900 Ladybrand deserted by the Boers.
Wednesday, May 9, 1900 Capture of Welgelegen. Mafeking Relief Force reached Vryburg.
Thursday, May 10, 1900 Battle of Zand (Sand) River. Occupation of Ventersburg.

[678] PF: Mr. Bosscher was the only advertiser, even though he need not have done so, since he still had the concession. See Hofmeyr, for particulars, and Rosslyn. The magazine did not begin publishing until May 1900 well after the prisoners had been transferred to a purpose-built prison outside of Pretoria.

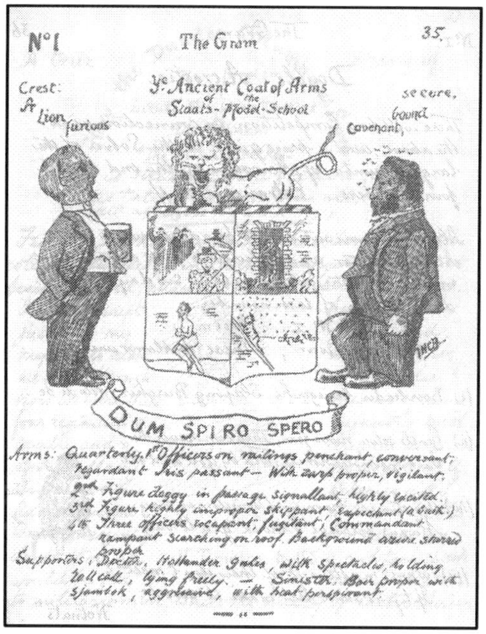

Image of "Captain Dunham" by Thomas Frankland[679] Page from the "Gram" Note THCF Author.[355]

Life has not been so bad lately. Buoyed up with hope of a speedy release, and occupied with the "Gram," time has passed, in my case, more quickly. We had a selling lottery the other day for the day of our release. The dates ranged from the 15th of May to the 15th of August. The Queen's Birthday was much in request, while "the field" (any day after August 15th) went for six pounds.

The "Volksstem," of course, progresses as usual. Having exhausted all other insults on England, they commenced lately on the Queen! During the present British advance the mendacious powers of the editor are once more brought to trial, and once more he has not been found wanting. The burghers are full of courage (running everywhere); even the women wish to fight! There was, indeed, a rumour that our present guard was to be commandeered and the women put here to look after us. Poor time for us! I fancy we should be all shot! The Volksraad sat the other day, and after Kruger and others quoting a few scriptures the session of 1900 was closed after sitting two days!

May 14th—So much news has arrived today, that I think I had better inscribe it, while I remember. This morning came the rumour that a good many Boers actually did get into Mafeking, but, being unsupported, still remain there. This evening's "Volksstem" is truly a wonder. It gives more news than it ever has given before. An attack was made on Mafeking. The Boers took a "fort," but were attacked by night, and lost seven killed and "some" wounded and prisoners.[680] At present Carrington[681] and Plumer[682] are proceeding to Mafeking by train, so that it must have been relieved. Everywhere the Boers fly, and the British troops entered Kroonstadt on the 11th inst. Hunter,[683] with his 25,000 men, drove the enemy back at Warrenton,[684] and "the Boers are unable to resist the advance of the forces at Vryburg."[685]

679 Image and text per Neville Chamberlain "Frankland drew the illustrations on pages 1, 9, 22a, 35, 36, 39 & 41 of 'The Gram' (7 of the 18 drawings)." Anglo Boer War - Armored Train Ambush 11-15-99 Chuchill & Haldane - Page 17 - Boer War Forum

680 Saturday, May 12, 1900 Commandant Eloff attacked Mafeking, and was captured by Col. Baden-Powell. Roberts occupied Kroonstad without resistance.
Sunday, May 13, 1900 General Buller advanced towards the Biggarsberg.

681 Frederick Carrington. In February of 1900 he was appointed to command of the Rhodesian Field Force. This was a force established to prevent attacks on Rhodesia from the south, and the soldiers were drawn from the various colonial contingents in the war, including the South Australian Citizen Bushmen. Carrington left for South Africa in March 1900 on the SS Dunottar Castle, and arrived in Cape Town the following month. Frederick Carrington - Wikipedia

682 Herbert Plumer. In 1899 Plumer went to Southern Rhodesia where he raised a force of mounted infantry and led them during the Relief of Mafeking during the Second Boer War. Herbert Plumer, 1st Viscount Plumer - Wikipedia

683 This is: Archibald Hunter - Wikipedia.

684 Warrenton is situated in the north-eastern part of the Northern Cape province, on the south bank of the Vaal River, approximately 70 kilometers (40 mi) north of Kimberley. Warrenton, South Africa - Wikipedia

685 Vryburg is situated halfway between Kimberley and Mahikeng Vryburg - Wikipedia

"But," says the "Volksstem," "the fact that Kroonstadt is in the hands of the enemy need create no alarm. [686] As we retire our line of defence becomes less and our commandos can be concentrated to resist more effectually the advance of the British forces. Besides, many things may happen which will put an entirely new face on the war. Our delegation has reached America, &c., &c. Lord Roberts's hastened advance is said to be caused by his desire to reach Pretoria on the Queen's Birthday, but might not the real reason be the fear of foreign intervention? Lord Roberts wishes to strike a decisive blow before his forces are needed elsewhere. Every day's delay is, therefore, an advantage to our cause. Courage is all that is needed, &c., &c."

The above is a *precis* of the "Volksstem" leading article. Still they harp on foreign intervention, but from what I gather from recent Continental criticisms on the war, I fancy their chances in this line are less than at the beginning of the war. As to the burghers' courage, I doubt if the majority of them have much left. For many months the Transvaal Government have whipped their subjects to the fight; but even the worm will turn, and to the simplest, or the most ignorant, the Government promises and hopes must seem vain.

The day of our release is, perhaps, approaching; but it does not do to be too sanguine; one never knows where a check may occur. Still I "plump" on the end of the present month.

May 20th—The month is drawing to a close, and the day of our release is still a matter of speculation. News is pretty plentiful; even the "Volksstem" tries to hide nothing. Roberts has made a great advance, but whether he has halted at Kroonstadt or not is uncertain. We all hoped he would not stop until he had reached Pretoria.[687]

We have been very much alarmed lately at the rumoured intention of the Government to move us to Lydenburg,[688] but at present it is only a rumour. If we are moved we shall have every prospect of being shunted about the country with guerilla bands of Boers who would keep us merely as hostages, if, however, we are kept here we shall have every chance of being released during the siege of Johannesburg. The Boers, it is said, have decided to hold that place and are not going to blow up the mines. The defence of Pretoria would be impossible with the troops at their disposal.

Life goes on as usual. The only diversion that has lately occurred was the athletic sports, which were got up by some energetic people. The event took place yesterday, and, on the whole, was a decided success. The chief feature, however, of the day was the betting. Several enterprising officers kept books, but Haig,[689] of the Inniskilling Dragoons, cut the best figure in that line, and it was chiefly owing to his amusing performance that the day was a success. White has made an excellent sketch of "Our Bookie" for the next "Gram" number.

The sermon this morning is worth recording. The Rev. Mr. Bateman delivered a most extraordinary speech as part of his service. Whether it was meant for our spiritual edification, or merely intended to convey news to us under the disguise of a text, was not quite certain; but, by preaching on the text that begins "as cold water is to the thirsty soul, so is good news, &c.," he led us to believe that we were to be released in a very short time. Roulette has been going very strong. Large sums have been lost and won.

May 25th—Yesterday we, prisoners of war, joined with the British Empire all over the world in the celebration of the Queen's Birthday. In our little enclosure we have quite a representative British Empire—English, Scotch, and Irish soldiers, Colonials, South Africans, Australians, and civilians, and, indeed, we only require a Canadian to complete the list.

Yesterday evening we drank the Queen's health in light port (rather nasty). The first drops of wine or spirit I had tasted since the 18th of November. This was followed by "God Save the Queen," sung by all with a heartiness and feeling that I never heard before. It must have sounded very well outside. To us it was as it were "giving

[686] Saturday, May 12, 1900 Roberts occupied Kroonstad without resistance. Commandant Eloff attacked Mafeking, and was captured by Col. Baden-Powell.

[687] Tuesday, May 15, 1900 Christiana occupied. Christiana, South Africa - Wikipedia

Thursday, May 17, 1900 General Ian Hamilton occupied Lindley. Colonel Mahon, at the head of the relief force, entered Mafeking. Methuen entered Hoopstad.

Friday, May 18, 1900 Occupation of Newcastle, in Natal by Buller's forces.

[688] Lydenburg, officially known as Mashishing, is a town in Thaba Chweu Local Municipality, on the Mpumalanga highveld, South Africa. It is situated on the Sterkspruit/Dorps River tributary of the Lepelle River at the summit of the Long Tom Pass. Lydenburg - Wikipedia

[689] Neil Wolseley Haig Lieutenant. Prisoner. Vredes Verdrag, 10 May 1900 Released Source: South African Field Force Casualty Roll 6th (Inniskilling) Dragoons MID LG: 10 September 1901, page: 5929. Source: Field Marshal Roberts. 4 September 1901. Re: General mentions Captain QSA (3). Source: QSA medal rolls.

vent" to our imprisoned feelings, while we also found in it a link with our country, from which we have for so many months been severed.

It is now pretty certain that Roberts is resting his troops, and rumours have it that the Boers have asked for an armistice. Whether Lord Roberts celebrated the Queen's Birthday by a victory or a peaceful armistice remains to be seen.

The "Volksstem" considers that it would be a graceful act on the part of the State President if he were to wire the Queen and offer her as a birthday present the unconditional release of all the British prisoners of war. As the "Volksstem" is the official organ, this may quite possibly be merely a feeler to the public (if public there be in this country). At any rate it would be an act worthy of the wily Boer. He finds it a source of trouble and expense feeding and guarding 5,000 prisoners, so he gives them away with a pound of tea—I mean as a graceful act. Whether the offer would be accepted is uncertain. But we at any rate will be very happy if the Transvaal Government puts us over the border.

The weather (by day) is simply perfect. Every morning the lovely air makes one long for a walk or ride, and causes one to chafe at the inability to roam beyond the one hundred yards' enclosure. We are henceforth to be allowed to have wine, but personally I shall wait for freedom before I indulge in that luxury again. The second number of the "Gram" came out yesterday, and, I believe, was much appreciated.

May 26th—Two prisoners of war arrived this morning. They were caught at Lindley, which the Boers have apparently reoccupied. They were taken across country to the Natal railway, and then conveyed straight to Pretoria. They say they have heard firing at the Vaal, so I suppose Lord Roberts is there. The Boers hold a strong position south of Johannesburg, and they also intend defending that town. One of the De Wets is still on the right rear of our army, but will be dealt with by Rundle's division which is coming up that way. It is said that De Wet at one time offered to surrender on condition that he himself should not be made a prisoner. But Roberts would receive none but an unconditional surrender. Buller has been ordered to force Laing's Nek at all costs. The "Volksstem" says that Lord Roberts's headquarters are at Honningspruit, some way north of Kroonstadt, but this is probably news of some days' standing. There is also a rumour that our troops have occupied Potchefstroom.[690]

May 29th—At last our release seems near at hand. Yesterday and to-day big guns were heard plainly in the direction of Johannesburg, which is now in our hands.[691] Boscher, the grocer, has just arrived, having come up by the last train. He says that the Dragoons were actually in the streets when he left. [Johannesburg] I fancy to-morrow or next day will see us out. Everybody is in the best of spirits and full of excitement.

[690] Tuesday, May 22, 1900 General Ian Hamilton occupied Heilbron after a series of engagements. The main army, under Roberts, pitched its tents at Honing Spruit, and General French crossed the Rhenoster to the north-west of the latter place.
Wednesday, May 23, 1900 Rhenoster position turned.
Thursday, May 24, 1900 British Army entered the Transvaal, crossing the Vaal near Parys, unopposed.

[691] Tuesday, May 29, 1900 Battle of Doornkop: Boers defeated. Battle of Doornkop - Wikipedia

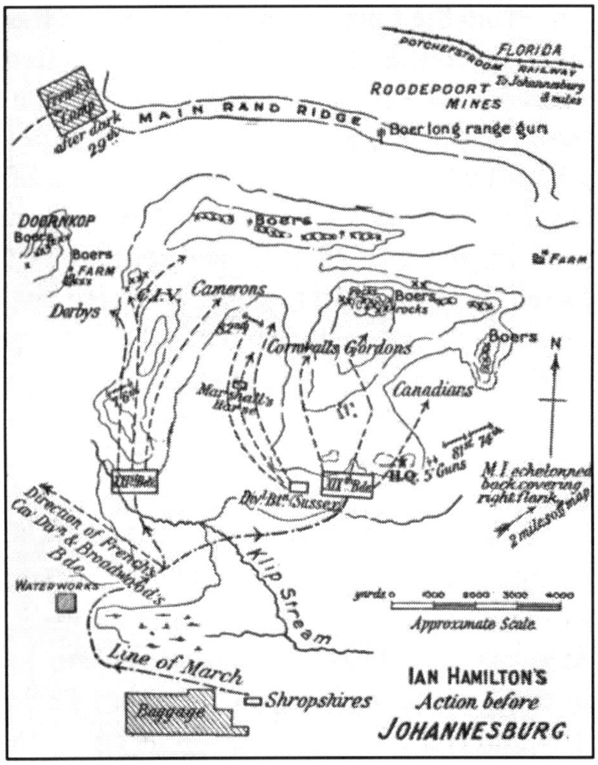

Ian Hamilton's Action Before Johannesburg.[356] Dorset Yeomanry entering Johannesburg, 1900[357] Also published as The City [London] Volunteers Entering Johannesburg.[358]

Greatest excitement during dinner. Mr. Hay and Mr. Wood came in and asked Colonel Hunt to send twenty-four officers to Waterval to look after the men.

Adelbert S. Hay U.S. Consul at Pretoria.[359 692] The Consul for The British Prisoners at Pretoria.[360 693]

[692] Adelbert Stone Hay (November 1, 1876 – June 23, 1901) was an American consul, politician and son of U.S. Secretary of State John Hay. Adelbert S. Hay - Wikipedia

[693] "Hon. Adelbert S. Hay, American Consul-General at Pretoria, who attended to the wants of the British prisoners and acted for British subjects during the war. He stands by the railway carriage, behind him is Mr. John G. Coolidge, Vice-Consul; they are bidding goodbye to attache's leaving Pretoria, and Mr. Hay is talking to Captain Slocum, United States Army, American attache.' Other travelers seen in the picture are the Japanese and Russian attache's" Hillegas, Boers in War, 1900.

Richard Harding Davis,[694] Adelbert S. Hay, [695] Consul and J.G. Coolidge Vice-Consul. Referred to as Gardner F. Coolidge of Boston. American consulate, Pretoria. [361 696]

Kruger has gone to Holland.[697] The British are expected here to-morrow, and we shall be free! We sang "God Save the Queen" and cheered Hay and the Commandant, who made a very nice speech, saying he hoped to shake hands with us outside. Oh! how I longed to see the old regiment once more! The Commandant says that there is still fighting at Klip drift, but a force of 4,000 men has broken through and come here. I believe there is a lot of looting going on in the town now. Roulette is at an end. I can scarcely write coherently, so excited am I. Fancy being free; I can scarcely believe it! Six and a half months' imprisonment, and about to be freed! Thank God!

May 31st—Too premature were our hopes. Yesterday and to-day have been spent in awful suspense. Distant guns have been heard, Boers have been seen riding about, and rumours of all kinds and descriptions are rife. It is too awful this final suspense. We do nothing in hope of a speedy release, and we pass the day anxiously scanning the horizon for the approach of troops.

All day commandos have gone through the town, and one was seen on the plain coming in from Mafeking. One commando came up our way, and we were rather surprised that they made no attempt to shoot us. Indeed there was nothing to prevent them. Three prisoners came in. They were caught in or near Johannesburg. That town was officially surrendered at 10 A.M. this morning. The Boers intend making a sort of stand (one of their usual ten-minute affairs I suppose) at Irene, a place six miles south of Pretoria, and a fight is expected there tomorrow. Their line of flight is past our abode and Waterval, and I should not be surprised if, unable to face and shoot armed men, some of these foreign ruffians shoot a few prisoners.

The town is evidently to be handed over quietly. The "Volksstem" is still covering a sheet of paper with print, but seems to take not the slightest interest in the war. They speak of giving up Pretoria as one of our papers might of a concert. Well, I suppose it will come at last, but I shall heave a sigh of relief when it does!

[694] Richard Harding Davis With Both Armies in South Africa. Charles Scribner's Sons, New York 1900 Richard Harding Davis (April 18, 1864 – April 11, 1916) was an American journalist and writer of fiction and drama, known foremost as the first American war correspondent to cover the Spanish–American War, the Second Boer War, and the First World War. His writing greatly assisted the political career of Theodore Roosevelt. He also played a major role in the evolution of the American magazine. Richard Harding Davis - Wikipedia

[695] Webster Davis. "[upon] my return to Pretoria from the visit to the battlefields and camps of the fighting Boers, I had the pleasure of meeting United States Consul Hay, who had just arrived. His predecessor, Consul Macrum, after having made his complaint to his government that his mail had been opened by British officials through whose hands it had passed, was relieved of his office and Mr. Hay appointed in his place." Quote from Davis' book, John Bull's Crimes. Hay was the eldest son of John Milton Hay, who served as United States Secretary of State and the United States Ambassador to Great Britain at the time. John Hay - Wikipedia

[696] During the time intervening between Mr. Macrum's dismissal and Mr. Hay's arrival, Mr. W. J. Hollis, United States Consul at Lorenzo Marquez, was stationed in the consulate at Pretoria. Webster Davis' John Bull's crime; or, Assaults on Republics, 1901 p. 23.

[697] Not Until Tuesday, September 11, 1900 "Kruger, having departed from the Transvaal, arrived in Portuguese territory, and proceeded to Lorenzo Marques."

June

June 1st—No sign of the British! But we expect to hear guns tomorrow. There are plenty of rumours about—Roberts captured, French killed, &c. There was a good deal of looting in the town yesterday, and five men were shot. Our hopes of a few days ago have been somewhat damped, and most of us put our release down at a week hence.

Boer Retreat from Pretoria.[362]

The "Volksstem" is remarkable. The editor is evidently wishful to avoid his tarring and feathering, and scarcely speaks of the war at all.

June 3rd—I have almost given up looking forward to our release, and have fallen back into the ordinary monotonous life. No guns have been heard, and therefore no serious fighting can have taken place anywhere near Pretoria. Rundle has been reported as having received a check in the Free State, and Lord Roberts is said to be still in Johannesburg; otherwise there is no news at all. Botha has taken matters into his own hands, has kicked out the officials appointed by Kruger, chosen a committee of his own, and has arranged the defence of the positions outside the town. He has therefore made himself practically President of what remains of the Transvaal. Kruger went off with a million of hard gold, paying the Government officials with dishonoured cheques on the National Bank, from which he has removed all the money. Every one of his ministers thirsts for the old man's blood, and perhaps it were best for him to go further than Middelburg.

President Kruger's private car used as the capital after evacuation of Pretoria.[363] Mr. Kruger…at Lorenzo Marques[364] (October 1900).

June 4th—At about 8.30 this morning firing was heard at no great distance, in the south-west direction—field-guns, "pom-poms," Maxims, and even musketry. At about nine o'clock a shell was seen to burst on an earthwork on a ridge of hills south of the town. Field-glasses and telescopes were immediately brought out, and we were well entertained for the rest of the day. Shrapnel burst all along the ridges, and presently lyddite[698] shells were planted on the hills. The firing seemed very unmethodical, and the Boers made little or no reply. On the western kopjes shrapnel was seen bursting all over the place, and we expected the Infantry to attack them. But the lyddite shells were certainly the most interesting. They burst with a tremendous noise, throwing up clouds of brownish earth. For some time the forts seemed the mark our gunners were aiming at, and these costly erections certainly received their share—four shells pitching well inside the west fort; but, later, the shells were directed on the eastern outskirts of the town. Whether these were intended for the railway station, we could not make out; but, otherwise, they seemed to have no object.

At about 4.30 the Boers were seen leaving the western ridges and trekking at a remarkable pace across the plain, disappearing along the northern road. The day's action was ended by a kind of *feu de joie*[699] of lyddite shells, which struck the two forts and the surrounding hills. Then peace ensued. The last few shots seemed to have been fired by guns which were much closer than at the commencement of the bombardment, and the flight of the projectiles, which we could distinctly hear, passed from west to east, so that we hope our troops have occupied the hills on the west.

The hills are burning to-night, and the scene is strangely illuminated in honour of our approaching rescue.

June 5th—A day of strangely mingled hopes and fears. This morning at about 1.30 the Commandant awoke us and ordered us to pack up at once and prepare to march to the railway, whence we were to be transported by train down the Delagoa Bay line to some station beyond Middelburg. All were filled with consternation. To be hurried away when release was so near at hand seemed too awful. Words cannot express my feelings. At last we decided to refuse to go. Let them massacre us if they dared. We reminded the Commandant of the promise made to the officers the week before that if they restrained the men in Waterval neither they nor the men should be transported. The Commandant replied that he had his orders and must execute them, and he rose to leave the building, but we refused to let him or his lieutenant go, and held them both prisoners. The Commandant said that the guards would soon come in to rescue him, but he eventually promised to do his best to save us from being deported, if we set him free.

Then, by Colonel Hunt's advice, for we did not know when a commando might appear, we returned to bed—you cannot shoot men in their beds. And so passed the anxious hours away till dawn. With the first streaks of daylight we scanned the hills anxiously for the British troops. We could see lines of men moving on the race-course, but it was impossible to make out what they were. Presently, at about half-past eight, two figures in khaki came round the corner, crossed the little brook and galloped towards us. Were they Boers come to order our removal?—The advance scouts, perhaps, of a commando to enforce the order! or were they our friends at last? Yes, thank God! One of the horsemen raised his hat and cheered. There was a wild rush across the enclosure, hoarse discordant yells, and the prisoners tore like madmen to welcome the first of their deliverers.

Who should I see on reaching the gate but Churchill, who, with his cousin, the Duke of Marlborough,[700] had galloped on in front of the army to bring us the good tidings. It is impossible to describe our feelings on being freed. I can scarcely believe it, after seven months' imprisonment; the joy nearly made up for all our former troubles, and, besides, the war is not yet over.

[698] A type of high explosive. Picric acid - Wikipedia

[699] Literally "fire of joy."

[700] Charles Richard John Spencer-Churchill, 9th Duke of Marlborough, (13 November 1871 – 30 June 1934), was a British soldier and Conservative politician, and a close friend of his first cousin Winston Churchill. He was often known as "Sunny" Marlborough after his courtesy title of Earl of Sunderland. Charles Spencer-Churchill, 9th Duke of Marlborough - Wikipedia

Waterval after the liberation of the prisoners.[365]

Boer Military Prison at Waterval.[366]

"Boer Prisoners in the Prison Prepared for the British Officers Just Outside Pretoria."[367]

To close the scene we hoisted the Union Jack which Burrows (one of the prisoners) had made by cutting up a Vierkleur[701], on the staff whence the Transvaal colours had so long reminded us of our condition.

I will not write about the triumphal entry of Lord Roberts and the army into Pretoria, because that has been already told by so many others.

The Dogman [Patterson] and Cullingworth shared our good fortune, both being speedily released from the gaol where they had languished since their attempt to get through to the British lines, and with this happy fact let me end my record of so many weary days passed in uncertainty, disappointment, and monotony, but borne, I hope, with patience, and ending at last in joy.

[701] PF: The flag of the South African Republic was a vertical green stripe with horizontal red white and blue stripes.
Flag of Transvaal - South African Republic - Wikipedia

First Troops to Enter Pretoria.[368]

Pretoria Square occupied by the British, June 5th 1900.[369]

Hoisting the British flag on the Raadzaal, Pretoria.[370]

The scene in the square on June 5th [1900], when the flag was raised on the Raadzaal.[371]

Colonials "Marching Past" at the formal Occupation of Pretoria, June 5th 1900.[372] "Lord Roberts and staff cheering the Queen--Raising Union Jack on occupation of Pretoria, June 5th, S.A."[373]

British troops just before raising the Union Jack.[374]

Raising the flag at Pretoria.[375]

The Model School, Pretoria, Now An Hospital For British Soldiers, Recently The Prison of Our Officers.[376]
Three Thousand British Soldiers Released From Prison At Pretoria Being Re-Armed In The Public Square.[377]

Biographical Summary

Thomas Hugh Colville Frankland[702] was born on 17 October 1879 in Cork, County Cork Ireland.[703] He was the son of Colonel Colville Frankland[704] and Mary Jay Dawson.[705] He attended Cheltenham College from September 1889 to July 1890[706] and continued his education at the Charterhouse School, Godalming from September 1893 to March 1895.[707]

He fought as a Lieutenant in the Royal Dublin Fusiliers during the Boer War between 1899 and 1902. He earned the Queen's South African medal with three bars and the King's medal as well. [708] His commanding officer, Captain Aylmer Haldane, described Frankland's actions during the ambush of the armoured train in his after action report as: "a gallant young officer who carried out my orders with coolness."

He served as adjutant to the Headquarters staff of the British Army in 1911 and was posted to Ahmidnagar, Deccan India that same year. He gained the rank of Captain and Brevet Major in the Royal Dublin Fusiliers.

He gained the rank of Flying Officer in the Royal Flying Corps.[709] He used a Vickers Biplane at the Vickers School, Brooklands to earn an aviators certificate on November, 5th, 1913. [710]

He most likely trained in a Royal Aircraft Factory F.E.2 of which 1,939 examples were constructed according to an evolving design.[711]

The first F.E.2 Pusher Biplane in 1911.[712]

F.E.2a with original undercarriage.[713]

[702] Person Page #53145 citing [S37] BP2003 volume 3, page 4292. Person Page (thepeerage.com)

[703] Source: 8476612 Source: 8476606 Life story: Thomas Hugh Colville Frankland | Lives of the First World War (iwm.org.uk) And: Brevet Maj T H C Frankland And Capt R C C Frankland | War Imperial War Museums (iwm.org.uk).

[704] Colonel Colville Frankland was born on 6 November 1839.2 He is the son of Sir Frederick William Frankland, 8th Bt. and Katherine Margaret Scarth. He married Mary Jay Dawson, daughter of William Dawson, on 22 September 1870.2 He died on 22 December 1913 at age 74. He fought in the Indian Mutiny. He gained the rank of Colonel in the 2nd Battalion, Royal Dublin Fusiliers. Person Page #36657 @ Person Page (thepeerage.com).

[705] Mary Jay Dawson was the daughter of William Dawson. She married Colonel Colville Frankland, son of Sir Frederick William Frankland, 8th Bt. and Katherine Margaret Scarth, on 22 September 1870.1 She died on 25 January 1914. Person Page #53414 @ Person Page (thepeerage.com).

[706] Source: 8510474.

[707] Source: 8510475.

[708] Frankland Thomas Hugh Colville Lieutenant QSA (3) bars Source: List of QSAs with the clasp Talana from the Talana book. Also: QSA and KSA 1st Battalion and 2nd Battalion medal rolls Royal Dublin Fusiliers. https://www.angloboerwar.com/name-search

[709] Person Page #53145 @ Person Page (thepeerage.com).

[710] Record Card #679. List of pilots awarded an Aviator's Certificate by the Royal Aero Club in 1913 - Wikipedia

[711] Between 1911 and 1914, the Royal Aircraft Factory used the F.E.2 ("Farman Experimental 2") designation for three quite different aircraft that shared only a common "Farman" pusher biplane layout. Royal Aircraft Factory F.E.2 - Wikipedia.

[712] An FE 2 pusher biplane, on the ground with pilot and passenger. File:Aviation in Britain Before the First World War RAE-O763.jpg - Wikimedia Commons

[713] F.E.2a with original undercarriage File:Aviation in Britain Before the First World War RAE-O597.jpg - Wikimedia Commons.

The following photograph was taken by the Royal Aero Club when he gained his flying license. Comparing the signature to the initials on his sketches, particularly the 'F' in Frankland it is his autograph.

He fought in the First World War, and was mentioned in despatches.[714]

He died on April 25th 1915 at age 35, killed in action at the Dardanelles, Gallipoli, Çanakkale, Turkey.[715]

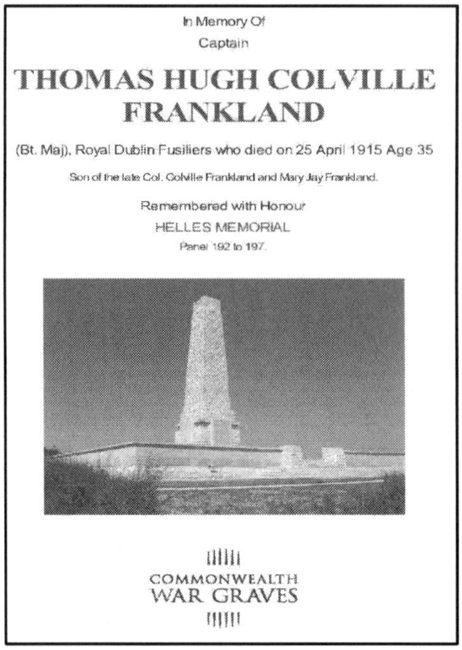

Thomas Hugh Colville Frankland.[716] Thomas Hugh Colville Frankland Captain (Bvt. Maj) Royal Dublin Fusiliers Died 25 April 1915 35 years old, Remembered with Honour - Helles Memorial Panel 192 to 197. Turkey (including Gallipoli).[717]

His older brother Robert Cecil Colville Frankland[718] was born on 7 July 1877. Robert fought in the Boer War between 1899 and 1901.[719] He continued in the military and fought in the First World War, attaining the rank of Captain in the 3rd Battalion, North Staffordshire Regiment. He died on August 7th 1915 at age 38, killed in action at the Dardanelles,[720] Gallipoli, Çanakkale, Turkey.[721]

Immediate Family: The children of Mary Jay Dawson and Colonel Colville Frankland:

William Jay Colville Frankland b. 14 Apr 1876, d. Nov 1896
Robert Cecil Colville Frankland b. 7 Jul 1877, d. 7 Aug 1915
Thomas Hugh Colville Frankland b. 17 Oct 1879, d. 25 Apr 1915
Eleanor Colville Frankland d. 10 Mar 1946
Katharine Marian Colville Frankland d. 17 Sep 1950
Margaret Lee Colville Frankland d. 26 Nov 1874
Beatrice Colville Frankland d. 11 Oct 1959
Mary Olive Elsie Colville Frankland d. 26 Mar 1960

714 Source: 8476604 of Commonwealth War Graves Commission.

715 Captain Thomas Hugh Colville Frankland | War Casualty Details | CWGC The Commonwealth War Graves Commission | CWGC | CWGC And: Thomas Hugh Colville Frankland (unknown-1915) - Find A Grave Memorial Memorial Id 56009464.

716 Photo Credit: Lives of the First World War Life story: Thomas Hugh Colville Frankland | Lives of the First World War (iwm.org.uk) Photo Credit to Royal Aero Club. Contributed by JOHN14663.

And Captain Thomas Hugh Colville Frankland Royal Dublin Fusiliers. Died Sunday 25 April 1915 - A Street Near You

717 The Commonwealth War Graves Commission | CWGC | CWGC

718 History Life story: Robert Cecil Colville Frankland | Lives of the First World War (iwm.org.uk) Also: Person Page #53415Person Page (thepeerage.com).

719 R C C Frankland 1st Battalion Source: QSA and KSA medal rolls South Staffordshire Regiment.

720 Imperial War Memorial #56284 Located on the south pillar of the chancel arch, St. John the Baptist Church Palmeira Square, Hove, East Sussex Brevet Maj T H C Frankland And Capt R C C Frankland | War Imperial War Museums (iwm.org.uk).

721 Helles Memorial Panel 171 to 173. Turkey (including Gallipoli) Captain Robert Cecil Colville Frankland | War Casualty Details | CWGC.

Robert Frankland

Thomas' older brother was Captain Robert Cecil Colville Frankland b. 7 Jul 1877, d. 7 Aug 1915 who also served in two of the same campaigns as his younger brother.

London Gazette: 2nd Lieutenant, 3rd & 4th Bn the South Staffordshire Regiment 14 July 1896, Lieutenant, 2 June 1897, Lieutenant, 3rd Battalion, the South Staffordshire Regt, Reverted to be 2nd Lieutenant, South Staffordshire Regiment 4 May 1898, 2nd Lieutenant South Staffordshire Regiment, 30 December 1899.
Resigned Commission South Staffordshire Regt, 25 February 1903.
Assigned to be Captain 10th (Service) Battalion, North Staffordshire Regiment, 24 October 1914,
Killed In Action: August 7th, 1915, No Known-Grave.[722]

Robert Cecil Colville Frankland Medals.[378]

1. QSA, Transvaal, Wittebergen, Cape Colony
2. KSA South Africa 1902 and South Africa 1901
3. 1914 - 1918 War Medal - Captain R. C. C. Frankland

[722] Data and image credit: QSAMIKE, Past-President Calgary Military Historical Society O.M.R.S. 1591: Anglo Boer War - Armored Train Ambush 11-15-99 Chuchill & Haldane - Page 18 - Boer War Forum.

The Story of My Captivity

During the Transvaal War 1899-1900

By

Adrian Hofmeyr

Dedicated
With Sincere Affection
And Heartfelt Gratitude To
The British Officers
My Fellow-Prisoners of War

Publisher Edward Arnold 37, Bedford Street, Strand, London, 1900
Printed by: Unwin Brothers, The Gresham Press, Woking And London[723]

Adrian Hofmeyr.[379 724]

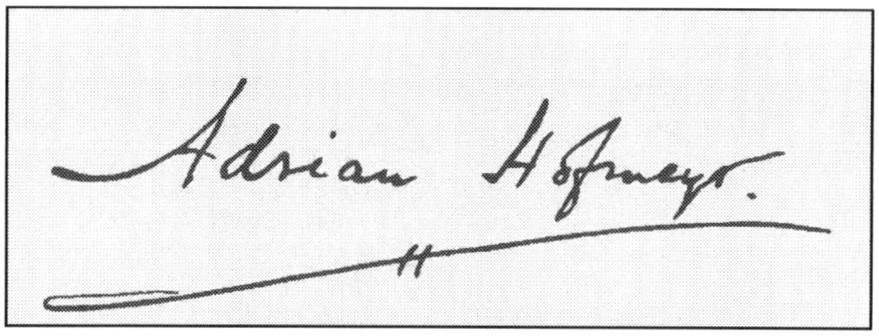

Adrian Hofmeyr's signature.[380]

[723] Adrian Hofmeyr The Story of My Captivity During the Transvaal War 1899-1900, Publisher Edward Arnold 37, Bedford Street, Strand, London, 1900 Printed by: Unwin Brothers, The Gresham Press, Woking And London The Story of My Captivity During the Transvaal War 1899-1900 - Google Books.

[724] Rev. Adriaan (Attie) Jacobus Louw Hofmeyr (Calvinia, 13 April 1854 – Bellville, 1 May 1937) was a beloved but controversial minister in the Dutch Reformed Church, known as one of the Church's most eloquent and fascinating ministers as well as a political activist. See Afrikaans Wikipedia for an extensive article. https://af.wikipedia.org/wiki/Adriaan_Hofmeyr

Athenaeum Review

THE ATHENÆUM N° 3813, Nov. 24, 1900

Mr. Edward Arnold publishes *The Story of my Captivity during the Transvaal War*, by Mr. Adrian Hofmeyr, who has suffered as men always suffer at the hands of their kinsmen in civil war, and who writes with bitterness not unnatural. He will not even believe in the piety of Mr. Kruger, of which some of his fellow-captives are still so sure, and writes thus of the President's telegram after the relief of Ladysmith: "Was it blasphemy, or am I mistaken? ......I know *God is not mocked with impunity*."

Citation [725]

Annotator's Introduction

Adrian's story is the first opportunity we have to for a native of South Africa to recount, as a civilian, the earliest moments of the conflict and how the Boer forces went about their military operations.

Subsequent to his arriving at the Officer's Prison his experiences become synoptic with those of Haldane and Frankland's, as well as Churchill's.

While it is important to faithfully allow him to chronicle his journey into and out of captivity there are a few chapters that have been excised on account of space considerations and relevance to the main stories in this volume.

The omitted chapters are:

Chapter IX Harry's Experiences

XIV Some More of the Same Kind

XXVIII Majuba Day-1900

XXVIX And After,

XXX A Last Chance, has a portion excised at the end.

I will note where the omitted material was removed but I have not changed the original chapter numbering so there will be gaps in the numbers.

Crossing A South African Drift.[381]

[725] Athenaeum Review Athenaeum 1900 #2 July to Dec. Index: https://babel.hathitrust.org/cgi/pt?id=chi.15885677&seq=8. p. 680 Just a note. Bottom Right column: https://babel.hathitrust.org/cgi/pt?id=chi.15885677&seq=692.

My little corner in the Staats Model School—8 feet by 6 feet—in a room occupied by eight officers. Kaffir blankets make a good tablecloth and carpet, and an umbrella-stand upside down a bookcase. The table I bought, and the flowers my sister sent me.[726]

Preface

Just a word to the reader. The greater part of my "Story" was written in prison. One feels then most keenly and poignantly, and suffers more acutely. I have given vent to such feelings, and this will account for many seemingly hard words.

Will the kind reader please remember this? Now that I am free I cherish no hard thoughts; I have forgotten all indignity heaped upon me. To pacify, to conciliate the genuine Boer, whom I love and respect, I would do anything. My ardent desire is that he, too, soon may experience the inestimable advantages of a beneficent British rule, as we all do in the Cape Colony. What a grand, happy country South Africa then will be—the brightest jewel in our gracious sovereign's crown. May I live to see my dream realized!

Adrian Hofmeyr.

Introduction

When the Transvaal War broke out, October 11th 1899, I happened to be at Lobatsi,[727] in British Bechuanaland,[728] a small railway station on the Buluwayo line, fifty miles north of Mafeking. I was on my way to Lake 'N Gami [Lake Ngami,[729] 365 miles, 590 km, to the northwest] to visit the trekkers who had settled there since 1894.

I was travelling by bullock wagon, accompanied by my young friend Harry Neethling[730] and four Kaffir [731] servants. We were well provisioned, and had sixteen bullocks and two riding horses. I delayed at Lobatsi a couple of weeks waiting for the drought to break up before crossing the dreaded Kalahari.[732]

[726] Hofmeyr The Story of My Captivity.

[727] 45 miles (70 km) north of Mafeking and 165 (265 km) west of Pretoria. Lobatse is a town in south-eastern Botswana, 70 kilometers south of the capital Gaborone, situated in a valley running north towards Gaborone. Lobatse - Wikipedia

[728] British Bechuanaland was a short-lived Crown colony of the United Kingdom that existed in southern Africa from its formation on 30 September 1885 until its annexation to the neighbouring Cape Colony on 16 November 1895. British Bechuanaland - Wikipedia

[729] Lake Ngami is an endorheic lake in Botswana north of the Kalahari Desert. It is seasonally filled by the Taughe River, an effluent of the Okavango River system flowing out of the western side of the Okavango Delta. Lake Ngami - Wikipedia

[730] PF: Unable to identify this individual.

South West Africa.[733] Lobatsi is just north of Mafeking but outside the bounds of this map.

[731] During the Dutch and English Colonial Period the word Kaffir was used to describe monotheistic peoples (Nguni ethnic groups in particular) of South Africa, who were not of a Christian or of an Islamic religious background, without derogatory connotations, until the early twentieth century. It appears in many historical accounts by anthropologists, missionaries and other observers, as well as in academic writings. See Also: Kaffraria and British Kaffraria (Queen Adelaide's Province) were historical regions along the coast north or Port Elizabeth. See: File:Eastern Frontier, Cape of Good Hope, ca 1835.png - Wikimedia Commons Subsequently it became a racial slur. Kaffir (racial term) - Wikipedia See also the British military unit Kaffrarian rifles which were white and black units which were still organized as late as 1947. East London, 1 March 1947. King George VI inspecting 'Kaffrarian Rifles' guard of honour. - Atom site for DRISA And: Kaffrarian Rifles AKA: Buffalo Volunteer Rifles Buffalo Volunteer Rifles - Wikipedia Also: a historical region in Afghanistan and the setting of most of Rudyard Kipling's famous 1888 novella "The Man Who Would Be King." Kafiristan - Wikipedia And: an Arabic and Islamic term which, in the Islamic tradition, refers to a person who disbelieves in God as per Islam. Kafir - Wikipedia PF Note: I have no authority to change the wording used by original authors. There are instances where a pejorative or admiring modifier is used. While this is racist in and of itself, to identify someone by their complexion, in this narrative, identifiers such as Dutch, Boer, English, Uitlander, Indian, and many others are used as means of identification sometimes with pejorative or admiring modifiers.

[732] The Kalahari Desert is a large semi-arid sandy savanna in Southern Africa extending for 900,000 square kilometres (350,000 sq. mi), covering much of Botswana, and parts of Namibia and South Africa. Kalahari Desert - Wikipedia

[733] General Botha, the Career and the Man, By Harold Spender, 1916 Google Books, UCAL-Davis: https://books.google.com/books?id=dxw1AQAAIAAJ

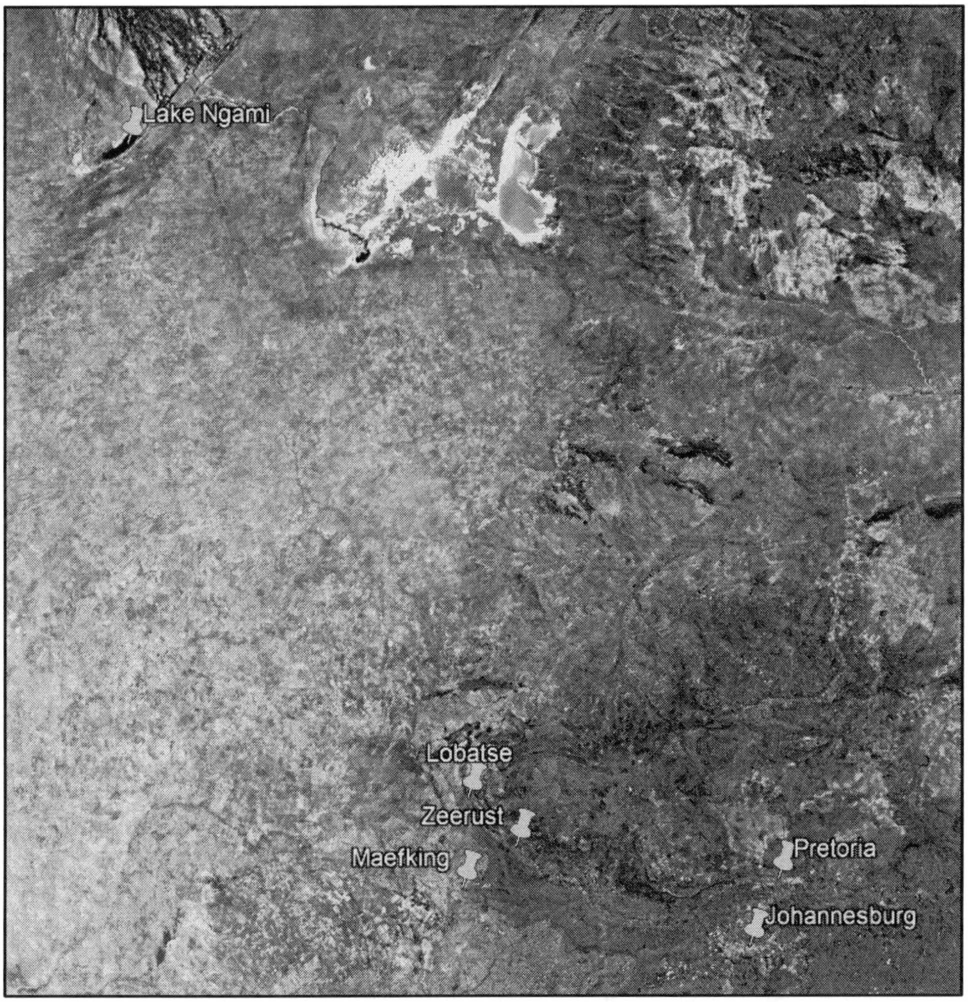

Western Africa Region where he is traveling. Google Image © 2023 Image Landsat/Copernicus.

Wagon on Pontoon over River.[382]

Ox wagon stuck in Crocodile River.[383]

All of a sudden, to our intense surprise, we heard of the Transvaal Ultimatum, and a few hours later of the declaration of war. We had seen and heard of commandos gathering and preparations going on apace, yet no one expected war—at all events not so soon. A couple of white families were living at the station, and round about some hundreds of Kaffirs. When rumours of war became rife, most of the women and children left. Thus it happened that on the fateful 11th [of October 1899] only a hospital nurse and the manageress of the hotel were at the station. Lieutenant Cole with twenty-five men and two B. S. A. police constituted the strength of the garrison. This was the position on Thursday morning, the 12th, when news of the declaration of war reached us. Let me take up the tale from there.

I. My Position with Regard to the War

I happened to be one of those colonial-born Africanders who believed that the Transvaal ought to have given in to the demand for a *five* years' franchise.[734] I could not see how that could have endangered Transvaal independence. I felt persuaded that a *refusal* to do so would *undermine* that independence. Why? Because, first of all, history teaches us that equal rights bestowed upon the inhabitants of a country never endanger such a country's safety. And secondly, because it is unnatural to expect a majority to consent to the supremacy of a minority. Wherever a minority rules it tyrannises, and when that takes place the fires of dissatisfaction are always smouldering, and sooner or later civil war ensues. I feared this for the Transvaal. I felt assured that giving in to the Uitlanders' *just* claims (all were not just) would be a masterstroke of policy.

And, therefore, I advocated the "climb down" on the Kimberley and Beaconsfield platforms[735] in the month of July [1899]. I then said that I knew I would be made to suffer for my opinions. I never dreamt, however, what horrors of vengeance awaited me. But more. The above being my humble view on the matter, I, all the same, was decidedly averse to the idea of having recourse to arms. In conversation with Sir Alfred Milner and the leaders of the "Progressives" in the Colony, I made bold to urge patience. In public and private I strenuously urged it. My devout wish was that England would resort to no other arms than diplomatic ones. Let it take three, six, nine, twelve months—never mind. Be patient—all will eventually come right. For this I pleaded. I begged our leaders to remember that climbing down is always hard work, and that therefore we ought to sympathise with Paul Kruger.

Give him time, act courteously, use every available weapon of diplomacy—only, *no war!* This was my honest opinion—to realise it, I spoke and wrote and did what I could. Imagine then my horror when I heard that the *Transvaal* had sent an ultimatum I knew what this meant. It meant burning the boats behind them; it meant seizing hold of their one opportunity of striking a blow at the Empire—their only one. Thus were my hopes dashed to the ground, and the dreaded war was upon us. I verily believe, to-day still, that there would have been no war, if the *Transvaal had not* declared it.

If the Imperial authorities had any such idea they would have flooded the Cape with troops long before. No, their hope too was, I am sure, that diplomacy would gain the day. Alas! the unwise Ultimatum dashed all such hopes to the ground. As an Africander,[736] born and bred, I shuddered at the very thought of such a war. But as a *loyal Africander* I did not consider it compatible with my duty to my Queen to give anyone reason to think that by word or deed I egged on anyone in the Transvaal to persist in refusal of the demanded reform.

I make bold to say that if all the Africander friends of the Transvaal had acted thus, had dissuaded the authorities from persevering in their headstrong policy, instead of allowing them to think that all the Colonial Africanders would fight for them in case of war, there would be no war to-day. No, their friends *would* tell them—"the Colonials will rise, European Powers will interfere, America is sure to help you, &c. don't give in." And they did not. But scarcely any Colonials have risen, no European voice is heard, and America remembers England's loyal attitude to her during the Spanish War. Oh, the pity of it all. Things might have been so different. And now, what will the end be? I am writing this in prison—I cannot foretell. I must wait! I only know this, the British arms will triumph.

734 That adult male immigrants would be granted voting rights after 5 years' residence rather than the 14 then in effect.

735 Beaconsfield is now a suburb of Kimberley in the Northern Cape Province.

736 *Africander*, earlier form of Afrikaner. Afrikaners are a South African ethnic group descended from predominantly Dutch settlers first arriving at the Cape of Good Hope in the 17th and 18th centuries. They traditionally dominated South Africa's politics and commercial agricultural sector. Afrikaners - Wikipedia

PF Note:

Some of the hostility that is directed towards Rev. Hofmeyr results from his being a "clergyman of the Dutch Reformed Church from Cape Colony." He is also "a cousin of the Bond[737] Hofmeyr" [738] Who is Jan Hendrik Hofmeyr (Onze Jan) [739]

Jan H. Hofmeyr.[384] During the last phase of the negotiations with the British, Paul Kruger offered to reduce the residency requirement from fourteen to nine years as shown in the cartoon below. President Kruger (to the British colonial): "There see what I am doing for you. Ain't you very grateful. Who is your true friend now?"[740] In the background is Joseph Chamberlin looking stern and not satisfied.

II. The First Days of the War

When on Wednesday [October 11th, 1899] we heard a vague rumour that an ultimatum had been sent by the Transvaal, we were naturally greatly excited. The question debated was, What shall we do? The hospital nurse was anxious to get to Mafeking, so also the manageress of the hotel, to help nursing there. I intended going back to the Cape myself. Unfortunately, on that very day the trains ceased running both southward and northward. I decided then to send my young friend Harry[741] on horseback to the Boer commando, ten miles off, to ascertain, if possible, definite news. At ten in the morning he rode off. How the day dragged on! We could obtain no definite news, and in the absence thereof we could not decide definitely what to do. Being a *Times*[742] correspondent, as also of the *Cape Times*, [743] I spent a great part of the day in writing to these papers and also some letters. How to get these communications away, however, I had no idea.

I remember that I felt low-spirited. Writing to a friend on business matters, I said, "If anything happens to me, do this, &c., please." I knew the Transvaal Government hated me for daring to differ from them. I knew that that

[737] The Afrikaner Bond (Afrikaans and Dutch for "Afrikaner Union"; South African Dutch: Afrikander Bond) was founded as an anti-Imperialist political party in 19th century southern Africa. While its origins were largely in the Orange Free State, it came to have a significant presence across the region, and especially in the Cape Colony and the Transvaal. It was established for "the States of South Africa to be federated in one independent republic" "However the Cape Colony branch of the bond was less extreme in its republicanism and more inclined to cohabit with an imperial policy of indirect rule. Afrikaner Bond - Wikipedia

[738] Rosslyn states that Adrian was a cousin of: "the Bond Hofmeyr." See Page 314 of Earl of Rosslyn, Twice Captured:

[739] Jan Hendrik Hofmeyr (4 July 1845 – 11 October 1909) was a South African politician. He was affectionately known as Onze Jan, "our Jan" in Dutch. As the recognized leader of the Cape Dutch, he protested against such abuses as the dynamite monopoly in the Transvaal, and urged Kruger even at the eleventh hour to grant reasonable concessions rather than plunge into a war that might involve Cape Afrikanderdom and the Transvaal in a common ruin. In July 1899, he journeyed to Pretoria, and vainly supported the proposal of a satisfactory franchise law, combined with a limited representation of the Uitlanders in the Volksraad, and in September urged the Transvaal to accede to the proposed joint inquiry. Jan Hendrik Hofmeyr (Onze Jan) - Wikipedia See also, a younger relative and family history. Jan Hendrik Hofmeyr (1894–1948) - Wikipedia

[740] Slow and Sure - JM Staniforth - Uitlander - Wikipedia

[741] Harry Neethling.

[742] The Times is a British daily national newspaper based in London. The Times - Wikipedia

[743] The Cape Times is an English-language newspaper published in Cape Town, South Africa. Cape Times - Wikipedia

particular Hollander[744] element which swayed the Government counsels, was a particularly narrow-minded one, vengeful and spiteful—an element which believed that it could only prove its true love for the Transvaal by exhibiting excessive bigotry of zeal against its enemies—by out-Heroding Herod, in fact. I knew that though I was a nobody, not even a public politician, it would make no difference. Little things delight little minds: they would simply delight in taking me, aye, and even shooting me, a nonentity, as if I were some great political power or military commander. After-events justified my opinion.

Evening came, but no Harry. Imagine my anxiety. Late that night I sat up writing in my lonely hut. A gloom was cast over all of us. About eight o'clock Thursday morning a Kaffir came up to my hut with a letter from Harry. He wrote that he had arrived at the "lager" of Commandant Swarts' company, but found him out, that his *locum tenens*[745] had received him kindly, but detained him till Swarts' return, and that on Thursday he would, therefore, be back again. He wrote under the sub-commander's instructions and at his dictation, amongst others, the following sentence: "Tell Mr. Hofmeyr that we do not intend to attack Lobatsi, and that women and children are quite safe." It will appear later with what a diabolical plan this was inserted. At twelve o'clock that same Thursday we received official intimation that war was declared. What now?

In consequence of the contents of Harry's letter we felt quite safe, and decided to arrange for departure to Mafeking[746] as soon as possible, whilst Lieutenant Cole, rightly judging that it would be useless to stay in a place which was not going to be attacked, evacuated Lobatsi that very same evening.

Another lonely evening and night, for Harry had not yet returned—in fact, he never was allowed to come back. The Boer commander had deceived him and us purposely. Friday passed quietly. In vain we tried to hire oxen from the Kaffirs to take my wagon on to Mafeking, with those of us who were left there. My oxen I had sent away a few days previously to a sale, and that was the last I saw of them. On Saturday morning we found that the wires were cut a mile north of Lobatsi, and three lengths of rail taken up.

The wires southward had been cut early on Friday morning. The stationmaster, the gangers,[747] and myself immediately set to with a will and succeeded in restoring connection on one line. I got through one message to the *Times*, but could gain no further information—nothing was known northward of us, except vague rumours. A large body of Boers had come during the night to cut these wires, and whilst repairing the line I more than once expected that we would be fired upon from the surrounding hills, where they perhaps were in hiding. Nothing happened, however. My letters and communications I had sent on to Mafeking with two Kaffir boys. Whether they ever reached their destination I know not. We were anxious now to get away from Lobatsi, and had already decided to run down to Mafeking on a ganger's "trolley," when we heard from other gangers further down the line that it had been wrecked in several places.

Hand Trolley.[385]

Engineers repairing a Railway destroyed by the Boers near Kroonstadt.[386]

[744] As shown here and in other narratives the term *Hollander* refers to recent Dutch immigrants to the Transvaal, a distinct class of individuals, many of whom have prominent places in the Government and are often associated with the "War Party."

[745] *A locum*, or *locum tenens*, is a person who temporarily fulfills the duties of another; the term is especially used for a physician or clergyman.

[746] Mahikeng, still commonly known as Mafikeng and previously Mafeking, is the capital city of the North West province of South Africa. It is about 45 miles (72 kilometers) from Lobatse. Mahikeng - Wikipedia

[747] Railroad Gangers who lay and repair railroads.

Perforce, therefore, we had to stay. That evening—Saturday evening—our whole Lobatsi population consisted of the two ladies, the stationmaster, a ganger (a Scotchman), his wife, her brother, and myself—seven whites all told—and one Kaffir boy.

I sat out on the platform late that night. The air was to me full of ominous forebodings. There was one saddle-horse, but how could I take it and escape, leaving the others? All seemed to look up to me for comfort and advice. It was a beautiful moonlight night. Aye, the thoughts, the hopes, the memories that crowded in on me!

I *knew* something was going to take place. And yet I knew not what! I read Harry's letter again, and said to myself, "They won't come, we are safe; see, they were here last night, a mile off only; no, I can believe the message!" I was sorely mistaken, but knew it not. Long I sat there. I can still picture the glorious scene—all the country around, the bold hills and open plains bathed in gleaming moonlight, and all so calm and quiet. Only under the crags, down in the river, and under the trees dark, ominous shadows lay. Overhead the majestic moon and silvered sky breathed messages of rest and hope, around me the hyena's howl and jackal's screech warned me that enemies were nigh. It was a picture of peace and war! A great peace had fallen on me. I felt no fear, and it struck me how quiet and calm the others also were. There was no nervousness, no hysteria; no, it was the quietness of courage! Thus ended the week—the first days of the war.

III. The Fateful Sunday, October 15, 1899

Can I ever forget that day—ever shut out its horrors? I was awake very early that morning, just as day broke. I went out, but not a sound. I went back to my room, to be startled a few minutes later by the report of a gun. Going out again, followed by some of the others, in a moment we heard the clatter of a horse's hoof, and up rode a man with a white flag. On his arrival he dismounted and came up, asking me if we intended to surrender or fight.

With a smile I said, "Well, my friend, the stationmaster and I are the only men here—what think you, is it worth while fighting? How many men have you?" said I. "We have about eighty," said he. "Well, we are unarmed—what think you, shall I hold the fort, or surrender?" He understood my little pleasantry and signaled to the others to come up. He told me, too, that when that shot was fired the whole commando had just drawn up in the river before the hotel, and when I came out on the platform they were on the point of firing on me, thinking they were going to be attacked. But as no more shots were heard they desisted. This was my first escape that day. Up thundered the eighty horsemen now. The Commanders, Swarts and Potgieter,[748] came up and asked for the keys of the station, &c. I met with dark looks all round, and felt that they were satisfied at the success of the trap they had laid for me. To wreak a petty vengeance they had plotted to capture me, and now they had me—one against eighty—a defenceless man, too. I remembered all their pious talk in their newspapers—papers which otherwise had no respect at all for religion.

I remembered their story of the President's three days' prayer, locked up in his room. I remembered their boast that they never fight on a Sunday. And so, in my indignation, I said to the Commander, "I thought you do not fight or make war on Sundays?" "No," said he, "we don't as a rule, but orders are orders." "Oh," said I, "the orders of man are more than the orders of God?" My noble warrior fell into the trap, and said, "Yes, in time of war!" And I began to understand how much these religious protestations were worth.

All was bustle and confusion now all over the station and the hotel. The ladies made coffee and breakfast for as many as possible, and the commanders held consultations. A trench had been dug some weeks before all along the rails, to drain off the rain-water, and during the digging process an old bit of dynamite fuse, buried under ground—who knows how long before—had been cut in two by the diggers' spades. A length of a couple of inches protruded from the side of this trench. This bit of fuse was spotted by one of the burghers, who reported it to the commander. What a scare it created! Immediately a guard was placed to warn every one off the spot, and terrible was their rage and deep their curses. Of course they concluded we had laid a mine to blow them up! I assured the men that it was not so. No use. "That's the accursed Englishman all over—he can't hit us with the rifle, now he wants to blow us up with dynamite." To prove that they were mistaken I suggested they should all withdraw to some distance, leaving me alone to fire the fuse. This ought to have shown them that they were mistaken, but no one was generous enough to acknowledge it. Indeed, one cowardly fellow cried out, "Don't

[748] There are two Potgieters listed in the Zeerust Commando at ABW.com. They are both listed as Privates. The is no Swarts. Both are common names with too many to sort through without a first name or definite rank.

trust the verd ..verrader[749] (the d ... traitor); he has some other plan." And so they made a pretense all that day of guarding the reputed mine, just to show the stationmaster and myself that they did not believe us.

I said above that the leaders were holding a consultation and overhauling the station papers. During this time I had a bath and dressed myself, being guarded all the while. To my astonishment I found that my guards—all honest patriots—were purloining everything of mine they could lay hands on unseen. Thus I found my beautiful gold watch, a valuable hunting-belt, gold links, &c., gone, when I re-entered my room. And of course no one had ever seen these! Oh no! And such good fellows don't tell lies! When finished, I appeared on the platform again, to find every one scowling at me. Mutterings and murmurings were heard on every side. I transcribe some literally, as I jotted them down at the time. "This is the man who made that disgraceful speech at Kimberley." "He is the traitor who wired those cables all over the world." "He told his hated Government that we were massing on the border."

What caused this sudden access of anger, I wondered? I sat down under the verandah and watched. I soon found out. With a pale, scared face Nurse Ruth came up and said: "Mr. H., they've seen the copies of your wires in the office; they know now you are a correspondent, and they are going to shoot you." And bursting into tears the kind-hearted nurse fled to her room. Imagine my sensations! I felt stunned, and had only begun to realise somewhat what was awaiting me when the stationmaster managed to get close beside me and whisper, "What a fool I was not to destroy the wires in the office! They know of the armoured train coming from the north, of which I was advised, and also of *your* wires; and ... Mr. H., the commander told me they were going to shoot you! "In cold blood?" said I. "Yes, just now!" He was called away then, and what passed through my mind I can only leave the reader to imagine.

There I saw before me on the railway platform eager groups talking and gesticulating fiercely. The first and second in command, Swarts and Potgieter, were close to me, some ten yards off, talking to an excited group. I could hear snatches of what was going on, especially the horrible curses flung at me. "Schiet hom dat hy vrek" ("Shoot him dead") said one. "We'll teach him to follow the English, and not stand by us Africanders!" &c. I felt sure now that my doom was sealed. Their scowling looks, the way they handled their rifles, their words—all proved it, besides what my friends had whispered to me. I was to die because I was *loyal*, and would not turn *traitor*. And these poor men had been drilled into the idea that one could be an Africander only by being a Krugerite. They fought for independence, as they averred, by trying to make others dependent.

So I really thought that I had but a few moments more to live. There was a pang of regret, a wistful look "over the hills and far away," a silent prayer for courage, and a quiet calm took possession of my soul. I felt that I was not afraid to die, though life was sweet. I remained sitting quietly. Furtive looks were cast my way continually. Oh, how I prayed for courage! How they would like to see me fall on my knees before them, howling for mercy! No, no, not that! O God! help me! help me to die as a brave man! Shooting me would be murdering me—that I knew. It would be an infringement of all international usage, for I was innocent and a non-combatant. But I had seen enough of these Commanders now to know that they were capable of doing it. My friends came out to me again, sad and sorrowful, as may be imagined. I told them I was ready and not afraid. Then that little meeting of the Commanders broke up, and they came towards me. "It has come," I thought. Just one more silent prayer, and I stood up and walked to the entrance of the verandah to meet them.

They came with stern, set countenances and loaded rifles. Quietly and calmly I said, "Gentlemen, I am ready to die!" Oh, it was a horrible moment. I am not ashamed to say I felt as if my heart could break, but, thank God, outwardly I was quite calm. There was an ominous silence, and then the Commandant said: "We'll take you prisoner to Zeerust, and shoot you there; we won't shoot you now." I afterwards heard that this was the resolution come to by the narrow majority of one vote. The second in command was determined to shoot me then, and when outvoted he mounted his horse and rode off in a rage, with a large body of men, northward, to break up more of the railway line.

I then quietly asked: "Why am I taken prisoner?" The answer was a shuffling one: "It is our orders." *Behind my back*, I was cursed and sworn at, and called traitor with ever so many adjectives thrown in. *To my face*, no one spoke out. Bullies are cowards! The real, honest Transvaal burgher—the thousands of them now fighting without pay, only out of patriotism—is no bully, and therefore no coward. Oh no; but those *Government minions*, lusting for loot, braggarts all—*they* were the bullies and the cowards. Into their hands had I fallen.

[749] Dutch for traitor or betrayer.

They could talk lofty patriotism by the hour, and yet send a treacherous message; they could make wonderful professions of honesty, and yet do a treacherous deed; they could talk of being civilised, and yet take women prisoners!

I knew now that the danger of immediate death had passed by me, but I knew too I had tasted its bitterness. A scene of bustle ensued hereupon. The Boers dismantled the station, hotel, and stores, and we packed our few belongings. Lots of my clothes had by this time been stolen. The seven of us before referred to were now prisoners, and told to get ready to start for Zeerust soon. It was Sunday, day of peace, but there was none there. It was a beautifully clear day, but dark clouds hung over our lives; the sun was bright above us, but for me especially a blackness of darkest night had come—I was riding on to meet my death.

IV. More About that Sunday

At about noon we started for Zeerust[750]. My own wagon and several others were loaded with loot, and under escort of four mounted men the seven of us left. I was searched for revolvers or poison—of all things in the world! Soon we left Lobatsi behind us, on the way to the lager. I was a *prisoner* now, and began to realise the meaning of the word. Our wagon was very full, and packed by the Boers without any thought of providing comfort. The stationmaster and myself had to sit the whole three hours' journey in the broiling sun—from twelve to three o'clock. We arrived at the lager then, getting there a cup of tea, and eating some provisions we had brought with us. We here rearranged everything on our wagon so as to get a little more comfort for the ladies of the party. I was greatly struck by the difference in treatment we all, and I especially, experienced here at the hands of the burghers in the lager—those who had *not* taken part in the Lobatsi expedition.

Those who had were of course still there. The lager was pitched at a place called "Schildpad" ("tortoise"), and all the men were most courteous to all of us. At Lobatsi all were scowling, and many a bitter word and curse were flung at me. Here every one was kind, and eagerly helped me to unload and reload the wagon. The Commandant, Swarts, had also joined us, having followed on horseback. He, too, was most courteous now, so much so that I was wondering what might have caused the change. When about to start he said to me—

"Mr. H., don't trouble yourself about anything. Your comfort will be well attended to at Zeerust. Tis much safer for you and all your party to be there now." To our escort he gave orders to take us to the hotel. He left me under the impression that we were really not prisoners, but were taken for protection. Alas, delusive hope! I did not know the man I had to do with. At the time I believed him; and so cheered we all felt, that when on the point of starting I thanked him and all the others, in the name of all our party, for their courtesy and kindness to us. More still. Belonging to the tent of my travelling wagon was a handy folding-table. I had put it up for the party's use—there it was, standing before us. And now, since my trip to the interior had to be given up, I asked Swarts to accept of it as a present, a token of our gratitude for his courtesy *now* extended to us.

The simple-minded warrior actually accepted of it and thanked us. And weeks afterwards, when I lay ill, sorely ill, in Zeerust gaol, and he was asked to give me parole on that account, the good Samaritan, who had listened to my words of thanks, who had accepted of my gift, gave answer and said, "Laat hem vrek" ("Let him die"). "Such men as he ought to be shot or sentenced to twenty years' imprisonment." Reader, I've given you a fair type of a Transvaal *Government minion!* Judge for yourself. The Boer, the real one, is a different man, however.

At about sunset on that Sunday we pursued the journey to Zeerust. I had arranged the front room of my comfortable travelling wagon for the use of the old ganger and his wife and her brother. The others occupied the room at the back. The stationmaster and myself occupied an outside seat. The weather was now very pleasant, and our spirits were comparatively buoyant again. Our escort consisted of four mounted men. Two, who were brothers, were young men of German descent, and had accompanied us from Lobatsi. They were better educated than the others, and were indeed interesting companions. They were most courteous to us all along. The one soon found a place inside our wagon. We were consequently fairly cramped for space, and as we travelled through all night there was not much sleep to be had.

[750] About 40 miles (60km) Southeast of Lobatse. Zeerust is a commercial town situated in Ngaka Modiri Molema district North West Province, South Africa. It lies in the Marico valley. Zeerust - Wikipedia

Loaded Ox Wagon.[387]

Civil Prisoners from Dundee entering Pretoria, November 10th, 1899.[388]

"The above picture represents a Train -load of the Wives of Boers who are still in Arms against us, and who have been Deported from the Towns we Occupy and sent by us to Rejoin their Belligerent Husbands. The Women are, apparently, as Hostile as the Men, and can Display their Rancour without Fear. It will be noticed that one has hoisted the Transvaal Boer Flag."[389] [PF: Also so they would not be fired upon when approaching Boer held territory.]

The younger of these two brothers has since died of fever, greatly regretted by many friends. Thus that fateful Sunday ended—a day never to be forgotten. Many a time during that night I shuddered when again I pictured some of the events I had passed through. At times it seemed to be only a dream, till the rumbling of the wagon and the gleam of the rifle barrels in the beautiful moonlight reminded me of the stern reality. Alas! It was no dream—'twas a stern reality. We were prisoners, but yet hopeful. Had not the Commandant given orders for us to be taken to the hotel, where we would be practically free? and had he not told me to have no concern, because the greatest care would be taken for the comfort of all of us?

He did; but I then, as I said above, did not know the stamp of man I had to deal with. Among civilised nations an officer's boast is that his word is his honour, a word given to a conquered enemy especially, or a prisoner. But here, where war was commenced with a great blare of trumpets, announcing to the world how much they had prayed to God to prosper their arms, and theirs alone—here an officer's word, an officer who is a *Government tool*, is *not* his honour, to put it very mildly.

This was not my only experience of the fact. It is some months now since I was taken prisoner, and during these weary weeks I have had to experience this characteristic tens and tens of times. I will try, in charity, to believe that they behaved falsely to me only—me whom they delighted in hating and humiliating. But there! as soon as I try to cherish this charitable disposition a new piece of fraud and deceit is practised upon the little colony of prisoners here, and even charity refuses to be charitable. Oh! what must I do?

The Government officials we are brought into contact with here, in our prison, cannot even buy me a little book *marked* 1/6, without charging 2/- for it—the extra 6d. naturally for their own honest pockets—a salve to the conscience, is it? But more of this later on. Monday morning at seven o'clock we arrived at Zeerust, hungry and thoroughly tired out by the excitement of the previous day. We were kindly received at the hotel, and a bath and cup of delicious coffee soon set me up again. And thus began another eventful day, which deserves a chapter to itself.

Zeerust lies in the center of the Marico Valley.[390]

V. Another Eventful Day

I soon discovered that, in Zeerust at all events, the state of affairs reminded one of some parts of America during the Civil War. Every other man I came across was some kind of official—a Commandant, or an Assistant-Commandant, or an Acting Commandant, or a Field-cornet, assisting or acting. The *acting* was always good—can't say much for the *assisting!* The dignity was always immense, the ludicrous never absent. And if a man was not a Commandant something or other, or a Field-cornet something or other, he was a gaoler or a Zarp,[751] you may be sure. If not that, then he wore a badge on his arm to show that he was an orderly, or an assistant-orderly, or a non-assistant-orderly, of the Red Cross Society. Some people were so uncharitable as to say that all who were afraid of going forth to war joined the Red Cross!

However it be, every Jew belonged to it. Thus it happened that before breakfast, an acting assistant-deputy-commandant-general of nothing in particular came over to the hotel to tell me that the whole thing was a

[751] South African Republic (Dutch: Zuid-Afrikaanse Republiek) Police; hence ZARP. This police force later made a gallant stand at the Battle of Belfast (Bergendal) holding a Kopje salient under intense fire. Pakenham p. 467. Battle of Bergendal - Wikipedia

mistake—*that I was free!* Imagine what a breakfast I took on this good news. After breakfast this man with the long title came again. I found that he was a Colonial man who, years ago, had come to the Transvaal as a schoolmaster. From that he had developed into a law agent, and from this, when the war began, into a Commandant something or other of Zeerust who had to stay at home. As I said, he came again, and brought his notebook with him. He took down my name, age, profession, abode, and asked me ever so many unimportant questions, then told me I was free, and asked me if I would like to go back to the Colony *via* Delagoa Bay. To this I agreed.

All seemed to be well now, and we all felt light-hearted again—all were to go back. Walking about in the streets of the little village during the forenoon, I was several times invited by men total strangers to me to go indoors with them into their own houses. They would then close the door, draw the curtains to, so that nobody could look in from the street, and say to me in a whisper that they had longed to meet and thank me for the sentiments I had uttered at Beaconsfield and Kimberley. "You do not know," said they, "how this country is governed, what a curse there is on us. No justice, detectives everywhere, no public opinion allowed, nor dare you differ from Government policy—all is monopoly. And our officials, such as Landdrosts, &c.—alas, it is a crying shame!" The same was said to me by several parties, and these were all *Africanders*, Transvaal born, or coming originally from the Colony.

I need not say that I was surprised, but also intensely grieved to find that matters were even worse than I had suspected. Aye, the people might well fight for liberty and independence. There was none of it in the country. Unfortunately they were fighting for it in a wrong way! Poor Transvaal—the yoke was indeed a galling one! Notwithstanding their outward union against what the people were taught to consider their common enemy—England—the mutterings and grumblings amongst themselves against their own wondrous specimens of Government officials were deep and strong. It was a pitiable sight! During this morning, too, I became acquainted with the doctor and his family. What have I not to thank that grand old gentleman and his wife and daughter for! But for them I doubt if I ever would have left Zeerust alive.

I feel almost sure I would not. His generosity, and invariable kindness and courtesy, his ever candid friendship and shrewd advice, God knows how much. I am indebted to him for all this, and what a load of gratitude I owe him and would gladly pay, if possible. As for his family, his kind, gentle wife and daughter, always so sympathising, so full of comfort—well, they were to me simply ministering angels! From the depths of my heart I say may God reward them, and help me always gratefully to remember their courageous friendship for me.

Remember, to treat a prisoner-of-war kindly meant then to lay yourself open to suspicion and consequent persecution. My noble friends braved all this. They were *white* people, as the saying goes. During all my stay in Zeerust not one of the people belonging to my own [Dutch] nationality[752] ever showed me the least kindness, or offered me a drop of water even, except the few who were suspected of English proclivities, and who were up to that time utter strangers to me, but good Samaritans all the same, who could not, and would not, see a fellow-being suffer unjustly, whilst priest and Levite coldly passed by on the other side.

Aye, Dr. Blake, how I revere and honour you, my noble-hearted friend, and your noble family! Shall I ever be able to repay you something of what I owe you—a great debt of deep gratitude? I do trust so.

Back to my story now. A pleasant forenoon passed quickly—alas! too quickly. After dinner we were all summoned to the Landdrost's office. The Public Prosecutor there told us that the ladies would be looked after in the hotel, but that we men would be considered as prisoners of war; these were the instructions just received from Pretoria. Imagine what a thunderclap it was! I had but just begun to enjoy liberty, and here I was about to lose it again. Why? The *Pretoria authorities* had been communicated with, and they had decided that we should be re-arrested. You see, this unhappy Government has such a bad conscience that it is afraid even of women, railway gangers, and correspondents. Everything was grist that came to their mill in the way of spite. And so we were marched off to prison, and thrown, the four of us, into cell No. 1, 18 feet by 9 feet. In another room seven more prisoners of war, gangers also, brought from near Mafeking, were huddled, and in another room four Kaffirs, two of them in chains. The misery of that first night in prison how shall I speak of it! But let me devote a chapter to something else first.

[752] As noted above he is a minister in the Dutch Reformed Church and comes from Dutch ancestry.

VI. Magna Est Veritas Et Prevalebit

Magna Est Veritas Et Prevalebit.[753]

This is a true word. Never mind what falsehoods you tell, but truth will out, sooner or later; only it is generally *later*, more's the pity!

Let me give a short summary of the untruths told me during the very first days of my captivity—untruths later to be overtaken by the truth. On the Sunday morning, after their first ebullition of temper was over, I asked several of the burghers, and especially Commandant Swarts, to tell me what had happened since war had been declared on the previous Wednesday evening. This is the answer I received: "There are 90,000 Transvaalers and Free Staters in the field all determined to die rather than give in; our commandos have rushed down Natal, from Laing's Nek, where we were massed *before* the war, *ready* for it (not like England, blustering but unready), *and reached Durban yesterday, and to-day Natal is ours!*"

When I expressed my doubts the answer given me, on this sacred day of rest, which the Transvaal Government says it respects so highly, was that "it is official news," sent to their lager late Saturday evening.

This is No. I., and a very good I. too—a regular AI. [This is how it is printed in the book.]

Now prepare yourself, reader, for the next; it might take your breath away otherwise. It is this: "The Free Staters *took* De Aar[754] *yesterday*, and combined Free State and Transvaal forces are besieging Kimberley now, which will surrender within a few days. We have swept everything before us, because we were *ready!*"

No. II. is a stunner, and no mistake about it. Now the next: "Mafeking fell *yesterday*, Cronje taking it with a rush; Baden-Powell and all his men being prisoners now, and on their way to Zeerust."

This is No. III., and a real, lively, frisky lie it is. Another one: "*The whole Colony has risen,*[755] the combined forces will sweep on to *Table Mountain!* [756] We will not rest till we are there. Already a body of between 7,000 and 8,000 Colonials has crossed over into the Free State to help us. One flag now from the Zambezi[757] to Table Bay."

No. IV. makes me grow dizzy, it is so tall to look up to. "And," said my truth-loving Commander, "our northern commando will this week take Buluwayo,"[758] bringing the total number of crammers up to five. And to every doubt I expressed the invariable answer, given with the simplicity of the "heathen Chinee," was, "but it is official, we got it last night." And this latter statement was quite true, let me say it here—I've seen myself many of these *official* wires. Thus a pure and honest Government delighted in gulling its simple subjects! Of course, *official* meant Paul Kruger says so, and even if he tells an untruth it must be true!

That was the general idea. I knew instinctively at the time that this was what the Yankee would call "awfully tall lying"; and now, when I look back—four months after these "official truths" were told me—I could laugh and despise these liars, only I feel more inclined to weep that men, guiding a State through critical times, could sink so low. I know quite well that false reports, outrageously false, were spread on our side, too; but then it was done by irresponsible correspondents and scare-mongers. But here every lie bore the official stamp. The good, simple burgher would not lie thus—his Government—wondrous Government!—taught him, and, alas, too many proved to be but too apt pupils. 'Tis February now, and their troops never got beyond Colenso, and are now retiring rather quickly; and they never got further south than our few Colonial towns occupied by them; and they have never taken Kimberley, which is now relieved, nor Mafeking, soon to be relieved, I hope; nor did they ever get further north than the Crocodile[759] they could not even take Fort Tuli.[760] Since then time upon time

[753] Truth is mighty and will prevail.

[754] De Aar is a town in the Northern Cape province of South Africa. It is the second-most important railway junction in the country, situated on the line between Cape Town and Kimberley. The junction was of particular strategic importance to the British during the Second Boer War. De Aar - Wikipedia

[755] Meaning the Cape Colony.

[756] Table Mountain overlooks Cape Town.

[757] The Zambezi River (also spelled Zambeze and Zambesi) is the fourth-longest river in Africa, the longest east-flowing river in Africa and the largest flowing into the Indian Ocean from Africa. Zambezi - Wikipedia Victoria falls are on the Zambezi River.

[758] Bulawayo is the second largest city in Zimbabwe, and the largest city in the country's Matabeleland region. Bulawayo - Wikipedia

[759] The Crocodile River, also referred to as Crocodile River (East), is a large river traversing Mpumalanga province of South Africa. Crocodile River (Mpumalanga) - Wikipedia

[760] Tuli is a village in the province of Matabeleland South, Zimbabwe. It is located about 90 km west of Beitbridge on the eastern bank of the Shashe River. The village grew around Fort Tuli, which was the first settlement built by the Pioneer Column in July 1890 at the

Government officials assured me that the combined troops of the two States amounted only to 50,000 men, and as for the 7,000 Colonials, why, they are still waiting for them here!

And Table Mountain is still as far off as ever. All this was official lying, of which the reader will hear more later on. And these very officials are the men who tell the world how they pray to God for success, and how sure they are that God will give them the victory, for their cause is so just! Does a just cause require to be bolstered up by such shameful lying? Are real God-fearing men boasters and idle braggarts? Alas! I have been terribly disillusioned. I had a different opinion of the President and his officials. I differed from their political opinions, but respected those among them who were *not* political Hollanders, for there are many other good Hollanders among them. And now?—well, never mind what I think—only I am disillusioned and bitterly disappointed!

So much for official lying—now a little private lying.

One of our escort, a Colonial-born German, who had had a little education, and been to Europe with his parents as a youngster, acquainted some of my fellow-prisoners on the way to Zeerust with several little interesting experiences of his. He told them that he had once met Sir Donald Currie,[761] who was so struck with his ability and general personality that he offered to adopt him as his own son, but "I preferred coming back to South Africa," said he. I wonder what Sir Donald will say to this! Further, he told them that he once travelled from Kimberley to Cape Town, and was mistaken for the Attorney-General of the Cape (was it Sir Thomas Upington,[762] I wonder?) and given a whole saloon carriage to himself! And then he informed the party that he never would have fought on the Transvaal side, but as he owned all the farms from Zeerust right away some thirty miles to the north-west, he had to do it, or forfeit all his property. And thus the tedium of the journey was pleasantly (?) beguiled. Arrival at Zeerust let in the glare of truth, and our imaginative hero dwindled down to a very commonplace little fellow—very commonplace, indeed, even in the estimate of his own compatriots. But I forgive him—I can even excuse his propensities. Does not his Government officially encourage him to develop his truth-disguising talents? But *magna est veritas et praevalebit!*[763]

VII. In a Prison

We were arrested on Monday afternoon at about four o'clock, and marched up to prison and lodged in cell No. I, as stated. It was a little room of 18 feet by 9 feet. At one end, high up in the wall, was a hole 18 inches by twelve inches, called by courtesy a window. In that hole is fixed a wooden frame with four panes of glass tilted at an angle of 45°. All the air one gets when the door is closed comes through this aperture. In the door of the cell there is a little square hole—a peep-hole for the warders—3 inches by 3 inches. Here you have the whole elaborate system of ventilation explained. Four of us were locked up in this little cell—the stationmaster and myself and the old Scotch ganger and his brother-in-law. Fortunately I had thought of taking my two mattresses from my wagon before it was sent away again. So I could use them and my own rugs too. Otherwise I would have had to satisfy myself with the two rugs, *and nothing more*, which the kind, humane authorities provide sentenced prisoners with.

On the hard boards of the floor, for no bedstead is allowed, I would have had to lie had I not my own bedding. No chair, no table, nothing was there but the four bare walls. I was not even allowed to keep my portmanteau in the cell nor my dressing-case, which I had managed to save. A bucket of water, a tin mug, and a towel constituted the list of our luxuries. Being allowed at first to take our meals in the hotel, we went down for dinner

place known as Selous Camp and used by Frederick Selous as a base for his hunting expeditions. Tuli, Zimbabwe - Wikipedia

[761] Sir Donald Currie GCMG (17 September 1825 – 13 April 1909) was a Scottish ship-owner, politician and philanthropist. In 1862, determining to start for himself, he established the 'Castle' shipping company, which consisted at first of sailing ships plying between Liverpool and Calcutta, owned by a circle of personal friends. Currie first introduced the plan of despatching sailing ships on fixed dates. In 1865, he made London the port of departure of his vessels and took up his residence there. Development of the Union-Castle line: The line grew steadily in strength and importance, and he resolved on a line of steamers from the United Kingdom to Cape Town Donald Currie - Wikipedia.

[762] Sir Thomas Upington KCMG (1844–1898), born in Cork, Ireland, was an administrator and politician of the Cape Colony. He was briefly Prime Minister of the Cape Colony, between 1884 and 1886, during a period of extreme turbulence in the Cape's history. The town of Upington in the Northern Cape is named after him, as was the short-lived Boer republic of Upingtonia. Thomas Upington - Wikipedia

[763] Truth is great and shall prevail.

under escort of the gaoler. Let me say here that this man, according to his lights, all along during those dreary weeks treated me very kindly. He meant it well, and I feel grateful to him.

After dinner this first night we were conducted back to gaol, and now came the ordeal. We were going to be locked in for the night. Shall I ever forget the horror of that moment? *Our* door was shut last of all. One by one the terrible bolts of the other cells were shot to. There were seven other prisoners of war in another larger cell, and some Kaffirs in a third, two of them in chains, as I mentioned before. And now our turn came. When the three bolts were closed and the double lock locked, a terrible sensation came over me.

The sound of these bolts was jarring and grating to an ear accustomed to beautiful music. The clank of the two prisoners' chains seemed to spell my doom—it came like an awful whisper to say," *You are a prisoner too*." It seemed to me as if now only I began to realise the full extent of the Government's treachery towards me. It was a bitter moment, one of the bitterest of my life. Once or twice already I had been brought very near the shadowland by fell disease. More than once have I passed through the dangers of the desert and the savage. The very day before I had escaped being shot in cold blood by a hair's breadth. But I doubt if ever I felt a more bitter pang than when those horrible bolts were shot home on me, and I knew that my liberty was taken away from me. I walked up and down in that cell like a caged animal. Now I felt a terrible rage taking possession of my heart, then again a feeling of despair threatened to overwhelm me. God alone knows how terrible that night was. It seemed to add years to my life. All my property stolen, all my plans frustrated, all my prospects blighted, myself a prisoner in a felon's cell—and the jarring of bolts and clanking of chains which continued to ring in my ears! Aye, I hear them still, those horrible bolts and chains; I still see that door shut firmly upon me.

I remember how I looked up to the star-studded sky with a bursting heart. Through that little hole of a window I could see a few, sparkling brightly, oh, so far away. How sweet their glimmer was, how they seemed to beckon to me! And I was impotent—a caged prisoner! I could not go to them. Then I heard our warders talk outside in the guardroom, and Kaffirs sing in a location near by. But I was locked in. I could not go out into the glorious night. Liberty, sweet liberty, how I longed for thee then! Lie down and sleep? I tried to, but how could I? The atmosphere was stifling, and I was overpowered by my own terrible emotions.

My brain was on fire, my lips dry and parched. The fever of indignation and horror, coupled to the consciousness of my own impotence, was devouring me. And for ought I knew I might be shot on the morrow, as I was told. It seemed to me as if now only I began to understand what suffering was, and what pity was. How many men, thought I, besides myself have not been unjustly punished and imprisoned! How many have not suffered innocently, whilst I, a free man outside their terrible dungeon, little thought of such, or pitied them. Never before had I realised what it meant to be a captive; now I knew and experienced it.

I felt as if I could go down on my knees beside my two fellow-prisoners, the Kaffirs in chains, to comfort them and speak words of encouragement, for soon they were to be mercilessly lashed—lashed as only a Transvaal gaoler can lash a Kaffir. I felt as if my heart could break with pity. All this happened months ago, but I thank God that I still have the same feeling. I have become convinced of many evils in existence here, and it will be my duty as a lover of my country to expose wrongs and seek to redress grievances when I am free again. As a man I cannot help despising the unmanly, mean things I have seen done, and that too in the name of patriotism; but sufferings appeal to me more strongly than ever before, and the cry of the miserable pierces my heart deeper than ever. That first horrible night taught me many a lesson—taught me to pity and feel compassion, but also to feel indignant when cruelty and meanness are practised.

Here where I sit writing in my spacious room in our Pretoria prison, surrounded by British officers, educated men and gentlemen, I cannot help comparing men and matters. But always, when I close my eyes, the experiences and scenes of that terrible night come back to me; I hear all those sounds again, the horrible grating of bolts and bars. I still hear the voices of my warders and the people outside—aye, and that one other terrible sound.

The poor Kaffirs, the chain-gang, could not move in their sleep without the iron chains clanking against each other. In the deathlike stillness of the night those ominous sounds, dread remembrancers of captivity, struck on my ear from hour to hour. It was horrible! How long I tramped up and down in my cage I do not know now. I only know it was near the dawn of day before some fitful moments of sleep brought temporary oblivion. But even these were disturbed by dreams. God only knows how a prisoner must suffer—prisoners who are the victims of petty spite and meanness, hated by the very men they had tried to benefit. That very night, as I learned next day, there was a prayer-meeting in the church. I had heard the bells ring the evening before, and I loved to listen to the sound; but what happened inside? Long, passionate prayers were sent up to heaven for success to their arms; but not a prayer, not a sigh for fellow-Christians, close by, treacherously imprisoned by the very Government they were praying for.

The very official who had locked me up was in that prayer-meeting. Did any thought of pity stir his heart? Nothing of the kind. An eyewitness informed me that bitter, spiteful things were said, only that, even in the house of God! I am afraid Matthew v.7 was sadly forgotten in the Transvaal during the war: "Blessed are the merciful: for they shall obtain mercy."

Is my accusation a hard one? No, and again no, for I have seen and heard too much here. Oh, the cruel things done by officials, and their frightful words of hatred, and their cursing and swearing! May God have mercy upon them! They had no mercy—they knew no *mercy*. What right then have they to expect a *blessing*? Echo answers, What?

The beautiful bright morning came at length, but not for us. It came for the free, for the oppressor, for the tyrant, for the merciless. The poor prisoners could only look on from behind prison bars and long for liberty.

Before I was taken to gaol the previous day my wagon and everything belonging to me, was taken—no, *stolen*—even my dear little folding Hamlin and Mason[764] organ, shut up in its box, with all my beloved music.

Mason & Hamlin Harmonium Pump Organ.[391]

What a comfort music would otherwise have been to me in my captivity! But no—these kind, patriotic officials had to prove and show their zeal by deeds of dishonesty and cruelty. Hundreds of pounds in value were thus "looted" from me. I saved only a portmanteau with some clothing and my dressing-case, and a few rugs and my two mattresses.

But then the Transvaal Government, *not the real burgher*, fights against non-combatants. And—the mockery of it!—they consider themselves particularly up-to-date in civilised warfare!

In our case, as I learnt afterwards, they really thought they had made a smart haul. The ladies, forsooth, were spies (to spy what?), and I had a Maxim concealed about my person, I suppose, and dynamite in my pockets, or poison, and I was going to send the whole Transvaal "up in the sky so high" with an explosion of liquid air. Oh, I was so terribly dangerous a man; eighty-four good men and true had to catch me by strategy and treachery. They succeeded, and the South African Republic was saved! Puny, insignificant little officials! Really, I write about you more in sorrow than in anger—or should I say, more in mirth than in anger? You do cut such a ridiculous figure! You capture railway officials and non-combatants, and then flash the story of your grand success all over the land.

Yet this first, horrible night remains a reality, a terrible reality. A lifetime will not wipe out the memory thereof. Nothing can; nor do I hope ever to forget the lesson it taught me, or to neglect practising it—"Blessed are the merciful: for they shall obtain mercy." The officials at Zeerust could not have displayed more venom towards us, their prisoners, and towards me especially, had we been criminals of the deepest dye.

[764] Mason & Hamlin is a piano manufacturer based in Haverhill, Massachusetts. Founded in 1854, they also manufactured a large number of pump organs (harmoniums) during the 19th century. Mason & Hamlin - Wikipedia

VIII. The First Week

As I said before, for the first few days of my imprisonment we were allowed some liberties. I have to thank the Public Prosecutor of Zeerust for this courtesy. This official was a gentleman, one of the very few I met with in the Transvaal among Government officials. I gladly make use of this opportunity to assure him of my gratitude for his great courtesy always extended to me. When I was harshly treated by others he, I know, did his best for me and the others. He was the man who allowed us the few liberties we enjoyed at first. The stationmaster and myself were allowed to take our meals at the hotel, and, accompanied by the gaoler, to go out at about four in the afternoon and stay out till after dinner, being back at 6.30, of course always attended by the gaoler. Every afternoon, therefore, till the Sunday following we went to the good doctor's [765] house, at his invitation, to stay till dinner time.

Those little visits were like glimpses of heaven. Shall I ever forget them? We had music and fruit and delightful talks, the doctor with his charming wife and daughter doing everything to make us feel at home and enjoy our short spell of liberty. Aye, the music especially—can I ever forget it? Lord Henry Somerset's[766] "Echo" and "Song of Sleep," and some other grand old favourites—how I sang and re-sang them! And the duets we had! Yes, those were glorious hours. Their memory made the locked cell bearable every evening. May I say again, God reward these good Samaritans for their infinite kindness and staunch friendship, notwithstanding the risks they incurred thereby, as we shall see later on.

On Wednesday my young friend, Harry Neethling, who had been taken prisoner, so treacherously too, *before* the war broke out, was also sent to the gaol. He, the stationmaster, and I then occupied one cell. The next chapter will tell about his experiences.

Our meals at the hotel, however, did not last long. This privilege granted us was too much for the charity of some of the Zeerust people. Our going to and fro every day for our meals, though we were guarded, was to them an eyesore. Pressure was brought to bear on the authorities, and on the Saturday this privilege was cancelled.

A later chapter—"The Man and the Woman"—will deal more fully with this part of our history. We were still allowed to visit the doctor every afternoon, but, alas! on the Monday morning following this privilege also was cancelled. The Government could not extend its drop of generosity any longer. They were so unaccustomed to act generously that this little trial hurt them too much! It had to be abandoned accordingly. The one official referred to before did his best for us, I know, but in vain. Naturally, this first and only comparatively happy week passed like a dream—all too soon.

When our hotel meals were stopped the kind doctor's wife came to our rescue. She offered to send us our daily meals to the prison. We took our prisoners' share of meat and sent it down to her house, where she then prepared our meals, adding ever so many nice things. Those meals—so appetising, and so daintily served—how they increase our load of gratitude! Imagine the three of us in our little cell around a little table, which was then allowed us, with snow-white cloth and napkins even! A beautiful bouquet of flowers came up every day; and many other little things a woman only thinks of were provided. We always said that our meals were little whiffs of home. Nothing was wanting except the pudding of liberty!

As the sequel will show, however, even this innocent pleasure was begrudged us, and a stop put to it very soon afterwards. For the first week, too, our fellow-prisoners, Nurse Ruth and Miss R., were allowed to visit us for a minute or two, and then that also was put a stop to. With great ostentation we were informed that a merciful Transvaal Government was going to treat the prisoners of war as *untried* prisoners. And their own regulations showed what liberties were extended to such prisoners. We, however, were *denied* all such privileges and treated as condemned felons. Transvaal justice all over! And yet the Government pretends to have the monopoly of Providential care. A friend, getting tired of waiting, when during the first few months everything seemed to go against the Imperial troops, expressed the situation very shrewdly: "The mills are grinding very slowly, and the Miller is commandeered and monopolised by the Transvaal." It is possible to entertain the thought for a moment that a just God would specially bless the efforts of an unjust Government.

[765] This is almost certainly Dr. Blake who is mentioned above and infra.

[766] Henry Richard Charles Somerset PC, DL, JP (7 December 1849 – 10 October 1932) was a British Conservative politician and composer of popular music. He served as Comptroller of the Household under Benjamin Disraeli between 1874 and 1879. Lord Henry Somerset - Wikipedia

I do not mean to say that England has never done an unjust deed, that during this war even nothing wrong was committed. But then our Generals did not quote Bible texts to cover a false report and an abominable falsehood. Think of those misstatements I referred to in a former chapter. And there are many more to which I will have to refer later on. Let me mention one now. *After* I had been lodged in prison one of their newspapers, the meanest rag I have ever come across, published an official wire from Zeerust informing the public that *I had been released on parole*. And in Pretoria the *Times* correspondent was similarly informed by the Government, of which *honest Paul Kruger* is the head, that I "had been taken prisoner, but immediately released"! Is it not an insult to the Divine Being to expect of Him *special* blessing on their undertaking? No wonder that a Government so hypocritical should have been a curse instead of a blessing to the people.

And now, I'm afraid, the culminating calamity is going to overtake them. And yet this calamity they so dread will only be a blessing in disguise. If the Augean stable[767] of the Transvaal Government is not cleansed now worse things will befall the country. There has been civil war once before—there can be one again. It surprised me to hear from so many sides expressions of deepest hatred against the Hollanders and their Mephistophelian influence. One said to me, "We have to thank the Hollanders for this war." Another, "They will try to get the war to last as long as possible; they sit at home enjoying the salaries of fat billets, whilst the burgher fights and gives his life at the front."

But more of this also later on. Our first comparatively happy week passed like a dream. Had the environment remained unchanged captivity would have been bearable. This my captors noticed, consequently, goaded on by a lust for cruelty and oppression worthy of the fiends of the Spanish Inquisition, they determined to make of my little prison cell a torture chamber. I had to be placed on the rack as far as was possible—as far as they dared go. These are bitter words, but the reader must kindly remember that I differentiate greatly between the good, genuine Transvaaler and his Government gang. The burgher is all right—brave and honest and kind-hearted till he falls under the baleful influence of his Government or becomes a Government official himself. Farewell, then, a long farewell, to all his simplicity and honesty of character. This has been too often experienced, and I have kept eyes and ears open.

Our first Sunday in prison was to us a very pathetic one. By that time there were in all seventeen prisoners of war, besides the criminal prisoners. We gathered in the entrance-hall and I conducted a short service. I spoke in Dutch to the few warders who cared to attend, and then in English to my fellow-prisoners. It was a solemn, touching service. How we prayed for dear ones faraway! And, reader, can you believe it? even that service was begrudged us, and put an end to a few Sundays later. The very Monday after this first Sunday the report was rife in the little village that I had conducted a service in the doctor's house; and surely no miserable prisoner ought to be allowed to do, this!

Just one more episode connected with this week before the chapter closes. I referred before to the Acting-Commandant (or whatever his full title may be) of Zeerust—the little agent fellow. He distinguished himself beautifully—thusly. A telegram had been received from the Mafeking besieging army to the effect that their General had sent a messenger, under a flag of truce, to Colonel Baden-Powell, and that he had not yet returned, though some hours had already elapsed.

People with evil consciences are naturally suspicious, and such was this self-important little official. Of course he at once jumped to the conclusion that a British officer had made a prisoner of a man coming up under a flag of truce—and this was an insult to the Transvaal flag. The little man went mad; he rampaged up and down the street, foaming at the mouth, telling all and sundry that he would wait two hours more and then, if no news had arrived of the safe return of the messenger, he would have all the Zeerust prisoners brought down to the market-place and shot! Aye, there was blood in his eye and blood in his words, but very little in his veins—the foolish man!

And the messenger? Do you know what happened to him? Colonel Baden-Powell had received him, as a British officer would receive such a messenger, courteously; had listened to his message, and taken him to the hotel and given him a rattling good meal. In fact the young man, according to his own statement, enjoyed himself so hugely that he was loth to leave. And here our little frothy Commandant, for no reason at all, went curvetting up

[767] The fifth Labour of Heracles (Hercules in Latin) was to clean the Augean stables. King Eurystheus intended this assignment both as humiliating (rather than impressive, like the previous labours) and as impossible, since the livestock were divinely healthy (immortal) and therefore produced an enormous quantity of dung. These stables had not been cleaned in over thirty years, and 3,000 cattle lived there. However, Heracles succeeded by rerouting the rivers Alpheus and Peneus to wash out the filth. Augeas - Wikipedia

and down the street, for all the world like a toy general on a tin gee-gee, thirsting for our blood! Were the sight not so degrading, its humour could make you hold your sides with laughter. Tis the old story over again, of little minds and little things!

Ford Der Boeren, Mafeking.[768]

Boer Fort at Mafeking. [769]

Major-General Baden-Powell and Major Hanbury-Tracy. The photograph was taken in Mafeking, during the siege, next to a shell-proof "dug-out." [392]

Mafeking Artillery.[393]

[768] Webster Davis, John Bull's crime; or, Assaults on Republics. 1901 New York The Abbey Press

[769] Webster Davis, John Bull's crime; or, Assaults on Republics. 1901 New York The Abbey Press

Colonel Baden-Powell and his military and civil staffs.[770] Mafeking Defenders. [394]

Bombarding Mafeking: Laying [aiming] the big Creusot gun, "Creechy."[395] The signal to seek shelter. "A Hindoo, on a heap of flour bags, announcing with a flat the firing of a shell from 'Long Tom.'" Gordons getting into shelter.[396]

Cronje and lieutenants around one of the Creusot guns. General Cronje near gun, with whip pointing down. [397] (L to R) Commandant Steenkamp, General Cronje and Commandant Van der Merwe identified. [398] Sarah Wilson at doorway[399] [771]

[770] Top Row: Left to Right: Major Panzera (artillery, Captain Ryan (Commissariat) Captain Greener, Edward Cecil (Chief Staff Officer), Captain Wilson, A.D.C., Lieut A. Hanbury Tracy (press censor, Military Intelligence Officer). Captain Cowan (Bechuanaland Rifles). Second Row: Major Godley (Commanding Western Outposts), Colonel Vyvyan, Mr. C. Bell (Civil Commissioner), Col Baden-Powell, Mr. Whiteley (Mayor), Colonel Hore (in command of the Western Fort), Dr. Hayes (Principal Medical Officer. In front, Lieut. Moncrieff. With the Flag to Pretoria, H. W. Wilson Vol 2 May 1900.

[771] Lady Sarah Wilson, (born Lady Sarah Isabella Augusta Spencer-Churchill; 4 July 1865 – 22 October 1929) a first cousin of Winston Churchill was recruited by Alfred Harmsworth to cover the Siege of Mafeking for the Daily Mail. Sarah Wilson (war correspondent) - Wikipedia He recruited Lady Sarah after one of its correspondents, Ralph Hellawell, was arrested by the Boers as he tried to get out of the besieged town of Mafeking to send his dispatch. She had moved to Mafeking with her husband, Captain Gordon Chesney Wilson, at the start of the war, where he was aide-de-camp to Col. Robert Baden-Powell, the commanding officer at Mafeking. Baden-Powell asked her to leave Mafeking for her own safety after the Boers threatened to storm the British garrison. This she duly did, and set off on a

Old Ship's Gun, 'The Wolf' made during the siege.[400]

Graves at Mafeking. [401]

IX. Harry's Experiences

This chapter dealing with Harry Neethling's brief imprisonment has been omitted. The chapter number has been retained as a placeholder.

X. The Man and the Woman

As promised before, I must now explain how it came about that our privileges were curtailed. Up to the Friday we three, inmates of the historical cell No. I, had our meals at the hotel, where the owners treated us with exceeding kindness and consideration. What put a stop to it? The heading of the chapter gives the answer to the question. Let me tell the story. I have mentioned already that scowling looks were continually turned on us, and muttered curses hurled at us as we passed through the streets to and fro from hotel to prison. The principal performers in this farce were the officials and their understrappers. It was so enjoyable, so manly, so courageous to jeer at an unarmed man, a prisoner, one who could not defend himself. In the hotel dining-room we noticed the same thing. Everyone who wanted to prove his great loyalty to the Transvaal—the Jews especially—believed that the only way of doing it was by behaving like a cowardly cad to a defenceless man.

To give as little cause of offence as possible we decided to take our meals before the usual time, and then, when others came in, not to speak a word. On the very day we agreed to do this, however, *the man and the woman came to lunch*. It was a neighbouring farmer and his wife. Both had been educated above the usual standard of burgher, and in the Colony, too, I believe. They knew what politeness and decent behaviour meant. I would otherwise not have remembered their exploits, nor have written about them. The man was a rather ugly specimen, decently clad yet ugly, even when he looked his best. Imagine his looks when angry! The woman was neatly dressed, was large and stout. There was a good deal more of her than of her husband. *Her* complexion was bright—*his* like yellow parchment. She could look pleasant, but the face was of that class which surliness and anger at once deform, making it repulsive. This couple had obtained all the necessary information in the village, and therefore knew all about us. They were in a towering passion, and when they entered the room I was attracted, or rather repulsed, by their dark, scowling looks.

The woman was livid with passion, the man green and yellow. Why? Because, as he whispered loud enough for everyone to hear, that "hated Hofmeyr, the accursed prisoner, is allowed to eat with decent people!" How complimentary! Now the *man* was a deacon of his Church, and wore a long face and longer frock-coat on Sundays, whilst the *woman* sat in her pew solemnly and attentively, praying, "Forgive us our sins as we forgive them that sin against us." Ah, but that was on Sundays; and when one prays "Forgive us," &c., one does not really mean it! And then the pair, thus contorted with rage, banged down on their chairs.

The *man* sat down somewhere on the small of his back, stretched out his legs as far as they would reach, thrust his hands into his trousers pockets, and stared at me, vulgarly and insolently, mumbling like a maniac all the time. The *woman* sat down motionless, boiling over with rage. All the while the poor waiter, a coloured boy, stood waiting for orders. The *man* turns to him now and gives his orders in a stentorian tone, venting his anger on that poor fellow, disapproving of every dish, condemning everything. Scowling, grumbling, and in a

madcap adventure in the company of her maid, travelling through the South African countryside. Eventually, she was captured by the Boers and returned to the town in exchange for a horse thief being held there.

towering passion they finish their meal. Had the spectacle not been so sad a one—for I knew what it all meant—and such a travesty of politeness and good manners, I could have burst out laughing.

And I am fond of a good hearty laugh. Why, I could have lifted the little puppy by the scruff of his neck, and swept the floor with him! Behaviour such as theirs is reserved only for the semi-barbarian, and such unfortunately this poor man was, this noble warrior, who *could* gloat over a conquered enemy; who *could* thus treat a defenceless man. What a noble-hearted brave! Aye, how this *man* and *woman* thirsted for my blood! how they longed for my scalp!

Unfortunately I could not then oblige the turtle-doves. Was it worth while writing down all this? *Cui bono*?[772] some one asks. Wait a moment, dear reader. The story has a sequel. The man went out after his lunch and painted the town red. *He* would see to it, he shouted, that no prisoner got his meals in a hotel, and that "this fellow H. should be locked up like a Kaffir!" The *man*, ably seconded by the *woman*, was verily on the rampage. Government officials were interviewed, the wires were used to communicate with Cronje's lager, near Mafeking, and after a day or two the noble *man* and his angel of a wife were made happy by hearing that the order had come to lock up altogether Mr. H. in his cell, as if he had been a murderer, instead of an inoffensive man and an honest *political* opponent.

How happy they must have been, and how soundly they must have slept that night in the consciousness of a deed well done! They had added to the load of a fellow-man's misery and sorrow. They had refused him the proverbial cup of cold water, as had the other civilised, brave heroes who so ably seconded their efforts. I feel sure that a barbarian enemy would have shown more consideration and mercy to a conquered foe. Ah, but these were civilised people, Transvaal officials who were anxious that the world should believe in their civilisation and chivalry! *Credat Judaeus!*[773] say I.

Thus were our privileges cut off. Such was the beginning; and then the curtailment came with a run, as the sequel will show. This *man* and *woman* were but types of many others, some even more hard-hearted and cruel. Alas! what does not a heart miss that knows not generosity, and a life that understands not that the greatest of all gifts is charity!

XI. Daily Routine in Prison

I may just as well tell here what our daily routine was in prison, and may premise by saying that I have never been much of a sleeper in my life, and that now since I am a prisoner I sleep less than ever before. In fact I have often for weeks suffered from insomnia. It is my great enemy—one I fear. In prison we were locked in at 6.30. A courteous gaoler, until our doors were ordered to be locked day and night, some little time after, allowed mine to stand open till seven o'clock, so that I might have as much cool air as possible, for the nights especially were stifling. When locked in we went to bed at once, for the chairs kind friends had supplied us with were taken out, as also our little table sent by our friend the doctor. Lying on the mattress on the floor we read or studied by candle-light till 9 or 9.30. Then the lights had to be put out, and we lay talking, of absent friends generally, till eleven or later. My friends managed to drop off then. Morpheus,[774] however, seldom took me in his arms before 12.30 or 1, and very often much later. As a rule I was up by six, when the doctor's delicious coffee was at the door. How we missed this luxury when it was after a while forbidden! How we enjoyed that delicious early morning beverage! Yes, there *were* some kind people in Zeerust who understood Christ's doctrine of the drop of cold water, and put their knowledge into practice, too, which says ever so much more! Shall I forget the hot scones that used to come up with this coffee, and the doctor's kind, good daughter, who purposely got up early in the morning to prepare these for us? Unselfish Daisy—forgive me for revealing your name: a name as pretty as you were, reminder of all your kindness and goodness. Please once more accept my warmest and heartfelt thanks. You will never know what you and your dear parents were to me and all of us—how you really made our sad fate bearable. May God reward you abundantly, and grant you great happiness, for you have tried to make others happy.

After this early coffee I had my bath in a *zinc bucket*, which was all the washing-stand we were allowed. Attached to the prison were two little yards, surrounded by high walls, thirty by twenty feet each. In one of these

[772] Who stands, or stood, to gain? (from a crime, and so might have been responsible for it).

[773] *Credat Judaeus*: the Jew may believe it, not I.

[774] Morpheus is a Greek god associated with sleep and dreams.

I used to sit till breakfast time (eight o'clock) reading or studying. After bathing and dressing Harry and I swept our room (several times we scrubbed it, too, ourselves) and tidied up, brought in our table and chairs, and above all our beautiful flowers. After breakfast, whilst our cell-doors were still allowed to be open, we had a walk up and down in the corridor, and then again reading and studying till one o'clock, the luncheon hour. The heat in the stuffy prison was by that time generally intense, and therefore after the meal we lay down in as little clothing as possible to woo a little sleep. This very seldom proved a success, for our warders always talked and laughed in the loudest tone possible, and their incessant march up and down soon banished all sleep.

At about four o'clock we again frequented the little yards to get some fresh air. The doctor used to come up to see us then, bringing news, cigars, and words of good cheer. How we looked forward to these valued visits, how we appreciated them! Ah, doctor, doctor, how good you were to *all* of us! Shall I forget how you dosed the prisoners less favoured than we were with quinine? Shall I forget the amused twinkle in your eye when you produced your large black bottle, containing one tiny drop of quinine and the rest—well, let us say water? Shall I forget *the at first* astonished and then *grateful* looks of the men when they tasted this wonderful quinine, first sipping a wee little, then tilting the half-filled tumbler to a generous angle, and caressingly coaxing the last lingering drop out of that tumbler's bottom!

You were district surgeon. You dared do such things. Had anyone else tried it—but the thought is too terrible! And then, too, you are a brave, independent American, and the poor Government officials of Zeerust at the time actually believed that the United States would not only interfere in their behalf, but also pay their debts. Can you imagine such folly?—and yet so it was. But the doctor could another tale unfold. During these happy afternoon hours we often, too, used to sing together songs we all knew. There were some good voices among us. How often the "Old Folks at Home"[775] brought a mist to the eyes, and "Whisper and I shall Hear," and the "Holy City,"[776] and "Sweet Marie," and "Mary of Argyle." On Sundays we sang hymns (Ancient and Modern). Ah, those old favourites, "Abide with Me,"[777] "Onward, Christian Soldiers,"[778] "Art thou Weary, art thou Languid?"—how they cheered and comforted us! God knows what happy, blessed hours these were.

Then, too, there were tales told and jokes cracked. One of our fellow-prisoners was a splendid *racouteur*. Lombardini was his name, and my good friend will, for the sake of those days of tribulation, forgive my mentioning it. The stories he could tell! I have never heard the like. One after the other they rolled out, each wittier than the other and smarter. He could keep us in roars of laughter. Yes, Lombardini, old chum, you contributed more than you know to our happiness and cheerfulness. You do not know how you helped me especially to ward off the demon of melancholia. Will I ever see you again? I trust so, for I will certainly try to find you "after the uproar is over." We must have another long chat together. At six o'clock dinner came, and the day was over. Thus passed every day, till our cells were locked, and then all communication between cell and cell was forbidden. The *man* and the *woman* did not rest till the cruelty practised upon me broke up my whole nervous system and shattered my health. I wonder if they will be glad to hear that I have recovered. If not, I am sorry that neither in this can I oblige them. They will just have to grin and bear it, poor things!

The gaoler and a few of the warders were very nice, and often one or two would come into my cell of an evening for a little chat. They really intended to be kind, and I am sorry to say that their kindheartedness later on brought them trouble, for they were warned most emphatically not to show us kindness on pain of dismissal. And then came again darkness and too often sleeplessness for hours. Ah, how one's thoughts did travel then! There was a good deal of rain, too, terrible thunderstorms, mostly in the night. I delighted in watching the flashes, and listening to the mighty cannonade of heaven, the voice of the Great Creator. Listening one night the following thoughts came to me, and herewith I close the story of the daily routine of our life. Let me call it—

[775] "Old Folks at Home" (also known as "Swanee River") is a minstrel song written by Stephen Foster in 1851. Since 1935 it has been the official state song of Florida, although in 2008 the original lyrics were revised. It is Roud Folk Song Index no. 13880 Old Folks at Home - Wikipedia.

[776] "The Holy City" is a religious Victorian ballad dating from 1892, with music by Michael Maybrick writing under the alias Stephen Adams, with lyrics by Frederic Weatherly. Its sheet music sales made it one of the most commercially successful songs in the UK and United States around the beginning of the 20th century, and also "perhaps the most pirated musical piece prior to the Internet," according to copyright scholar Adrian Johns. The Holy City (song) - Wikipedia

[777] "Abide with Me" is a Christian hymn by Scottish Anglican Henry Francis Lyte. It is most often sung to the tune "Eventide" by William Henry Monk. Abide with Me - Wikipedia

[778] "Onward, Christian Soldiers" is a 19th-century English hymn. The words were written by Sabine Baring-Gould in 1865, and the music was composed by Arthur Sullivan in 1871. Onward, Christian Soldiers - Wikipedia

The Prisoner's Reverie.

I heard a voice, a gentle voice, soft coming
Through the storm.
The rain was fierce, the roaring winds were chilling,
Yet breathing warm.

Of skies so blue, my sunny homeland lying
In peace beyond,
This gentle voice, unto my voice replying,
Made answer fond.

And then again, like crash of armies meeting
In yonder vale;
But through it all that sweet voice kept repeating,
"God shall not fail!"

XII. Traitors

I have come to a very dark page of this little history. The episode I am about to relate made a deep impression on me, revealing depths of meanness to which I did not think a man could sink. One can expect anything from an enemy, but when a friend turns against you there are no words which can adequately describe one's feelings. I have suffered from such false friends before this, but this act of treachery involved a question of life and death, and was, therefore, perhaps my worst experience of human nature. Let me tell all that happened. I have before referred to the ganger and his wife; perhaps in not too courteous terms, but the reader will excuse these when he has read this story. This man and his wife came up to Lobatsi station on the Friday morning before our capture from their cottage down the line.

They seemed so helpless. The stationmaster therefore allowed them to put up in the ladies' waiting-room, and we all showed them every kindness. On the way to Zeerust I, as an experienced traveler, made them as comfortable as possible—they *were* so helpless. During the night, as mentioned before, I managed to let them have the half of our wagon. Arrived at Zeerust, they were still more helpless, and confidentially informed me that they had no money with them at all. I myself had only a few pounds—everything else had been taken from me—but yet I said I would manage to help them. I paid for them that Monday, till we were arrested in the afternoon. From then up till Friday I paid for the meals of the man and his brother-in-law sent up from the hotel to the gaol. I was indeed sorry for them. On Friday I heard that the woman had told the nurse that they had their bank-book with them (Savings' Bank), and had a sum to their credit four times larger than all I possessed at the time. I accordingly told the man that I could no longer pay for him, but would help them get their money from some one in Zeerust; and I succeeded, too, in doing it. However, this action of mine gave umbrage to the stingy Scotchman. I received no word of thanks, and he began grumbling over the few privileges granted to me, and not to him.

In the village his wife all at once turned Transvaaler and complained bitterly that I, their benefactor, mind you, enjoyed privileges which he did not. This pair thus ably abetted the other *man* and *woman* I told about. But worse: from questions put to her this woman gathered that the Government was eagerly looking for proofs of charges they were anxious to make against me, so as to justify my capture and detention. The officials were thirsting for my blood, and were determined to have me court-martialed and shot. Talk to this effect was soon prevalent on the streets, and one of their miserable papers had a paragraph to that effect, too, gloating over my coming end.

This woman was allowed to come and see her husband every day in prison; she told him about it, and to revenge themselves on me for my generosity to them, I suppose, they concocted a vile plan. What added to their spite was the following. This ganger used to walk up and down in the prison corridor every day after lunch with a pair of frightfully creaking boots. He was kindly asked to desist, but in vain. I then spoke to him too, for I knew not that he bore me a grudge, but also in vain. Then I told him, as also did the other prisoners, that we would have to complain to the gaoler, as everyone wanted a moment of quiet after lunch. This caused the old curmudgeon to fly into a passion and swear at all of us and me in particular.

The gaoler, hearing our story, ordered quietness every day from one to three o'clock. And now, when the woman appeared on the scene with her suggestions, the man was ready to assist. What a degrading sight!—two subjects of our beloved Queen plotting against another loyal subject! And they were clever about it, too, as I will try to show if you will accompany me back to Lobatsi for a moment. On the Friday night when the couple came to the station and took up their quarters in the waiting-room, the following happened, unbeknown to me. The police-officer in charge wanted to put a box of cartridges, for safety, in the stationmaster's room.

The latter objected, however, thinking it would be too dangerous, but advised him (this is the stationmaster's own story) to put the box in the lavatory behind the waiting room in which the ganger and his wife were. This lavatory could be reached *only through the waiting room*. The police-officer called the Kaffir porter to take up the box and carry it into the lavatory. He and the stationmaster and this woman (who was in the waiting-room at the time) saw the porter do it. On Saturday morning the stationmaster had the box taken out again by the porter, but the woman happened to be absent at the time.

Let us go back to the prison now. Here the two plotted and arranged that she should go to the officials and tell the following story—namely, that I had brought in a box of cartridges into the lavatory, had dug a hole in the ground, and then hid it; and that *she had seen me do it*, without my being aware of it, and that my intention was to *arm the Kaffirs* with these cartridges. The chivalrous Government officials greedily devoured the story of this traitor and traitress. The gaoler is ordered to leave early the next day for Lobatsi, but he does not know for what purpose, as he privately informs me. Next morning early he is told to take the woman there, and is then informed of their vile charge against me. The good fellow afterwards told me that he did not believe it, and that when the woman was told, unexpectedly, to go with him to show the spot he could see her weaken perceptibly. Off they went, accompanied by a couple of officials. On the way the woman grew more and more uncomfortable. "But he may have taken them away," said she. Said the gaoler, "But you said he dug a hole, and that he had not come back for them on the Saturday, and that early on Sunday morning you were all captured, so they must be there, and if taken away we can see traces of digging."

The traitress, however, grew more and more restless. Arrived there, the party made her point out the spot, and, armed with spades and pickaxes, they entered the place to find—a *boarded floor* and not a vestige of a cartridge-box! To do their work very thoroughly they even broke up the floor, but not a trace. Then the men burst out in revilings and the woman into tears. And then she confessed that *I did not hide* the box; I had only *put* it there. And, later on, when confronted with the stationmaster's affidavit, she confessed that *the porter* had brought it in, and that I had nothing to do with it at all—in other words, that she and the man had *lied, knowing* what the result might have been for me.

The creature confessed, too, to the two other ladies that she had been promised release for herself and husband if she could help them prove anything against me! And to this they had lent themselves! Another case of warming a serpent in your bosom—only this time it was two serpents. The story had a very depressing effect upon us all. Here we were, a very small band of loyal subjects, but prisoners in the enemy's hands. And one of our party was a traitor! No wonder that after this no one ever spoke a word to the man again. He became a veritable pariah amongst us. And the woman—when the people in the village heard of her treachery—was she avoided and shunned?

Oh, no—Landdrost and Commander continued calling on her, and condoled with her, and grieved over the plan that failed! No, no, treachery is not repulsive in the sight of these officials; they practise it themselves! Some six weeks later this miserable couple were released, with the other prisoners, who were railway men. They have gone back to the Cape Colony now, but where they are to-day I don't know. The taint of their dastardly treachery will cling to them all their lives, however. One more remark: what harm is there, when a war is declared, and you are on your own soil, and not in the enemy's country, in hiding a box of cartridges, supposing that I had done it? Surely no harm at all. So blind were these people, however, with hatred against me, and so cruel, that they magnified mole-hills into mountains, and eagerly snatched at every rumour, even the word of a traitress. I suppose they made sure that I had hid a Maxim, too, in my portmanteau, and a bomb in my hat, and that I was just awaiting a good opportunity to blow up the whole country. Among such officials had I fallen.

XIII. Some Episodes—Ludicrous and Disgraceful

All the time I was in the Zeerust gaol I obtained but very scanty news, and that little was excessively bad. Every day an official telegram—the inevitable *official*—was posted up at the Landdrost's court. These words were masterpieces of words, *et praeterea nihil.*[779] At first when I read them myself, and later on, when told about them, I did not know whether I ought to cry or laugh. They were so funny, so false, so flighty—exquisite poetical productions. And, in addition to this official gospel, the world-renowned *Standard and Diggers' News* regularly appeared on the scene. What a beauty it was! According to these two authorities the whole English army was killed or wounded during the first week or two, whilst after every fight the Republican forces had the inevitable "five killed and seven wounded." I think the headline was afterwards stereotyped.

One Bank Holiday, in London, I spent at the Crystal Palace. Returning home, I got off a bus at Piccadilly Circus to walk up Regent Street to my club. At a corner two little newspaper boys were having a heated argument, which threatened to end in a fight. A third stood grinning by. Buying a paper from him, I said, "What are the youngsters wrangling about?" With a twinkle in his eye, his answer leapt forth: "Oh, sir, tis only the exuberosity of the verbosity of the 'oliday!" I cannot describe these official telegrams and this newspaper better than in my young friend's terms—any amount of *exuberosity of verbosity*, and nothing more. But seriously, deceiving the people like this certainly was not a feat worthy of a Government always professing its dependence upon God and His providential care. It was a shame.

[779] Voice and nothing more: sound without substance.

It was amusing to read continually "The fall of Mafeking" imminent! "The last hour has struck." "The doom is sealed." At first I felt nervous and thought—will plucky little Mafeking be taken after all? Then this news became monotonous, and afterwards ludicrous. This is the month of February, and Mafeking has not yet fallen, nor is it likely that it ever will fall. Their papers to-day never even mention the name any more. Are they ashamed of their bluster and brag? I hope so.[780]

The same thing was done in the case of Ladysmith. Its "last hour" struck continuously; twice a Government official in Pretoria solemnly informed me that *"Ladysmith has surrendered this afternoon, and the President has sent off an official to take over the booty!"* Well, Ladysmith is still standing, and from all accounts it will be relieved soon. Kimberley has already been relieved. How small our braggarts must feel to-day if they have any sense of shame left! The Government have claimed the monopoly of God's blessings and help; and newspapers, which in time of peace knew not God and His service, now talk piety by the column, and in the next lie and brag. Is this excusable, or am I mistaken? I had an idea that in Matthew v. 5 we read "Blessed are the meek: for they shall inherit the earth." Evidently I must be mistaken, for these people here say they are right in their view— "piety and brag" go together.

After a while it became a standing joke in Zeerust that every Sunday every burgher was to be back in the lager, [at Mafeking] for on Monday Mafeking was going to be taken. When on Wednesday or Thursday I inquired what had become of Mafeking, the answer was, Oh, it could easily have been done, but Cronje decided to wait, for reinforcements were coming up to help Colonel Baden-Powell, and then he could make a grand scoop of them all! Evidently they are still waiting, and I am really very sorry that Colonel Baden-Powell cannot oblige them. Aye, the bounce of these "officials" is simply ludicrous.

Another story is told of a well-known burgher, who, after a little bit of an attack on Mafeking something they called "a charge on Mafeking" was found *behind* a little hillock holding some horses. When laughed at and asked what he was doing, his answer was, "I am holding the position"—mind you, the position *behind* the hill! And he was one of the bullies who was hurrying through Zeerust to the front, because "they were going to take Mafeking, but were anxiously awaiting his arrival before charging. They could not get on without him."

Another youngster came home on furlough, after a few weeks of doing nothing, and bragged about the streets of what he had done. One of his yarns was that the Lord had worked a special miracle for them, for when they were fired on by the terrible Maxims, the bullets, though pattering on their breasts, fell harmlessly to the ground, not one being wounded. "Well, all I can say is," said I to my informant, "that the Lord I know is not in the habit of working miracles in *favour* of an Ananias,[781] but rather the other way."

Another one rushed back with the story that Mafeking was burning in ten different places, and that he had seen the smoke ever so far off. I suspect the brave defenders were cooking their dinners, and the truth-loving warrior had allowed his imagination to run off with him. I do not wonder at their failing to capture any of the besieged towns—to put it plainly, the dense cloud of lies told obscured their vision and warped their judgment, and the result is *failure* writ large!

Another episode of a more painful nature was told me by the chief actor himself. It was a clergyman, and on the first Monday after war had been declared, he had gone to the scene of the fight there had been outside Mafeking on the previous Saturday. There he found, he tells me, the dead body of an English sergeant who had been shot, and whose body had not been found by the burying party. He then cut off several of the buttons of the dead man's coat, to keep as souvenirs of the war. All this he related to me with the greatest gusto. He evidently enjoyed the *fun*. I said: "Of course, you and your companions buried the dead man then, with a prayer for the relatives left behind." "N—no," stuttered my friend, "we—you see—we—er—had no time really. We were afraid they might come out of Mafeking and shoot us!"

Aye—time to desecrate a dead body but not to bury it! And the man in charge of the souls of so many people had never thought of it! No wonder this dear man could play the leading part in the following little tragedy—or comedy, if you prefer it—a few days after!

[780] Mafeking will not be relieved until May 17th, 1900.

[781] Ananias was a disciple of Jesus who was sent to Damascus to restore the sight of Saul of Tarsus (who later was called Paul the Apostle), When Ananias laid his hands on him, the "scales" of dead tissue on the surface of his eyes fell off, and he looked up at Ananias. After additional instruction, Saul was baptized. (Acts 9:18; 22:16 Ananias of Damascus - Wikipedia

Shortly after the famous charge on Lobatsi—eighty-four bold warriors to capture some unarmed men—several Zeerust people drove out to see the battlefield! This was the ostensible reason; the real one was to pick up loot, for the Commandant was busy carrying off everything belonging to the traders there. My friend the clergyman also drove out to join in the fun. He got what he wanted and drove back to Zeerust with his loot. Imagine the picture—our clergyman has a jolly, fat, rubicund countenance. I can see him, sitting in his little buggy, driving his horses, trotting through the main street, bowing right and left, calling out, "I've got it! I've got it!" *It*—what was that *it*! It was a *Berkshire pig*[782] in a box tied on to the back of the cart!

Champion Berkshire boar at 2005 Royal Adelaide Show.[402] A Goods Shed at Viljoen's Drift.[403]

Tell it not in Gath![783] On they went in triumphal procession—a shouting parson and a squealing pig! What a pair! What company for a parson to keep! Over the box every one could see the pig peep, squealing like one of his Gadarene [784] relatives, as if possessed. On the seat sat the jolly, self-complacent "Domine,"[785] a hero in his own estimation, a dead soldier's coat-buttons in his pocket, a stolen pig in his cart!

People laughed, but not all; they approved, but not all. Parson and pig—had I but a Cruickshank's pencil to picture the scene, or a Dickens's pen to describe it! And this dear man was the party who, after my capture, wrote to my aged father in the Cape Colony that I was taken because I was loyal to my Queen, an act unworthy of *a true Africander!* Yes, my reader, there are such men, even at the end of the nineteenth century. In prison we were subjected to some unpleasant experiences, too. It often happened that Pretoria youngsters, who had joined the bicycle corps of despatch-riders, passed through Zeerust. Of course, they were anxious to see the prisoners of war, and consequently often favoured us with their company. There were gentlemen among them, but alas! also arrant[786] cads, capable of insulting a prisoner any day. Thus, for instance, some four came up one day "to see the animals feed." They were dressed to kill, and *looked clean*, which, unfortunately, they not always are. But clean clothes do not make a man a gentleman. So it was here. These youngsters were foul-mouthed and vulgar to an extent. They made a degrading spectacle of themselves. They looked into the cells where the prisoners were having their evening meal, jeered at them, mocked them, and swore horribly. They deplored the fact that they were not allowed to shoot down these "dogs of prisoners," for they did not deserve to live.

Then they came to our cell and found us, inmates of No. I, sitting round a little table enjoying our meal. This was too much for these beardless ruffians. "Prisoners eating round a table, sitting on chairs!" Turning to the gaoler, they said, "Such good food for prisoners? The dogs ought to have only mealies and bread, like Kaffirs." My young friend Harry now looked indignant, and said, "Look here, our food is paid for, and that is very likely more than you can say of your fine clothes." (It was a known fact that many Transvaal youngsters proved their patriotism by requisitioning clothes galore for the front, where they remained to guard the rear! Such were these little creatures!) This called forth some more cursing and swearing, till the indignant gaoler asked them if they knew whom they were speaking to. He then told them who I was. I spoke never a word, just looked them up and

[782] The Berkshire is a British breed of pig. It originated in the English county of Berkshire, for which it is named. It is normally black, with some white on the snout, on the lower legs, and on the tip of the tail. Berkshire pig - Wikipedia

[783] Gath, often referred to as Gath of the Philistines, was one of the five Philistine city-states, established in northeastern Philistia. Gath is often mentioned in the Hebrew Bible and its existence is confirmed by Egyptian inscriptions. Gath (city) - Wikipedia

[784] The exorcism of the Gerasene demoniac, frequently known as the Miracle of the (Gadarene) Swine is one of the miracles performed by Jesus. The story shows Jesus exorcising a demon or demons out of a man and into a herd of swine, causing the swine to run down a hill into a lake and drown themselves. Exorcism of the Gerasene demoniac - Wikipedia

[785] A clergyman; especially a settled minister or parson.

[786] Alteration of *errant*. Originally meaning wandering, the term came to be an intensifier due to its use as an epithet, e.g. in the phrases arrant thieves and arrant knaves (i.e., "wandering bandits").

down. They were silenced, perhaps ashamed, and slunk away. And in Pretoria they are big swells. What a miserable outlook for the future of a country when its rising generation is of such *genre!*

Even our simple-minded, and in some instances kind-hearted, warders were shocked at their behaviour. Let us be charitable: these young fools were out without their mother's leave, and being thirsty after a dusty journey had put more fire-water behind their waistcoat buttons than they were licensed to carry. Hence their disgraceful behaviour. How proud their sisters will be of them!

On another occasion, a Sunday morning, a big burgher honoured us with a visit. It was very early still, but yet even then the odour that radiated from him betrayed the fact that he had been very thirsty that morning, and that certainly not because he had abstained all Saturday evening! I was busy washing and dressing when this burly Bacchanalian stalked into my cell, without so much as by your leave even, and stood staring at me. He was the son of a commander, but had not gone forth to war, for which some very funny reasons were given. One was, he is never sober; another, he can't see any more, and will shoot his own people; and a third, uttered in a stage whisper, he is *afraid.* Never mind what was correct or not, but there he stood staring at me through bleared eyes. "Good morning," said I, but no answer. "It is going to be a hot day," said I, but yet no answer. "I am busy dressing," I pursued, but still no answer. "Have you never seen a man wash himself?" I asked. Not a word. You see I began thinking that the man was lost in amazement, seeing another, bigger and stronger than he was, wash himself.

I could see that he and a bath were very distant relatives; in fact months must have elapsed since last they had met. I could see that, but after a while the fact made itself known to another of my five senses too. As my sphinx would not or could not speak I too now preserved silence. After a few minutes the gaoler touched him on the arm and intimated that it was time to go. Off he walked, and when he had proceeded as far as the big door of the gaol, where he knew no prisoner could get at him, the coward broke forth in the most horrible invective against me, cursing me roundly and swearing at me horribly. I could hear it all, and could not help thinking what a parody! Here you have a man blustering about his love of independence, bouncing about his prowess in war, informing the world that his life's blood may flow before he would surrender home and hearth. And yet, he is an abject slave to his passions and vulgar habits and a moral coward. What a degrading sight!

A Government boasting of its belief that God was going to give them the victory, and no one else, and here stood one of the defenders of that faith, maudlin drunk and swearing, allowed admission into prison to jeer at a prisoner and curse, because he was a son of one in authority, and therefore *dare* not be refused. Such is the Government of this Republic! The man's last words when he stalked away, were, "I am going to church now, and will tell all the people that I have seen this cursed Africander!" This was the burden of his revilings. "He is not a true Africander!" Just fancy, a loyal subject of his Queen, who preferred remaining true to his oath of allegiance and not break his word, by doing which he might have gained any amount of popularity and influence among these people—fancy such a man not a *true* Africander, but that drunken sot one!

Away with the name, if that be true! Tis Africanders such as these that imperil our future as a people, that degrade us in the eyes of civilised Europe, that make one despair. Often in my cell I lay thinking of the situation. Often I mourned the mistakes made by our leaders. Verily some of them have given sharp knives into the hands of children, who will now either injure themselves or others. To be Africander is vastly different to being a blustering braggart who knows not and understands not what he is talking about. I could have wept at the degrading spectacle before me. It is the firm conviction of my heart that the influences at present paramount in the Transvaal—that especially of the blustering, irreligious Hollander—are admirably calculated to produce such Africanders!

XIV. Some More of the Same Kind

This Chapter has been omitted. The chapter number has been retained as a placeholder.

XV. All Privileges Curtailed

During the beginning of my imprisonment several privileges were accorded me, but, as mentioned before, they were cut down gradually. The *man* and the *woman*, ably backed by others, among them dear friends of mine, men who had enjoyed my hospitality, and never scrupled to ask my help in former days, rested not till the cell door was closed upon us day and night.

I have referred to the doctor's family and their infinite kindness. There were several other kind people too. The English clergyman and his wife often sent us food and books and periodicals. Several other people, strangers to us but good Samaritans, sent flowers, coffee, potatoes, sometimes whole dinners. "How dare they!" said the great Commander Cronje. How dare anybody feed my prisoners—my dogs! Let them eat prison fare! To-day Mr. Cronje is a prisoner himself, fortunately for him not in the hands of men like himself, but of British officers and gentlemen who know how to treat a fallen foe.

Meeting of Lord Roberts & General Cronje.[404] February 27th, 1900.[787] General Cronje and his officers on St. Helena.[405]

But the Government officials here know not what magnanimity means. And so the good doctor first got a most impertinent letter forbidding him to give me or any prisoner food or any luxuries.

Another gentleman and his kind wife had continually sent us nice things. Poor man, he was dragged off to the Hoofd-Lager to give account to Cronje and Snymann of his evil, treacherous deed. He escaped by the skin of his teeth. A wire was sent to the gaoler, forbidding him to treat me in any way different to the usual criminal prisoner. I read this telegram myself. These poor men were so afraid somebody might show a kinder spirit than they themselves had exhibited. We thereupon took prison fare, but were allowed by the gaoler, surreptitiously, to get up one decent meal from the hotel.

The persecution, however, was not over yet. The order came that no one was to be allowed to enter another cell to have a chat. Then again, all of a sudden, came the cruel order, "Cells to be locked night and day"—and this in the summer, with its stifling heat![788] It was a cruel thing to do.

Three times a day we were allowed out—cell by cell—in that dirty little yard, for a quarter of an hour at a time. The inmates of one cell were not allowed to talk to the inmates of another cell. One would say we were condemned felons or murderers. Then the righteous, patriotic Government was happy. What was worst is this, that I was distinctly informed that this cruel treatment was inflicted to humiliate me and break my spirit. I am thankful to say that they never succeeded in this noble attempt.

All visits, too, were forbidden; I was not even allowed to have the short Sunday service we all enjoyed so much. Outside, in the church, the people prayed, and in the lagers, we were informed, they sang psalms and prayed. But their hearts were meditating cruel deeds. Remember, reader, this does not refer to the genuine burghers. No, *their* hearts were tender. They protested both in the village and in the lager against this cruelty practised upon

[787] 3/2/1900: Cronje and staff, moved to Simonstown under a guard and put on board H.M.S. Doris, and sent to St. Helena.

[788] Summer in the Southern Hemisphere.

us. No, it was the *patriot*—I beg his pardon, the *traitor* to his country, the Government official who did all this. The people considered by them dogs of Englishmen were the good Samaritans—they themselves nobly and heartily played the parts of Priest and the Levite. No comment is necessary. Yea, "the ways of men are narrow, but the gates of heaven are wide." Many was the time then that I softly crooned to myself the well-known "Ora Pro Nobis"[789]:—

"Out of the dark and dreary street,
Out of the cold and driving sleet,
Into the church the folk had gone,
Leaving the orphan child alone.
Tatter'd and so forlorn was she,
They cross'd themselves, as they pass'd,
To see so frail a child In that grievous plight
On such a relentless and stormy night!"

Our only relaxation now was a swim in the river every Saturday. It was real enjoyment. We were marched down, the stationmaster and myself leading the van, and guarded by fourteen armed men. There were then nineteen prisoners. The village people gaped and stared, but made no remarks. How I enjoyed my swim! Ever since my youth swimming has been a delight to me. And now it seemed to me as if I could drown my sorrow in that glorious swim. I must say I wished that our guards could have had, if not a swim, then a wash too. They sorely needed it. Every one of us longed for those Saturdays, and was sorry when this one solitary hour of pleasure was over, for then came the terrible "cells" again, so hot, so dreary. The stationmaster, who had been ailing for a long time, could not stand this confinement and now fell ill.

He stated his case to the authorities and asked for parole. To his own and our surprise, it was granted him, and he was taken to the hotel. I then wrote to Cronje, in the name of all the others, to ask that our cell doors might be left open, as the heat was intolerable. Our petition was curtly refused. The doctor and several kind people in the village interceded for us, but all in vain.

The enemy was glad to hear that we suffered so dreadfully. Some six months before I had been at death's door with a sharp attack of pleurisy. My lungs needed all the fresh air they could possibly get, and this was refused me. I felt my strength going daily, and in addition the dreaded insomnia invaded my narrow cell. Oh, those nights—stiflingly hot, restless, and miserable! In the distance we could hear, every morning, the cannon bombarding Mafeking,[790] and then, during the afternoon, the sensational reports began coming in, all retailed to us in the evening by gaoler and warders. No wonder that health and nerves failed! To crown our misery another tragedy must needs be enacted. One frightfully warm afternoon I was lying down to get a little rest when in a passage just outside my cell I heard preparations go on of which we knew not the meaning. Soon enough, however, we did. It was a poor Kaffir, condemned to twenty-five lashes, who was being tied up to a ladder with bared back.

Then the blows descended. Oh, that horrible thud, and the yell and scream that followed the first five or six lashes! Then the poor victim just moaned like an expiring animal. And all the while the lash rushed through the air with its horrible swish, whirled by an arm that did its work with gusto. It was horrible! I can't think of it, or write of it, without again experiencing the horrible sensations of that day. I had never witnessed anything of the kind in my life before. I could not bear to hurt another myself. And here, a captive, I had to be a silent listener. To crown all, a day later two more Kaffirs were lashed on the very same spot, each receiving twenty-five lashes. It was terrible.

What happened to me I was told of afterwards, for long before it was over I was unconscious. Imagine poor Harry's sensations! He and I were alone in the cell then. There I lay unconscious, the door was locked, and outside the frightful shrieks and yells of the poor cruelly lashed Kaffirs filled the air. When it was over Harry bombarded the cell door till the guards came up. They opened the door and carried me out into the gaoler's office. Fortunately the doctor was nearby. He did for me what he could, with a heart full of tender friendship and care. In about half an hour's time I revived a little, and recognised all again.

[789] Pray for us.

[790] Thirty miles, 50km to the Southwest.

My head, however, was buzzing round and in frightful pain; I knew of nothing, and could not talk coherently. The doctor feared that my brain had given way and I was mad. Thank God! that did not come over me, though it would have delighted the hearts of mine enemies, I believe. Towards evening I was carried to my bed, and the doctor ordered the cell to be left open for the night. The mischief was done, however, and during the next two days the fits returned continually, whilst I, when conscious, felt as if I was really going mad. One afternoon, they tell me, I started up, calling out to my little boy, my Willie, whom, I said, I saw standing in the door stretching out his arms to me. I started up to go to him. Harry called for help, and he and some four or five guards held me down. I fought them, I struggled fiercely.

I begged them in piteous accents to bring my darling boy to me, or let me go to him. So piteous was the scene, I was afterwards told, that Harry and the guards themselves were all in tears. Again I was carried out to the cooler and airier office. The good doctor came in haste, and what love and gentleness could do he and Harry did. I became conscious for a little, and then was mad again, talking to my little boy, trying to break away from them to go to him. At last the doctor managed to get me to sleep with morphia.

In the meantime he had wired to the lager—to Cronje—and to the government in Pretoria, apprising them of my condition, and telling them that he feared the worst if I were not removed from that cell. And can you believe it, gentle reader, the *authorities absolutely refused*. The next day I was no better and getting very weak. And then the brave doctor, as district-surgeon, showed his teeth. He simply *ordered* my removal to the hotel, where a generous hearted host and his wife were only too willing to play the part of good Samaritans. The doctor was ably assisted by an official I have before referred to, the Public Prosecutor—a gentleman who, I'm sure, would all along have seen to my being properly treated if only he had *dared*. He countersigned the doctor's order and had me removed. I thank this official for his courtesy shown me that day and on other occasions.

Poor Harry was now left alone in our cell, but a considerate gaoler put two other prisoners, friends of ours—Messrs. Howard and Fischer—with him to keep him company. I was now installed in a cool, comfortable room in the hotel, but so weak that I had to keep my bed for several days. The noble-hearted commanders, frightened by the doctor's reports and threats, now perforce acquiesced in the action taken by the Public Prosecutor, and I was left in peace.

A dastardly thing was done, however. An *official* telegram was published to say that I was somewhat indisposed, "and had been taken to the *hospital*." There is no hospital in Zeerust, and my hotel bill I had to pay myself, the kind owners making it as low as possible. The Government officials deliberately told a falsehood and then took credit for a charitable action never done by them! The next chapter will tell about my "hospital" experiences. As I said, I was laid up, but still guarded—one guard sitting outside in the passage at the door, and another outside before the only window of the room. In fact, the doctor had great difficulty to keep them from sitting down day and night *inside* my room.

XVI. In Hospital

I could breathe in my new quarters, and see kind faces again. I was not allowed to see visitors, but the people of the hotel and the doctor were with me continually. The nurse and Miss R. were allowed to come in for a minute just to say goodbye, for the day after they were taken to Pretoria by coach, whence they were sent over the border, *via* Delagoa Bay.

Great kindness was shown me, and soon I began to be myself again, the attacks and frightful headaches leaving me altogether after a few days. My window looked out on a beautiful garden. How I enjoyed the green trees and flowers, and the general air of peace, amid all the turmoil of spirit within and the fierce roar of war without! Even in my bedroom I could hear the boom of distant cannon. After a while I could read a little every day, and then sit up in my room, and finally go into the garden for a little change, as I was not yet very strong.

Every day the good doctor paid me a long visit—how I enjoyed these visits, how I appreciated them! I had not been long in prison when information came from Pretoria that all the prisoners were to be sent thither. One half was sent off first, and the other half had to follow. I too was informed that I would be taken there as soon as I felt strong enough. The thick-skinned Government was evidently beginning to feel a bit nervous regarding the opinion of its people and others outside the State about the barbarous treatment accorded us. Hence this change. A little paragraph was inserted in the two papers to say that I was simply a prisoner of war, a political prisoner, giving people to understand that I was treated as such—another falsehood, perpetrated just to allay the fears of relations and friends in the Transvaal and everywhere else. Yes, it is but too true "man's inhumanity to man makes countless thousands mourn."

My guards were an object lesson. There they sat before the window and the door, staring at me, afraid that if they removed their eyes for a moment I might vanish into thin air. One pair of guards never ceased talking, either to me or to one another; another pair never talked a word—they deemed it part of their duty to preserve the silence of the grave.

During the night they slept by turns, but usually made such noises, both asleep and awake, that it took me several nights to get accustomed to the groaning and the grunting and the snoring, and then again the jabbering. And all along—the farce of it!—each one was armed with a rifle and a revolver and fifty cartridges. I never knew before that I was such a dangerous animal. I ought to be proud of the estimation I was held in and the value put upon me. Really, it was too funny! I verily thought that I must be the "Prisoner of Zenda," or "Rupert of Hintzau,"[791] whom they feared so terribly.

In a fortnight's time I was able to travel, and so decided to go with our second batch of prisoners. The doctor was anxious for me to go. He hoped that I would be quartered with the officers at Pretoria, but knew not what might happen if I stayed at Zeerust—very likely solitary confinement would be my share. And so it was decided that I should go. When all this had been arranged I received a visit from the Landdrost, greatly to my surprise. I had known him before and he me, but he had pretended not to know me. Now that I was going away, however, he came to see me and condole with me, and assure me that, on arrival at Pretoria, I would either be released or quartered with the British officers. Well, it was kindly meant, and I freely forgive him any indignity put upon me before. He was the man who soundly rated the gaoler for allowing me to take my mattresses into my cell. Like other prisoners, said he, I ought to have been made to sleep on the ground. But I forget all that now. He meant to be kind when I was ill. And on the day I left he again came to say goodbye very heartily.

On the day of my departure a doctor, whom I had known years before, and who had just come in from the lager, came to see me, very kindly, and with him came the doughty stay-at-home Acting-Commander, who wanted to shoot us the day their messenger was so right royally treated by Colonel Baden-Powell in Mafeking. He was so polite and affable—one would say he had all along treated me as one gentleman treats another. My friend the doctor would never have dreamt that this was the man who had grudged me a little food sent by friends, and some fresh air. And yet so it was. He was a man who had to prove his patriotism by his ruffianism, and—whisper it gently!—he was the Landdrost-elect of Mafeking, the Mafeking they had still to take, the Mafeking they were taking every Monday but have not yet taken and never will, though its brave defenders number only some 600 trained soldiers to their 6,000!

My great sorrow was that I could not say goodbye to the doctor's home circle before my departure. I felt that it might compromise them if they came to see me. After my departure obloquy would be heaped upon them for going to say goodbye to a prisoner. Nor could I go to them—it would cause the same trouble; so the doctor and I concluded it would be best not to meet. And I saw them no more, only a glimpse as the coach drove past. The doctor saw us off, but came to my room for a last long talk early in the morning. Tis no shame to say that our last hand-grip brought a mist to the eyes of both of us. Knowing how I was situated, and had been robbed, the dear friend provided me even with money. "You may need it," said he. Aye, and I did, for did not the Philistines in Pretoria despoil us? But this is another story. When we got into the coach several friends and loyalists pressed near, surreptitiously to shake hands and say, "God bless you." And then we were off.

[791] *The Prisoner of Zenda* is an 1894 adventure novel by Anthony Hope, in which the King of Ruritania is drugged on the eve of his coronation and thus is unable to attend the ceremony. Political forces within the realm are such that, in order for the king to retain the crown, his coronation must proceed. Fortuitously, an English gentleman on holiday in Ruritania who resembles the monarch is persuaded to act as his political decoy in an effort to save the unstable political situation of the interregnum. A sequel, Rupert of Hentzau, was published in 1898 and is included in some editions of The Prisoner of Zenda. The Prisoner of Zenda - Wikipedia

XVII. The Journey to Pretoria

I was imprisoned at Zeerust on Monday, October 16th, and left for Pretoria on Tuesday, November 26th. The coach started at twelve o'clock. Quite a crowd saw us off, and there were a few very hearty goodbyes. The coach was crowded, and as I was very shaky still, the journey thoroughly wore me out. We had to sit three on a seat—twelve inside and ten outside. Among the passengers were two Transvaal artillerymen, young Colonials, whom I had known years ago. One was an old school-comrade, whom it was a pleasure to meet again. He and I were on the back seat, and also a hospital nurse, a Transvaal lady, whose relatives I knew in the Colony. This good nurse looked after me with great kindness. I will not forget that. I thanked her then; I do so now again. We travelled through the whole night, arriving at Krugersdorp on the Wednesday morning at nine o'clock, and during all that time she tried to make me comfortable with pillows and rugs.

A Mail Cart Leaving Johannesburg Post Office.[406]

Whenever there was a relay of horses she managed to get me a little milk or soup. She was really a good Samaritan, and naturally I was very grateful. I was very weak and in great pain, and not very happy. God bless the kind nurse and other kind friends who helped me during that wearying journey. Arrived at Krugersdorp, the Landdrost there, too, an old friend from the Colony, showed me and Harry much kindness.

Gathering at the Paardekraal Monument near Krugersdorp.[407 792]

At two o'clock we left by train for Pretoria. At the Krugersdorp station a large crowd was assembled. Among the people I recognised many well-known faces. Many came up to shake hands and say a kind word—but many also turned their backs upon me. At about 7.30 in the evening we arrived at Pretoria, and were all taken to the common gaol, and locked up in the room where the Reformers,[793] after the Jameson Raid, had been locked up. I was kindly treated by most of the officials there, as well as at the station.

[792] The Paardekraal Monument is situated in Krugersdorp, South Africa. The monument commemorates the original site of vow made by Transvaal Boers on 13 December 1880, prior to the armed rebellion known as First Boer War, when they vowed to regain their independence from the British Empire. Paardekraal Monument - Wikipedia

[793] The Reform Committee was an organisation of prominent Johannesburg citizens which existed late 1895 / early 1896. They demanded a stable constitution, a fair franchise law, an independent judiciary, a better educational system, etc. Some members of the Committee had been instrumental in the preparation for the Jameson Raid. For conspiring with Jameson, the members of the Reform Committee were charged, confined in deplorable conditions, and finally found guilty of high treason against the Transvaal. Many received severe sentences, including death; however, the majority of the membership escaped with high fines in January 1896. Johannesburg Reform Committee - Wikipedia

A crowd had assembled to see us arrive, and one could not help overhearing many remarks. A woman said when some one had pointed me out to her, "And what will they do with him? I am so sorry for him." A big bully standing close by said thereupon, "Give him to me and I will soon show you what to do with him!" "What?" said several voices eagerly. "I will hang him on the nearest tree till he is dead," said the gallant champion of Liberty! He fights for liberty—his own—so he says, but no other man may be free or dare think for himself! This is their idea of liberty. Poor deluded people. I can assure my reader I was glad of a bed that night, and had a fairly good rest. In the *Diggers' News* of that day I found three little bits of news, which I give below. They show the exact character of the paper—dislike of truth and narrow-minded hatred.

First of all there is a telegram from Zeerust: "A second batch of prisoners, amongst whom is the Rev. Adrian Hofmeyr, left by to-day's coach for Pretoria."

Then comes this telegram from Pretoria: "Nineteen civilians detained (that is very good) by the burghers on the western border, and subsequently sent to Pretoria, were released this afternoon, and were supplied with passports for Delagoa Bay (without food, however). The Rev. Adrian Hofmeyr is seriously ill, and is suffering from melancholia. He will be quartered in the State Model School." The funny part is that I have never in my life suffered from melancholia. I do not know what it means.

Now comes the editor's own bit of news—first the falsehood, then the spite. Here is the tit-bit:—

A Minister in Prison

Adderjan [Adrian] Hofmeyr, formerly clergyman, but now sleeping partner of the firm Rhodes, Chamberlain & Co., has arrived at Pretoria. "His ex-Reverence is accommodated at the expense of the State in the Grand Hotel of 'Oom Koos' (the gaoler), on the corner of Koch and Skinner Streets, where all the cells are provided with iron bars, whilst a gallows to accommodate visitors, is kept in readiness."

How charmingly delicate! what an example of ultra-refinement in literature! what a proof of the broad, yawning chasm there lay between the writer and barbarism! Poor man! and he is a renegade Scotchman, I hear, on trial for patriotism—Transvaal Government double-barreled patriotism. How hard it is for him to hide the stuff he is made of! I do feel sorry for him. My only comfort is he is in congenial society, very congenial—German Jews and Transvaal liars!

On the morning after my arrival Harry and I were placed in a little room, which I thought was going to be my home for the future. Everything we had had been taken from us—knives, scissors, note-books, even our money. When I asked the acting-gaoler—a very kind young man—why that too was taken away, his immediate answer was, "Ah, sir, you see if we leave money on prisoners they might bribe the gaoler or the warders!" You see, reader, bribery and corruption are not unknown here—oh no. The Reformers who were imprisoned here in '96 could a tale unfold.

A real little bit of happiness and joy fell to my share at about twelve o'clock, when three Government officials, appointed to look after the British officers, visited us to liberate Harry and the other prisoners who were railway men. I was indeed happy and glad that after *seven* weeks of injustice and cruelty the Government had at last awakened to a sense of decency. We then heard that the ladies and the stationmaster and the other railway men, among them our friend Lombardini, had already been sent to Delagoa Bay. I was so glad for Harry's sake that I forgot all my own misery. The glorious constitution of this Free Republic, having the *shadow* of liberty, without any of the *reality*, had made it possible for its minions to capture one man *before* war was declared, and then three inoffensive women shortly after, one a hospital nurse too, *and keep them prisoners for seven weeks*. Nurse Ruth had a good sum of money on her. This was taken—I beg pardon, *stolen* from her at Zeerust. I wonder whether it was ever restored to her.

The liberated prisoners were sent to Delagoa, and from there they had to manage as best they could. I had to borrow £20 to pay Harry's expenses to the Cape. And then the dear President still tells his people that they are a chosen people! Comment is not needed! I was now informed that the Government had decided to quarter me in the Model School, where the British officers had been accommodated. A meeting of the Executive Council had been called to consider the question, I was told, whether I should languish in gaol, or be sent to the Model School.

Yes, the whole machinery of the State was stopped, even the war was forgotten, while this momentous question was being discussed! Up rose one of the fathers and proposed that I should be sent to the racecourse, where the privates were imprisoned. "Don't we hate this man and, though we are perpetrating an injustice, showing all the world that we are unfit to govern and have no idea of the true meaning of the word liberty, should we not, having treacherously captured him, treat him as cruelly as possible?"

"No," says another Civis Romanus, a noble patrician—"no," says he; "the racecourse is too good for him. True, we have acted disgracefully, but we are accustomed to such action; we have disgraced ourselves by committing an injustice, but this is not the first time—we speak a great deal about right and liberty, but we never allow it or recognise it in another—therefore, I say, let this man lie in prison. Has he not dared—aye, dared—to differ from us, and, more horrible still, to say so in public yonder in Diamondopolis?[794] To prison with him!" (Sensation.)

The vote was about to be taken, when slowly the President rose to speak. He had a Bible under his arm. "*Patres conscripti*,[795] fellow-slaves of a Hollander-ridden and Leyds-guided Government, monopoly-granters and spider-receptivants, listen! This Book says, 'Love your enemies.' We have disgraced ourselves by taking him and his companions prisoners; let us rectify our error, after seven weeks of *unavoidable* delay." (He coughed slightly over the word, but a firmer grip of the Book restored confidence.) Let us now show our love to our enemy, not by doing justice to him—it is against our principles so to do; we can't, even if we tried to—no, but by quartering him with the officers and not sending him to the racecourse, or to gaol, where he has already been for seven weeks.

Listen to my inspired words of wisdom. And now let us rejoice that we are not like other men. *I* do not *say* we are better than they, I leave that for you and my trusty printers of the *Volksstem* and *Diggers' News* to say!" And then they sent three officials to take me to the officers' quarters, whilst they themselves went home, having strained so mightily the quality of mercy that they were quite exhausted!

I demanded from these three gentlemen the reason of my capture and detention. Their unblushing answer was, *"Your Kimberley speeches!"* O land of liberty and free speech, thy name is not Transvaal!

And so I entered upon a new phase in the story of my captivity.

XVIII. The Staats Model School

This building is a very fine one, consisting of sixteen large rooms, twelve of them of one size, about 30 feet by 40 feet, and four of a size twice as large. A long corridor, some 9 feet wide, runs from end to end along the length of the building, and a shorter one, some 12 feet wide, at right angles to it, through the width of the building. One large room is excellently fitted up as a gymnasium. The building stands in a plot of ground of about the size of an acre. Outside this we were not allowed to go. We were fenced in, on two sides (on the street) by iron railings, and the other two by iron fencing, and were guarded by nine armed policemen—Mauser and revolver. A change of guard takes place every four hours. Just next door is a Red Cross Hospital, in which building our gaoler and his assistant have their guard-room.

State Model School.[408]

Model School (back view).[409]

Our gaoler is a Mr. [R.W.L] Opperman, one of the Transvaal burghers. In time of peace he is a criminal magistrate (Landdrost). Now he goes by the title of Commandant van de Wacht (Commander of the guard, of the prisoners). He is a short, thick-set man, generally very curt and uncourteous, and a terrible hater of the English. He shows his lack of intelligence and commonsense by believing that if the country is conquered by us

[794] Kimberly, where the diamond mines were mostly located.

[795] "Roman senators," literally "enlisted fathers."

he will be shot and his wife degraded to the rank of a servant. He consequently talks of shooting himself and her first before it comes to that. He has never bestirred himself to show the prisoners any legitimate kindness. He always tried to look down upon men infinitely his social and intellectual superiors, and naturally always failed. Nobody ever took any notice of him. His assistant was a Dr. Gunning,[796] a Hollander, who had lived in the Free State for many years, and a thorough gentleman. He is the curator of the Staats Museum. He ought to have been our head gaoler; I am sure things would then have run much more smoothly. The police who guarded us were as brutal a lot of men, with very few exceptions, as one could find in a day's march. They deserve a separate chapter, and will get it.

This will give the reader some idea of our prison and its surroundings. Our walk was a tramp round and round the building. In one-half of the backyard rounders were played, in the other half the tents of the police and our servants (privates from the different regiments) were pitched. Later on, in the month of March, these police tents were taken out and placed in the street, to our great relief. The close proximity of these men was anything but pleasant or edifying. The building itself was rather badly ventilated for a school, but fortunately the rooms were very lofty, about 16 feet, and thus by leaving doors and windows open night and day, we really did not feel the summer heat severely at all.

When I arrived I was prisoner No. 54, and was put into one of the rooms all by myself till the middle of December, when the Stormberg disaster[797] took place, supplying me with fellow-prisoners as companions in my room, to my great joy and comfort. I had often felt very lonely in that big schoolroom. A considerate enemy, gloating over our capture, provided us daily with food about enough for half a day—a half-pound of meat and one pound of bread, and in addition we were supplied with tea or coffee and potatoes and salt. Everything else had to be bought, such as milk at 6d. per bottle, and sugar at 1s. 6d. to 2s. 3d. per pound.

An officers' mess was consequently formed. To have anything like a decent and healthy table our expenses amounted to from 2s. 3d. to 3s. per day. On arrival every prisoner was provided with a free issue of clothes, bedding (a mattress and two rugs) and a towel. Our gaoler, however, never hurried over this; the comfort of an officer who arrived here after a long journey, tired and hungry, was a matter of no consideration. Officers have arrived who had to wait as long as *two* weeks before *anything* was given them, not even bedding. *Seven* weeks passed before I received my free issue of clothes. Our captors took a delight in inflicting these little annoyances on us. It did their little souls good. Everything we wanted had to be bought through a certain appointed party. He, of course, had to distribute "palm-oil" very freely. Ugly whispers were heard as to the parties who stretched forth hands to be oiled. But you see, there was no disgrace in pilfering defenceless prisoners. And as we bought monthly for as much as £800, the clean-handed officials, you may be sure, made a tidy penny by us. One room was reserved as a store, and two of the officers kindly acted as storekeepers, thus sacrificing a great deal of their time. It was indeed a generous action on their part.[798]

The building, facing west, was provided with a large verandah, both in front and at the back. These places naturally were the daily resort for readers and writers and chess-players, &c. We were provided with two bath-rooms only, and when our populations reached the number of 120, plus the twenty-three servants,[799] these baths were kept fairly busy.

Two pumps in the yard supplied us with fresh water, and washing-day was quite a feature—colonels and majors and captains, in all sorts of costumes, with buckets and tubs and soap, washing away for dear life! Some of our guards, showing a lamentable ignorance of the advantages of clean water and cleanliness, considered all this a terrible waste of soap and energy. Why could we not go about in one suit of clothing for three or four weeks? One does not need a change oftener! This was their creed. The authorities evidently had a hankering after the same opinion, hence their unwillingness to supply more bath-rooms. By crowding the bedrooms somewhat we

[796] Dr. J. W. B. Gunning is one of the administrators of the State Model School. Dr. J. W. B. Gunning, per Brian Davis, Churchills in Africa p. 201 See also: The Strand Magazine, No. 112, April for an article on the Staats Museum of which he is the director.

[797] The Battle of Stormberg (December 10th, 1899) was the first British defeat of Black Week, in which three successive British forces were defeated by Boer irregulars in the Second Boer War. 696 British soldiers were captured. Battle of Stormberg - Wikipedia

[798] Hofmeyr note: The monthly expenditure later on amounted to £1,600.

[799] "Batmen" A batman or an orderly is a soldier or airman assigned to a commissioned officer as a personal servant. Before the advent of motorized transport, an officer's batman was also in charge of the officer's "bat-horse" that carried the pack saddle with his officer's kit during a campaign. Batman (military) - Wikipedia

were able to reserve one large room as a fives-court.[800] This game, with rounders[801] and the gymnasium, provided athletic exercise for all who desired it. The fives-court was the scene of many an interesting tournament between representatives of the different regiments and public schools. We managed to get up a great deal of excitement over these games, making the days pass fairly pleasantly. On my arrival we were allowed the use of the school harmonium—a very good instrument. I can play and sing a little, and several of the officers too sing. And thus many a pleasant evening was spent. My relatives and friends sent me lots of music, and many were "the songs that reached our hearts."

Alas! this was happiness too great to last. On a certain day, quite unexpectedly, our harmonium was taken away. The people in the town, we were told, the bigoted ones, objected to hearing us sing and having a little bit of sunshine in our lives. Some cantankerous individuals had even spread the report that I had played "God save the Queen" and "Rule Britannia," and that too on Christmas Day. A friend, who had come from the Ladysmith lager, told me that the report had been spread even there, and that I had been roundly cursed and sworn at.

Can one imagine anything smaller, more childish than this?—and, worse still, that Government officials should lend an ear to such nonsense? No wonder they are fighting for liberty; the pity only is that they have mistaken the enemy—they are fighting the wrong party. On Christmas Day we had a solemn service, and in the evening sacred music. Several songs were sung. Last of all, at the request of the hospital people next door, I sang Lord Henry Somerset's "Song of Sleep," and then we all retired. Somewhere in the vicinity of the school, in the house of English people very likely, some one—a child perhaps—had played "Rule Britannia." A truth-loving, patriotic passer-by then immediately rushed off to the President and told him that I was playing patriotic airs, when, as a prisoner I ought to sit down in sackcloth and ashes. *Hinc Illae Lachrymae!*[802] The Duke of Plazatoro[803] could not have acted more foolishly and with less dignity. Verily there are little-minded people in this world of ours, with souls so small and mean that only microscopic inspection will reveal them. Ever since my arrival here I have had the pleasure of conducting a service on Sundays. The officers honoured me by asking me to give them a short address every Sunday. It has been an honour and a privilege to me to do so. Many solemn but also pleasant hours were thus spent. Let me tell about them in another chapter. So much for the present of life in the Staats Model School.

XIX. Sunday

How shall I describe our Sundays and our pleasant services? There are memories, sad and sweet, connected with those short hours which will never be effaced, I am sure. Till my arrival a clergyman of the English Church used to come of a Sunday to conduct a service, and after my arrival he still came for a few Sundays, but on Churchill's escape (of which we will hear later on) he wrote the officers a very curt and curious letter to say that in consequence of that escape he could no longer conduct a service, as he had to guard the honour due to his position.

J. H. Godfrey [804]
To the British Officers, being Prisoners of War.
From Rev. Mr. Godfrey
St. Albans, Pretoria, Z.A.R.
Dec. 15, 1899.

Gentlemen - By the kind courtesy of the Government I have been permitted to hold services for you in connection with the Church of England, which services I have felt it a privilege on my part to conduct. After what has recently occurred—viz., the escape of Mr. Churchill from confinement—I exceedingly regret that, in

[800] Fives is an English sport believed to derive from the same origins as many racquet sports. In fives, a ball is propelled against the walls of a 3- or 4-sided special court, using a gloved or bare hand as though it were a racquet, similar to hand-pelota. Fives - Wikipedia

[801] Rounders is a bat-and-ball game played between two teams. Rounders is a game that involves hitting a small, hard, leather-cased ball with a rounded-end wooden bat. The players score by running around the four bases on the field. Rounders - Wikipedia

[802] Hence those tears: that is what those tears were for.

[803] "The Duke of Plaza-Toro," Poem by William Schwenck Gilbert. W. S. Gilbert - Wikipedia

[804] Per Haldane: "Up to the time of the incident I have just recounted, the Reverend Godfrey a Church of England clergyman, had held a service in the schoolroom on Sundays; but after Churchill's escape he deemed it wiser to forsake us."

consideration of my duty to the Government, I must discontinue such regular ministrations, as I desire to maintain the honour due to my position. Of course I shall always be glad to minister to you in any emergency with the special permission of the authorities, who will, with their usual kindness, duly inform me.—With my best wishes, I am, Gentlemen, yours sincerely,

J. H. Godfrey.[805]

I will not criticise the reverend gentleman's motives nor his letter. I can understand neither. So let it pass. The officers then desired me to continue the services, which I gladly consented to do. Our hour of worship was eleven o'clock, and till Christmas Day, as we have seen, we had the use of an American organ. I acted as organist too. We turned our large mess-room into a temporary church. A schoolmaster's desk did duty as pulpit. We spread a table-cloth over it and put a flower, growing in a pot, on it, and I am sure no domed cathedral could have solemnised hearts and minds more than our homely arrangement did. The singing was hearty, and my short addresses had no other merit than that they were simple, straight talks from the heart to the heart, words of good cheer and comfort. At the request of several of the friends I am preparing these short addresses for the press, just as a memento of hours pleasantly spent here.

After Christmas Day the organ was peremptorily taken away, and so I led the singing every Sunday, and we enjoyed it, and praised God heartily, to the disgust of some of the officials, I am afraid. Why? Let me tell you. After my first service, conducted November 26, 1899, the question was debated amongst the authorities whether I should be allowed to continue such ministrations. A relative of mine was made use of to hint to me that I should discontinue this Christian work, as otherwise I would imperil my chances of release. When I desisted not I was informed by an official that the Government had decided to forbid me to preach, nor would it allow the English Church clergyman to come either.

The day after I was again informed that I might conduct a service, but if I prayed for her Majesty the Queen I had also to pray for his Honour the President. I said to my informant that I had had no idea that the President was at all desirous of my prayers, for I knew he hated me, but that *my* religion taught me to pray even for those "who despitefully use you," and that therefore I was quite prepared to grant his request. Possibly the letter written by the English Church clergyman was the outcome of a similar talk had with him. I can't say. About a week after the same official again expostulated with me for taking the service. Amongst other things he had the bad taste and disgusting animosity to say, "You ought rather to pray that every d .... d Englishman might be sent to h … l at once, than to preach the gospel to them." In my wrath I made answer, "You might perhaps find a clergyman in the Transvaal capable of entertaining so *cowardly* an idea, but *nowhere else*." And then I turned my back on him and walked off. And these are the people who speak of liberty. Alas! they have not the slightest idea of the fundamental meaning of the word. Does not its most glorious application and intensest signification arise from a clear understanding of our grand Christian maxim, Do unto others what you wish that they should do unto you? I am sore afraid that this grand truth is a neglected and forgotten one in some Transvaal circles.

I continued my services, now knowing that a bad mark had been placed behind my name. An Africander, how dare he preach to accursed Englishmen! Did this feeling die away? By no means. As late as the month of March an official told me that great disgust was expressed outside, in the town, because I had not refused to continue these services. One remarked, "It is clear that his imprisonment has not yet made a Transvaaler of him—his spirit is still unbroken!" Aye, I was imprisoned because they wanted to cow and frighten me; and to be Transvaaler means to *hate* your enemies and do them evil. I am afraid, if this is true, that the last thing I would like to be is Transvaaler.

Thank God the *genuine*, real Transvaaler is not of this calibre. He is a brave man and therefore generous. It is the town-bred burgher, the younger upstart class, the man who has imbibed infidel ideas and learned drinking and swearing habits from the worst Hollander importations, who speaks thus, poor fellow!

[805] The letter as reported by Rosslyn. And quoting from his work: "Needless to say, no reply was sent to this epistle. This gentleman, I was also told, did not pray for the Queen! On the 15th February a letter signed by the committee of officers to the State Secretary, Pretoria (owing to the death in jail of Lieut. Tarbutt from enteric fever), "requested the Government to reconsider the position of the prisoners of war who were former residents in the Transvaal, and who were then in prison, and to treat them in a similar way as the British officers were treated, seeing that they were not subjects of the Transvaal, but of the Queen." From: Twice Captured: A Record of Adventure During the Boer War. Earl of Rosslyn, James St Clair-Erskine, 5th Earl of Rosslyn. Wm. Blackwood and Sons, 1900. Page 320 See: Twice Captured - Google Books And: James St Clair-Erskine, 5th Earl of Rosslyn - Wikipedia.

After our organ had been taken away we petitioned again and again for an instrument, even though it were allowed on Sundays only. We offered to hire one ourselves, if the school harmonium[806] were begrudged us. I too wrote again and again from New Year's Day onward. On March the 8th I received an answer for the *first* time, saying that we would be allowed to hire one, and use it *for sacred music only*. How glad we were of this boon at last granted unto us, as a man throws a dog a bone. But even after that it took us days to get in the harmonium, which we had to buy, as none could be had on hire. And thus on the next Sunday we had an instrument again, and how we enjoyed our singing every lover of music will understand. How heartily the "Church's One Foundation" pealed forth, how tenderly our "Hymn for Absent Friends,"

Hymn for Absent Friends.

Holy Father, in thy mercy
Hear our anxious prayer;
Keep our loved ones, now far absent,
'Neath Thy care.
Jesus, Saviour, let Thy presence
Be their light and guide.
Keep, O keep them in their weakness
At Thy side.
When in sorrow, when in danger,
When in loneliness,
In Thy love look down and comfort
Their distress.
May the joy of Thy salvation
Be their strength and stay.
May they love and may they praise
Day by day.
Holy Spirit, let Thy teaching
Sanctify their life;
Send Thy grace, that they may conquer
In the strife.
Father, Son, and Holy Spirit!
God; the One in Three,
Bless them, guide them, save them, keep them,
Near to Thee."

Aye, our hearts were mightily stirred. Yes, I am certain of it—we may forget many days and many incidents of our prison life; but few, if any, will ever forget our happy, blessed Sunday services. How they cheered our hearts, giving us new courage and faith; how they helped to still the tumult of impatience and anger within; how they softened our hearts, especially when the prayer "for the absent, for those near and dear unto us" solemnly ascended on high. God did indeed bless us richly.

Several of the officers belonged to the Roman Catholic Church, but no priest was allowed to conduct a service for them. The clergymen of my own Church, the Dutch Reformed, I blush to relate, simply ignored the fact that any of the prisoners might perchance have a soul worth caring for; not even our sick and wounded. More about this, however, later on.

From the beginning of February we have been singing every Sunday the following [previous] hymn. The wife of one of the officers here, with kindly forethought, sent her husband a copy of it, as also of the prayer daily used in Westminster Abbey. We made copies of that hymn, and used it, as well as the prayer, every Sunday.

806 The pump organ is a type of free-reed organ that generates sound as air flows past a vibrating piece of thin metal in a frame. The piece of metal is called a reed. Specific types of pump organ include the reed organ, harmonium, and melodeon. The idea for the free reed was imported from China through Russia after 1750. Pump organ - Wikipedia

They always stirred our hearts mightily. Our copies of Hymns Ancient and Modern[807] did not contain this hymn. The services I was thus privileged to conduct will always remain to me the bright spots, the happy hours of my prison life. I can never forget them, nor the kindly words of appreciation so often spoken by the officers, my fellow-prisoners.

Note.—Let me add here that on Sunday morning, March 11th, before service, the following note was put in my hands by Colonel Hunt:—[808]

"Dear Mr. Hofmeyr,—Your fellow-prisoners of war beg that you will accept the recently purchased harmonium as a small acknowledgment of the help you have been to them both by holding the Sunday services and by your ever cheerful sympathy. They hope that, though the time for its use here may be short, you may be able to find some useful purpose for it in connection with your future work.

"Yours very sincerely,

"H. Hunt, Lt.-Cl.

"For the officers at the Staats Model School."

I need not say how surprised and overwhelmed I was by this generous gift and great kindness of the friends here. I could not but most sincerely thank them, and wish them in turn God's richest blessing. It was then we sang so heartily and feelingly, "The Church's One Foundation is Jesus Christ Her Lord."

XX. An Escape

Looking through my little memorandum book I find entries like the following, which will give an idea of what our life here was like up to the evening of "the great escape." I read this, for instance: "Since Saturday, 27th Nov., newspapers are allowed us. These say that I am going to be released soon. Is it true? I doubt it very much." "Several of my relations call and send me fruit and flowers." "My health is improving, though my head still troubles me very much." "American Consul[809] called, but says he never received my letter of appeal addressed to him as protector of British subjects. The authorities here are quite capable of suppressing a letter."

"Music is my great solace. What happy, happy hours we pass, singing the old songs over again! How nice our Sunday services are too!" "Had fine rains of late. Heat has practically passed away. Hope it won't return again." And then I come to this extract:

"Wednesday, Dec. 13 [1899].—Great excitement. Churchill escaped last night." Yes, it was a great to-do; it stopped the whole machinery of state. It paralyzed the officials. It seemed to me that even the war was forgotten. Let me tell the story. On their arrival here the officers had asked for parole. This was refused them. It was reported at the time that this refusal was mainly due to the representations of one Consul, representing a European Power, who informed this Government that "a British officer has no honour," and that therefore parole should be refused. Naturally such a statement was eagerly seized upon, and action was accordingly taken. I cannot reveal the name of the Consul, for I am not sure of the fact. Some officials here think nothing of telling a falsehood. I can therefore only after the war verify the story, and expose the "gentleman," if it is true. On account of this refusal escape talk became rife. Many were the plans concocted. During the first week of December two of our servants had managed to escape, no one of the authorities discovering it for a fortnight at least.[810] This added fuel to the escape flame.

In a short time the plan took shape and form, and it was known to all our little Community that Mr. Winston Churchill and three officers had made up their minds to escape. Their plan was to try during the dinner hour—

[807] Hymns Ancient and Modern is a hymnal in common use within the Church of England, a result of the efforts of the Oxford Movement. Over the years it has grown into a large family of hymnals. Hymns Ancient and Modern - Wikipedia

[808] Mr. H. Hunt Lt Col. Royal Regiment of Artillery. Severely wounded at Colenso. 15 Dec 1899. Source: Natal Field Force Casualty Roll, page 27 line 34 PF: Col. Hunt was probably associated with the artillery unit that was devastated and abandoned during the battle and he remained on the field after the retreat of the army to be captured by the Boers. Anglo Boer War - Name search

[809] Mr. Hay.

[810] Because of lax roll calls by the guards.

seven to eight—to elude the guard in the back yard, scale the fence behind the offices,[811] and then try to get on a goods train. Fortune favoured Churchill that Tuesday evening. Unnoticed by the guard he got clear away over the fence.

Unfortunately, the guard turned too soon for the others to get over also. One had already begun to climb the fence when the guard noticed some disturbance and came nearer. He therefore had to come down. This was a great disappointment to them, for no other chance offered that evening nor after. Churchill had provided himself with the Transvaal colours, which were easily smuggled in, so vigilant was our gaoler! These he put on his hat—it was my soft felt hat, too, for he had only a pith helmet—and thus armed he coolly walked the Pretoria streets till about 10.30 that night. Then he went to the railway station and managed to get on a truck of a goods train just drawing out. He travelled on this till near daybreak, got down and hid himself till night came on again. Then he again boarded a train and thus, travelling by night, and hiding by day, he reached Delagoa Bay and was safe.[812]

The next morning—Wednesday—it was discovered here that our friend had got away. It came about thusly. A little barber man used to come once or twice a week to shave and cut the hair of members of our little colony. Unfortunately, Churchill had made an appointment with this little man for Wednesday morning. At eight o'clock the tonsorial artist, a little Hollander, appeared on the scene accompanied by a policeman. He knocked at the door of the empty cage, in vain, of course. Sorely afraid of losing his sixpenny fee the little man ran up and down the building, interrogating every one he met. Some gave him no answer, just looked him up and down, the little there was of him; others referred him to the most unlikely corners; a third said, "In his bath." And so outside the bath-room barber and bobby took up a strong position, holding it against all comers for fully half an hour. Then it struck the little man that perhaps they had been guarding an empty bath, so he knocked gently and apologetically. No answer. A louder, rather self-assertive knock then. Still no answer. And so a loud, peremptory, taking-no-refusal knock came, but yet no answer. Bobbie then ventured to turn the knob and open the door inch by inch, peeping in gingerly. Is the man perhaps dead? Has he cut his throat? What ghastly sight am I doomed to see?

He opens the door a little more. There is no one. It is a little room of about 8 feet by 3 feet. Yet he examines carefully. Under the bath? No. Behind the door? No. Under the chair? No. He gets wild and excited. He examines sponges and towels. Under these, perhaps? No. Where, then? Bath and water, soap and towels are there, but where the bather? Consternation is now changed into panic. The gaoler is called; the guard is alarmed; there is bustle and confusion. We are all asked when we saw Churchill last. "Last night" was naturally the universal answer. So cocksure had the authorities all along been of the utter impossibility of escape, and of their superior intelligence, that they never even called a roll. Our chief gaoler, [Opperman] who naturally was in a great stew, was sure that he had seen the fugitive out under the verandah as late as 10.30 the evening before, which, to put it mildly, was a cram, for the grand *coup* had come off at 7.15.

The fact was recognised now that our friend *had escaped*, and the higher authorities were notified. The Field-cornet, the inspector of police, and some more big bugs came down to investigate the matter. The Field-cornet, a big, burly fellow, was in a great rage. He came to my room—I was still alone then with our first and second gaolers to ask me when I had seen Churchill last.

Of course my answer, too, was "Last night." The big man then said to the two trembling gaolers: "Well, you must produce Churchill; if not General Joubert will hang you!" He looked daggers at them, and they shook. And now the play moved on right merrily. Wires were sent all over the country, and the houses surrounding the Model School were searched. Even the house of our doctor gaoler[813] was searched, forsooth, because his wife is an English lady. Right airily the officials informed us that "Of course he can't escape; he will be caught soon enough, and then put in prison, and very likely all the officers too." The big, blustering Field-cornet went so far as to say, "Churchill is a blackguard; he gave his parole, and that is the way he keeps his English word of honour." Well, *Churchill* was not the blackguard. Somebody else was—a baffled official who takes refuge in

[811] A euphemism for the bathroom stalls outside in the back.

[812] Hofmeyr note: Mr. Churchill tells his story in his very able book on his war experiences. PF: London to Ladysmith via Pretoria is a book written by Winston Churchill. It is a personal record of Churchill's impressions during the first five months of the Second Boer War. It includes an account of the Relief of Ladysmith, and also the story of Churchill's capture and dramatic escape from the Boers. London to Ladysmith via Pretoria - Wikipedia

[813] Dr. Gunning.

lies! The idea! A man gives parole, and yet he is guarded by men armed to the teeth! But, you see, with some of these officials anything will do for a fig-leaf. Public opinion, they knew, would point to them for remissness, and therefore any cowardly lie would do for a subterfuge. Vengeance was now to be taken on us—newspapers were stopped, beer also, the roll was to be called every day, no one allowed to sleep outside under the verandah on a warm night, and lights to be put out at 9.30 (though this latter was never done).

Then *everybody* was refused admittance to the building; not even our caterer was allowed to come in for orders. Oh, there was a to-do and a fuss! Everybody and everything was suspected. A judge was appointed to hold a court of inquiry; the police on duty were changed, some being dismissed; parties were arrested on suspicion and examined—for our astute gaolers were sure that the fugitive was still in Pretoria. A nurse in the hospital, a young lady from the Cape Colony, who had never even seen Churchill, was suspected of having supplied him with a nurse's uniform. She was put over the border and sent home. Another young lady, who had passed our prison on several occasions, was arrested and examined on suspicion of having smuggled letters in or out. Yea, even a little dog was suspected of having brought in a letter in its mouth! A policeman was suspected of having sold Churchill a uniform of his, thus enabling him to escape to Delagoa Bay as a Zarp. And the newspapers came in with their fine detective work—each one was a veritable Sherlock Holmes. The Hollander (*Volksstem*), at Pretoria, and the Scotchman (*Diggers' News*), at Johannesburg—well, they were going to capture him in no time.

He would never escape the Argus-eyes[814] of Transvaal Zarps—no, never! They had his picture in their windows—yea, and about fifteen times in ten days they reported his recapture, cleverly, marvelously done; and now this scion of a noble family would have to sit behind iron bars and weep. Then came the awful news, conveyed in a telegram sent by Churchill himself to one of the officials, to say that he had really escaped to Delagoa Bay, and was off to Durban. How dared he! How could he be so sharp as to elude their sharpness; so bold as to fool all his brilliant trackers, who were so hot on his scent, knowing exactly where he was every night, and going to lay capturing hands on him every morning? Aye, how dared he! The authorities pooh-poohed the news, said it was a bogus wire, the man was still in Pretoria, and would soon be laid by the heels.

For us the whole thing was a grand pantomime—a screaming farce. The authorities did act so foolishly over the whole matter, and their two brave handy men, their beautiful newspapers, making them out to be even greater fools than they were. Our poor gaoler could not, and would not, believe that the escape was a *fait accompli*. Long after Churchill had publicly addressed the public in Durban and Maritzburg he said to me, "I don't believe that he has escaped from Pretoria. Some of the accursed English are hiding him here somewhere."

Winston on deck.[410] "Arrival of Mr. Winston Churchill at Durban from Pretoria."[411] "Arrival of Mr. Winston Churchill at Durban from Pretoria."[412] PF: This image has been reversed.

[814] Argus Panoptes (all-seeing) is a many-eyed giant in Greek mythology. Argus Panoptes - Wikipedia

Winston was importuned to deliver two speeches that day before he was allowed to depart from Durban. The first was at the dockside and the second in front of the 'Town Hall.'

Winston's first speech by the dockside in Durban.[413] Second speech in Durban.[414]

This poor man could not appreciate Churchill's dash, nor swallow the bitter pill of his escape like a man. One of their men [a Boer] escaped in Simon's Bay from the [HMS] *Penelope*,[815] and on arriving in Pretoria was led about like a tame lion. He was brought in to see us, too, and proudly shown off. We all admired his pluck and gallantry, and nobody here or in Cape Town, I'm sure, called him names or bothered much about his escape. But then here, you see, little things trouble—well, never mind. Only it takes a brave man to admire another brave man! This Churchill escape inquiry dragged on its childish existence for months, and it was only in the month of March that a report was made by the Committee.

According to one official it stated that the railway company could prove that Churchill *did not* escape by goods train. And what the railway company here (the Netherlands Company) says is gospel. But, says the report, if my official informant speaks the truth, he got off in this way very likely. An Englishman in Pretoria had asked for a railway pass for his wife, but after Churchill's escape it was found that his wife had not gone away. And when called upon to give up the pass given her, the lady had answered that she had burned it, having no use for it.

The brilliant conclusion come to, therefore, was that the English gentleman had given the pass to Churchill. But and there are ever so many "buts"—how about the lady's name written in the body of the pass? Where did Churchill find this man, when he knew not a soul in Pretoria, and when (as we all know) no one outside was aware of his intention to escape that night. Ah, but the honour and the vigilance of gaoler and Zarp, and detective and official, in fact of the whole State, had to be vindicated, by putting obloquy on an accursed Englishman, even though they stultified themselves in the attempt, as they glaringly did in this case. In the meantime Churchill is free and safe, and his escape is a very bitter pill for them. Our uncourteous gaoler now became, if possible, more uncourteous still and more morose. He no longer shaved, seldom brushed hair or clothes—in fact, he looked as fierce as the famous "Wild Man of the West."

XXI. Our Time of Tribulation—And After!

Our darkest days, I think, came with the middle of December, and lasted till about the end of January. War having been declared before Great Britain was prepared for it, the Republican forces, as was expected, had a long run of successes at the start. When, however, Sir Redvers Buller came to Natal every one hoped that the tide would turn. Imagine, then, the disappointment when the British forces met with that fatal repulse at Colenso,[816] and when, a few day before, that great disaster befell our troops at the Stormberg. Hundreds were killed, and hundreds were taken prisoner. Our little colony of 54 [officers] at once grew to 94, and several of our wounded officers were brought to the hospital. On top of this came the Magersfontein[817] catastrophe, when the brave Highland brigade, caught in the relentless barbed wire, was so terribly cut up.

[815] HMS *Penelope* was being used as a prison ship at the time. HMS Penelope (1867) - Wikipedia

[816] The Battle of Colenso was the third and final battle fought during the Black Week of the Second Boer War. It was fought between British and Boer forces from the independent South African Republic and Orange Free State in and around Colenso, Natal, South Africa on 15 December 1899. Inadequate preparation, lack of reconnaissance and uninspired leadership led to a British defeat. Battle of Colenso - Wikipedia

[817] The Battle of Magersfontein was fought on 11 December 1899, at Magersfontein, near Kimberley, South Africa, on the borders of the Cape Colony and the independent republic of the Orange Free State. British forces under Lieutenant General Methuen were advancing

We received the news of these catastrophes in the order I have mentioned them. Since Churchill's escape we were not allowed to buy newspapers, and were thus dependent for information on news casually heard, and a newspaper now and then smuggled in. Our gaolers delighted in telling us of reverses, but we then already had discovered that one could rely on their tales even less than on broken reeds. I will, however, afterwards devote a chapter to the methods we resorted to for obtaining news.

Let me merely state here that our tribulation was increased tenfold during these dark days, by the fact that it was so impossible to obtain reliable information. On the two newspapers one could not depend. They were simply impossible, to put it mildly. For *suppressio veri* and *suggestio falsi*[818] I have never read the like. They could give heavy odds to the French newspapers, and still win hands down. The editors could only insult the enemy, and give garbled accounts of everything. When the natives on the western border rose, the leading article hurled miserable, cowardly insults at our gracious Queen. When every generous Transvaaler—fighting at the front—admired the indomitable pluck and courage of the British soldier, these mendacious scribblers would inform the public that the Tommies refused to fight any more, that British officers fired on their own men with their own cannon, when they were afraid to advance and refused to obey.

Reports of fictitious interviews with soldiers taken prisoner were published under startling headlines, and splendid theories anent[819] the demoralisation of the British army were built up. Once a week, regularly, the gullible public was informed that the Cape Colonials were going to rise *en masse*, that European complications were imminent, that America were going to interfere, that Russia was marching on to India. The effect of all this on a prisoner may be better imagined that described. Speaking for myself, I am not ashamed to say that it required all my courage and determination to keep up a cheerful appearance, and always hope for better things on the morrow. There was many a sleepless night, and many a weary sigh when no one was near. When is the tide going to turn? We all along knew it would turn, no one doubted that, but often one's faith threatened to be obscured. Many were the quiet talks we had, and the discussions. Then, too, during these dark days the authorities were particularly nasty. The Zarps were impudent and insulting, but no notice was taken of it. Every week we were told that Government was going to move us to another prison, an iron building. Our mess committee could write letters, asking for articles promised us, necessaries of life, measures to be taken to ensure health—drainage, disinfectants but no notice was taken as a rule, and if an answer was vouchsafed it was only after the request handed in was weeks old.

To men of refinement and education all this was galling. Each one tried to fill up the time as well as possible. I studied hard—law, French, Portuguese—and taught Dutch often as much as five hours a day. More prisoners came in, too, so that at the end of January we had about 120 all told. And then the tide began to turn. I must not forget to mention, however, great kindness shown us at Christmas time by some Christian friends—some of the hospital nurses and outside friends. Remembering this time of peace and goodwill, their kind hearts and willing hands provided us with plum-pudding, cakes, and ever so many nice things for our Christmas dinner. Everything was nicely done, and tasted doubly well, for we partook thereof with the consciousness that nothing was begrudged us, and that other hearts were happy because we enjoyed these nice things. Yes, after all, there were some bright spots in the darkness, and some kind hearts, too!

Towards the end of January, after their Platrand repulse,[820] and after our Spion Kop disaster, we somehow all began to feel that better news would come soon. I dreamt of it, and thought of it, and afterwards felt unaccountably sure of it—so sure, that on the last Sunday of January I said to the friends, during service, that I firmly believed that the tide had turned, and that we would not be here much longer. And the tide did turn with a

north along the railway line from the Cape to relieve the siege of Kimberley, but their path was blocked at Magersfontein by a Boer force that was entrenched in the surrounding hills. Methuen failed to perform adequate reconnaissance in preparation for the impending battle and was unaware that Boer general De la Rey had entrenched his forces at the foot of the hills, rather than the forward slopes, as was the accepted practice. That allowed the Boers to survive the initial British artillery bombardment, and when the British troops failed to deploy from a compact formation during their advance, the defenders inflicted heavy casualties. The Highland Brigade suffered the worst casualties, and on the Boer side, the Scandinavian Corps was destroyed. The Boers attained a tactical victory and succeeded in holding the British in their advance on Kimberley.

818 For *withholding the truth* and *propagating the false* …

819 *Anent*: with regard to, about, in respect to.

820 1/6/1900 Battle of Wagon Hill on the Platrand near Ladysmith.Boers repulsed on every side with heavy loss. General Buller made a demonstration in force to aid General White.

vengeance. Good news came pouring in. Lord Roberts and Lord Kitchener, who commanded round Kimberley, had arrived, and on Thursday night, February 15th, Kimberley was relieved.

General Buller, too, took possession of the position between Colenso and Ladysmith, and on Majuba Day, February 27th, the latter was practically relieved. On the same day Cronje and his thousands were taken prisoner, as also sixty-one others near Ladysmith. And history telleth what happened further. How cheerful we all were now! In fact, there were hours when I, and I am sure many others, could forget that we were prisoners. Time seemed to fly. And the attitude of the authorities underwent a marked change, too. Towards the end of February a newspaper was allowed us again—true, it was the miserable *Volksstem* (People's Voice), and I pity the people whose voice it is—yet its caricatures and travesties of truth amused us.

Its marvelous stupidity astounded us; its horrible selfishness disgusted us. Even the wounded burghers in hospital were heard to say: "Confound that Hollander editor fellow! We have got to fight and shed our blood to preserve the country for him and those like him. All our present misery we owe these Hollanders!" Yes, the poor fellows were quite right. Had Paul Kruger, after 1881, surrounded himself by others than these Hollander importations, there would be no war to-day. Then, too, we began to get answers to our letters and requests, and, *mirabile dictu*,[821] a prisoner who came in shortly before Majuba Day got served with a bed and clothes on the day of arrival.

All the others who had arrived during the dark days had to wait for days and weeks. And our gaoler, who used to go about like a roaring lion, seeking to devour us, now was intensely meek and mild. No one, however, will soon forget those dark days. How we tried to hide our miserable feelings from one another! how we tried to cheer one another; to look round the corner to see the bright side of things; to catch, like drowning men, at any straw of good news; and to be sure that the "tomorrow" would see the aspect of affairs changed!

Aye, and it did. Right merrily the sun shone again, and our hearts were bright. As I sit down writing this the news has just come in, "The British flag is waving over Bloemfontein"!—March 13th. Our excitement can only be imagined, not described. We knew that this was coming, only the brilliant editor of the *Volksstem* could not see it. In his paper of the 12th he calls the attempt of Lord Roberts a "vain hope." Oh no, this Mr. Stay-at-home was not going to allow it!

But now that it is no vain hope, but stern reality, of course this courteous, true, and imported patriot will put the blame on the Free Staters, just as he tried to put the blame of Cronje's surrender on General de Wet (of the Free State), who, he said, ought to have relieved him (Cronje). I can't help it—I just can't help it—I must again say, all this would never have happened if Paul Kruger had listened to better advice, given him by his best friends, and not surrounded himself with such friends—such time-servers. When his vessel goes down they will be the first to leave him—the very first to accommodate themselves to the new regime, whatever it might be. The genuine Transvaaler knows this, and despises the put-on patriotism of the Hollander, self-advertised and belauded. Thus the days of tribulation passed away, making place for better and brighter ones. We passed through the deep waters bravely, I trust; and we now listen to good news gratefully, hoping that dark days will come no more again.

XXII. The Zarp

The official name for the Transvaal police is "Zuid Afrikaansch Republiek Politie." If you take the initials of these four words you make of it the word *Zarp*. This has already become a notorious word, for the man it describes is so notorious. We have had ample opportunity of studying him here, and I must say he fully comes up to all I have before heard said to his discredit. A more uncouth and cruel body of men I have never seen. For months now we have had them with us. Some thirty have been living in the back yard in six tents. They have been on guard around us day and night.

We could study them closely, and a striking object-lesson they were. Let me say, at the very outset, that a few among them were fine fellows indeed, of soldierly bearing and intelligent. They were men who had seen service in Bechuanaland and Matabeleland. But these men, though their best servants, were not trusted by the officials here. They were too courteous to us, and treated us as if we were men too and not animals. Unfortunately they did not stay here long. They were sent to the front, or to do duty elsewhere in the town.

[821] *Mirabile-dictu*: wonderful to relate.

The famous Johannesburg Police, Passing Through Johannesburg.[415] "The Johannesburg Police (Van Dam's Commando) Off to the Front, 26 October 1899. No commando behaved more gallantly in the war than the Johannesburg Police, which, with the other police and the State Artillery, formed the standing army of the country. Originally the Johannesburg Police numbered over a thousand men. After the Battle of Dalmanutha (23—27 August 1900) their strength was reduced to less than 200."[416] Boer Mounted Police.[417]

I must also add that to me, personally, all were respectful—through habit, for the people of our country are taught from youth upwards to respect clergymen. But it counts not how they treated me. Here were British officers—gentlemen—their superiors at all times and under all circumstances, and the point is, How were these treated?

The generality of Zarp is drawn from what is known as the *poor white* class. The "poor white" is, as a rule, a man who has failed at everything, either through lack of ability, or more often through laziness or other evil habits, and who now lives in the towns waiting for something to turn up, Micawber-like.[822] These men or their sons, who closely follow their fathers' footsteps, become Zarps. Their physique is of the poorest—never an Adonis among them. Of training there is no trace, of education as little. Their personal habits are disgusting. They affect the tobacco-cud much more than the bath. They are loud-voiced and foulmouthed. On their beat they slouch about, stand at ease, lean lazily on the rifle or up against a tree.

Smartness of walk or action and the Zarp are as yet total strangers. When on duty, standing as much as fifty yards apart, they talked to one another, bawling as loudly as possible. And the talk!—vulgar, often obscene, nonsensical, impertinent, bountifully interlarded with curses. Oh, the language they used was horrible. I have already noticed the fact that many of the officers were surprised at the bad language they heard in the lagers where they were kept as prisoners. The major of a large party of captured officers and men thought it his duty to warn the men not to use bad language, for Mr. Stead [823] had written, said he, somewhere that it would shock the burgher horribly to hear bad language.

It did not take the major and officers long to find out that their captors could swear as glibly, and even more so, than Tommy himself. Alas! so it is. Years ago the genuine old Transvaaler would have abhorred a curse or a foul word. Now too many of the younger generation are adepts in the art. Where did they learn it? Some shrewd guesses may be made. In the Zarp this habit has reached its acme, it seems. With him it is a refined art. He cultivates it carefully and assiduously. He swears at the officers, at the gaoler, at his companions, at his gun, at his boots, at his mackintosh, when it is wet, at the streets, at the dogs, at the Hollanders, at the authorities, at the sun, at the moon.

[822] Wilkins Micawber is a clerk in Charles Dickens's 1850 novel *David Copperfield.* He is traditionally identified with the optimistic belief that "something will turn up." Wilkins Micawber - Wikipedia

[823] William Thomas Stead (5 July 1849 – 15 April 1912) was a British newspaper editor who, as a pioneer of investigative journalism, became a controversial figure of the Victorian era. Stead published a series of hugely influential campaigns whilst editor of The Pall Mall Gazette, and he is best known for his 1885 series of articles, The Maiden Tribute of Modern Babylon. These were written in support of a bill, later dubbed the "Stead Act", that raised the age of consent from 13 to 16.
Stead's "new journalism" paved the way for the modern tabloid in Great Britain. He has been described as "the most famous journalist in the British Empire." He is considered to have influenced how the press could be used to influence public opinion and government policy and advocated "Government by Journalism." He was known for his reportage on child welfare, social legislation and reformation of England's criminal codes. Stead died in the sinking of the RMS Titanic. W. T. Stead - Wikipedia

William Thomas Stead.[418]

Yes, and when there is nothing more for him to swamp with his stream of abuse it seems to me he is quite capable of swearing at himself. And what the Zarp is on duty, he is also off duty, only more so! And then their cruelty! This, for instance, was something they would delight in. Our prison was a corner house, and the two streets meeting at right angles on the corner were roped off as far as the grounds extended on each side. Nobody on horseback or on a bicycle was allowed to pass through these enclosed portions of the streets, only pedestrians, though not coloured ones, and then they had to confine themselves to the opposite side-walk. This was what often happened.

Mr. Zarp stood on duty. A little dog comes up the street. He waits till the hapless little animal has passed, and then roars to the next sentry, with a curse, "Look out!" With that he begins hurling stones at the poor inoffensive creature, which runs off howling till it gets opposite Zarp No. 2, who stands ready with his handful of stones. With a laugh and a curse he lets the little fugitive have it. Poor thing! it has to run the gauntlet of a third humane Zarp still before it can find shelter for its battered and bruised little body.

And the three guardians of the law roar with laughter, and bawl to one another and consider themselves smart. What a pity the Royal Humane Society never hears of these dear, gentle creatures! Won't somebody inform the Society? How our blood boiled when we were forced to be passive spectators of this Zarpian ruffianism!

And then a poor, luckless Kaffir comes up the street. Poor fellow! he does not know that the street is not wide enough for his Majesty the Zarp and himself, and thus walks on with that apologetic air which every Kaffir in the Transvaal wears, till he has passed the Rubicon—that rope which excludes him and forces him to walk a block round. Now you'll hear choice language, whilst with loud voice this polite Zarp explains to this benighted creature that no black man dares pass that way. My reader, this Zarp is one of the brothers Mr. Stead is so concerned about. He was our *neighbour* here, but, alas! not our *brother*. He was a being who would have delighted in shooting us down in cold blood, any day, *when our backs are turned*. With Mr. Stead it is distance that lends enchantment to the view. Had he been prisoner of war—he the champion of purity and justice—I am sure there would have been very little of the fraternal in his criticisms on them, and less in his relations towards them.

Am I too bitter in my criticisms? Then listen again. An officer, a captain, stands in the shadow of a tree waiting his turn for the bath. He stands near to a sentry, who seemed to think that the tree and its shadow were his exclusive property. How dare an officer stand in that shade too! Up he comes to him and aims a blow at him with the butt end of his rifle. The captain walks off to complain to the gaoler, who vows dire vengeance on the Zarp—he was to be hung, and then quartered, and then burned, and then killed! "What," say we, "will justice be done?" Alas! our gaoler had but gassed—the Zarp was not even reprimanded.

Officials in this country have to walk very circumspectly. You see, they might give umbrage to a man whom one day they might have need of, when another—a Hollander, very likely—through jealousy is trying to create a vacant billet into which he might step himself. Such was our gaoler—afraid to do right, for there were so many others ready to step into his shoes. Another officer was treated in the same way by this same Zarp, and again nothing was done to him. On another occasion a lieutenant, a giant of a young soldier, stood leaning over the iron railings before the door.

A Zarp had been making a fool of himself in the street, acting in a particularly uncouth and uncourteous way, whereat the lieutenant smiled. The guardian of the law sees the despising smile and feels mightily insulted. With a stride he is at the officer's side and aims a blow at him with his rifle. Had the officer not been a prisoner, of course that fool Zarp would have measured his length in the dust, and very likely his own mother would not have known him that evening. As it was, our young giant just smiled blandly at the irate preserver of the peace and asked him if he felt better.

The authorities were informed of the outrage. The Zarp, if not yet promoted, will soon be—indeed, for a time he was acting-corporal. He is one of their best men now. Did not he, a policeman, aim a blow at a defenceless man, a prisoner? Was not this a chivalrous, noble action? Ought not such an admirable Crighton[824] to be encouraged?

More still. Standing inside the railings, at our street corner, were several officers. Some inoffensive children were passing under the rope and were soundly rated by the Zarp on duty for some imaginary misdemeanor. This brought up a look more of pity than of anger on the faces of the officers. It was too much for his insulted majesty. He yelled out to another Zarp, ostentatiously *loaded* his rifle, *left his beat*, and rushed up to the fence. He foamed at the mouth, and swore horribly, again yelled, and then, drawing his cleaning-rod, aimed a vicious blow at one of the officers, threatening to shoot him.

The blow was warded off, but the officer's thumb severely injured in the attempt. Again a complaint was lodged, and to our surprise the Commissioner of Police came a few days later to try the case. I acted as interpreter. The accused Zarp lied as if he had no immortal soul, but the Commissioner was shrewd enough to find that out, considering him guilty of a "disgraceful, mean, and cowardly assault upon a prisoner of war." His sentence was "instant dismissal, a fine of £5, and off to the front tomorrow morning." Several days later this particular bobby passed in the street before our prison looking very happy. Query: Was he ever dismissed, did he pay his fine, for he was not sent to the front?

But then this is the Transvaal, and the pure fountain of justice is here—well, never mind. (Later on we heard that the man never was made to pay his fine, nor was he sent off.) There is not a little village in the Cape Colony, to speak of no other British possession, where the policeman, be he white or black, is not taught by his magistrate to be courteous to everybody and to remember that, as preserver of the peace, he has to help the public, and not to bully it.

Natal Native Police.[419]

Zulu Native Police.[420]

Uniformed Boer Artillerist.[421]

[824] No definition found.

Natal Native Volunteer Corps.[825] [826]

In this free country, terrorised over by a tyrannical Government, as unjustly as Russia is, and more so, it is the other way round. The greater bully you are, the better servant of the Government so it seemed, especially if you bully the defenceless—though it be a little dog only. I imagine the feelings of the British officer, with vivid remembrance of a courteous London policeman, when he is informed that the Zarp here represented is the best they had in the whole Transvaal.

Were they not *the* police of the capital? Were not these the men who constitute the President's bodyguard? Surely they must then be fit guardians of the State and of the safety of its inhabitants! Yes, and yet they were not trusted, not a soul of them. But, like master, like man. What else can you expect when justice is not a fixed quantity, but a variable factor? But enough of the poor Zarps—the awkward squad, the battalion of bullies! Let me remind the reader again, there were good men and true among them, courteous and brave, but they were men who had been British subjects once in the Cape or Natal, and were therefore not trusted here. If this Government but knew; *they* alone could be trusted, not the cowardly bullies left to guard us, every one of whom could have been bribed, if an officer had cared to condescend to such an attempt. Poor Zarps—verily one can afford to pity them!

XXIII. Dead

Right in the beginning of the war a Transvaal Government official said to me, "We'll show the world that we are a fighting nation as well as a civilised nation." Let us discuss the latter qualification to-day. First of all let me again remind the reader of my distinction between the real, honest Transvaal burgher and the Transvaal Government. Some eight weeks ago Lieutenant Tarbutt, [827] of the Imperial Light Horse, was taken prisoner near Colenso.

Because his father and himself had been inhabitants of Johannesburg, the Government hated him and sought to humiliate him. This was done by putting him in the common gaol in Pretoria, on his arrival. The same was done to other officers and men of that regiment. You see, this civilised Government holds peculiar ideas. One is that if you have lived in the Transvaal, though all political rights were denied you, and though you were made to pay directly and indirectly into its treasury, without getting a vote, or any political status, yet when war came on you are expected to forswear your own country and even, though you are a British subject, to fight for the Transvaal.

[825] Page 160 South Africa; its history, heroes and wars by Mackenzie, William Douglas (William Douglas), 1859-1936; Assisted by Alfred Stead, 1877-1933, joint author. Published 1899 by The Co-Operative Publishing Company, Chicago, Ill. South Africa; its history, heroes and wars : Mackenzie, W. Douglas (William Douglas), 1859-1936 : Free Download, Borrow, and Streaming : Internet Archive.

[826] See: Natal Native Contingent: Natal Native Contingent - Wikipedia.

[827] Charles A Percy Tarbutt, Lieutenant: He died of typhoid fever at Pretoria, February 13th, 1900. He was the eldest son of Percy Tarbutt of the Consolidated Gold Fields of South Africa. Lieutenant Tarbutt was twenty six years of age, and was educated at University College School (where he was in the football XV), and at the Bedford Grammar School. When Lieutenant Tarbutt landed in South Africa he was given a commission as Lieutenant in the South African Light Horse. After the battle of Colenso he was reported missing, but it was subsequently discovered that he had been wounded and captured. Source: Donner. Anglo Boer War - Name search.

Imperial Light Horse.[422]

Woe betide you if you do not! Its President loves to preach the gospel of peace and justice. A portion of that gospel is love your enemies; forgive them. Ah! but perhaps that portion is not found in the Government version of the Bible! And so peculiar is the sinister influence of this body upon the members, that men I've known in the Colony and Free State as gentlemen of refinement and education, but who have joined the Government here, have within a year or two lost all that refinement of spirit and instinct of justice, and become as cruel and merciless as the other members of the Government, which, again, is permeated by the worst imaginable spirit of the Hollander adventurer class. When I looked upon the actions of these Colonial gentlemen I was surprised and intensely grieved. "Can these," said I, "so soon have forgotten and killed all the nobler instincts of the real Africander? Is it possible so soon to deteriorate?" Alas! it is but too true—some of them even out-Heroding Herod.

Patriotism with these, and other Transvaal Government officials, means hatred, bitter hatred of everybody who dares to differ from them. It is with them *aut Transvaal aut Nullus*.[828] What a travesty of patriotism! And thus it was that every member of the Imperial Light Horse,[829] because most of them had lived at Johannesburg before, were special objects of this hatred. And so poor Lieutenant Tarbutt was sent to the common gaol, to be locked up in a miserable cellar cell fit for a condemned criminal, a thief, or a murderer, but certainly not for a gentleman, though he was a prisoner of war. And this they call civilisation—chivalrous action towards a fallen foe! Weep, guardian angels of Transvaal! For weeks Lieutenant Tarbutt was detained there, and, as could be expected, he fell ill of fever. You see, sanitation is everywhere at a discount here, and especially in the prison—I was there myself, I speak from experience. Imagine a fever patient in a miserable stuffy cell, in summer! But the officials gloated over it. "We'll show him what it is to fight against the Transvaal; we'll pay him out!" Such were the comments.

Meanwhile our young friend grew worse, and only when after a fortnight's illness it was proved that he was suffering all the time from a virulent attack of typhoid fever did the Government remove him to a decent hospital. And he had a father and mother in the old country, and some of these men also have children. And one could expect them to have a parent's instincts and a Christian's tenderness—that their hearts would go out to his poor mother far away, and that for her sake they would be merciful. No, their peculiar patriotism killed all such instincts. They forget to be merciful as their Father who is in heaven is merciful.

Lieutenant Tarbutt now grew rapidly worse, and, after about twelve days in hospital, died yesterday, February 13th. What a shock the news of his death was to all of us (the hospital was just next door to our prison)! So young—I believe about twenty-seven years—and now cold in death.

And the horrible thought it might have been otherwise, if just and Christian treatment had been accorded him, would come again and again. If you neglect your duty to a man for whom you have to care, even though he be your enemy, and he dies, is not this called murder in every Christian country? What, then, is Lieutenant Tarbutt's death? Is it wonder that many an officer was heard to mutter the dreadful word "murdered"? It was a cruel thing to do, to cast him into that hole of a prison, and, worse still, to leave him there when he fell ill till it was too late.

[828] From "aut Caesar sut nullus" Either a Caesar or a nobody: either Emperor or nothing: all or nothing.

[829] Adam Brockie was part of this force and Churchill's valet, Thomas Waldon, joined them after Churchill's capture. See: Johannesburg Light Horse Regiment - Wikipedia.

The funeral took place to-day, at four o'clock in the afternoon. Ten of us were allowed to attend it. We walked down to the English Church and took our seats round the coffin. Several people were in the church—people with charitable, kindly hearts; and as we walked up the aisle we heard many a sob, and met looks of intense sympathy. It was a solemn scene, and from our very hearts we sang—

"When our heads are bowed with woe,
When our bitter tears o'erflow, When we mourn the lost, the dear,
Jesus, son of Mary, hear!"

The sad procession then moved off to the English Church cemetery, followed by the nine officers and myself, and a goodly number of private carriages. Arrived at the cemetery we carried our comrade's body to its last resting-place, the coffin being covered with beautiful wreaths, sent by several friends in the town. The officers of the Model School had also sent a beautiful one in remembrance of their dead comrade. When it was all over we drove back again to our prison home. It was a sad experience for us. The horrible "it could have been otherwise; it might have been prevented," would boom in our ears. Bitter thoughts would come. I had continually to quell my rebellious heart with words spoken on higher authority, "Vengeance is Mine, saith the Lord."

Aye! but the poor mother's heart and father's heart at home, and their bitter grief when one day they learn the whole truth! I could not help thinking of them. May God help them to say, "Thy will be done." We noticed how respectful and reverent every one was as the hearse passed down the street! Hats were doffed by every passer-by. But these were not Government officials. Only one woman shook her fist at me, gesticulating fiercely. Even death could not cool her hatred. But then she is a woman, and I forgive; I have no harsh words for her.

With solemnised hearts and minds we returned to our prison. Death has entered our circle—grim, horrible death, and there are hearts that mourn for our young comrade, but a few weeks ago so strong and full of life, a veritable athlete! Now dead—dead! Was it murder?

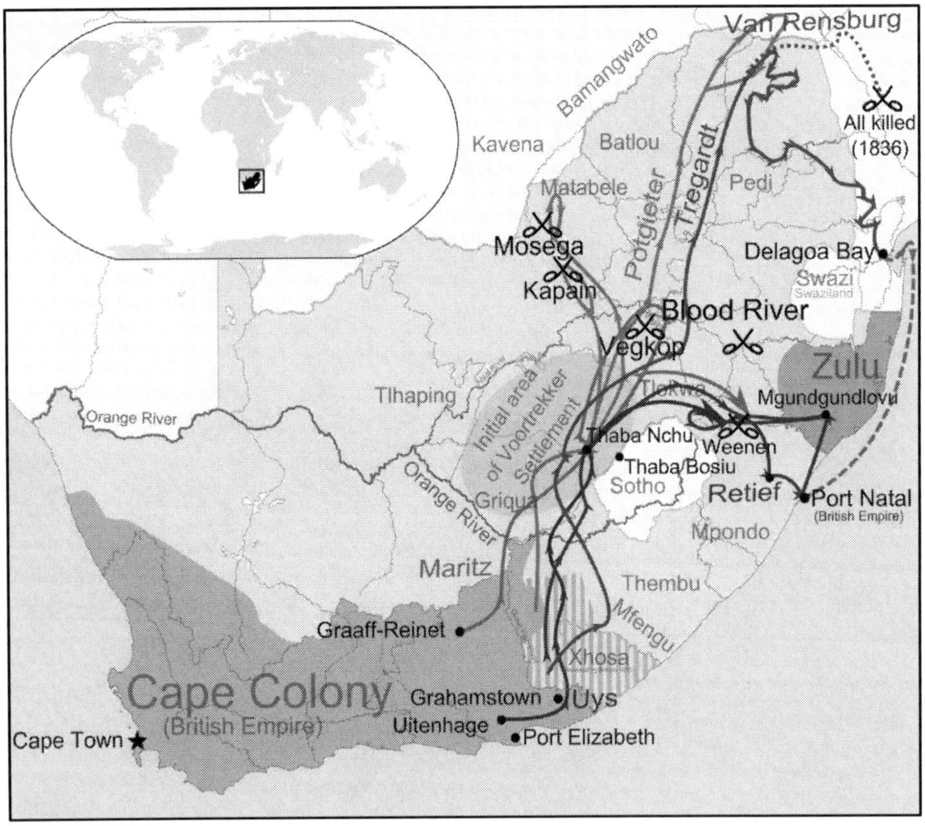

Great Trek Map[830] A map charting the routes of the largest trekking parties during the first wave of the Great Trek (1835-1840) along with key battles and events.

[830]Solid Blue: Louis Tregardt's route (1833 to 1838) Dashed Line: Survivors of Tregardt's trek evacuated by sea, 1839 Dots: Van Rensburg's route, after it separated from Tregardt's Red: Hendrik Potgieter's trek, campaign and scouting routes. Green: Gerrit Maritz's

XXIV. Was It a Mistake?

To while away the tedium of prison life I opened a Dutch class for several of the officers who were desirous of learning the language. Many pleasant and instructive hours were thus spent. As reading books we used, amongst others, a little work by my brother, "Kijkjes in de Geschiedenis,"[831] and another work by him and our brother-in-law called "De Voortrekkers."[832] The contents of these books, especially of the latter, gave rise to many interesting discussions, and to much thought during many lonely hours. As a son of the soil I have always been an admirer of the old "South African Pioneers" and of their prowess. I sympathised with them and their troubles and misfortunes. I must honestly confess I never allowed myself to hear the *alteram partem,*[833] or think for a moment that there might have been a mistake, and that to-day everything might have been otherwise and for the better in South Africa if these good old people had not "trekked."

Hence the question that heads this article—"Was it a mistake?" I have been reading a good deal about the subject, and I have come in contact here with a good many of the descendants of the first trekkers. I have been more than four months now a prisoner in this unhappy country, and cannot help instituting comparisons and forming conclusions. Let me reason the matter out calmly. There is no doubt that the grievances of the first emigrants were real. Lord Glenelg[834] and many other politicians of his day blundered stupidly, so stupidly that we can to-day only think of it with amazement. Could men of common sense really have acted like that? Yet so it was.

Then, too, the Colonial farmers suffered much from the cruel, marauding Kaffirs and the frequent Kaffir wars. These grievances were real, and there were more, several more, very galling too. I do not dispute the sufficiency or the validity of the *cause*. What I ask myself is, Was the cause intended to have this *effect*! Ought their trials and tribulations to have driven them to this extreme measure—the leaving of their country? Let me reason again, Where was there ever a country that never was misruled? Had the people of Holland and England not much to complain of, and that often too?

Did not the Dutch, ancestors of many of the old trekkers, suffer much and often under the rule of the East India Company in the Colony? Read the history—what great discontent there often was, how strained relations frequently were! And yet those people did not seek relief in emigration. And more: many of the trekkers were descendants of the Huguenots, who, though oppressed, left their country only when prevented from worshipping their God according to the dictates of their consciences.

And how did the Dutch treat them in the Colony—the East India Company people? The use of their language was forbidden them, they were compelled to live only in certain districts, and many more petty annoyances were heaped upon them. And yet *they* trekked not. With patience, through constitutional agitation, all the friction and irritation were removed. Again, the well-meaning trekkers, the pioneers, did they escape Kaffirs and Kaffir wars by emigrating? How is it possible that they could have expected such a blessing? True, their avowed intention was not to quarrel with the Kaffir, but to buy or barter his land and live peaceably. This they honestly meant to do, I really believe; but did they not yet know the Kaffirs?

Did they not know that he was *jealous* of, and would act *treacherously* to every white man entering his country? Surely they knew that the Kaffir was convinced that the advent of the white man meant his possessing the land, and that, therefore, at all risks he must be stopped. They knew all this. Why, then, trek to escape Kaffir wars? Verily it was leaping from the frying pan into the fire. Even more: they knew already that the Zulus of Natal and

route. Burgandy: Piet Retief's route, including missions with his entourage. Blue: Piet Uys's route. Afrikaans: Kaart van die eerste uittog van die Groot Trek (1835-1840), met die roetes van die grootste trekpartye van die Voortrekkers aangedui. Gebaseer op 'n kaart in "The Great Trek" (1985) deur B.P.J. Erasmus en die "Illustrated History of South Africa: The Real Story" deur Reader's Digest (1988), bladsy 115. Author Discott File:Great Trek map full.png - Wikimedia Commons.

[831] Glimpses into history.

[832] Searching HathiTrust: HathiTrust Digital Library | Millions of books online For author "Hofmeyr" prior to 1900 returns a dozen books by several authors. None matching this title however. Possibly: De voortrekkers: eene reeks van voorlezingen, gehouden in verband met de Kimberleysche Christ. Jongelings-Vereeiniging door J.D. Kestell. Catalog Record: De voortrekkers: eene reeks van... | HathiTrust Digital Library

[833] A Latin phrase meaning "listen to the other side", or "let the other side be heard as well."

[834] Charles Grant, 1st Baron Glenelg PC FRS (26 October 1778 – 23 April 1866) was a Scottish politician and colonial administrator. Charles Grant, 1st Baron Glenelg - Wikipedia

the men of Moselikatse[835] were ten times more cruel and warlike than any tribe of the Colonial Kaffirs. And yet they trekked to escape trouble with the Kaffirs. Surely something was wrong. It cannot be that their troubles were sent in order to drive them to this. I said that they leaped from the frying-pan into the fire. Alas! so it was: witness the frightful massacres in Natal by Dingaan[836] and his hordes! Again, they emigrated, as every one said, to find liberty, to rule themselves in their own way, to escape all oppressions. Have they obtained their wish? To answer the question properly, let me first glance at the old Colony.

Ever so many more farmers remained behind [in the Cape Colony] than those who left the Colony, as we know. They had the very same grievances. They remained, however, overcame *all* their troubles, and are to-day the happiest people imaginable. We (I am one of them) have perfect political freedom, no Kaffir troubles, and English and Dutch live most peacefully together, loyal to their Queen.

Our most distinguished clergymen, lawyers, politicians, doctors, teachers, &c., are pure Africanders, sons of the soil. We have made marvelous strides in every way—educationally, socially, in religion, in politics, everywhere. No one can visit the Colony and become acquainted with the people without noticing how happy, hospitable and contented they are. Remember, now, this condition of affairs has followed on the very same circumstances which caused the old pioneers to emigrate, *forced* them to it, as *they thought*. Here you have *cause* and *one effect* in the case of the Colonial people.

Let us now examine the effect in the case of the trekkers. They began their nomad life with trouble and disaster. Take, for example, Trichardt's[837] expedition and Vechtkop.[838] Then came the frightful chapter of Natal history written in blood.[839] And then, alas! jealousy caused disagreement amongst themselves, and there came divisions. Natal was annexed[840] and a number of them remained there, becoming soon, as they are to-day, prosperous, happy farmers and citizens. A number, however, joined the Free State and Transvaal trekkers, and the history of these people, what does it teach us? Take the Transvaal, for there most of them settled. [841]

No sooner had they made a republic of the country than they began quarrelling amongst themselves. Political strife and religious dissensions rent them in twain. Even civil war ensued. Yea, even with the Free State they quarreled! It was a most unhappy country. The Church factions were at one another's throats, liberty had degenerated into licence. "No taxes!" was the great cry. Poverty reigned and bankruptcy stared the country in the face, just at the time of the annexation in 1878.

The War of Independence in 1881 reunited them, and especially through religious influence, brought to bear by Colonial clergymen, the two factions in the Dutch Reformed Church were reconciled. Did it last long? No. At the time of the Raid the fire of religious dissensions was again raging fiercely. Verily the people were only

[835] Mzilikazi Khumalo (c. 1790 – 9 September 1868) was a Southern African king who founded the Mthwakazi Kingdom now known as Matebeleland, in what became British South Africa Company-ruled Rhodesia and is now Zimbabwe. His name means "the great road." He was born the son of Mashobane kaMangethe near Mkuze, Zululand (now known as KwaZulu-Natal in South Africa), and died at Ingama, Matebeleland (near Bulawayo, Zimbabwe). Many consider him to be the greatest Southern African military leader after the Zulu king Shaka. In his autobiography, David Livingstone referred to Mzilikazi as the second most impressive leader he encountered on the African continent. Mzilikazi - Wikipedia

[836] Dingane ka Senzangakhona Zulu (c. 1795–29 January 1840), commonly referred to as Dingane or Dingaan, was a Zulu chief who became king of the Zulu Kingdom in 1828. Dingane came to power in 1828 after assassinating his half-brother Shaka with the help of another brother, Umhlangana, as well as Mbopa, Shaka's bodyguard. They were traditionally said to have killed Shaka because of his increasingly brutal behaviour after the death of his mother, Nandi. The assassination took place at present-day Stanger. In November 1837 Dingane met with Piet Retief, leader of the Voortrekkers. In return for their recovering some stolen cattle, Dingane signed a deed of cession of lands (written in English) to the Voortrekkers. It is generally believed that Dingane knew what he was signing although he could not have had any formal education, have read the contents of the document or have understood the concept of permanent land ownership since it was not a custom of the Zulus to assign land to individuals permanently. On 6 February 1838, after two days of feasting, the chief had Retief and his diplomatic party killed. They had been told to leave their firearms outside the royal kraal. Suddenly, when the dancing had reached a frenzied climax, Dingane leapt to his feet and shouted Bulalani abathakathi! ("Kill the wizards!") The men were totally overpowered and dragged away to the hill kwaMatiwane, named after a chief who had been killed there. Dingane - Wikipedia

[837] Louis Johannes Tregardt (10 August 1783 – 25 October 1838) was a farmer from the Cape Colony's eastern frontier, who became an early voortrekker leader. Louis Tregardt - Wikipedia

[838] The Battle of Vegkop, alternatively spelt as Vechtkop, took place on 16 October 1836 near the present-day town of Heilbron, Free State, South Africa. Two farmers and 184 Matebele died. In spite of their loss, the spoil was great: The Matebele plundered 50,000 sheep and goats and 5,000 cattle, including all their draft cattle. Battle of Vegkop - Wikipedia

[839] PF: I believe he is specifically referring to the Battle of Blood River in 1838. Battle of Blood River - Wikipedia

[840] In 1843 the British annexed Natal and ended the Republic of Natalia. See: Natalia Republic - Wikipedia.

[841] Great Trek - Wikipedia

united when there was a common foe. Otherwise, internal strife and religious squabbles pervaded the whole country. Just before the [Jameson] Raid Church lawsuits were threatening everywhere. Expensive ones had been entered upon, and fanatical zeal and hatred were rampant. And again, only a common foe united them. All these vicissitudes, have they favourably influenced Transvaal character? The immediate descendants of the old trekkers are still men of sterling character and religious convictions, but a restless and nomadic people, no tillers of the soil, by no means industrious, always ready to resent any encroachments on their rights even by their own Government!

When Burgers[842] became President some hundreds trekked to Humpata,[843] because they approved not of his religious views. Common sense will tell you what the descendants of such men are like to-day. They are uneducated, for only lately has education begun to be recognised; the poor amongst them are frightfully bigoted; the rich with no ballast to counteract sudden accessions of wealth.

Through contact with the irreligious Hollander, young Transvaal in the towns has lost that veneration for religion which characterised their fathers. I know the Colony and its people very well, and make bold to say that this description is quite inapplicable there. Of course, we are not saints there, but sinners too, only there is a mighty difference between sinner and sinner. Again, the political and religious quarrels of the country have tended to form fanatics. From the President downwards everyone is but too ready to find a Bible text wherewith either to prophesy future greatness or to justify any action of theirs which is criticised. This peculiarity has been exemplified to an alarming extent during the present war. Newspaper men, even well-known irreligionists, utter Bible texts and pious sentiments *ad libitum*.[844]

But, see you, this goes down with the poor, simple, back-country burgher. Once more: capacity for hatred has been developed in this country to a pitiful extent. I have never seen or heard the like. What I have heard here, now and often too, has made my blood run cold. Expressions of intensest hatred and cruelty—words the Colonial educated, refined Boer would not dream of uttering. Their history has tended to make the people quarrelsome, spiteful, suspicious, self-righteous, in fact to a great extent even sanguinary.[845] A Christian, civilised people ought not to cultivate such a character, and the Colonial Africander, I trust, is not such a man. Has there, then, not been a mistake? Would it not have been better for these good old men to have waited patiently and laboured industriously, rather than trek about and fight and nurse their hatred, learning to despise everyone daring to differ from them. The *Cape Colony* and the *Transvaal*, or *Happiness* and *Discontent*, or *Love* and *Hatred*—good names for the two pictures. Look upon that picture and then on this, and tell me, *Was it a mistake?*

[842] Thomas François Burgers (15 April 1834 – 9 December 1881) was a South African politician and minister who served as the 4th president of the South African Republic from 1872 to 1877. Thomas François Burgers - Wikipedia

[843] Humpata is a town and municipality in the province of Huíla, Angola. Humpata was the primary destination of the Trekboers on the Dorsland Trek in the 1870s. These Afrikaners formed the majority population in the area for some time, before leaving to South West Africa after World War I, though some stayed until 1975, Humpata - Wikipedia

[844] "at one's pleasure" or "as you desire," often shortened to *ad lib.*

[845] An action accompanied by bloodshed or bloody violence. Sanguinary - Wikipedia

Thomas Burgers

Thomas Francois Burgers (15 April 1834 – 9 December 1881) was a South African politician and minister who served as the 4th president of the South African Republic from 1872 to 1877.

In his role as a minister of the Church, Burgers became involved in a controversy with the Dutch Reformed synod over his alleged liberalism and disbelief in the literal truth of the Bible. He was critical of traditional culture and strongly emphasised knowledge and rationalism. In 1862, his unorthodox doctrine brought on him an accusation of heresy, (an imbalance of faith) and in 1864, he was found guilty by the Synod and suspended. The Supreme Court overturned the decision, and in 1865, he was readmitted to the ministry.

In 1872, the burghers of the South African Republic urged Burgers to stand for the presidency, and he was elected by the considerable majority of 2,964 to 388 in 1872. The South African Republic's first coins, the Burgerspond, were introduced by Burgers in 1874. Some people in the South African Republic objected to the issue of the Burgerspond, because the portrayal of the president on coins likened him to a dictator.

One of his plans, inspired by the neighbouring Cape Prime Minister John Molteno's massive railway programme, was to build a railway linking the Transvaal to the sea. In 1875 he traveled to Europe to raise funds, but his plans were thwarted by the Pedi chief Sekhukhune, whose lands lay in the path of the proposed railway.

Burgers tried to force the issue with a military campaign against Sekhukhune[846] the Bapedi leader. The siege of Sekhukhune's stronghold failed because commando members objected to Burgers's theology, calling him a heretic, and abandoned the siege in droves. Burgers resigned and left the country. This gave Sir Theophilus Shepstone the pretext he required to annex the republic on 12 April 1877. Sir Garnet Wolseley, High Commissioner for South East Africa, declared war on Sekhukhune. With British troops and allied troops (including the Swazis who had supported Burgers), in 1879 Wolseley defeated Sekhukhune and imprisoned him in Pretoria. With the Bapedi threat removed, the burghers were no longer so amenable to British rule.

On 13 December 1880, the members of the last Volksraad were summoned to a meeting at Paardekraal. The meeting placed authority in the hands of a triumvirate: Paul Kruger, Piet Joubert, and Marthinus Wessel Pretorius. They hoisted the Vierkleur there on 16 December. The war – not officially declared to the occupying British troops – opened with an irregular attack on a British regiment on the march at Bronkhorst spruit.

The Boers defeated the British at Laing's Nek and Ingogo, and on 27 February 1881 at Majuba, where General Sir George Pomeroy Colley fell at the head of his troops. While the British would in other circumstances have sent more troops and defeated the rebels, Prime Minister William Gladstone chose to make peace. Drawing up of the detailed peace treaty (the Pretoria Convention) was left in the hands of a royal commission comprising Sir Hercules Robinson, General Sir Evelyn Wood and Justice John de Villiers of the Cape Colony.

The ZAR became fully independent on 27 February 1884, when the London Convention was signed. The country independently also entered into various agreements with other foreign countries after that date.

Dr. Thomas Francis Burgers.[423]

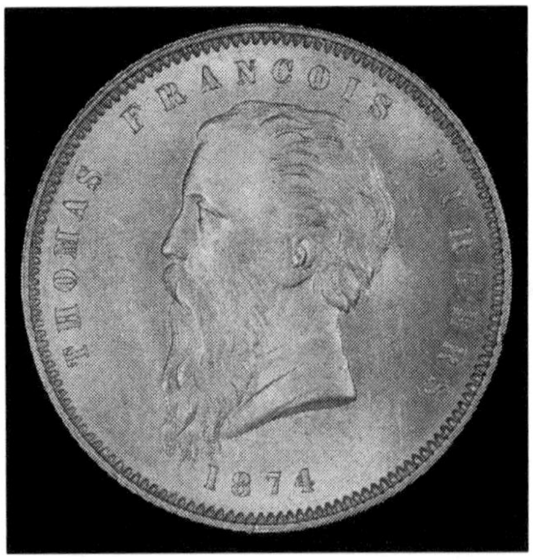

Transvaal One Pound Burgerspond 1874.[424]

Sekhukhune.[425]

[846] Sekhukhune - Wikipedia And: Pedi people - Wikipedia

XXV. No Doubt About the Following Mistakes, However!

I have been a prisoner so long now that thoughts and ideas I had about men and matters have had time to fructify. Illusions I had have had time to be dispelled; suspicions to be confirmed or otherwise. What I will attempt to state now is what I believe to be the solemn truth. I am trying to put into words prevalent ideas here which I have verified as being prevalent. And, as I write, I cannot help saying, oh, the pity of it! All this need not have been if wiser counsels had prevailed, and false hopes and delusions had not held sway. In the former chapter I referred to the capacity for hating prevalent here. Let me refer to it again for a moment. Ever since the Great Trek many parents here have considered it their duty to impregnate their children from their very youth with a bitter hatred against the English and whatever is English.

Gradually this hatred extended also to Africanders in the Cape and in neighbouring British colonies. Suspicion against Colonial-bred and educated clergymen was widely disseminated. Indeed the "Hervorende"[847] Church attempted to pass a resolution which would exclude any but Hollander clergymen from their pulpits. Their Politicians warned against the "Engelschgezinde" (English-minded) politicians and ministers of the Cape Colony. As I said, independence to them meant hatred of what is English and suspicion of British subjects. To be true Africander meant in their eyes to be Transvaaler.

The moralist will tell us how detrimental to strong, sturdy, noble development of national character such sentiments are. But to proceed. Remembering the above, I can now understand how the more bold among them, since the retrocession of 1881, have begun to dream of a united South Africa. By that they understand a Republic from the Zambezi to Table Mountain, under its own flag. In itself there was no harm in this dream. Then came the discovery of gold and eventual prosperity, and this dream began to assume proportions. [Orange] Free State politicians had an honest conception of this great idea. They entered into a Customs Union, for instance, with Natal and Cape Colony, and at the same time lived in close union with the Transvaal.

They believed in waiting and not forcing the hand of history, till a Steyn[848] became President. The Transvaal, however, proud because it had gold, discovered and worked for them by Uitlanders and English enterprise especially, blinded through success in their struggle for freedom in 1881, began plotting. Let us have a united South Africa, under our own flag, said they; all the time they meant it to be the Transvaal flag. Stupendous ambition! And now began their policy of propaganda in the Colony, Natal, Free State, and in Europe. Secret service money is freely spent.

The Bond leaders are got hold of in the Colony, and every little village is filled with insinuations and ideas which, if allowed to fructify, would mean simple rebellion against the Queen. All the time internal development in the Transvaal was disgracefully neglected, or muddled. Take, for instance, education and State finances; notice the galling monopolies! Then came the unfortunate Raid, lending colour and show of reason to all Transvaal pretensions and claims. Oh, that Raid—it spoiled so much. Well, sins are punished! So England must suffer to-day for what some of her children have wrongly done through mistaken zeal. Since the Raid the Transvaal propaganda were prosecuted with redoubled zeal.

Promises of reform remained unfulfilled, and, as a high official here told me, "we were determined to let it drift to a rupture with England. Then," said he, "our dreams would be realised—a Republic right south to Table Mountain." *Sotto voce*[849]— "under Transvaal flag with Kruger as President." Said I, "But England will conquer you." "No," said my informant; there are three factors we rely on" (and many other talks I have had during the past four months confirm what this gentleman said); and *these* are the three he mentioned as promising them ultimate success:—

1. "The Cape Colony will rise to help us as soon as we declare war."

When I looked surprised, he said, "The Bond leaders have assured us of it." "What," said I, "Mr. Jan Hofmeyr[850] and Mr. Schreiner?"[851] Then came the unwilling answer, because he saw that I knew something—"No, not *the*

[847] The Dutch Reformed Church. (redirect from Nederlandse Hervormde Kerk).

[848] Martinus (or Marthinus) Theunis Steyn [needs Afrikaans IPA] (2 October 1857 – 28 November 1916) was a South African lawyer, politician, and statesman. He was the sixth and last president of the independent republic the Orange Free State from 1896 to 1902. Martinus Theunis Steyn - Wikipedia

[849] Under one's breath. Intentionally lowering the volume of one's voice for emphasis.

[850] Jan Hendrik Hofmeyr (4 July 1845 – 11 October 1909) was a South African politician. He was affectionately known as Onze Jan, "our Ja" in Dutch. His chief importance as a public man was, however, derived from his power over the Dutch in Cape Colony, and his control of the Afrikaner Bond. But it remained an organization for obtaining the political supremacy of the Cape Dutch. His control over

leaders, but leaders in the north of the Colony, even members of Parliament, in disaffected districts" (and then he mentioned names, which I will not put in print, for I cannot believe these gentlemen guilty of such traitorous conduct). Aye, so sure was the Transvaal of the Colonial Africanders joining them, that on October 15th, when I was treacherously taken prisoner, the Commandant solemnly assured me, as I mentioned before, that between 7,000 and 8,000 Colonials had already joined the Transvaal forces. "And soon the rebellion will have spread right to Table Bay. We ourselves have now already," said he, "90,000 men in the field."

Thus the Transvaal Government had filled the poor, good burghers with their dreams and delusions. That commander solemnly believed in the Colonial rising. I am glad that *then* even I said to him, "I do not believe it," and that my trust in my countrymen has been justified. Since then several others have said, again and again, the Colony will soon rise—"thousands are coming over every week." Alas for them, the Colony has *not* risen, and now there is bitter disappointment.

An official said to me a little while ago, "You see the Colonials, of course, could not openly join; we can understand that they must remain loyal, naturally!" But such was not the talk at the beginning. No, they try to whittle down the fact, but their disappointment and chagrin are immense. The first week of the war the papers stated that Jan Hofmeyr and Schreiner had been arrested on account of Transvaal sympathies and treason. Then the bubble burst, and the truth was heard—these men are both loyally doing their duty!

Today they excuse and explain the failure of their hopes of a Colonial rising, but their hearts are bitter against Jan Hofmeyr and other Colonial leaders. One said, "We would not have fought had we not been assured of a Colonial rising to help throw off the hateful yoke." Aye, how dared they hope that we would prove to be traitors and rebels in the Colony? But such is the Transvaal Government thus the poor burghers are deceived. Bubble No. 1 is pricked, and there is wailing and gnashing of teeth.

2. "Europe and the American States will interfere."

Thus ran the story. Up to very lately I heard it again and again: "The United States are going to help us, and have offered already to pay all debts for us. Germany will, after a month or two, interfere. France will not allow England to conquer us. Russia will cause trouble in the east, in India; and then England must withdraw all her forces!" This was firmly believed in. And a dishonest Government had no other ground to go on than club-talk and the gutter-press, of France especially.

Thus the real Transvaaler, the burgher, was gulled, buoyed up with expectations which their Government had no positive proof of. During the long term of Transvaal successes every one here was patiently waiting for the European or foreign intervention that was sure to come. Now that the tide of reverses is setting in, and yet no foreign intervention, uneasy thoughts are rampant. An official said to me a few days ago, "I can't understand it; America and Germany promised intervention" (hear that!) "and now not a word, not a sign—nor anything of the kind from anywhere else. I can't understand it. It seems to me that every country is afraid of this hated England, except the Transvaal!" Poor deluded fool, thought I; what right had you to go to war on the strength of suppositions and expectations? "Now," said he, "if there is no intervention we can't go on. Our ammunition will run out, and we were sure Germany or France would supply us." And thus another bubble is burst.

Poor Transvaal burgher, your shrewd President and his myrmidons have cruelly deceived you, and your beautiful newspapers have ably seconded their nefarious objects! What will the reckoning be? Let me here say what struck me as particularly nefarious in the whole delusion—A man is known by the society he keeps. Behold, then, a friendship between righteous Transvaal and Dreyfus France! Unholy alliance, forsooth! General Joubert's ancestors were driven out of France, because no liberty of conscience was allowed them then. Today that general and ever so many descendants of those Huguenot refugees, champions of liberty; mind you, and righteousness, hob-nob with France, buy her cannon, employ *her* gunners, applaud *her* gutter-press, forgetting that St. Bartholomew's night still clings to France like a foul taint, and that the shadow of the Dreyfus story darkens her history to-day. Pity the poor, deluded Transvaal burgher; but then, what about the Government that so willfully misled him with its sophistries? Aye; a terrible blunder has been committed—a crime!

the Bond enabled him for many years, while free from the responsibilities of office, to make and unmake ministers at his will, and earned for him the name of Cabinet-maker of South Africa. Jan Hendrik Hofmeyr (Onze Jan) - Wikipedia

[851] William Philip Schreiner CMG PC KC (30 August 1857 – 28 June 1919) was a barrister, politician, statesman and Prime Minister of the Cape Colony during the Second Boer War. William Schreiner - Wikipedia

XXVI. The Biggest Mistake of All

It is quite possible that the Transvaal's expectations, mentioned in the former chapter, especially Nos. I and 2, might have been realised, but their own blundering made it an impossibility.[852] I felt that and said it from the beginning. I am more than ever convinced of it now.

As I pointed out before, the Transvaal was really wanting war. During the days of their triumphs, one said to me, "When our President met your Governor at Bloemfontein, we were ready for war and had decided to have it." Let me say here that I am as positive that neither then, nor up to the last moment, was England desirous of, or ready for, war, nor did she wish to be plunged into it. "England wants to annex your country, to take your farms away, to make servants of your wives and daughters," this was the bogie-man a dishonest Government was making use of to whip up the burghers to a fanatical enthusiasm. But to return. The Transvaal was ready and wished for it, and then did the best thing it could do. It published its Ultimatum which had been lying ready for use ever so long. It was a master stroke. England, of course, could not accept of it; war was therefore inevitable, and yet England was unprepared for it in the Colony and in Natal. What ought the Transvaal to have done now?

1. *To have left the Free State out of it.* It had no quarrel with England, and to drag it in was most selfish conduct on the part of the Transvaal. They knew it meant making a buffer of the Free State. They knew that it would have to suffer most severely when the invasion began. It was cruel, selfish conduct. The Free State President, I think, has committed political suicide by allowing himself to be dragged into this snare.

2. *To have defended its own country against any invasion*, but not to have *invaded* Natal or the Colony. This was a tremendous blunder. Of course the greatest mistake was—wanting to make war at all. The Ultimatum ought never to have been sent, and there never would have been war! I am convinced of this, and have good grounds for my faith. But having set the ball rolling, the next great mistake was the invading of the Colonies.

What was the result? England was naturally unprepared and suffered reverses and severe losses. This of course exasperated public feeling in England and alienated sympathy. Foreign countries, such as Germany, commented severely on the impertinence of sending an Ultimatum and the further blunder of invading the Colonies.

Moreover, by invading the Colonies and looting and destroying property, they alienated all the sympathy obtained even there.

This was the case especially in the Cape Colony. It was a blunder taking Aliwal North, Colesberg, and Dortrecht, and Burghersdorp, and commanding the inhabitants to join their forces.

Commando Crossing Orange River to Aliwal, North.[426]

Boer Commando Arriving at Upington Courthouse.[427]

It was playing a low trick upon them. These poor men became rebels by doing so, and will have to be punished as such when the Republican forces are forced to retire to their own territories—and then? And just fancy taking little places like Kuruman, and Barkly West, and Herbert! What was there to be gained by it, except empty glory for a little while! I cannot understand such blundering; but, you see, they believed so religiously what I mentioned before—that the Colony would rise, and foreign Powers interfere. And how could they but believe?

Did not their prophet-President tell them so? Had they defended their own country, bravely, desperately, as General Joubert wished them to do, then they could have expected sympathy and foreign intervention even;

[852] Hofmeyr note: Written February 19, 1900.

then, in case of defeat, there would not have been any humiliation. Now what a spectacle! England is unprepared and is taken at a disadvantage and suffers reverses. The burghers, maddened by successes, deem that they can carry everything before them. They rush on into the Colonies. They not only alienate sympathy; but now that the British soldiers have arrived, the story of humiliation begins. Kimberley is relieved. The Free State is invaded.

They cannot take Ladysmith, nor "plucky little Mafeking," notwithstanding the vulgar, insulting boasting of their bright specimens of newspapers—I beg pardon, *lies*,—not *news*-papers. They will soon be driven out of the Colonies, and forced to defend their borders. And then? Alas! then, too late, they will understand the full extent of their biggest blunder. The burgher was goaded on to fight for his country, his farm, whilst the Government all along knew that its idea was a larger one; it was going to drive England out of South Africa, forgetting that there were more Africanders preferring English rule to Transvaal misrule than there were Transvaalers preferring Transvaal rule to English domination.

To cloak their designs, it was given out that if war came on it would be one caused by the capitalists. The poor capitalist was made responsible for everything that happened. Well, perhaps he is very guilty, but then he is also responsible for the Transvaal's prosperity. His enterprise saved the country from financial ruin and provides the magnificent salaries of the Transvaal President and Government myrmidons to-day. But any stone will do to throw at a dog, and so the capitalist was made a target of. And thus the biggest blunder was committed. It is the old story over again, "Whom the gods wish to destroy they first dement." Have pride and prosperity demented the Transvaal Government? They have played their cards well on many occasions, but this time pride, I fear, has come before the fall.

A grand opportunity was lost, and the end is near. But then one can't play ducks and drakes[853] with impunity with truth and honesty; and this leads me to another false step taken by the Government here. Let me speak of it in a following chapter.

XXVII. Another False Step!

To whip up the good honest burgher into fanatical enthusiasm for the war and implacable hatred against the English, the Transvaal Government and its agents all about the country had to resort to every possible means.[854] In doing this they took many a false step—one of the principal was the following. I call it a false step, not only because it was based on falsehood, but also because, when one day the burgher discovers that it was so, a reaction will set in, in a direction the Government does not dream of now. To proceed, it was assiduously given out that England not only meant to annex the country, but also drive every burgher from his farm, compelling him and his wife to earn their daily bread by their own hands. This was told and readily believed, and had its due effect.

The real old Transvaal burgher who knows little of the outside world and whose little knowledge is one-sided, has his peculiar ideas of independence and liberty. He understands by these terms more or less the following: the possession of his farm, no direct taxation, and no equality for the Kaffir or any coloured person. (The latter grievance was one of the greatest influences causing the Great Trek.) Amongst others he was diligently taught that the Englishman makes no distinction between white and black; and that if once England ruled the country, then the conquered Boers and the blacks would be placed on the same level.

Imagine the effect of such a doctrine on the Boer mind! "In addition," said the propagandist, "the country will be taxed and burdened most frightfully to indemnify England's war expenses. Therefore, if you wish to retain your farm, if you wish to remain dominant as regards the black man, if you desire to maintain your independence (a mild kind of license) fight the English to the very death." All this was told me before the war began by many a burgher, and reiterated since by many more. The result is that the universal bluster one hears now is—"We will fight to the end," and, as one man said to me, "England will have to pass over our dead bodies to seize our farms!"

[853](Redirected from "Ducks and drakes.") Stone skipping and stone skimming are considered related but distinct activities: both refer to the art of throwing a flat stone across the water in such a way (usually sidearm) that it bounces off the surface. The objective of "skipping" is to see how many times a stone can bounce before it sinks into the water; the objective of "skimming" is to see how far a bouncing stone can travel across the water before it sinks into the water. Stone skipping - Wikipedia

[854] Hofmeyr note: Written February 21, 1900.

Another one, a Government official into the bargain, said to me, "If we lose, then before I give up my property and submit to English rule, I'll shoot my wife and children and then myself. I am not going to be made a slave of, nor allow my wife to be someone else's servant." Can you imagine a man of ordinary intelligence indulging in such rhodomontade![855] And yet such is the case. Something more, I referred to equality of white and black which the Transvaaler dreads so much. You can have no conception of the annoyance and harshness wherewith a Kaffir or any coloured person here is treated. It pained me continually, when on the western border, to hear the Kaffir speak of the Boer—the cruel, hard ones amongst them. One said, "Kicks and cuffs they give in plenty, kind words never." Another, "The Boer makes me pay as much as £1 17s. 6d. hut tax and work for him at a very low price, whilst he himself pays next to nothing in taxes."

After I was taken prisoner I was struck by the disdainful way their own Kaffir servants were spoken to. Many officers here attest to similar experiences. Treating a coloured servant kindly and courteously, they call putting him on the same footing as yourself. Treating him as if he were but a dog, and ought to be thankful that he is allowed to live at all, is what they call keeping him in proper subjection.

I saw myself with indignation and horror how Kaffirs, my fellow prisoners, some even in irons, were treated. Oh, the brutality of it! it makes my blood boil to think of it. A magistrate said to me, There is only one way of punishing a Kaffir. I find out how much money he has, and then sentence him—a fine of about all he has on him, and twenty-five lashes! Tell these people now that to give the Uitlander the franchise is to stop such proceedings and introduce the hated equality, and you can imagine how he will object to such a step. Tell him that to allow England to conquer means not only that, but in addition the loss of his beloved farm and his reduction to a state of personal servitude, and you can imagine with what dervish fanaticism he hates, and fights.

As I write this the tide of war seems to have turned. What will happen now? When defeated the burgher will find that England proclaims amnesty to everyone who lays down his arms and goes back to his farm within a certain time. He will be told that no one ever intended to drive him off his farm, or equalise him and the Kaffir, or in any way deprive him of liberty. And then he will round on his mis-informers, who primed him with such stories, and I would not like to stand in the shoes of the Government minion they get hold of on that day—no, by no means!

But what a pity all this is! Ah! I cannot help feeling sorry for the real Transvaaler, the burgher, I respect and like him. But who can respect a Government that rules thus and uses such officials? I am sorry for this deluded country. There will be a rude awakening when one day the burghers find out that they have been misled. Verily, this move of the Government was a false one—stupidly false! It will never recover from this false step so ingloriously taken.

XXVIII. Majuba Day-1900

This Chapter has been omitted. The chapter number has been retained as a placeholder.

[855] *Rodomontade* [a mass noun]: boastful talk or behavior.

XXIX. And After

I have in the former chapter referred to what happened on Majuba Day—Cronje surrendered with over 4,000 men, and the Boers evacuated their positions round Ladysmith, starting in full retreat. Ladysmith was thus practically relieved on that day, Tuesday, the 27th, whilst our forces really entered the town on Wednesday. *All this was known to the Government and many other; even we prisoners knew it.* But it was kept from the people by telling the most cowardly falsehoods. In the very churches, where the President and other officials were present, they prayed, knowing the above, that Cronje might be rescued.

With these data given me, I ask how must I understand and interpret the following? This is what the *Diggers' News* of Wednesday, February 28th, published, knowing that it was a lie:—

"ANOTHER BOER VICTORY.

Prince Alfred's Guards Engaged and routed. 200 Killed and Wounded."

Also this, ten times worse:—

"THE BATTLE OF PIETERS.

BULLER's FOURTH ATTEMPT

and how it failed.

GLORIOUS BOER VICTORY.

Two Regiments Cut Up.

FORLORN HOPE FAILS.

British Positions surrounded.

Guns taken over River again."

Is this blasphemy, or am I mistaken? But there is worse. On their sacred Majuba Day, the 27th, a wire—inspired, of course—is sent to this beautiful paper, *from Pretoria*, as follows:—

> "Pretoria, 27th (special). General Botha wires from Tugela to-day that the British forces are retreating south of the river lock, stock and barrel, and that all their movements since Sunday are inexplicable."

On the *very same day* [the 28th the following telegram from the President to his people is published, a telegram written *after* the President *knew* of the double event of Majuba Day, a telegram bristling with Bible texts, a telegram full of passionate appeal to the God of hosts, who, however, as they seemingly forget, is also the God of Truth who hates a lie, as well as the *suppressio veri.*[856] But any argument, and any attitude of hypocrisy serves these officials evidently, fighting for *liberty*, whilst it *oppresses* others, who dare differ from them in opinion.

I say candidly that when I read the following wire a cold shudder of horror crept over me. Was it blasphemy, or am I mistaken? Are the Transvaalers God's chosen people more than the English or any other Christian people? They God's chosen people, when the same President, Paul Kruger, fought against his own people in their civil war, animated by religious hatred and political jealousy. Oh, the horror of it all! I can only say this: I know *God is not mocked with impunity*. Majuba of 1900 is proof enough thereof. Read the following now attentively:—

"THE PRESIDENT'S TELEGRAM.

"The following telegram has been addressed by the State President to all officers and burghers of the Republic:—

> "SIRS AND BROTHERS,—
>
> "So soon as you cease to stand fast and to strive in the name of the Lord, then has unbelief appeared, and then follows cowardice, and when you turn your backs on the foe, there is no hope of further protection, for then have we ceased to trust in the Lord.
>
> "No, no, brethren, let it not be so. Has not the Almighty already given us a double proof that He is on our side? Where our burghers have stood fast, no matter how severe the struggle has been, the Lord has

[856] Suppressing the truth.

driven back our enemies by a small number of our forces. Brethren, is it not the same God who cleft the Red Sea asunder, and punished Pharaoh with all his hosts when Moses remained firmly trustful? Is it not the same God who enabled thousands to drink of the stream which: He brought forth from the rock?

"Is not the same God who walked upon the sea, and stilled the storm and the tempest, and to whom the winds were obedient? Is it not the same God who called Peter to Him upon the waves of the sea, and who supported him as he walked thereon, and when his faith wavered, grasped him by the hand, while He rebuked him for his want of faith? Is it not the same God who has promised to us through the Christ, born of the tribe of Judah, and who encourages all who trust in Him, with the words, 'Ye who believe in God, believe also in Me, I will not leave you nor forsake you. Lo I am with you always, even unto the end of the world, for I have overcome the world'?

"It appears to me from a study of God's Word, that we live at a time which is spoken of in the Revelation, where the Beast has received the power to persecute the Church of Christ, and to try her as gold is purified when placed in the furnace. The enemy has not the securing of victory in his own hand, but the victory is with the Lord. Up to the present time the Lord has shown that He is on our side, having given us so many victories in times when we have remained steadfast in the faith. No, no, brethren, let us not fall into unbelief, by forsaking Him, but stand fast in the faith to the death. The Lord says that they who fight in faith, and who fall by the sword, are dear in His sight, and that their death is a sacrifice upon His altar. This is the strife by means of which we may hope to win the crown, both in a material and a spiritual sense. Read Psalm xxvii. verse 7. God there says: 'Be of good courage, ye God fearing host, for the Lord is faithful. In your weakness He will make perfect His strength.'

"Read Psalm xxxiii., from the seventh verse to the end, where it says that the victory is in the hand of the Lord, and not with the multitude of horses or carriages. Read Psalm cviii., and especially the last two verses, which tell us that they who strive in the name of the Lord, shall perform mighty deeds, and shall destroy the enemy. Read Psalm cxviii.: 'They had surrounded me like bees, but in the name of the Lord I slew them, and they were as thorns in the fire.' Read also the tenth verse of the same Psalm. Nay, brethren, let us not bring our whole progeny into destruction, but stand fast in the faith, and continue to fight, and we shall be convinced that the Lord will stand up in His might and scatter His enemies.

"Read also Psalm lxviii. The faith of the burghers is at its extreme point, and the test is most severe, but the Lord will shortly show that He alone lives and reigns. The young men rather chose death in the fiery furnace than forsake their faith. Our forefathers chose rather to suffer at the stake than to give up their faith, and the Church still enjoys Divine protection, and those who choose death rather than deny their faith are as a sacrifice upon the altar.

"Read this telegram to all the officers and burghers, and my belief and prayer is that the Lord will strengthen all our burghers in the faith, even though they have no earthly rock behind which to seek protection, but on the open plains must fight for victory. At Modder River our burghers fought against thousands, who completely surrounded them, on an almost open plain, but they drove back the foe, and destroyed them by hundreds."

On Friday, March 2nd, the *Volksstem* had the following acknowledgment of what happened on the previous Tuesday, Majuba Day. It makes most curious and instructive reading, when compared to the deliberate misstatements of the previous Thursday and Wednesday. I give the reader first *their* account of what happened, and then the leading article, of which note the concluding paragraph especially. To me it is inexplicable that, when on one of their sacred days such a double and terrible catastrophe takes place, its officials should still allow such utterly mendacious reports about it. The reader knows exactly what happened, and a summary of the news we received day by day, given later on, will give him an idea of the utter horror of the position here. From my heart, I say again, I pity the good, real Transvaaler, thus sacrificed on the altar of mendacity by a hypocritical, tyrannical Government. This is their account:—

PF Note: The remaining portion of this chapter, reciting other newspaper articles has been omitted.

XXX. A Last Chance

The reader will remember the chapters on the "Mistakes"—which I tried to point out. These chapters were written *before* Majuba Day—this one *after*. Taking the *before* and *after* together, I can't help feeling that there is a last chance still for the Transvaal. What has happened since I was taken prisoner, and the revelations of this last Majuba week prove to my mind most incontestably that under the present Government the Transvaal never will be a free, independent country, even if no war had taken place; but also that it is doomed now that the war is there. They cannot prevail in the end. The revelations of the former chapter are enough to convince the sober thinking mind that the country has no right to expect the Almighty to work a miracle to save it. And yet I believe a chance has come now still to save something valuable out of the sinking vessel, and gain that independence which is real, and which will make this a happy and prosperous country—a last chance! I give my opinion for what it is worth.

The remainder of this chapter is an exposition on this theme and has been omitted.

Concluding with:

They have imperiled the independence of the Free State—they are now cutting their own throats. Will they awake ere it is too late! Were I free to-day, I would give them a mighty shake and cry, "Give in; tis your last chance!"

XXXI. Another Escape

In addition to all the bad news of Majuba Day, the poor Government officials had another bitter pill to swallow. On Tuesday morning early,[857] about six o'clock, when our gaoler came round to count us, he was horrified to find that three officers had escaped—Captain Haldane, and Lieutenants Le Mesurier and Brockie.

Aylmer Haldane.[858]

Frederick Neil Le Mesurier.[859]

Adam Brockie.[860]

[857] Tuesday February 27th, 1900.

[858] General Sir James Aylmer Lowthorpe Haldane, GCMG, KCB, DSO (17 November 1862 – 19 April 1950) was a Scottish British Army General. Aylmer Haldane-Wikipedia

[859] Frederick Neil Le Mesurier Captain QSA (4) QSA known to exist Source: List of QSAs with the clasp Talana from the Talana book Royal Dublin Fusiliers. MID LG: 10 September 1901, page: 5951. Source: Field Marshal Roberts. 4 September 1901. Re: General mentions Anglo Boer War - Name search Page 5927 | Issue 27353, 10 September 1901 | London Gazette | The Gazette.
He died at Ypres in 1915. See: Frederick Neil Le Mesurier Captain 2nd Bn. Royal Dublin Fusiliers Died 25 April 1915 Ypres (Menin Gate) Memorial Panel 44 and 46. Belgium The Commonwealth War Graves Commission | CWGC | CWGC PF: He may be related to Gerald John Le Mesurier DFC was a South African flying ace of World War II, credited with 3 'kills.' See Gerald Le Mesurier - Wikipedia

[860] Adam Brockie is listed in four places at Anglo Boer War - Name search My supposition is that they all refer to the same individual. He may have been promoted because Haldane states that he is tasked as a recruiting officer in Johannesburg after his escape.
Brockie Adam No. 489 Sergeant QSA (5) Source: List of QSAs with the clasp Elandslaagte Imperial Light Horse.
Brockie A. Lieutenant QSA (5) Source: List of QSAs with the clasp Elandslaagte, Johannesburg Mounted Rifles.
Brockie A. Lieutenant QSA (5) CC Eland DoL Tr SA01. QSA (5). Officially renamed. Sotheby October 1987 Provisional list of recipients Source: Ladysmith Siege Account and Medal Roll Johannesburg Mounted Rifles.

How shall I describe the scene of wild excitement and confusion that ensued! Of course we were all as sober as judges, and … well … we knew nothing about it. But let me tell the tale. I had just got up that morning and had begun dressing, when our gaoler came to my little corner. I saw at once, from his appearance, that something was wrong. I thought at first that he was ill. He looked so ill, was deathly pale, whilst beads of perspiration stood on his forehead, though the early morning was very cold.

"What is the matter?" I asked. In solemn voice he answered, "Three more fellows have escaped." "What," said I, in apparent astonishment, "three more officers gone away, without your leave?" "Yes," came the lugubrious answer. "How did they manage that?" asked I. "They cut the wires of the electric light, and in the darkness they got off, how I know not." Now the reader must know that the previous evening the electric lights were found to be out of order. Not a light was burning, neither those in the house, nor the big ones in the yard. When it was discovered this official at once sent for an electrician to rectify the mistake. The man could not find the cause of the collapse, however.

All that could be definitely discovered was that during the afternoon, whilst our servants were kicking their football about, it hit the wires, and the conclusion was that this shock had severed some connections up in the roof. One day I will tell the true inwardness of this remarkable story! Howbeit, the electrician could do nothing. We lit up our prison with candles, and the yard was lit up with the bright presence of fifteen extra Zarps, all on the *qui vive*.[861] The authorities naturally were satisfied. No British officer can get through a cordon of fifteen Zarps, be the night as black as Erebus,[862] so thought they. Alas! vain are the hopes of man, especially when he relies on a Zarp. The next morning three British officers had disappeared. And the night was not so dark after all. How? Well, this also is another story. It will come in its place. I dare not reveal it now. My papers may be seized any day.

Let me tell what happened. Remember, it was Majuba Day. At 8.30 we breakfasted, and had just received news that Cronje's surrender was expected every moment, also Ladysmith's relief. Naturally this triple event caused us to feel greatly excited. Our guards and every one else seemed to be completely down on their luck. Like Milton's "Il Penseroso,"[863] they were "sober, steadfast, and demure." At about 9.30 a *posse* of corporals and police invaded the place. The inspector of police, a little fussy man, was dancing about the show. He had been here all last night from nine o'clock to three in the morning. And he was ready to lay £100 to Is [1 shilling] that the officers did not escape during the hours of his vigil. Vigilant, was he? Oh! vigilant was not the word for it. He was Argus-eyed; he was here, there, and everywhere *at the same time*; nothing could escape his notice, not even a mosquito. Nothing, except a British officer! some one volunteered to say. This set the little man off into a war-dance and a fresh panegyric on his mighty prowess and lynx-eyed qualities. He could lay an oath that they did not escape whilst his Ubiquitency bossed the show.

Unfortunately for the lynx-eyed one, they … but that is another tale, as I said.

And now we were all requested to come outside the building, while it was being searched. What a hiding of diaries and letters there was! The pages of my manuscript I hid in a school-cupboard, in between the leaves of a number of Bibles. Police were placed inside the building to keep us outside. And then the search began. Under the roof the searchers found where the wires broke, being cut as they said. They also found a knife turned into a saw, wherewith a trap-door had been cut in the ceiling of the gymnasium. Through this door, according to their opinion, the officers must have crawled to cut the wires. The big dormitories, too, were searched for the missing men or for clues—beds and mattresses and cupboards and portmanteaux.

But in vain. And all the while the fussy little man danced about in great excitement, looking for tracks and footmarks about the building. Supposing he had found these, what would he have done with them, I wonder? They would have brought him no nearer the fugitives. But, you see, it is always good to lock the stable-door after the horse is stolen!

Brockie Adam Corporal BSACM Mashonaland 1897 (0). Source: BSACM rolls British South Africa Police Roll.
The South African National Archives website shows evidence that Brockie died in 1904 as there is a deceased estate file for him. He died in the Transvaal. http://www.national.archsrch.gov.za/sm300cv/smws/sm30ddf0?20210926024800064684B63&DN=00000003 Image source: Vera Gallony.

861 On the alert or lookout.

862 One of the Greek primordial deities who was the personification of darkness and shadow, brother-husband of Nyx and son of Chaos.

863 Il Penseroso (The Serious Man) is a poem by John Milton. Il Penseroso - Wikipedia

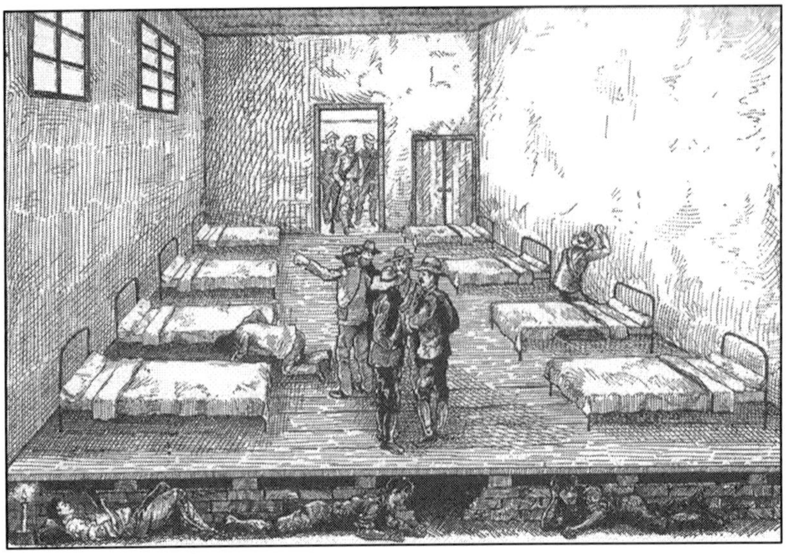

Hiding under the floor. "We hold our breath as they approach the roof of our humble dwelling."[428]

The next day an official told us that they knew where the officers were, and could lay hands on them any time—they were only waiting for daylight. How smart! Next day they apparently must have forgotten where to lay their hands so as to include the officers in that mighty grasp, for no capture was made. And let us hope that no capture will ever be made.

After Churchill's escape the papers blustered considerably. "We will soon have him"—"he will never elude the vigilance of our police"—and more such froth was uttered. *Now*, very curiously, all they say is "they do not believe *all* will manage to get out of the Transvaal." Also, say they, a place has been found where evidently they had passed the night and taken food. As if our hunted friends would pass the night in any other way but in making, "a forward movement." By day they might rest, but not by night. A great fuss was made over a Hollander town guard (all the town-guards seem to be Hollanders) who reported that during the night of the escape he had chased a suspicious-looking party, but was prevented from catching him by a wire fence. But why did the valiant guard not jump off his horse and over the fence and catch his quarry?

Yes, echo answers, Why not? Of course he was not afraid! No Hollander ever is! But let him take this consolation to his heart—it was not an officer he had spotted—not at all. Then our fussy little inspector heard of the "lair" referred to above. Bravely he rode out with a company of his merry mounted men. After a long search they found two young bank clerks on bicycles. Poor unfortunates, they were promptly marched off to prison on suspicion of having helped the officers to escape—and they had never even heard of the escape. And for this they were fined £10 each.

Our fussy little officer was now happy. Some days have elapsed since the escape, and every one—even our little man—is quite subdued now. After Churchill's escape we had to suffer for it, as told before. Now a corporal of police, the only gentlemanly fellow they had, was dismissed, and other changes were made in the *personnel*. The yard of our present home was cleared of all police tents and paraphernalia—so much the better for us, from a sanitary point of view especially—and trees and shrubs cut down, behind which we could possibly hide. And, says the paper, a thorough investigation is being made. Poor police, and nurses, and ladies! I suppose they will all be suspected again as on the occasion of Churchill's escape.

But now comes the amusing part. On the Thursday after the escape, a letter was published in the daily *Volksstem* purporting to have been written by a "mother and sister of one at the front." (There's a conundrum for you.) This irate lady goes hammer and tongs for our gaoler.[864] "Here he sits in luxury," says she, "whilst our relatives risk their lives at the front to catch the prisoners. Could he not guard them better?" She was sure that if the women had to do the guarding no officer would ever escape—and more to the same effect. This grieved the

[864] She is referring to "Commandant Opperman, who was directly responsible for our safe custody and was in times of peace a Landdrost or Justice. He was too fat to go and fight, but he was an honest and patriotic Boer, who would have gladly taken an active part in the war. He firmly believed that the Republics would win, and when, as sometimes happened, bad news reached Pretoria, Opperman looked a picture of misery, and would come to us and speak of his resolve to shoot his wife and children and perish in the defence of the capital." (WSC London to Ladysmith.).

poor man sorely, for he had done his best. Had he not treated the officers as uncourteously as possible? Had he not sworn at them (behind their backs, of course)?

Did he not lie on his back all day, in his hammock, and say "Not at home" when anyone wanted him, or there was any duty to perform? Didn't he tell people, every one who wanted to know, whatever private news there was in the officers' letters, which he had to censor? Didn't he on all occasions forget to tell the truth, till he no longer could distinguish between it and falsehood? Didn't he do all this? And now to be blamed and exposed and suspected! And then the poor little fat man, badly in want of a bath and clean clothes, blubbered, and said: "I know I've got the accursed Hollanders to thank for this. They just covet my billet, and now want to get me kicked out." Poor little man!

Yes, chickens come home to roost. As you treat others so you will be treated yourself. They were a couple—this little dancing dervish of an inspector, spare as a walking-stick, and this gaoler, small, fat, forty, phlegmatic, and—but never mind. Drop the curtain. Only, from my heart I say, may our fugitive friends escape detection! Another chapter will tell of their adventures. I cannot write them down now. We might be searched again. Who knows?

XXXII. Our Great Trek

Ever since December last we had been at stated intervals informed by our gaolers that the Government was going to move us to new quarters built for us north of the town, just outside it.

Nothing came of this move, however. Later on we were informed that the new place was reserved for the Ladysmith prisoners—the Ladysmith they were *going* to take every day of the last four months, but never did. Since the last escape of officers, however, there was such agitation in the town, amongst the women, so we were told, that on Thursday, March 15th, we were informed that on the following morning we would be moved to the new quarters.

What bustle and excitement there was that night; what packing and planning! Early on Friday some donkey wagons were loaded with our luggage and sent to the new quarters, about a mile and a half away. We were sent off in cabs. The Commissioner of Police, the same man who had sentenced the vicious policeman I told you of before, *without, however, carrying out that sentence*, came to me twice to ask me kindly to inform the officers that he had purposely provided cabs, so that no crowd should gather to follow and annoy us.

"We have to show the world," said he, "that we are a civilised people." He expatiated on this sentiment at length. I thought at once that surely the Majuba of 1900 had already begun to make its influence felt.

No such consideration would have been shown a despised prisoner before. Alas! I was reckoning without my host! My dream was doomed to a rude awakening. The awakening began immediately on arrival at the new premises, when we were invited to pay up for the cabs, offered to us, at 4s. per head, and look bright.

Officers leaving the Model School.[429]

My friend the Police Commissioner had kept up his Government's traditionary inability to tell the truth. The act was not one of courtesy—by no means. It was a little business transaction! The whole thing was a *show* of civilisation, there was no *reality* about it.

It was barbarism with a gloss over it—and but very little of that too. How little will appear if you read further. On arrival at the new quarters we naturally had a look round at once to see what they were like.

Imagine our disgust when we found the building one fit only for cattle—in fact, a large cattle shed. This they call civilisation here! Let me give a description. The building was a long oblong shed of iron, with thin boarded ceiling and *mud* floor, small windows far up near the ceiling like those of a prison, and no ventilation at all. The building stood *down-grade*, too, one side being about 10 feet higher than the other. The roof was terraced, but not the floor—it just sloped down; beds and chairs had thus to be propped up to get anything like a level. Oh, it was a magnificent fraud! At the lower end of the building, which was 40 feet wide, were five bath-rooms, 8 feet by 8 feet each. Then came one tremendous bedroom of 80 *yards* by 40 *feet*, in which our whole colony of 120 prisoners slept, the beds standing in four long rows. Then followed the dining-room, 60 feet by 40 feet, with four rows of tables and benches fixed in the mud floor. A room of 15 feet by 40 feet was a further extension, subdivided into a kitchen, pantry, and servants' room. The latter occupied the higher level of the building, and some fool of a sanitary inspector had so arranged that all the scullery water run down the full length of the building, down the higher side of it, in a small furrow. The building ran from east to west, sloping down that way as also from north to south. The whole was enclosed by barbed-wire fencing, the circuit being fully 600 yards. It was lit up at night by thirty-two large electric lights.

Waterval Camp

The Birdcage, British Officer's Prison, Pretoria.[430]

How the British prisoners were guarded at Waterval.[431]

Camp Interior.[432]

Inside the Waterval prison.[433] Note the extensive electric light poles.

The Bathing Tank for the Privates at Waterval.[434]

The building itself was well lit by electricity—this being the only good feature about it. The other sanitary arrangements were such as any civilised community would be ashamed to use for prisoners. It was a disgrace! And then that official's proud dictum would ring in my ears, "We must show the world that we are a civilised people." Civilised? save the mark! We were all simply horrified and disgusted. A strong protest was at once drawn up and signed by all of us; we feared, however, that no notice would be taken of it. The prospect of a prolonged stay is a horrible one. We are still in the rainy season. Our floor is now alternately mud and dust. The health of more than one will break down, I fear.

We try to make ourselves comfortable with pieces of wood on the floor, and spare blankets and bits of green baize. The Government have provided strips of linoleum to serve for carpets before each bed. Linoleum and dust, rugs and mud, beds and chairs at all imaginable angles, zinc-plate walls through which one could see the sun, moon, and stars, a variegated ceiling 11 feet high—imagine all this, and you have our cattle-pen.

Why, the Kaffirs at Mafeking, for instance, are more comfortably housed.

Even our chief gaoler professed to be terribly disgusted, and so too, we were told, the State Secretary, Mr. Reitz. I have ma doots, [my doubts] however.

I mentioned above that we were informed that the women of Pretoria were anxious to have us thus incarcerated, so that no one should escape. I do not believe that, however. What I suspect is this.[865] The Hollander editor of the paper keenly advocated our removal. The Hollanders in the town had petitioned the Government, we were told, not to press the sentence passed upon the brutal Zarp I told of.

These Hollanders, too, filled the place with insinuations and innuendoes against the officers, frightening everybody. And then they spread the report that the women—the Boer women—wished our removal. I gathered all this from many a hint dropped by some of the officials. A hundred Hollanders had come back from the Natal front to refit, or *refill*, I forget which! They were just the class of men capable of meanness like this, and they had the ear of the Government. Poor Hollander ridden Government! And here we are now, in a place hardly suitable to stable horses in, and the civilised barbarism of the official is satisfied, quite satisfied, and he is happy. They have revenged themselves on helpless prisoners! I have *heard* that Kaffrs do the same thing! I have *seen* it done here!

We have been here now for about three weeks. Eight days after our protest against this "hole" had been sent in the Committee in charge of this place visited us to investigate matters. The Landdrost of Pretoria, Mr. Schutte, is their chairman. He is a very venerable-looking man, and, I believe, a straight man. I have a suspicion that he has been made a tool and a fool of by the younger and more hot-headed patriots on the Committee and round the President, always fawning on him and flattering him all day long. This Mr. Schutte asked to be introduced to me, and was very courteous. He told me that he and the Committee had condemned our new quarters, but that the officials of the Public Works Department had over-ridden their objections.

He promised then that they would make several most urgently necessary improvements. This was two weeks ago, but *nothing* has as yet been done. Transvaal Government "to-days" are always "tomorrows," and consequently never come. Their way of doing business reminds me of a story that used to fetch me when I was a little boy. There was a certain shopkeeper who did what he could to attract customers. Amongst others he hit upon the following device.

He hung up cards everywhere in his shop on which was printed the following legend, "To-day you pay, to-morrow you don't." This drew large numbers of people "to-day" to make small cash purchases. "To-morrow" they returned again to make their "gratis" big purchases. Alas! there the legend stared them in the face: "*To-day* you pay, *to-morrow* you don't"—it was always *to-day*, of course. So it is here. The *to-morrow*, the *next week* of their *Government* promises, *never* comes! In our case they will very likely wait till some of us are down with typhoid fever, and then, too, till the disease has ended fatally for one or more, just as was done in the case of *several* prisoners of war who had been cast into the common gaol.[866]

It is indeed an ill-wind, however, that blows nobody any good. And this we too have experienced now. The wind was not so ill but that it blew us some good. I mentioned before that the Hollander machinations resulted in our being shifted to this place. But it also resulted in more. Since the last escapes these dear patriots, who pretend to love the Transvaal, whilst they are seeking only *self* and self-aggrandisement, have been dropping insinuations that our gaolers were not the right men in the right place. They were quite right as regards the chief gaoler, Opperman, but not as regards the second, who was a cultured gentleman.[867] But the result of these

[865] Hofmeyr note: This suspicion was afterwards verified.

[866] Hofmeyr note: After three weeks some drainage pipes were laid, and a few sanitary changes made—not much of an improvement, however.

[867] Dr. J. W. B. Gunning is one of the administrators of the State Model School Dr. J. W. B. Gunning, per Brian Davis, Churchills in Africa p. 201 See also: The Strand Magazine, No. 112, April 1900; University of Michigan and Google Books for an article on his Staats Museum of which he is the director.

insinuations was that our gaolers were dismissed, and also our guards, composed of the Zarps I have before described, whilst a *Hollander* gaoler and *Hollander* guards were appointed.

Nobody was sorry to see the first go, but all grieved for the loss of the obliging doctor.[Gunning] The new gaoler [Westernant] has hitherto been exceedingly courteous and gentlemanly. If he continues in the course adopted he will be an immense improvement upon his predecessor.[868]

This little incident, however, illustrates clearly how things are done here. *Transvaalers* are kicked out of good positions to make room for intriguing *Hollanders*. The Transvaal leaders know it, and yet they do not; they see it, and yet they see not. If this war had not taken place there would in a few years' time have been not a *Transvaal* Republic, but a *Hollander* Republic with a Hollander (Leyds?) President.

Dr. Willem Johannes Leyds.[435 869]

Francis William Reitz.[436 870]

Poor, deluded Transvaal! This same gaoler, Opperman, apeing the President, often told me how useful the Hollanders were, in fact, a pillar of the State; *now*—well, his language won't bear repetition.

Let me say it again: had this objectionable Hollander element not been introduced into the country by Kruger and his satellites, this war would not have come, and Transvaal independence would never have been endangered. I but endorse the opinion of many and many a sober-thinking, genuine Transvaaler. But there are none so blind as those who will not see. And for such willful blindness one always has to pay dearly. Here the payment will be a terrible one—a degrading one. Friend Opperman has now gone to the front, accompanied by his friends the Zarps. His rival, the Hollander, enjoys the *otium cum dignitate*[871] here. The former risks his own life at the front; the latter sits in safety and spouts patriotism and bunkum. Opperman deserves all he gets, and yet the pity of it!

Poor Transvaal—deluded Transvaaler!

[868] Per Frankland March 25th: "This evening the new Commandant had roll-call. We call him "Pyjamas," because he wears a suit of clothes for all the world like a pair of pyjamas. His real name is Westernant."

[869] Willem Johannes Leyds (1 May 1859 – 14 May 1940) was a Dutch lawyer and statesman who served as state attorney and state secretary of the South African Republic. From May 1 1889 – May 31 1898 he was succeeded by Francis William Reitz. He went to Brussels and until the conclusion of the war he was the Republic's special envoy and minister plenipotentiary accredited to several European states. Willem Johannes Leyds - Wikipedia

[870] Francis William Reitz, Jr. (Swellendam, 5 October 1844 – Cape Town, 27 March 1934) was a South African lawyer, politician, statesman, publicist, and poet who was a member of parliament of the Cape Colony, Chief Justice and fifth State President of the Orange Free State, State Secretary of the South African Republic at the time of the Second Boer War, and the first president of the Senate of the Union of South Africa. Francis William Reitz - Wikipedia

[871] Leisure with dignity.

XXXIII. A Lost Chance

The Transvaal Government did not make use of its last chance referred to in a former chapter. There was an attempt to make peace, but it was an impertinent, bungling one, and of course failed.[872] In the chapter on Majuba I tried to point out how little right the Transvaal Government has to expect special Divine interference in their favour. My opinion has been strengthened since then by after events. Let me refer to a few of these. They throw light on "the lost chance."

The President visited the Free State after Majuba Day, and addressed the burghers at Bloemfontein. In the course of his speech he said: "It was on Majuba Day that General Cronje surrendered. It was on that memorable day that Ladysmith had to be given up. Was it not because God wished to punish us for making that day a day for boasting instead of returning thanks for the great victory given by Him to our armies?" (*Standard and Diggers' News*, March 8, 1900).

This is fine *talk*. But what does the President *do*? To-day (beginning of April) his burghers are still kept in ignorance of the number taken with Cronje, and of the real events of that memorable fight. When Lord Roberts was marching victoriously on Bloemfontein *official* telegrams informed the people that the British were fleeing in all directions, after Abraham's Kraal. The President *knew* of these lies, but never said a word! The Boer forces are driven out of the vicinity of Colenso and Ladysmith; they *retreat* in an awful hurry to the Biggarsbergen.

There was some terrible fighting in those days in Natal. The English losses they put down at thousands. And their own? Listen: the official report said fifteen killed and twenty-five wounded—forty casualties. They left all their positions because they suffered forty casualties! Is this true? Then certainly they were arrant cowards. If not, then they are abominable liars. Cowards or liars! Not much to choose from. This is what the President and his Government *do*. How different to his *talk!*

More still. He appeals to the American people for sympathy on November 10, 1899. His appeal begins thus: "The main question in dispute between this Government and that of England was in regard to the *Franchise*," This is what he says. What does he *do?* He allows his mouthpieces, the two newspapers and the officials and the Commandants everywhere, to proudly aver that they fight because of the "Age of Injustice that is past," because of their hatred of England, because they want to drive the English out of South Africa, as the insulting leading article of the *Volksstem*, March 8, 1900, told us.

Oh, what a degrading spectacle to find a President who *approves* of misstatements (let us call them so) whilst he stands up thanking God that he is better than the English Government and people!

No wonder, then, that the infatuated Presidents, Kruger and Steyn, could pen the following proposal to Lord Salisbury: "The blood and tears which have been shed by thousands in connection with the present war, and the prospect which prevails of the ruin of all moral and economic conditions in South Africa, make it necessary for both the parties to the war, in the strongest and most emphatic manner, and as in the sight of the Triune God, to ask why they are at war with each other, and whether the objects they have in view justify them in the continuance of such a terrible and heartrending conflict.

For these reasons, and seeing that a number of her British Majesty's statesmen have declared that this war was commenced and has been carried out with the express purpose of undermining her Majesty's rule in South Africa, and in order to establish a Government wholly independent of that of her Majesty throughout the whole South Africa, we consider it to be our duty to solemnly declare that this war was commenced only as a defensive measure, with the object of preserving the threatened independence of the South Africa Republic, and is only being proceeded with a view to the securing of the unquestionable status of the two Republics as sovereign international States, and for the preservation of that status and for the purpose of securing that those of her Majesty's subjects who have taken part with us in this war should not suffer, either in person or property."

Is it possible to imagine any statement more misleading? The States declare war, and then they say they did not want it. They say they undertook the war purely as a defensive measure, and at the very start they invade Natal and the Cape Colony, whilst their battle cry is "Table Mountain!" Then they want protection and immunity for "her Majesty's subjects who have taken part with us in this war"—for *rebels*, in other words—men whom they

[872] 3/5/1900: Overtures of Peace made by Kruger and Steyn to British Govt.
3/11/1900: Presidents Kruger and Steyn received reply from the Prime Minister refusing to entertain their overtures for peace.

have invited to become rebels by spurious promises, or compelled to become such by force. These foolish people must take Lord Salisbury for a fool; in this supposition, however, they found themselves vastly mistaken.

But let me proceed with the despatch. "In this way, and in this way alone, we are still, as we have been in the past, desirous of securing the restoration of peace in South Africa, and of bringing to an end the calamitous war which at present rages in South Africa. Should her Majesty's Government, however, have definitely decided that the two Republics are to be destroyed then there remains for us and our people nothing but to continue in their present struggle to the end in spite of the overwhelming might of the British Empire, in the confident trust that the God who has implanted in our hearts, and in the hearts of our fathers, the inextinguishable love of freedom will not forsake us, but will complete His purposes in us and our posterity."

Love of freedom! Alas! the Government has no idea of the meaning of the word freedom. Licence—that is what they understand. Liberty—there is not a man in the Transvaal to-day who dares differ from the high officials in his opinion of men and matters. If he does, and is a non-burgher, he is put over the border at once. Is he a burgher, not belonging to a big family clique, he is ordered to shut up. If he is an influential man he is bribed. And then they dare say this is the love of liberty God has implanted in their hearts! But it was a slip of the pen, I suppose. They meant love of tyranny and oppression. And to ascribe this to the Almighty is blasphemy. Last Sunday one of their clergymen spoke from the pulpit about the burghers' disobedience and disorderliness in the field. He told them that even a youngster of *sixteen years* had dared to insult the late General Joubert in the lager. *And nothing was done to him*, for was he not a burgher, and are the burghers not free? Chickens come home to roost. This is the doctrine Krugerites loved to preach, applying it to non-Transvaalers, to whom they could do what they liked. England I call a free country; and we Cape Colonists enjoy real liberty, but not the unhappy, misguided Transvaalers.

But to continue. "We have hesitated to make this declaration to your Excellency at an earlier date, because we feared that, so long as we were victorious, and our army continued to occupy defensive positions at a distance within her Majesty's territories, such a declaration would have wounded the honourable feeling of the British nation.

"Now that the prestige of the British Empire may be considered as having been preserved by the fact that a portion of our troops has been captured by the forces of her Majesty, and now that they have consequently been compelled to evacuate the positions which have been held by them, now that that difficulty is removed, we are not justified in longer hesitating to make clear to yourself and your people why we fight, and on what basis we are prepared to restore peace!"

How kind and condescending! And while this communication is on its way to Lord Salisbury the Transvaal Government's puppets and mouthpieces are publishing the most disgustingly false and insulting statements regarding our Queen and Empire, and particularly regarding Lord Roberts and his army. Aye, their hope will be justified, I fear—"God will complete His purposes in us and our posterity." His purposes are that they must be honest and upright as were their fathers, as the real, genuine Transvaaler to-day is still, but not hypocritical and false as the Government clique to-day is. We all know how dignified and decisive Lord Salisbury's answer was to this impertinent peace proposal. Let me quote one sentence:

"This great calamity has been the penalty which Great Britain has suffered for having in recent years acquiesced in the existence of the two Republics.

"In view of the use to which the two Republics have put the position which was given to them, and the calamities which their unprovoked attack has inflicted upon her Majesty's dominions, her Majesty's Government can only answer your Honours' telegram by saying that they are not prepared to assent to the independence of the South African Republic or of the Orange Free State." And thus Kruger *lost his chance!*

Bloemfontein is already occupied by the British troops and the greater part of the Free State is conquered. Soon Lord Roberts will advance and the same fate overtake Pretoria and the Transvaal. When it is too late this willfully blind Government will rue its folly and dishonesty. Their punishment will be a hard one, but fully deserved. Then peace and prosperity will come to South Africa, under a flag which guarantees *real liberty*. England, I know, will not repeat the mistakes of 1879–1881.

No, the right men will be appointed to rule. The appointment of Mr. J. G. Fraser[873] as an official at Bloemfontein guarantees this. And the result will be that the sensible Transvaaler and Free Stater, experiencing what liberty really is, will soon become a most loyal subject, and then feel grateful that Paul Kruger has now *lost his chance*.

XXXIV. In the Catacombs

I can now finish the story of the escape of our three friends, as I promised to do in a former chapter. The reader will remember that on the morning of Majuba Day[874] three British officers were missing, and great was the search made for them inside the building as well as over the surrounding country. All in vain, however, and very naturally so, as will appear from the following. What really happened was this. Some of the officers had cut the electric wires in the roof of the Model School down to a single strand. A piece of twine was attached to the wire where the incision was made, and let down through the roof to the bedroom below. The plan was that at about ten o'clock that evening, the 26th, an officer in the room would jerk the string, thus breaking the wire and putting all the lights out. In the ensuing darkness, before the guards could be increased, some ten or twelve would then try to escape.

This special evening was fixed on, because we expected that our removal to the new quarters prepared for the officers would take place the next day. It was believed that the triple combination—viz., the Majuba Day reverses, the officers' escape, and the removal next morning of so large a number and its consequent bustle and excitement—would facilitate the escape of the fugitives. Unfortunately the football, as related before, spoiled the *coup* that had to come off at ten o'clock on the night of the 26th, and the guards were accordingly strengthened before anyone could get away.

The three officers who eventually got away were, however, determined to make a try. And this was what they did. The Model School is a big building and fairly high, resting on foundations about 2 1/2 feet above the ground. Under every one of the large rooms, therefore, there was an opening of about 20 by 20 feet and 2 1/2 feet high. Our three comrades opened a screwed down trap-door which led into these *catacombs* below, from their own room, and descended during the night, before the discovery next morning took place. Another officer in the room drew his bed over the door, and everything in the room proceeded as if nothing had happened.

Next morning, accordingly, when the discovery of the escape took place, and the hue and cry was raised, our three friends were comfortably, or rather uncomfortably, housed in the catacombs under the very feet of their eager, anxious would-be captors. The reader can imagine our keen enjoyment of the situation. It was as good as a farce to see our gaoler and the police rushing up and down in great excitement, peering behind doors and under beds, searching for the fugitives in all possible and impossible places, whilst all along, like the "beautiful stars" of the song, they were "so near and yet so far." Verily this was our Christmas pantomime! But now cometh the tragedy. During the day we discovered that the report of our contemplated removal was a premature one. The new quarters were not ready yet, and would not be for some time to come.

This meant that our friends down below would have to wait and live in the catacombs for an indefinite period. How could they come up and surrender? But, also, how could they live in the darkness and circumscribed space and damp for any length of time? What now? Let me mention here that our friends were fed through the trap-door, and that we held communication with them mostly by writing notes. And so it came about that I received a letter from below, after a few days, from Captain Haldane, asking me to ascertain from the authorities if we were going to be removed, and if so, when. I was also requested, in case the first query was answered in the affirmative, to try to hurry our departure and consequent release of our entombed friends.

[873] Sir John George Fraser (17 December 1840 – 22 June 1927) was a prominent Orange Free State lawyer, politician, statesman and member of the Volksraad. Fraser was a key player in the negotiations with the South African Republic and President Kruger in the late 1880s. He provides much detail in his book Episodes in My Life (Cape Town: Juta, 1922) about negotiations regarding the proposed Customs Union, the building of railways and defence pacts. Fraser distrusted Kruger. Fraser stood for election to the position of President of the Orange Free State in 1896 and was defeated by Marthinus Theunis Steyn (the husband of his niece Rachel Isabella Fraser). He resigned his position as President of the Volksraad in 1896. He remained in Bloemfontein during the Anglo-Boer War. Following the departure of the Boer Army from Bloemfontein and the imminent arrival of the British Army, on 15 March 1900 John Fraser, accompanied by the Landdrost, Mr. H.D. Papenfus and the Sheriff, Mr. Raaff, went out and surrendered the city to Roberts. John Fraser was knighted in 1905 and elected as a Senator in the first Union Parliament in 1910 John G. Fraser - Wikipedia.

[874] February 27th, 1900, the day Cronje surrendered.

Well, the answer I obtained was most positively affirmative: "You will be moved as soon as the new place is ready, for there you can be guarded much more securely." We knew already what "soon" meant in the vocabulary of the authorities, so I set about finding some means of accelerating their movements.

I took our gaoler apart and said to him, very solemnly and seriously, "Remove us as soon as possible."

"Why?" asked he. "The lights in the new place are not in yet, and we can't get workmen enough."

"Well," said I, "hurry up—that's all I can advise you to do."

"But why?" asked he again. I looked solemn and gloomy, and said, "Do you want to wake one morning and find yourself dead?"

"What do you mean?" asked he, as his face visibly paled.

"Half a word is enough to him of good understanding," I answered.

"But do you know or hear of anything—any conspiracy, any plot?" and beads of perspiration stood on his forehead.

"Certainly not," said I, "for if so I would assuredly not tell you. Only I judge by the signs of the times—four officers gone, reverses for your troops everywhere—look out—move us at once."

Our guardian angel was now thoroughly frightened, and immense was the pressure brought to bear on the authorities and workmen. And we too were anxious to be away, as our quarters were becoming dangerously crowded, and we were told the new quarters were so comfortable. What otherwise would have taken four or five weeks certainly was now accomplished in about seventeen days.

Daily our poor friends down below were fed and cheered by good news from the front, and words of comfort for themselves. But gradually their position and condition were becoming unbearable. Imagine, therefore, their joy and relief when it was notified that the removal would take place on Friday morning, March 16th [1900].

And so it did. Our friends were provided with as much as they could carry for a long, weary march towards Delagoa Bay, and our last good byes were said. I still have in my possession, and treasure highly, Captain Haldane's last note to me. And thus full of expectation and suppressed excitement we were off at about eleven in the morning. That night, as we know now, our three friends coolly walked out of their living tomb, and out of Pretoria, and out of captivity. After innumerable hardships and many dangers and narrow escapes they reached Lorenco Marques, tired and ill and worn, but full of pluck still, and rejoicing in their regained liberty.

Needless to say we heartily rejoiced with them when some weeks later we heard of their good fortune. But yet, when we think of it all, we cannot repress a smile. It was all so ludicrous—our little dancing Dervish Police Superintendent poking his little stick into every corner, and our perspiring, trembling gaoler following him everywhere and inciting him to still closer search and bolder efforts. And all the while they were literally standing on and walking over the heads of the very fugitives they were looking for—heads bowed low underneath in the catacombs!

How sick and disgusted these officials must have been when they learned the true story and whole inwardness of this second escape! I can still hear the little Police Superintendent (a German) say, "Eschape is impossible; I and mine polices are too vide avake!" One could not teach anybody anything in this country. What they did not know was not worth knowing. Did they not know that England could never win, and that no foreign Power would allow them to be conquered? This is what they knew. What they knew *not* was that their leaders had deceived them as to England's power and the attitude of the other countries. And was not this worth knowing?

Aye, had they but known! If they could but have had honest and true rulers all this misery and bloodshed would not have come over the country. Verily the ex-Transvaal Government has much to answer for! But to return to our three escaped friends. Captain Haldane and Lieutenant Le Mesurier managed to keep together, but lost Brockie after some days.

Let me tell what happened to the latter before I close this chapter—the story of the former has been told in several papers already. After losing his companions Brockie made his way out alone to the Portuguese border. The last station on the Transvaal side is Komati Poort, the first on the Portuguese side is Ressano Garcia, and in between is the bridge over the Komati River. The two stations are little more than a mile apart, the river being the boundary between the two countries. Brockie eventually arrived at the Komati Poort Station, looking for all the world like a dilapidated tramp. He felt convinced that only bounce and the proverbial tramp's cheek could now help him out of as tight a corner as one could well get into. He had a fund of both, fortunately, and so marches up boldly to the manager of the refreshment room, and playing the *role* of tramp in the uttermost stages of destitution begs to be taken on as barman.

The manager has compassion on him, and soon our hero is installed behind the bar, where he does the honours worthily. Inside the bar-room there is a Government notice on the wall—it is a description of the three fugitive officers, and an offer of reward for anyone recapturing them. There are men reading this notice and discussing it. Brockie joins the group and launches forth, as a good and loyal *burgher* of the State, in fierce invective against the English in general and the escaped officers in particular! He tells all and sundry what he would do in case he caught them, and promises drinks all round when he takes the reward. Little did the others surmise how near "the reward" they were standing themselves, and how easily they could have earned it.

For a few days Brockie played his part of barman to perfection, selling wines and beer, the sale of spirits being prohibited in the Transvaal during the war. Then, all of a sudden, the manager noticed that his model barman was seized with ague, a precursor of fever—and Komati Poort is a frightful fever hole. What now? With chattering teeth Brockie himself makes a suggestion. "Give me some money," said he, "and I'll go over to the Portuguese side to buy whisky." "But the bridge is guarded," said the manager. "I'll square the sentry," answered Brockie. And off he crawls didn't he have the ague terribly! Moaning and miserable he approached the sentry and begs to be allowed over the bridge, just to get a little whisky for himself and his manager, who was seized with ague even worse than himself. He begged so piteously and promised the sentry such a grand share for himself, that with an injunction to return at once he was allowed to cross over.

Brockie, be it known, had timed his ague and expedition very nicely—just when he knew that a train on the other side was ready to start for Lorenco Marques. And so over the bridge he crawled, and over the intervening ground, and when the sentry was out of sight, he ran, fresh and strong and healthy, till he reached the station. Never was man so quickly relieved of ague—it was a marvelous and quick cure. It must have affected Brockie's mind, however, for instead of returning at once and bringing the thirsty, longing sentry the whisky he had promised him, this absent-minded beggar of a fugitive took a seat in the train, and was whirled away to Delagoa Bay! How could he be so forgetful!

They say that oft in the silent midnight watches a sentry's impatient steps can be heard on the Komati bridge, and in between the curses deep and heavy, over the waters of the river ever floats the sad refrain of the song that sentry sings: "Oh, where and oh, where has my whisky-bottle gone?" And the crocodiles crawl out on the banks and each one wipes a tear from the other's eye. Thus Brockie escaped.

Crossing the Komati River.[437]

Komati Poort Railway Bridge.[438]

XXXV. Our Information Bureau

It is hard enough to be a prisoner, but when an unfortunate prisoner, thirsting for news from the outside world, is compelled to be satisfied with the most unreliable and willfully perverted statements, then indeed his lot is a most pitiable one! And such in very deed was our condition. The Transvaal Government allowed the publication of no paper, except its own inspired organs. And these were two in number—the *Standard and Diggers' News*, and the *Volksstem*, published at Johannesburg and Pretoria respectively.

These two papers were simply Government organs, and dared publish nothing the President or the Executive disapproved of. No independent judgment was allowed to be exercised in this unhappy country, no one dared offer adverse criticism. There is no free Press and no free speech. Anyone who dared say of and to the Government what, for instance, the *South African News* or *Ons Land* in the Cape Colony published in every issue, would find himself within the four walls of a prison cell in no time.

This muzzling of the Press was an egregious error on the part of the Republics, and a fruitful source of frightful evil. Let me mention only one result, already referred to. Every one knew that the statements of the above-mentioned papers were approved of by the authorities. Every intelligent reader soon found out that the majority of these statements—the war news especially—was frightfully and willfully misleading, if not purely false. Both the papers and the Government continually invoked God's blessing, and expressed their belief in the righteousness of their cause.

To what conclusion had the intelligent reader to come now? "Our cause is just," so the authorities and the papers say—and yet these papers tell such horribly false tales, and the authorities allow this. A *just cause*, bolstered up by *disgraceful lies*—surely something is wrong somewhere. No wonder that one result was this— *public confidence* was shaken; no one knew what to believe. A medical gentleman—one of the Transvaal—said to me: "I do not read the papers any more. One cannot believe anything they say." "But how can a professedly Christian Government allow such lies to be published, and, what is more, allow the papers to call their news *official?"* said I. "That is what I cannot understand," said the doctor; "something is radically wrong somewhere." And this opinion I have heard many a one express. No wonder that we poor prisoners pined for true news. And we got it too. Let me tell the reader *how.*

During the first weeks of our imprisonment we were permitted to buy only the *Volksstem*, from whose disgraceful columns nothing could be culled but insult to our Queen and abuse of everything English. As for the war news—why, it was simply a farce. In a few weeks' time they had killed off thousands of British troops, whilst their own losses amounted to next to nothing. As an officer expressed it in verse—

"Five thousand British now lie low,
One Boer is wounded in the toe.
Such is the news you learn to know
In prison!"

I do not mean to say that our English and Colonial papers never told tall yarns. Oh no! But when they published a telegram from "Roberts to High Commissioner," giving the news of the war, then every one knew it was absolutely correct. What was *official* was *true.* And this was *not* the case in the Transvaal.

Extracts from the papers I have given before prove this sufficiently. After Churchill's escape the paper was forbidden us, and the idea was that we should be cut off entirely from all sources of information. What now? Well, all we could do was to practise self-help. And it was done, too, most successfully. Let me mention instances.

Now and then a relative or friend visited me, and from them I managed to get some information which I knew was always absolutely correct. Thus, *e.g.,* in December already we learned that the Transvaal had to supply the Free State with food and money, as the resources of the latter State were practically exhausted.

Up to the time of our removal to the second prison the Government allowed us to hire a cook, a Colonial coloured boy, who, unknown to the authorities, was a British subject. We provided him with money, and the old boy faithfully smuggled in the papers for us, even a Natal paper occasionally.

How he managed to escape detection is a marvel, as he was searched both when he came and departed. One thing in his favour was this: he could put on a look of angelic innocence such as I have seldom seen before. And should a Zarp or gaoler warn him, or suggest only anything of this kind on his part, a look of injured innocence so perfect, so pathetic crept over old Johnnie's features that one felt inclined to apologise for ever having been so cruel, so mean, as to suspect him.

Old Johnnie was an old man, and used to slouch wearily and heavily through the streets as he came and went, and he pretended to be so deaf and stupid! Why? Oh, that was his cuteness, to listen to the conversation of the passers-by, and thus pick up threads of information. Arrived at our prison he detailed to us and others what he had culled during his peregrinations.[875] And how we snatched at every straw! Yes, old Johnnie, you were a most successful newsmonger, and we always looked forward eagerly every morning to your Kaffir-grams!

Another source of information was the Zarp himself. Some of us were always prowling about hoping to pick up some of our guards' conversation and thus gather news. As the men spoke generally in Dutch, however, I was, perhaps, the most successful in obtaining Zarpigrams.

I used to go out occasionally when I saw them talking excitedly and slowly walk past with a look on my face as if I were weighed down by deep thought on the most abstruse of philosophical problems, trying to solve the vexed question of the exact number of spirits that could find dancing space on the point of a needle, or some other problem of similar practical importance. All the time my ears were strained to catch up every word, as only a news-hungering prisoner can strain his ears. And very often I did pick up some important bits of news. Let me give a few. It was late one Sunday evening when I overheard a Zarp telling his companion that he had been in the church that night, where all the people were so deeply moved when the minister told them that Government had informed him that "up to Spion Kop the Boer losses in *Natal* were 500 killed."

"What!" said the other, "500 killed in Natal alone? Is it not horrible?" "Yes," repeated the first one, "500 in Natal. And then the minister prayed so tenderly for the widows and the orphans." Now the papers had led us to believe that the losses had been very slight. Naturally my repetition of the news to the officers caused great sensation. Then I heard this too one day. Early in the year there were reports that the great General Joubert had been wounded, or was killed, or was dying. There was no certainty, it seems. Discussing the matter I heard a Zarp say to another, "Won't the Hollanders be glad if the General dies?" "Yes," said the other, "they never liked him because he never could stand their underhand ways." Said the first speaker: "If General Joubert had had his way we never would have had the Hollander here, nor this terrible war."

Another source of information was the hospital. There was always a sick or a wounded officer in there, and till we were moved to the other place friends were allowed to see them. The matron and nurses and several of the attendants in hospital were exceedingly kind, and many of them sympathised heartily with our cause. They gave the invalids news and also papers, which we, when we visited the hospital, again casually became possessors of. And thus we were supplied with hospital-grams. And they were always more reliable than the newspaper telegrams—their sources were purer. But not always, as proves the following instance. At the time of the Spion Kop disaster I was in hospital myself unfortunately. When the first bits of news began to come in our gaoler came to my room in attendance on a visitor. When the latter had departed he told me, with a very serious face, that he had just come from the President, who had read out to him the following telegram (I took it down at the time): "British fleeing in all directions. Have killed 3,000; taken 2,000 prisoners; still disarming; will take 6,000 at least. Whole British army back over the Tugela; 1,000 drowned in the river." Well, I challenge any American paper to tell a taller yarn. Of course I was greatly moved by the frightful news.

"And now," said I, panting with excitement, "what will Buller do?"

"Well," said he, "the soldiers won't fight any more. The officers have turned their cannon on them, threatening to shoot if they do not advance. It is all up." "And how does the President take it?" asked I anxiously, dreading the effect that this tale might have on the old man. "The tears are rolling down his eyes," said my gaoler, "and all he says is just this: 'God is good; God is good.'"

"Yes," thought I, for I dared not say it, "good in that He does not let an Ananias fate overtake you."[876] Next morning I saw my friend again, and when I opened the subject of this Munchausen[877] story he said, with a hangdog look, "There was a mistake in the wire; about 200 prisoners only were taken." "And the killed and the drowned?" I asked innocently, "was that correct?" But he heard me not. A look so far, and withal so innocent, stole over his features that I could not repeat my question when he asked me so kindly, "And how are you this morning? Much better?" And thus the subject dropped. Only—and the reader anticipates my question—who

[875] A journey made by a pilgrim; a pilgrimage. A journey or trip, especially by foot.

[876] Ananias and his wife Sapphira were reportedly struck dead after lying to the Holy Spirit about the amount of the proceeds of the sale of their farm they were donating to Peter and the Church. Acts of the Apostles chapter 5. Ananias and Sapphira - Wikipedia

[877] Baron Munchausen tells tall tales. His fictional Baron's exploits, focus on his impossible achievements as a sportsman, soldier, and traveler; for instance: riding on a cannonball, fighting a forty-foot crocodile, and travelling to the Moon. Baron Munchausen - Wikipedia

told the fairy tale originally: the gaoler, the telegraphist, or the Pr——? But stop! Solve the riddle at your leisure. Let my story move on.

Another source: a magistrate had been captured by the Boer forces in Zululand and brought to Pretoria, accompanied by his wife and little daughter of about four years. Mr. Hignett, the magistrate, was quartered with us, whilst his wife and child stayed in a private boarding-house. A generous Government allowed Mrs. Hignett to come and see her husband once every two weeks in the gaoler's room and under his supervision. The brave lady, therefore, thought out a plan. She managed to convey to her husband in one of her visits a code of signals. She wrote out carefully what the different colours she should happen to wear would mean—colours of flowers and ribbons. For instance, *white*, with certain combinations, would mean good news from Ladysmith, and with other combinations bad news, and so on. And thus she had planned a code for Natal, Free State, Kimberley, and Mafeking—a simple and effective one. Then, as she walked past the prison, her husband made a note of everything she wore, and decoded the message. Aye, and how we grouped round him to hear the news! The brave lady and her dear little child passed by every day. How we looked forward to that coming! Verily her visits were an angel's visits. But there was a sad side to it all the same. The sweetly little girlie was passionately fond of her father, and as they were such utter strangers her mother could not leave her alone at home, but always had to bring her along.

And then, whenever they passed our prison, the little one would burst out crying and stretch out her arms to her father so pathetically that we on-lookers, who had dear little ones far away, could not bear to witness the sight. It was heartrending. On the day we were taken to the new prison I was standing close to Mr. Hignett. His wife and child had managed, taking advantage of the crowd, to sidle up close to him. Before a Zarp could interfere the little one sprung forward and flung her arms round her father's neck, sobbing bitterly, moaning with broken voice, "Come home, Daddie; come home with us." It was a piteous sight. It brought a mist before many an eye. The guard grumbled a little, but allowed the father to hold his girl in his arms for a while and say a few words to the mother. Then came the goodbye and the tearing asunder. And as we moved off the last we heard was the bitter wail of the dear little girl. How our hearts ached!

After a while, especially after our quarters had been shifted, did not the plucky butcher- and baker-boys smuggle in papers, and mumble the news to our soldier-cook as they delivered their wares? All honour to those plucky fellows. Their butcher- and baker-grams were sources of great comfort and excitement too to us.

But now I must tell of the grandest achievement of all. It was in the month of December already that we noticed a gentleman, always accompanied by a couple of large dogs, passing our quarters regularly every day. He usually seemed greatly excited, whirling his walking-stick about his head most energetically. He was quickly known as the dog-man.

At first we could not interpret his manoeuvres; but soon we did. He wanted to draw our attention, and when he perceived that he had succeeded in doing it he sidled up as near the iron railings of our prison as he dared, and tried to impart information by emitting short, disjected sentences. Thus one would hear him grunt: "Keep heart; things not so bad." "Battalions of troops coming." "White doing splendidly." "Can't touch Kimberley." "Little Mafeking makes them sit up," &c. Naturally after this discovery we eagerly watched for the advent of the dog-man and his doggrams every day. But he grew bolder. In a little while he began signalling as he walked past with his stick and handkerchief, giving us the latest bits of news, informing us, too, who he was. Let me tell about this brave gentleman. His name is Patterson, and he was one of the chief clerks in the telegraph department. He is an old resident in the Transvaal and a burgher, but remained an Englishman in sympathies, and would not fight against his countrymen.

As one of the chief clerks in the office he had ample opportunity of becoming acquainted with the authentic news that came in from the front. He himself often had to decode cipher messages. And now his heart just yearned to give us the news. He began in the way I indicated, but did not stop there. Just opposite the Staats Model School, our prison, is another corner house, such as ours was. We knew not who lived there, but we could see the inhabitants were friendly, at all events. Imagine our surprise, then, when one day we saw our friend Patterson standing in the long passage of the opposite house signalling vigorously with a little white flag. How excited we were! An officer—Captain Burrowes[878]—immediately went inside and signaled back through the upper panes of the opposite window.

[878] Possibly: Arnold Robinson Burrowes. Captain, (Princess Victoria's) 1st Battalion Royal Irish Fusiliers QSA (4) QSA Source: List of

The lower window was of frosted glass, as it was a school. And this fact was of immense importance to us. Captain Burrowes could now get on a box and signal through the upper window, and thus escape detection by the guards. Aye, it was an immense scheme and an immense success! Our friend used to come three times a day, generally before our meals, to signal the latest. Captain Burrowes would then write the "wire" and hand it round as we sat at table. How our hearts beat!

What suppressed excitement reigned supreme! And when the news was good, as it generally was then (from February 1st), how happy every face seemed! Verily, many a day we almost forgot we were captives; and there were days when the excitement caused by specially good news was painfully intense. Let me give an example of how well Patterson served us. Cronje surrendered at 7 a.m. on Majuba Day, February 27, 1900. At 7 a.m. our friend had signaled the news to us! We knew it practically as soon as the President, and before the world knew anything; yes, and days before the Transvaal people were told of it. Was it not grand? We dared not let out that we knew, of course; that would create suspicion. Nor could we openly show our elation. But in our rooms and corridors many a pathetic scene was enacted.

We pressed one another's hands silently, but firmly. Tear-dimmed eyes looked into others suspiciously misty. "God save the Queen" we said as we joined hands, and hummed softly as we walked about. Aye, such experiences must be gone through, lived through. One cannot describe them. Reader, you cannot understand what all this was to us, for you were free. We were prisoners, and that made all the difference. I can still see my good friend, Captain Burrowes, quietly, solemnly stalk up to our little prison corner, where I used to sit at my little table, and with solemn voice tell me the latest before it was made public.

How happy, how amusing he always was! Yes, Captain, we will not forget one another in a hurry; will we? Let me give one of Patterson's telegrams. On Friday, February 16th, he signaled: "Kitchener relieved Kimberley. Hurrah! Last night at midnight our (Boer) telegraph clerk reported 'British are on us and we fled with General de la Rey.' Hurrah! Just had bottle of fizz on it!" And this we received at eight o'clock in the morning! But better still. Patterson was getting so reckless in his desire to help us, that we had to ask him to be more careful. His coming up so often, during the day especially, we thought would create suspicion. Imagine what the brave fellow did then?

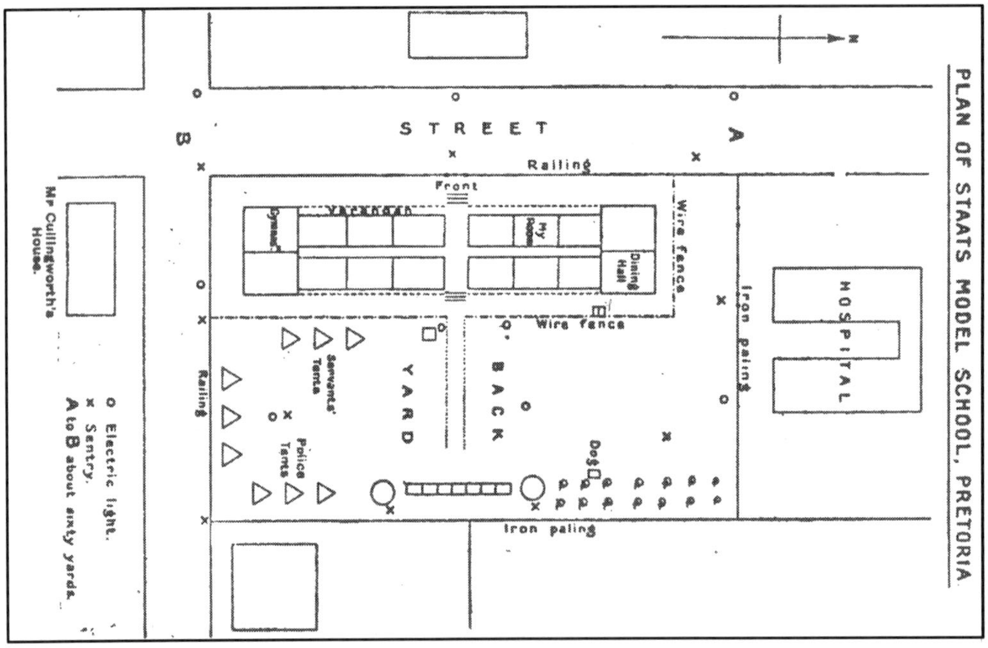

Plan of Staats Model School showing the location of the Cullingworth's house at the bottom.[879]

He taught the daughters of the gentleman living in the house, Mr. Cullingworth, flag signalling. And every one can understand our enthusiasm when one morning we saw the ladies signalling away bravely. Their little sister

QSAs with the clasp Talana from the Talana book Anglo Boer War - Name search.

879 PF: Note Cullingworth's home on the lower side of the south side of the 'school.' Also an overlooked point is that no one attempted to involve them in any escape plan. Nor did Churchill or Haldane and Brockie seek asylum or assistance once they had escaped. What noble men. As noted above both Patterson and Cullingworth were called up to the front and Burghers and, went to prison rather than engage in hostilities. Image: Haldane, How We Escaped from Pretoria.

stood outside in the garden on the *qui vive* for Zarps and detectives or other passers-by. And it was most interesting and amusing to us to watch how quietly and cutely the little girl gave warning of the approach of any one.

Then came a sad day. Just when we were being shifted to our new quarters both Patterson and Cullingworth were commandeered to go to the front. We afterwards learned that both were imprisoned for refusing to fight, and released only when Lord Roberts entered Pretoria. The young ladies signaled before we left that they would still try to give us news. And several times, when there was good news, they strolled past the prison fences, signalling with their handkerchiefs. Soon, however, this became impracticable, as nobody was allowed within half a mile of our premises. The end had then fortunately already come within sight, however. An amusing episode every day was the arrival of the *Volksstem*, which we were now again allowed to buy.

We found that as a rule the Dutch part contained more and better war news than the English columns of the paper. On its arrival, therefore, a large circle was generally formed around me, whilst I translated the Dutch columns. This was always an interesting event, causing great amusement—especially the correspondence column—but also disgust and indignation. The doctor who visited us in our new prison was a Hollander, a courteous and kindly gentleman. From him I managed to draw scraps of information whilst talking about mutual acquaintances in Holland, and in Utrecht especially, and whilst consulting him about a headache that was going to come, or an indigestion that might ensue given certain conditions, or a certain undetectable and unlocalised pain. Yea, I did often succeed in pumping the good medico thoroughly dry. Such is the news one learns to know in prison!

All honour to the brave Patterson and the brave Cullingworths. Is it wonder that the officers made them a handsome presentation after Pretoria's occupation; that Lord Roberts publicly thanked them for their brave deed, and took tea with the Cullingworths in their home? Is it wonder? Were they not most brave and daring? Remember detection would have meant horrible punishment. Once more, then, all honour to these brave hearts![880] They will never know how they helped us to bravely wait and hope till the sun of victory once more brightly illumined us! Such was our information bureau!

XXXVI. The British Officer and His Pastime

Ever since I was a college lad, many years ago, I have been a keen cricketer. In those days the matches of *the* season were the Western Province v. Garrison matches. How well I remember them; how keenly we enjoyed them! And ever since then I have known military officers. In late years, up in the north, in Bechuanaland and Rhodesia, I learned to know more of the brave men who helped open up the country and build the Empire there, and many were the friendships I formed. And now I have for many months been a fellow-prisoner of the British officer. I have lived with him under the same roof, I have come in daily contact with him, and learnt to know him intimately. And from the very bottom of my heart I say I have learnt to respect and value and love him. The British officer is a gentleman and a fine one withal. His self-restraint, his resourcefulness, his in many cases splendid intellectual development, his invariable courtesy, his kindness, his courage, have taught me honestly and fervently to admire him.

What the officers were to me during my term of imprisonment I can never forget and never repay. Is it wonder that bonds of firm and intimate friendship were laid between us? Aye, I love to recall their names and faces, and the many, many happy and profitable hours we spent together. I came among them a stranger, but in a few minutes' time I felt I had been admitted into their circle and been made one of them. I am not an Englishman; I belong to a different race. But the bond of a common loyalty and a common suffering drew us together and soon made us firm and fast friends. Thrown together day and night, character study becomes inevitable, and I believe that a prisoner sooner lays bare his inmost peculiarities, exhibiting his weaknesses especially, [more] than any other man does.

We thus had ample time to learn really to know one another. And there is now but one conviction in my mind—one impression—the British officer is a fine man! One cannot but admire and appreciate his loyalty to his Queen, his commanding officer, his regiment, his profession; nor his pluck and daring under the most trying circumstances. And yet he is kind and tender like a woman. I have seen him with his sick and wounded

[880] Hofmeyr note: I have since learnt that her Majesty the Queen has suggested to Lord Roberts that their brave act shall be officially taken notice of.

comrades in hospital; I have heard him speak with deep concern of his captured men, and seen him contribute most liberally, so that the suffering soldier might obtain some comfort and food. (The officers collected amongst themselves more than £800 for this purpose.)

I have been ill myself when amongst them, and experienced their kindly solicitude and care. I thus speak from experience. And I say again unhesitatingly—and I hold my opinion against all comers—the British officer is a splendid man. I do not say that he is perfect, that he never makes a mistake. No! But as a man he takes a tremendous amount of beating, and as an officer I know of no one I would follow more readily or trust more entirely.

The mistakes made, as far as I could gather, must be put down as the result of a defective military system, and not as the fault of the individual. Almost invariably the captive officer was the victim of such defect, and not the originator of his misfortune. For all this I honour the British officer. Like a brave man he spoke kindly and appreciatingly of a brave foe. There never was a word of bluster or brag. When Commandant-General Joubert died there was as genuine mourning and sympathy amongst the officers as if he had been one of their own officers. The relatives of the great general felt this and appreciated it highly. I have heard critics who pass a different verdict on the British officer. For such I have but one answer: You have not been his fellow-prisoner for seven months; you do not know him.

How such men spent their time in prison it would be interesting to know. Let me add to what I have written before on this point. In the Model School we had a beautifully fitted out gymnasium. In this classes practised regularly every day, whilst outside there was just room enough for a game of rounders in the afternoon. Walks round and round the building made up the balance of our physical exercise. The new prison had larger grounds outside, so that there was room for a little cricket with empty tins for wickets and walking-sticks for bats. Many and grand were the matches we played. There was also space for a game of hockey in the afternoons, a game which afforded the onlookers more pleasure, I think, than the participators.

Here, too, the walk round the yard was one of fully 600 yards, so that the exercise was a pleasurable one. In the Model School, as well as in this place, we had classes—for instance, I conducted a large Dutch class, whilst others of the officers had French classes. There was even a law class, and I am proud to say that with the kindly and able help of Lieutenant Adv. Gallwey,[881] of Natal, I finished Justinian's "Institutes," and did not find it dull reading at all. I am proud too of my Dutch pupils—many a one of the officers having obtained great proficiency in that language.

Lieutenant Adv. Gallwey is probably related to Colonel Sir Thomas Gallwey.[882] [883]

[881] Only return is for Gallwey E. J. Lieutenant Colonel. It does not appear he was ever a prisoner.
Entered 1870; Lieutenant Colonel, 1898. Staff service: Adjutant Auxiliary Forces, 1885-90; Commandant School of Inst. for Military and Volunteers, Aldershot, 1891. War service: Boer War, 1878-79; Sekukuni and Zulu Campaigns; Ulundi (medal with clasp); Boer War, 1899-1900. Anglo Boer War - Name search Source: List of CB recipients. Various sources

[882] Buller's Campaign with the Natal Field Force. 1902 Photo by W. W. Robertson. "There also marched by the Natal Volunteer Ambulance Corps, 1,400 strong, who had been embodied by Colonel Gallwey, Principal Medical Officer of Natal."

Oh, the happy hours we thus spent in study and sport! Shall I ever forget them, or the officers, my pupils? No, my dear friends, no! We also had a great deal of music, especially in the evening, after we were allowed to buy the organ I referred to. Some of the officers and myself provided the music, whilst the others, seated in the large dining-hall, read or studied, or wrote or played whist and chess, &c. There were moments, during such nights, when our songs, old friends of days ago, called up memories sweet and tender, memories of home and loved ones far away. In our new premises we had much less comfort than in the former ones, consequently, as far as possible, we lived outside, and the improvised tents made of blankets and large umbrellas dotted over the yard presented a scene most unique and interesting.

No longer having the use of the gymnasium we formed a Sandow class for dumb-bell practice. Our instructor, a very efficient and stern one, was Lieutenant-Adjutant Hill. Some fifty of us belonged to this class, and the advantages we derived from this exercise were inestimable.

Many of the officers pursued other avocations also. You would, for instance, see one with a rod and a thread and a bent pin, patiently sitting over a hole in the ground, fishing for mice or spiders or snakes. Many a snake was caught, under tremendous excitement, but never a one by the military fisherman. The skinning of the snake and the curing of its skin afforded pastime and excitement for many a one during many a day. During the night, too, there often was excitement. An officer would dream of rats, jump up, stick in hand, ready for the fray, and kick up such a general row that the whole show was awake in no time. No rat was to be found, of course, and all went to bed again, but this time without saying good-night to one another. In fact, dire threats of vengeance were uttered on that dreamy officer should he see a rat again. Thus, with laughter and song, though these came from heavy hearts, and were but a cloak to cover our deep-felt misery and dejection, we tried to cheer one another and get the heavy wheels of Time's chariot to roll on.

Our greatest excitement, however, came towards the end of our imprisonment. We started a newspaper. Lord Rosslyn[884] was appointed editor, several of us receiving appointments on his staff. We called our paper the *Gram*.

Why? asks the reader. For this reason. In a former chapter I related what sources of information we had. The reader will remember the telegrams we got—the Kaffir-grams, the butcher-grams, the baker-grams, the hospitalgrams, &c. This is why we called our paper the *Gram*—the name carried so many suggestions, and brought back so many pleasant memories of most exciting experiences.

Cover of the "Gram"[885] 'The Gram.' [886] [887]

James Francis Harry St Clair-Erskine, 5th Earl Rosslyn.[439]

[883] See Gallwey W. J. Lieutenant and Gallwey E.J. Lieutenant Colonel. Anglo Boer War - Name search

[884] James Francis Harry St Clair-Erskine, 5th Earl of Rosslyn (16 March 1869 – 10 August 1939), styled Lord Loughborough until 1890, was a Scottish soldier, author and aristocrat. From 1890 to 1897, he was Captain of the Fifeshire Light Horse Volunteers, serving with Alexander Thorneycroft's Mounted Infantry during the Second Boer War where he was taken prisoner twice, which he wrote about in his book Twice Captured. After his bankruptcy, he resigned his commission in the Fife. Lord Rosslyn was war correspondent for the Daily Mail in 1900. In 1904, he was Secretary to the Secretary of State for Scotland and from 1915 to 1917, he was Major in the King's Royal Rifle Corps. James St Clair-Erskine, 5th Earl of Rosslyn - Wikipedia See also his book: Twice Captured - Google Books.

The production of this paper gave tremendous work to the publishers, and great excitement to the public—the prison public. We had to buy paper and hectographs, and then distribute the work amongst the assistants and thus produce the paper. The first number appeared on the 12th of May, and was really a most creditable production. The illustrations, I make bold to say, will compare favourably with the cartoons and pictures of any published paper I know. Two full numbers of the paper were published, and the third number was well in hand when the memorable 5th of June came—the day whereon Lord Roberts entered Pretoria. I am glad to say that the hard work of editor and writers will not be lost, for Lord Rosslyn intends producing a facsimile of the published numbers. The smartest bit of work in connection with our paper was this, I think, and it says a great deal for the business instincts of the officers. We made our caterer advertise in the paper.

He was the only advertiser, and we gave him a full page for the sum of £5 sterling. And the only readers of that advertisement were the prisoners of war, who knew more than enough about our caterer. The good man, however, tumbled to the idea, when we proposed it and, I trust, has thus immortalised himself.

I must not forget to refer to another pastime we had. It was the chess tournaments. Many of the officers were keen chess-players, but with the arrival of Captain Duhan[888] we received in our midst a veritable demon of chess. He played three or four partners blindfolded, and beat them all. In fact nobody could touch him. His being so keen on the game, inspired others also, and many was the chess tournament organised and great the excitement caused.

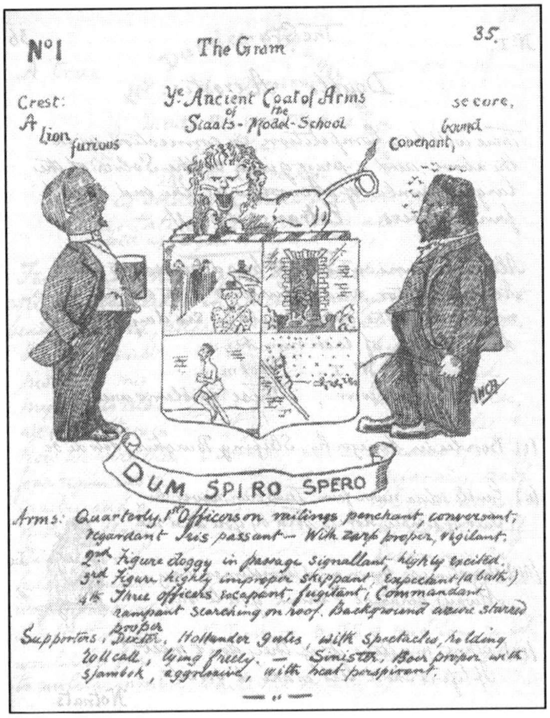

Image of "Captain Dunham" by Thomas Frankland[889] Page from the "Gram" Note THCF Author.[440]

[885] Image Credit QSAMike Anglo Boer War - Armored Train Ambush 11-15-99 Chuchill & Haldane - Page 17 - Boer War Forum.

[886] PF: Mr. Bosscher was the only advertiser, even though he need not have done so, since he still had the concession. See Hofmeyr, for particulars, and Rosslyn. The magazine did not begin publishing until May 1900 well after the prisoners had been transferred to a purpose built prison outside of Pretoria.

[887] "The Gram" A social Magazine Founded by British Prisoners of War in Pretoria, Facsimile Edition Eyre & Spottiswoode, Her Majesty's Printers, London #1 - The Gram ; a social magazine founded by British prisoners ... (no.1-3). - Full View | HathiTrust Digital Library Scanned from the University of Minnesota Copy. Image: #60 - The Gram; a social magazine founded by British prisoners ... (no.1-3). - Full View | HathiTrust Digital Library Cover Image: #61 - The Gram; a social magazine founded by British prisoners ... (no.1-3). - Full View | HathiTrust Digital Library.

[888] Probably: Duhan H. J. Lieutenant Prisoner. Near Paardeberg, 18 February 1900 Source: South African Field Force Casualty Roll Unit: Kitchener's Horse. See Also: Duhan Harry Major Source: Nominal roll in WO127 Warren's Mounted Infantry. Anglo Boer War - Name search

[889] Image and text per Neville Chamberlain "Frankland drew the illustrations on pages 1, 9, 22a, 35, 36, 39 & 41 of 'The Gram' (7 of the 18 drawings)." Anglo Boer War - Armored Train Ambush 11-15-99 Chuchill & Haldane - Page 17 - Boer War Forum

The reader will thus see that we did not fritter away our time, nor were we lazy boys. I am sure many an officer with me during those tedious months remembered with gratitude a father and a mother who had incited us to study when we were youngsters, or a teacher and college professor who had inspired us with a love of learning. Verily, but for the literary attainments possessed by so many of us, the scholarship and love of knowledge, time would indeed have been a chain dragging us down into the depths of despondency and hopelessness. Now we could set at defiance carping care and horrible despair. The poets and the novelists, the historians and the philosophers, all brought of their rich stores to comfort us and make us patient and strong and brave. Thus the British officer, gentleman and scholar, passed his time in prison. And I shall always account it a great privilege and an honour that for many months my lot was cast in with such men—so loyal, brave, and true.[890]

XXXVII. Free

I am coming near to the end of my story now. On the 12th of May, at ten o'clock in the morning, our gaoler came to our quarters accompanied by two policemen, and beckoning me aside, said to me, "Would you like to change your quarters, Mr. Hofmeyr?" I answered, "I am very comfortable here, and do not know if the change you propose will be an improvement." "Well," said he, "would you prefer going home to staying here?" I looked at him and said, "Don't mock a man who has been a prisoner seven months. Don't speak to me of home, when I have almost begun to think that I will never see it again." "No," said he, "I am not mocking you. You are going to be liberated to-day—now, to-day; and I have come to pack your luggage for you and to take you out."

I felt the man was speaking the truth, and I can never forget the wistful feeling that rushed over my heart. I sat down on my little bed and verily gasped for breath. My head was in a whirl and I felt as if I could fall down. One of the officials offered me a glass of water, and in a kindly voice said, "Come, Mr. Hofmeyr, we will help you pack and take you away."

The news spread like wildfire, and in a moment my little lager, as we called it, was surrounded by the officers, every one congratulating me with the good news. And now a scene of bustle ensued the policemen packing everything belonging to me and examining everything to prevent my taking out sealed letters and documents. Perceiving this, I really thought that now my newspaper cuttings, my manuscripts, and other trifles precious to me though obnoxious to the authorities would be taken away from me. Fortunately, through a little proper manipulation and juggling, I managed to get all my valuables safe in the pormanteau. But now came another trouble; round about me I saw on the faces of many an officer an eager look, interpreting the question, "Can you take out letters for me?" The most eager amongst these was Lord Rosslyn, whose entreating look, as he came up to congratulate me, told me how anxious he was to get away his copy.

What to do now? They planned and I planned. I managed to signal to several of the friends that they should place their letters and documents under the pillow on my bed. They did so without being noticed. And then, when the packing was over, I said to my guards, "Gentlemen, I am going on a long journey, you will allow me to change first before I start?" they kindly acquiesced, and turned their backs on me whilst I withdrew into my little red-baized recess to change. Whilst doing so I put on my football sweater, though the day was a fairly warm one. But it was done with a purpose. As I drew it over my head I slipped underneath the letters and papers under my pillow, thus finding a secure hiding-place for them between my body and the garment. I emerged rather stouter than when I entered my little sanctum, but fortunately nobody noticed it, and I escaped detection.

Then up came Lord Rosslyn with a biscuit tin: "My dear Hofmeyr, do take these biscuits with you, you'll need them on the way." I hesitated at first, not suspecting the dodge. But when he so kindly pressed the gift on me, his eyes winking most mysteriously, I suspected, and most gratefully accepted the welcome present. Yes, it was a biscuit tin, but a good deal more document than biscuit inside.

But the noble lord's kind attentions did not cease there. He would re-arrange my dressing-case for me, paying particular attention to my slippers. And whilst doing so he, inadvertently of course, slipped a box of matches into one slipper! Now, no guard would object to your putting a box of matches into a portmanteau! And then my friend came to me, managing to whisper into my ear whilst others talked around me, "Look in your slipper." I

[890] Hofmeyr footnote: Since writing the above I have been informed of an article recently written in *Scribner's* (I think) by *Mr*. Richard Harding Davis. I blushed for very shame when I read it—shame that a well-known writer could descend to such meanness, and tell such abominable falsehoods. The charges he brings against the British officers—prisoners of war—are absolutely false. I say this most emphatically!

caught the whisper; I understood his meaning. There were matches inside that box, but more letters and papers cleverly rolled up into the smallest space, and thus I emerged from the prison in possession of many home messages and much precious copy.

The parting on that happy though sad day I shall never forget. When all the packing and smuggling was over, we gathered in our big dining-room. In a broken voice I tried to convey my heartfelt thanks to my fellow-prisoners. They had all been so kind to me, so sympathising, that when I look back upon those months to-day, I feel I can never sufficiently value their kindness nor ever repay my fellow-prisoners what they had done unto me. I spoke a few words of goodbye, and then we shook hands. Every one had a kind word and a good wish. I could not speak any more. When the hearty handshaking was over the officers filed out to the gate on the other side of which lay freedom. And when I emerged, the last of all, they had formed up in a double line, and in between I walked through. There was a hearty "God bless you," from every one, and as I passed through the gates a ringing chorus was raised, "For he's a jolly good fellow."

I had my last look, as I turned round once more, my last wave of the hand to the friends I loved so well, and off we drove to the railway station. It was a glad day for me, and yet I felt very sad. Truly I have much reason gratefully to remember the British officers.

Having signed a declaration that I would not take up arms against the Republic, I was taken to the railway station and provided with a ticket for Delagoa Bay—a ticket and nothing more—and the journey was one of thirty-six hours. And this brings me to the relation of a most pleasing incident. I have all along said how I admire and respect the genuine burgher. Let me give an illustration confirming my belief. I was put into the train, as I mentioned, without a penny in my pocket and without food for a long journey. I sat in the corner of the carriage trying to realise that I was free, when an old burgher, whose name I do not know, passed by me, and to throw the bystanders off the scent pretended to be in a great rage with me, speaking gruff and harshly.

But in between came this question, "Have you any money with you?" My answer was, "No." He walked away, came back again, and, speaking as before, again managed to drop the question, "Have you any food?" Again I answered, "No." The old gentleman walked up the platform, spluttered and raged, all to avert suspicion, of course, and then passed me again, very deftly dropped two sovereigns into my lap.

View from Pretoria railway station.[441]

1842 Queen Victoria "Shield reverse" sovereign.[891]

That is just the genuine old burgher all over, and I need not say how deeply this kind deed touched me. A few acquaintances were on the platform, and after a last goodbye the train steamed away.

Let me tell of another pleasing experience. I was sent out in charge of a policeman. I may state here that Mr. Hallewell, a newspaper correspondent, was also released that day, and that we travelled together. The policeman in charge of us, in the course of conversation, found out that I was an old friend of his [the policeman's] father, who lived in the Cape Colony. Years ago I had been able to show his father some trifling kindness, and now the son tells me that his father has always spoken of me with great gratitude and love. "And for his sake now, Mr. Hofmeyr," said the policeman, "how can I look upon you otherwise than a friend? I am your servant, not your guard."

And indeed the man was true to his word, showing me the utmost courtesy and kindness, till he left me at Komati Poort on the border. How wonderfully the bread one casts on the water sometimes comes back again! The same thing happened at Komati Poort in the Customs Office. To our surprise we heard that there too we were to be searched, and I made sure that all our maneuvering was in vain, for they would now discover

[891] At 20 shillings per pound this is a generous gift of 40 shillings. 1842 Queen Victoria "Shield reverse" sovereign File:Sovereign Victoria 1842 662015.jpg - Wikimedia Commons Attribution: Classical Numismatic Group, Inc. http://www.cngcoins.com.

everything. But what happened? When I entered the office the official behind the counter looked up and greeted me by name with a kindly smile, telling me how glad he was of my release.

I apologised for not remembering him, and asked his name. "No, you won't remember me. It is so long since we met, but you will remember the little boy, R., who was your pupil twenty-five years ago in the South African College, Cape Town. You were kind to me then, and I have never forgotten it." I remembered, of course, and thanked him for his kind words, and then came up for examination. "Search you, Mr. Hofmeyr?" said my pupil of years ago. "No, certainly not!" and with that he jumped over the counter, chalked my boots, and made me pass out with all my luggage and papers. I am sure he could not understand how heartily and gratefully I then thanked him for his courtesy.

Our train then crossed Komati Bridge, passed the White Beacon—the boundary between the Transvaal and Portuguese territory—and I was free. Oh, the sensations of that moment! Seven months of durance vile, months of disappointment and uncertainty, but months too full of experience and instruction and useful information—these had now passed by. I could once more go where I pleased; no guard, no gun barred the way. How blue the sky seemed, how glorious the sun, how green the fields, how sweet the running waters—for I was free! But pen cannot describe a liberated prisoner's sensations. Enough that Hallewell and I grasped one another's hands and from the bottom of our hearts said, with loud and joyful voice, "God Save the Queen!"

Lorenco Marques from the east.[442]

Delagoa Bay. Showing the Old Portuguese Fort on the Left. "Delagoa Bay has been described as the 'Key of South Africa,' but we allowed it to slip through our hands."[443]

We reached Delagoa Bay on Sunday night the 13th, receiving great kindness from many of the English residents there, and in a few days' time sailed for Natal on the *Umbria*.[892]

[892] RMS *Umbria* was one of the last Cunard Line ocean liners that were fitted with auxiliary sails. RMS *Umbria* was built by John Elder & Co. at Glasgow, Scotland in 1884. War broke out in South Africa on 12 October 1899, and two months later on 22 December *Umbria* was chartered by the government and was prepared to carry troops and armaments to South Africa. She set sail on 11 January 1900 on her first voyage in her new role. On board were troops of the Warwickshire, Derbyshire and Durham Militia. They arrived in Cape Town on 29 January and after calls at Port Elizabeth and other ports she returned to Southampton with wounded soldiers. In April she again was back in South Africa and during the relief of Mafeking she was in Port Natal (now Durban) for the celebrations. She left Cape Town for the last time on 7 June, carrying 600 wounded soldiers. She arrived back at Southampton 19 days later on 26 June, and she was then returned to Cunard to resume her normal role. She was given a complete refit and returned on the New York run on 21 July. RMS Umbria - Wikipedia

RMS *Umbria*.[444]

I'll never forget the afternoon whereon we sailed. The big steamer was crowded with refugees from Johannesburg days before the hour of sailing. We went on board on Wednesday afternoon. It was a beautiful day, and the sea was gloriously calm. The big steamer was riding at anchor some eight or ten miles out in the bay. Many passengers left by tug-boat that afternoon, and I, accompanied by several friends from Delagoa Bay, sailed with the last little steam tug that went out to the vessel. The few days I had spent at Delagoa Bay were to me like a dream. I had telegraphed to my fellow-prisoners in Pretoria, and written to them, and tried to realise in every way that I was free. And yet it seemed impossible. The first morning I awoke in Lorenco Marques I looked out through the window for the guards with their guns. I found them not, and it was strange to me. When you have lived seven months surrounded day and night by armed guards the change from captivity to liberty cannot be taken in all at once.

So it was with me. I was as in a dream when we sailed for the *Umbria* on the Wednesday afternoon, and I am not ashamed to say that, as our tug-boat steamed under the Union Jack, gracefully clinging in that bright afternoon to its flagstaff, my emotions overpowered me. I seized hold of my companion's arm, and as tears trickled down my face, I huskily said to him, "Thank God, I am free once more!"

XXXVIII. Afterwards

On Friday morning [May 15th, 1900] before the break of day, the *Umbria* dropped anchor in Port Natal, and soon afterwards I was on shore and amongst friends again. I need not say what a pleasure it was to hear the news, to talk over the events of the war, and to discuss matters in general. The dinner given me by the friends in the Club that evening, and the pleasant hours we spent together went far to compensate me for the trouble and trials now fortunately over. Amongst the friends I then was introduced to I cannot forget Mr. James Bailey, M.P.[893] He and his wife had come out on a visit to the Colony, drawn thither by the fact that two of their sons were fighting for their Queen and that one had just before been made prisoner (I may here mention that the other son had also been wounded very severely). I met Mr. Bailey afterwards again in Cape Town, and am glad to say that when I went north to the Transvaal I managed to get letters through for him to his prisoner son.

On my arrival in Cape Town I was immediately and very kindly asked by his Excellency the Governor[894] if I would like to go back to the Transvaal to assist in the work of pacification. I at once consented to this, for my heart yearned for the people who were fighting under wrong impressions, urged on by false statements.

Entering the Imperial Service, I was then immediately sent to the Western Transvaal, to join General Baden-Powell. I may here state that on my arrival at Zeerust, the very place where I was imprisoned, Lord Edward Cecil,[895] the Imperial Commissioner of the Western Transvaal, appointed me his Assistant. He very kindly soon

[893] Sir James Bailey (10 November 1840 – 12 October 1910) JP, DL, MP, was a British Conservative Party politician who served from 1895 to 1906 as Member of Parliament (MP) for Walworth in South London. He was also a successful hotel developer, most notably for establishing the Bailey's Hotel in Kensington, and the founder of a London gentlemen's club, the Constitutional Club. Baileys Irish Cream is named for the hotel that bears his name. James Bailey (British politician) - Wikipedia

[894] Alfred Milner. Alfred Milner, 1st Viscount Milner - Wikipedia See: List of governors of British South African colonies - Wikipedia

[895] Edward Herbert Gascoyne-Cecil KCMG DSO (12 July 1867 – 13 December 1918), known as Lord Edward Cecil, was a distinguished and highly decorated English soldier. As colonial administrator in Egypt and advisor to the Liberal government, he helped to implement Army reforms. Lord Edward Cecil - Wikipedia

afterwards granted me three months' leave of absence. And when these have expired I hope to resume my duties, having only one object in view—the pacification of the conquered burghers. My one hope is that under the beneficent and generous rule of England all of them may very soon be happier and more prosperous than ever before. Before I detail any of my experiences in Zeerust, however, I would like to relate a few other incidents. The last year has been to me so full of coincidences and experiences, so full of woe as well as weal, that I sometimes really fancy that I am dreaming.

I left Natal for the Cape on the little steamer *Umvoti*. We met with exceedingly rough weather immediately after leaving on Sunday morning. Almost every passenger had fallen ill, and I was one of the few who escaped. I was sitting in the smoking-room that Sunday evening after dinner chatting with the ship's doctor. The ship was plunging and rolling madly, and we were alone. We must both of us have dozed away at about eight o'clock, I think. At nine I woke and saw the doctor, who was sitting next to me, apparently sleeping in his corner of the sofa. I dropped off to sleep, again to be rudely awakened at about 10.30 by a rush of water through the door of the smoking-room right over us. I started up, shook the salt water off my overcoat, and looked around me. The doctor was still sleeping. I wondered at this, and instinctively put out my hand to touch his. It was icy cold! He was sleeping his last sleep! With a placid smile on his face the old gentleman lay back in his corner. Never a muscle had moved, and no struggle had taken place, as his spirit passed away. I called the captain and the first officer, and gently we laid him aside, till on the following morning the body was solemnly committed to the deep. I was spared, he was taken! God knows best!

Naturally this event cast a gloom over all the passengers, a gloom that was dispelled partially only when we dropped anchor in Table Bay. There I left the vessel, and it was then that I was asked to proceed northwards again. Passing through Kimberley, visiting the scenes of the great battles fought in its vicinity, I reached Mafeking on the 2nd of July.[896]

Table Bay filled with steamers.[445]

Devils Peak, Table Mountain and Lion's Head from Table Bay.[446]

Table Bay with Alfred Docks and transports.[447]

Cape Town Harbor.[448]

[896] PF: Belmont, Modder River and Magersfontein were all fought in the immediate vicinity of the railway line. He did not have to go out of his way to make these observations; they were right there for all to see.

Mafeking

Col. Plumer's Column Entering Mafeking.[449] [897] Note Dixon's Hotel and Dixon & Company on the Building.

Panorama of Mafeking, Besieged by the Boers.[450]

The Victoria Hospital at Mafeking.[451]

The R.C. Convent as hospital showing shell holes. [452]

And here let me relate a most pleasant coincidence—let me call it the "Story of my watch." As I mentioned in the beginning of my book, it was looted from me on the day when I was made prisoner, almost eight months before. Of course I never expected to see that watch again. But what happened? While still in Cape Town I received a letter from a gentleman in Port Elizabeth, asking me for information about some of my fellow prisoners who were friends of his. I answered him, but before my letter reached Port Elizabeth this gentleman, who was a Transvaal refugee from the district of Zeerust, had left for Mafeking, whither my letter followed him. This is the first chapter. Now for the second. I arrived at Mafeking late on Monday night, July 2nd. My good friend the host of the Mafeking Hotel informed me that his place was crowded, and that he could give me a room to myself only on the following day, if I did not object to share a room for the night with another guest.

Of course I had no objections, and my friend showed me my room. As he left me he wished me good-night, mentioning my name. My room-mate hereupon at once sat up in his bed and asked me if I were a relative or an acquaintance of Mr. Adrian Hofmeyr. I informed him that I happened to be the very man. Imagine my surprise when he then told me that he was the man who had written to me from Port Elizabeth and that my answer, forwarded to him from there, had reached Mafeking that very morning.

Of course we were at once old friends and related our experiences to one another. I told him all about my capture, mentioning what was taken from me that day, including my gold watch. Then we slept. The next morning this gentleman, who owned property in Mafeking, visited the people who were living in his house, and amongst others told them of our wonderful meeting the night before. When he mentioned my name, the lady of

[897] 5/17/1900 Colonel Mahon, at the head of the relief force, entered Mafeking.

the house said, "But I have seen that name lately somewhere. Where can it be?" Thinking the matter over a little while it all came back to her, and this is her story.

"On the day of the relief of Mafeking, when the soldiers were pursuing the Boer commando, a Kaffir boy in my husband's service picked up in the Boer lager a gold watch. The boy was a driver of one of the gun-wagons, and in the hurry of their departure one of the Boers—very likely the one who had taken it originally—must have dropped the watch in his flight. The boy brought it to me and asked me to keep it for him. And this is how I remember the name, *it is in the watch*."

My friend came back to me about luncheon time, with a broad smile on his face, saying, "Open your hand, Mr. H., and I'll drop something into it." I did so, and the reader can imagine my surprise when once more I saw my valued watch, a present given me twenty years before, *on the 2nd of July!* I need not say that many and hearty were the congratulations, showered upon me by my friends, on this extraordinary piece of good luck. But my good fortune did not end there. On my arrival at Zeerust the very next day, and on my telling the friends at dinner this remarkable story, Lord Edward [Cecil] said to me, "Extraordinary! Have you lost anything else we might find, perhaps?" Jokingly I said, "Oh yes, there is my beautiful little American organ and my music, which I would like to see back again."

Lord Edward then remarked, "Well, we'll tell a detective about it to-morrow and send him out with a party to scour the country." I looked on the matter as a joke and never thought of it again. Imagine my surprise, therefore, when on the very next evening a detective officer came to me, followed by two men carrying my little organ. His story was that he had been sent out that morning by the Commissioner to see what he could find out. He proceeded at once to a lager of tents occupied by women whose husbands were still fighting. On arrival there he made investigations and found that the occupants of the lager were meditating a big bonfire of all looted articles in the camp, for the purpose of destroying all trace thereof.

The detective then searched, found my little organ intact, and thus saved it in the very nick of time. Many were the friends standing around me when I opened the box, set up the instrument and caressingly touched the keys. I need hardly say that the very first chords interpreted the tender refrain of "Auld Lang Syne." I now felt sure that some day or other I should get my music back as well. But I was not prepared for the following. Driving out in the district on official work the very next day, I came upon a deserted farm. Whilst my horses were resting I entered the house and found that one room had evidently been used as an office during the siege of Mafeking. Papers and telegrams were lying inches thick on the floor. These I turned up and read just to while away an idle half-hour. Imagine my surprise, therefore, when, on turning over a big heap of papers in the corner, I came upon a large packet of music, and on the top corner of every song there stood my initials clear as daylight!

Having found these valued articles again, I can honestly say that I regret the loss of others, infinitely more valuable, but very slightly. My return to Zeerust, however, afforded me much greater pleasure than this even. Did it not bring me back to the friends who so generously stood by me during the dark seven weeks of my imprisonment there? Above all, did it not give me the opportunity of once again gripping the hand of my dear friend, the brave Doctor Blake, I arrived in Zeerust late in the evening, but it took me a few minutes only to run up to his house—that house that held so many touching memories—to see him and his home circle. The small hours of the morning had already come before we parted for the night, and it took us days to say merely the half of what we wanted to say to one another. Yes, it was to me indeed an inexpressible pleasure and honour once more to be the guest of my kind friends. And once more must I say, "God bless them."

I found in the Zeerust prison many Boer prisoners, and amongst them my gaoler and several of my guards. The tables were now reversed. I visited them, I asked them what I could do for them, and gave them the luxuries they wished to have. They did not expect it, and were a little nervous about the future when they heard I had come back to Zeerust.

Then I talked the matter over with Lord Edward, and we came to the conclusion that if these men would take the oath of neutrality they could be released. I saw them again, told them of this, and asked them if they were willing to take that oath. They all consented, and then heard the glorious news, "You are free." I could see that the men felt thoroughly astonished. One turning to me said, "We expected that you had come to be avenged on us for our cruel treatment of you." My answer was, "You are right. I have come for that purpose. I have taken my revenge, and it is sweet! It is this, you are free! Thus a British subject takes vengeance on his enemies. Remember that."

Aye, reader, this is what I sincerely trust the great British Empire will do all over the Transvaal and Free States—show the people that we are not their enemies but their friends. My heartfelt prayer is that England may so rule the newly conquered territories that from north to south, we all, Dutch Africanders or whoever we are, may fall in love with the grand British constitution, the shadow of whose flag spells the magic word Liberty. That time, I fervently trust, is not far distant. It will come very soon if all her Majesty's subjects, both those in the old country and those in the South African Colonies, join hands and labour with heart and soul in the grand, the divine task, of pacification.

XXXIX. Conclusion

I have been free some months now. I look back on the time of my captivity with mixed feelings. There were moments, whilst in prison, when bitter feelings of anger took possession of my heart. But they passed away very quickly. I could not but feel intense sympathy and keen sorrow for the terrible mistakes that had resulted in that frightful catastrophe for the two Republics. I have written about my experiences, and sometimes used hard words. I did so because I had to tell the truth, and because I wrote when the impressions made upon me by what I had undergone were fresh in the mind. More often I wrote with a heart full of sorrow for a misguided people, for I had learned to know how shamefully the burghers had been misled and deceived. I cannot pity the Government that caused all this. It ought to have known better.

It ought not to have allowed foreign, inimical, selfish counsels to sway its deliberations. Oh, the mistakes that were made and the chances that were lost! Now I am persuaded there is but one course to be adopted in the future—our gracious Queen must rule from the Zambezi[898] down to the Table Mountain. And my heartfelt prayer is that the British Government may send out to rule over the country such exponents of British fair play and justice, of British rule and power, that the subjects of the newly conquered territories, like those of her Majesty in the Cape Colony and Natal, may soon find out that after all the greatest blessing that can come over us in South Africa is to be subjects of our Empress Queen.

South Africa is a beautiful country. Its climate is unequalled, its resources illimitable, its possibilities unfathomable. And therefore we stretch out hands across the sea and ask for the love and sympathy and the generosity of the conqueror to the conquered. Then, very soon the day will come which will make South Africa the brightest gem in the Queen's crown. Then the horrors and the sufferings of the present war will be forgotten, and from Rhodesia in the north to Cape Colony in the south, from Natal in the east to Bechuanaland in the west, in the Vaal River and Orange River Colonies, everywhere peace and harmony and brotherhood and prosperity will reign. May that glorious morning dawn soon.

[898] The Zambezi River is north of Botswana and Zimbabwe. It borders Mozambique. Zambezi - Wikipedia

Bibliography

Many works have been footnoted in the text, enthusiastically if also somewhat idiosyncratically in form. Restating those references again here would extend this work unnecessarily.

Anyone who has read this work will easily come across the major historical works which chronicle this era and conflict as they follow the leads I have cited. I would rather cite material directly in the location where it is most relevant than bury it at the end of the work or in the endnotes where I primarily include citations to images.

The full and equally idiomatic bibliography for this series is over sixty pages long and will be included in an Appendix volume to be published at the end of the series.

For this volume an abbreviated bibliography is as follows:

James Aylmer Lowthorpe Haldane, How We Escaped from Pretoria. "New Edition, Revised and Enlarged with Numerous Illustrations, Plans, and Map." William Blackwood And Sons, Edinburgh And London, 1901 William Blackwood And Sons, Edinburgh And London, 1901 How We Escaped from Pretoria : James Aylmer Lowthorpe Haldane : Free Download, Borrow, and Streaming : Internet Archive

Adrian Hofmeyr, The Story of My Captivity During the Transvaal War 1899-1900, Publisher, Edward Arnold 37, Bedford Street , Strand, London, 1900 Printed by: Unwin Brothers, The Gresham Press, Woking and London .The Story of My Captivity During the Transvaal War 1899-1900 - Google Books

Lieutenant Thomas Hugh Colville Frankland: Held by the Enemy: Extracts from the Journal of Lieutenant Thomas Hugh Colville Frankland, Royal Dublin Fusiliers, lately prisoner of war at Pretoria. Originally Included in Winston Spencer Churchill's book; Ian Hamilton's March, 1900 The Copp, Clark Company, Ltd. Toronto. Ian Hamilton's march : Churchill, Winston, 1874-1965 : Free Download, Borrow, and Streaming : Internet Archive

Biography of Frederick Neil Le Mesurier published by the Bushey Museum. https://busheymuseum.org/ I have not been able to actually locate this work but I did see that it was compiled.

Randolph Churchill. Official Biography Winston S. Churchill Volume I by Randolph S. Churchill (Houghton Mifflin Co.) 1966. The abbreviation OB is used to refer to this volume.

Official Biography Companion Volume(s) I and II by Randolph S. Churchill, Published by William Heinemann Ltd.[899] , 1967. Copyright by C & T Publications Limited. The abbreviation OBC is used to refer to these volumes.

License Links

My many thanks to all the authors, libraries, conservators, and scanners/uploaders who have allowed me to populate, illustrate and annotate this compilation. Responsible efforts have been made to ensure that material in this work public domain material and otherwise is governed by the following licenses.

[899] William Heinemann - Wikipedia And: Heinemann (publisher) - Wikipedia

[900] Creative Commons License: Creative Commons license - Wikipedia And About CC Licenses - Creative Commons

Some documents are reproduced pursuant to the UK Open Parliament License and this applies not only material in which either House owns the copyright but also material published before 1 August 1989 which is the operative Commencement date of the Copyright and Designs act of 1988; and 50 years having elapsed subsequent to such publication. The same terms applying to previously published material that at one time subsisted under Crown Copyright. [901]

Also, this work contains public sector information made available under the South African Promotion of Access to Information Act, 2000 ("PAIA"). https://www.gov.za/documents/promotion-access-information-act

Google Earth License Links

Google Earth License: https://www.google.com/intl/en-GB_ALL/permissions/geoguidelines/

Sales of these annotated, illustrated and supplemented historical narratives are anticipated to be less than the 5,000 copies permitted. If that is not the case the author will contact the Google Cloud Customer Team for additional permission(s). https://cloud.google.com/contact/

Copyright Discussion

This work is a comprehensive republication of the included works which have been compiled, annotated, illustrated and expanded in order to provide historical context and clarity.

Images and text used by permission or by law are clearly identified by citation and should not be reproduced; as to the remainder there is no copyright claimed by myself on any public domain images, text or any common property information contained in this work.

Future creators should continue to follow the permissions, guidelines and requirements from the original source and author/uploader who is cited. Links to Creative Commons, Google Earth and other licenses are noted in the license links section.

So far as can be determined, as of 2025, the text and images used to illustrate the first person accounts included in this volume reside in the public domain in the United States having all been first published prior to January 1st, 1930 as well as in all other jurisdictions which employ any of the following copyright regimens:

1. A copyright regimen that grants an author's works subsisting copyright for seventy (70) years pma[902] or less.

2. A copyright term subsisting for 120 years from the date of creation or first publication.

3. And such jurisdictions that apply the Rule of the Shorter term to an author's pma date or publication date.

This work will not be marketed or deliberately sold in any jurisdiction with a longer copyright term.

Copyright for the assembled compilation of this material, or substantial portions thereof that would comprise separate articles or works unto themselves is retained. This work and the series in sum comprise a derivative work. Literary, dramatic, and adaptation rights of the saga demonstrated and explicated by this series is retained.

Otherwise, with respect to the underlying texts and images recast in a more permanently accessible form I choose to pay forward my creative energy to the people who lived through these experiences and it has been a pleasure to find and assemble their collective words and images in a new synoptic and synergistic arrangement.

Except where noted otherwise, every effort has been made to ensure that no material has been used without an attributed source. However, if any non-attributed source, incorrect source, or copyrighted material used without permission or justification, has been included then I would be grateful for information that would enable any errors or inappropriate inclusions to be corrected or omitted in subsequent impressions.

Starting from creation, to publication, through to purchase, archiving, and digitizing, this annotated, illustrated and expanded edition is the sixth step in the journey of these works. My thanks go to all who have helped preserve and perpetuate these important historical records.

'Thank you, patient reader.'[903]

[901] If the work was commercially published within 75 years of the end of the calendar year in which it was made, Crown copyright will last for 50 years from the end of the calendar year in which it was published.
https://copyrightservice.co.uk/copyright/p01_uk_copyright_law Flow Chart: Crown copyright flowchart 2015

[902] post mortem auctoris or "pma" [post mortem author].

[903] Will Durant, in his eleven volume work 'The Story of Civilization', Simon & Schuster, 21st Edition. 1972 ISBN: 0-671-41800-9 concludes, Vol. 2 'The Life of Greece' with the statement, 'For those who have come thus far: Thank you again for your unseen but ever felt companionship.' Vol 3 concludes with my favorite: 'Thank you, patient reader.' Vol 4, concludes: 'Thank you again, friend reader.' Vol 6 Concludes: 'Courage reader: We near the end.'

Endnotes

[1] Frontispiece Photo, Revised Edition. Aylmer Haldane, signed photograph taken around 1900. File:Aylmer Haldane 002.jpg-Wikimedia Commons

[2] Adrian Hofmeyr The Story of My Captivity During the Transvaal War 1899-1900. The Story of My Captivity During the Transvaal War 1899-1900 - Google Books

[3] Photo Credit: Thomas Hugh Colville Frankland (unknown-1915) - Find A Grave Memorial Memorial Id: 56009464 Photo added by Nancy Wright: Member Profile: Nancy Wright - Find A Grave See Infra his Biographical summary and full photograph.

[4] A sketch by Mr. T. H. C. Frankland, from: Winston Spencer Churchill by A. MacCallum Scott, Methuen, London 1905

[5] Haldane, How We Escaped from Pretoria.

[6] Image source: With the Flag to Pretoria, H. W. Wilson Vol 1 1900 and britishbattles.com

[7] File:The Second Boer War, 1899-1902 Q72855.jpg - Wikimedia Commons

[8] With Roberts to the Transvaal. C.N. Robinson [1900]

[9] Anglo-Boer War Album, Cape Town, D. Edwards & co. [1902?]

[10] Anglo-Boer War Album, Cape Town, D. Edwards & co. [1902?]

[11] With Roberts to the Transvaal. C.N. Robinson [1900]

[12] Boer Commando assembled from Pretoria. 1899. File:1899 Boer Commando - Pretoria.jpg - Wikimedia Commons

[13] Cunliffe, History of the Boer War Vol 1 (NY Library Scan)

[14] Anglo-Boer War Album, Cape Town, D. Edwards & co. [1902?]

[15] File:The Second Boer War, 1899-1902 Q72296.jpg - Wikimedia Commons Author: William Skeoch Cumming

[16] File:The Second Boer War, 1899-1902 Q72152.jpg - Wikimedia Commons Author: William Skeoch Cumming

[17] Anglo-Boer War Album, Cape Town, D. Edwards & co. [1902?]

[18] File:The Second Boer War, 1899-1902 Q72258.jpg - Wikimedia Commons Author: William Skeoch Cumming

[19] See: Roll film - Wikipedia

[20] Presently the definition of Kodaking is the: 'present participle of Kodak.' A participle is: A form of a verb that may function as an adjective or noun.

[21] "Solomon's pools becomes a picnic & swimming resort. View of the middle pool [spouse] (kodaking husband & dog)" Source Library of Congress File: File:Solomon's pools becomes a picnic & swimming resort. View of the middle pool (kodaking husband & dog) LOC matpc.20398.jpg - Wikimedia Commons Catalog Record: https://www.loc.gov/pictures/item/2019711166/

[22] Presently the definition of Snapshooting is the 'present participle of snapshot.' A participle is: A form of a verb that may function as an adjective or noun. (I only mention this because I did not know it myself.)

[23] Film speed is the measure of a photographic film's sensitivity to light. See: Film speed - Wikipedia

[24] Photojournalism - Wikipedia

[25] The Navy and Army Illustrated was an illustrated glossy journal or magazine that covered historical and contemporary military matters from 1895 to 1915. The Navy and Army Illustrated - Wikipedia

[26] Black & White (magazine) - Wikipedia

[27] Black & White Weekly Vol 1-2 1899-1900

[28] Hillegas, Boers in War, 1900.

[29] Credit Neville Constantine: https://www.angloboerwar.com/forum/21-uniforms-badges-and-equipment/31231-artillery-and-ammunition?start=222

[30] File:Brownie Camera (1905) (ADVERT 428).jpeg - Wikimedia Commons June 1905

[31] File:Dorset Yeomanry Boer War.png - Wikimedia Commons Credit: Axel Ravera. Public Domain.

[32] Navy & Army Illustrated Vol 11 November 17th, 1900.

[33] With the Flag to Pretoria, H. W. Wilson Vol 1

[34] With the Flag to Pretoria, H. W. Wilson Vol 1

[35] Cunliffe, History of the Boer War Vol 1 (NY Library Scan)

[36] Mauser Model 1889 - Wikipedia Lee–Enfield - Wikipedia 6 Rifles Used by the Afrikaners During the Second Boer War – Tactical Life Gun Magazine: Gun News and Gun Reviews (tactical-life.com)

[37] With the Flag to Pretoria, H. W. Wilson Vol 1

He still has 5 volumes to complete. Publishing Vol 1 in 1935 he concludes Vol 11 in 1975 with his spouse and life companion Ariel as coauthor of the last five volumes.

[38] File:SouthAfrica1885.svg - Wikimedia Commons

[39] Navy & Army Illustrated Vol 9, October 7th 1899 and Creswicke, Louis South Africa and the Transvaal War, Vol. 1

[40] South African Railway System circa 1900: File:SouthernAfricaRailwayMap 1901.png - Wikipedia

[41] SA Map Roberts, Churchills in Africa p. 120

[42] WSC, My Early Life

[43] Frontiers and Wars p.360 https://archive.org/details/frontierswarshis0000chur

[44] Crop of map from My Early Life.

[45] Haldane, How We Escaped from Pretoria. Image credit: Neville Constantine, Neville C. at Angloboerwar.com Anglo Boer War - Armored Train Ambush 11-15-99 Chuchill & Haldane - Page 13 - Boer War Forum

[46] Haldane, How We Escaped from Pretoria. Image credit: Neville Constantine, Neville C. at Angloboerwar.com Anglo Boer War - Armored Train Ambush 11-15-99 Chuchill & Haldane - Page 13 - Boer War Forum

[47] Doyle, Arthur Conan. The Great Boer War: Vol 1 Smith and Elder Co. Stanfords.

[48] SA Boer-British War Vol 1. Hopkins, 1900

[49] Cunliffe, History of the Boer War Vol 1 (NY Library Scan)

[50] James. Stuart, A History Of The Zulu Rebellion 1906 And Of Dinuzulu's Arrest, Trial And Expatriation, Macmillan And Co., Limited St. Martin's Street, London 1913. History of the war in South Africa: containing an authentic account of the great struggle between the British and the Boers by James H. Birch, Jr; Henry Davenport Northrop, 1836-1909; and Robert Wilson, 1833-1912. R. A. H. Morrow Co. St. John, N. B. 1900 PDF 820 page version. P.VI
William Harding War in South Africa and the Dark Continent from Savagery to Civilization 1899 Dominion Company (The strange story of a weird world from the earliest ages to the present, including the war with the Boers) Page 66 History of the War in South Africa in four Volumes. By: Major-General Sir Frederick Maurice, K.C.B, London Huest And Blackett Limited 1906. Vol 1 p. 492-494

[51] The Boer war: its causes and its interest to Canadians by Emerson Bristol Biggar. 3rd.Ed. 1899.

[52] See Also: Dictionary of South African English https://www.upgradeyourbrowser.dsae.co.za/

[53] South Africa; its history, heroes and wars by Mackenzie, William Douglas, 1859-1936; assisted by Alfred Stead, 1877-1933, joint authors. Published 1899 by The Co-Operative Publishing Company, Chicago, Ill.

[54] History of the Boers in South Africa, Theal, George McCall. Sonnenschein & Co. London 1887 British Library HMNTS 9061.eee.18. Identifier: 003607007 https://access.bl.uk/item/pdf/lsidyv2afa3205

[55] File:SouthAfrica1885.svg - Wikimedia Commons

[56] War in SA, Harding 1899.

[57] War with the Boers, H. Brown, London [1902?] Vol 1

[58] Heroes of the Boer War. Review of Reviews, 1903.

[59] Black & White Weekly Vol 1-2 1899

[60] Vredefort crater cross section 2 - Vredefort crater - Wikipedia

[61] Cropped: File:Witwatersrand NS cross section.png - Wikipedia User:Oggmus - Wikimedia Commons

[62] File:Ferreiraskamp, winter 1886.jpg - Wikimedia Commons

[63] War with the Boers, H. Brown, London [1902?] Vol 1

[64] War with the Boers, H. Brown, London [1902?] Vol 2

[65] SA War Vol 8 p. 154-155

[66] File:The Second Boer War, 1899-1902 Q68483.jpg - Wikimedia Commons

[67] After Pretoria: The Guerilla War Vol 3 Jan., 1901

[68] After Pretoria: The Guerilla War Vol 3 Jan., 1901

[69] South Africa; its History, Mackenzie 1899

[70] With the Flag to Pretoria, H. W. Wilson Vol 2 June, 1900.

[71] History of War SA, Birch, 1899 See: Big Hole - Wikipedia

[72] South Africa; its History, Mackenzie 1899

[73] Heroes of the Boer War. Review of Reviews, 1903.

[74] South Africa; its History, Mackenzie 1899

[75] File:Kimberley Mine shaft.png - Wikimedia Commons

[76] SA Boer-British War Vol 1. Hopkins, 1900

[77] ABW Source

[78] Black & White Weekly Vol 1-2 1899-1900

[79] Neville Constantine 3rd: Pretoria_Commando_Crop_3.jpg

[80] War in SA, Harding 1899
[81] Richard Harding Davis With Both Armies in South Africa. Charles Scribner's Sons, New York 1900
[82] Photograph of one of the Transvaal Staatsartillerie 75 mm QF Creusot guns (Thanks to MC Heunis) Anglo Boer War - Artillery and Ammunition - Page 17 - Boer War Forum
[83] Portrait photograph of Lieutenant and Adjutant Carl Roos (with thanks to MC Heunis). Anglo Boer War - Armored Train Ambush 11-15-99 Chuchill & Haldane - Page 21 - Boer War Forum
[84] War with the Boers, H. Brown, London [1902?] Vol 5
[85] War with the Boers, H. Brown, London [1902?] Vol 1
[86] Creswicke, Louis South Africa and the Transvaal War, Vol. 1 1900 Photo by Elliott & Fry, London.
[87] Times History War in S.A Vol 1 (1901)
[88] Creswicke, Louis South Africa and the Transvaal War, Vol. 1 1900 Photo by Elliot & Fry, London.
[89] Cartoon from Der wahre Jacob, 5 December 1899. File:England und der Krieg in Südafrika - Rata Langa 1899.jpg - Wikimedia Commons
[90] Richard Harding Davis, With Both Armies in South Africa. Charles Scribner's Sons, New York 1900
[91] Davitt, Boer Fight for Freedom, 3rd ed. 1902. (Duke Scan)
[92] Cunliffe, History of the Boer War Vol 2 (Composite of Mich. St. & U. Cal. Scans)
[93] Creswicke, Louis South Africa and the Transvaal War, Vol. 2 1900 Photo by Knight, Aldershot.
[94] Black & White Weekly Vol 1-2 1899-1900
[95] Anonymous (1910) A Handbook of the Boer War - With General Map of South Africa and 18 Sketch Maps and Plans, London & Aldershot: Gale & Polden Ltd File:NorthNatalWarTheatre, colour, crop.png - Wikimedia Commons Also: Black and White Original: NorthNatalWarTheatre - File:NorthNatalWarTheatre, colour, crop.png - Wikimedia Commons
[96] Davitt, Boer Fight for Freedom, 3rd ed. 1902. (Duke Scan)
[97] Black & White Weekly Vol 1-2 1899-1900 by René Bull. From about one month later, but exodus from the Transvaal and northern Natal was already substantial. Titled: "Colenso captured by the Boers – Flight of Indians and Natives."
[98] And: No. 27157. The London Gazette. 26 January 1900. pp. 497–498. https://en.wikipedia.org/wiki/The_London_Gazette
[99] Cunliffe, History of the Boer War Vol 1 (NY Library Scan)
[100] File:Battle of Glencoe Map.jpg - Wikimedia Commons
[101] Image Source: Britishbattles.com\Talana Hill Battle.
[102] File:Telfer-dunbar J Q72268.jpg - Wikimedia Commons
[103] Hillegas, With the Boer forces, 1901 (Photograph by Reginald Sheppard, Pretoria.)
[104] Hillegas, With the Boer forces, 1901 (Photograph by Reginald Sheppard.)
[105] Hillegas, Boers in War, 1900.
[106] Black & White Weekly Vol 1-2 1899-1900
[107] Hillegas, With the Boer forces, 1901 (Photograph by Reginald Sheppard.)
[108] War with the Boers, H. Brown, London [1902?] Vol 2
[109] War with the Boers, H. Brown, London [1902?] Vol 2
[110] War with the Boers, H. Brown, London [1902?] Vol 2 Window & Grove Photo
[111] The Sphere - Saturday 24 November 1900 p. 13 https://www.britishnewspaperarchive.co.uk/viewer/bl/0001861/19001124/016/0013
[112] BritishBattles.com https://www.britishbattles.com/
[113] BritishBattles.com https://www.britishbattles.com/
[114] BritishBattles.com Map by John Fawkes BritishBattles.com analysing and documenting British Battles from the previous centuries
[115] Cunliffe, History of the Boer War Vol 1 (NY Library Scan)
[116] File:Battle of Elandslaagte Map.png - Wikimedia Commons
[117] BritishBattles.com https://www.britishbattles.com/
[118] Heroes of the Boer War. Review of Reviews, 1903.
[119] Cunliffe, History of the Boer War Vol 1 (NY Library Scan)
[120] With Roberts to the Transvaal. C.N. Robinson [1900]
[121] Heroes of the Boer War. Review of Reviews, 1903.
[122] File:Hollanderkorps Memorial at Elandslaagte.jpg - Wikimedia Commons Author: David Biggins
[123] File:Boer War; searching for the wounded from the battlefield at Wellcome V0015550.jpg - Wikimedia Commons Watercolour sketch by F. Craig, 1899.

[124] Please Note Indian Stretcher Bearer at battle site: File:Boer War; removing the wounded after the Battle of Elandslaa Wellcome V0015512.jpg - Wikimedia Commons

[125] Black & White Weekly Vol 1-2 1899-1900 Photo by D. Barnett. This must have been early in the battle or the next day since the battle ended at night.

[126] Image included to show R.A.M.C. personnel working with Indian bearers. "Removing the wounded after the battle of Elandslaagte. This operation had to be performed in a pouring rain and in almost total darkness." With the Flag to Pretoria, H. W. Wilson Vol 1 1900 Sketch by W. Dewdr.

[127] After Pretoria: The Guerilla War Vol 3 p. 51, 1900

[128] Hillegas, With the Boer forces, 1901 (Photograph by R. Steger.)

[129] Elandslaagte bridge. Damaged during war. - Atom site for DRISA P1813.jpg (1400×947) (drisa.co.za)

[130] Hillegas, With the Boer forces, 1901 (Photograph by Van Hoepen.)

[131] Image by R. Caton Woodville File:Elandslaagte 1.jpg - Wikimedia Commons

[132] BritishBattles.com Map by John Fawkes

[133] File:Battle of Elandslaagte Map.png - Wikimedia Commons

[134] Times History War in S.A Vol 2 (1902)

[135] File:The Second Boer War, 1899-1902 Q72417.jpg - Wikimedia Commons

[136] Creswicke, Louis South Africa and the Transvaal War, Vol. 2 1900 Photo by Wilson, Aberdeen.

[137] Navy & Army Illustrated Vol 10 "Our special Correspondent who was locked up in Ladysmith sends an illustration of the balloon being sent up to observe the Boer positions."

[138] Black & White Weekly Vol 1-2 1899-1900

[139] Image source: britishbattles.com

[140] Hillegas, Boers in War, 1900.

[141] Davitt, Boer Fight for Freedom, 3rd ed. 1902. (Duke Scan)

[142] Times History Vol 2 p. 216.

[143] Black & White Weekly Vol 1-2 1899-1900

[144] Black & White Weekly Vol 1-2 1899-1900

[145] Anglo Boer War - Pictures from Under the Union Jack - Page 3 - Boer War Forum Credit: Neville Constantine

[146] Heroes of the Boer War. Review of Reviews, 1903.

[147] War with the Boers, H. Brown, London [1902?] Vol 5

[148] File:Sir Ian Hamilton.jpg - Wikimedia Commons Photo by Elliot & Fry

[149] Promotional Picture. Reference uncertain. See: Rare British South African war song | marching onto Pretoria. - YouTube

[150] Image Credit: Rob Davidson Anglo Boer War - Small Arms and Ammunition - Page 6 - Boer War Forum

[151] BritishBattles.com Map by John Fawkes.

[152] File:Melton Prior00.jpg - Wikimedia Commons

[153] Hillegas, Boers in War, 1900.

[154] Cunliffe, History of the Boer War Vol 1 (NY Library Scan)

[155] Creswicke SA War Vol 2

[156] File:Positions round Ladysmith - November 1899 - Project Gutenberg eText 16466.jpg - Wikimedia Commons

[157] Creswicke, Louis South Africa and the Transvaal War, Vol. 2 1900 Photo by Cribb, Southsea.

[158] File:Devonshire Regiment-Ladysmith.JPG - Wikimedia Commons

[159] Navy & Army Illustrated Vol 9 Photo by: T. H. Murray, Pieter Maritsburg.

[160] Image source: With the Flag to Pretoria, H. W. Wilson Vol 1 1900 and britishbattles.com.

[161] Creswicke SA & Transvaal War. Vol 3 front. Edinburgh and London: T.C. & E. C. Jack.

[162] Navy & Army Illustrated Vol 10 p. 36.

[163] File:British Army, Boer War, Q 71941.jpg - Wikimedia Commons

[164] Webster Davis, John Bull's crime; or, Assaults on Republics. 1901 New York The Abbey Press

[165] Anglo-Boer War Album, Cape Town, D. Edwards & co. [1902?]

[166] War with the Boers, H. Brown, London [1902?] Vol 2

[167] Anglo-Boer War Album, Cape Town, D. Edwards & co. [1902?]

[168] Hiley and Hassell, The Mobile Boer, Grafton Press, New York, (c. 1902)

[169] Courtesy of Neville Constantine: Anglo Boer War - Artillery and Ammunition - Page 25 - Boer War Forum

[170] Image Source: BritishBattles.com

[171] File:The Boer Lilliputia(n) - the British Gulliver - J.S. Pughe. LCCN2010646243.tif - Wikimedia Commons
[172] Frontispiece. Haldane, How We Escaped from Pretoria. Photo by Elliot and Fry.
[173] Black & White Weekly Vol 1-2 1899-1900 Photo by D. Barnett.
[174] Black & White Weekly Vol 1-2 1899-1900.
[175] Image source: With the Flag to Pretoria, H. W. Wilson Vol 1 1900 and britishbattles.com Horace Nicholls photograph of Armoured Train No. 53 – "Embarking on a perilous journey. (Armoured train destroyed at Chieveley)." Copy of an original print held in the archive of the Royal Photographic Society, Bath. Credit: Neville Constantine Neville C. at Anglo Boer War - Armored Train Ambush 11-15-99 Chuchill & Haldane - Page 8 - Boer War Forum
[176] Black & White Weekly Vol 1-2 1899-1900 (Sandys p. 45. 3 officers, 117 men from Durban Light Infantry and Dublin Fusiliers, 5 sailors with a muzzle loading cannon.)
[177] Navy & Army Vol 9
[178] Black & White Weekly Vol 1-2 1899-1900 Photo by "Our Special Correspondent"
[179] Cunliffe, History of the Boer War Vol 1 (NY Library Scan)
[180] Times History War in S.A Vol 2 (1902) Photo by C. Knight, Aldershot.
[181] Image Credit: britishbattles.com Battle of Colenso.
[182] Anglo-Boer War Album, Cape Town, D. Edwards & co. [1902?]
[183] Cape Town. Table Bay Harbour with many sailing vessels and wagonloads of bags in the foreground. - Atom site for DRISA P0692_02G.jpg (1400×1080) (drisa.co.za)
[184] Black & White Weekly Vol 1-2 1899-1900 Photo by "Our Special Correspondent" Also: National Library of Scotland (Sandys)
[185] October 21 Folder Anglo Boer War - Small Arms and Ammunition - Page 4 - Boer War Forum
[186] October 21 Folder Anglo Boer War - Small Arms and Ammunition - Page 4 - Boer War Forum
[187] SA War Vol 3 1900 Photo by Vickers Sons & Maxim, Limited.
[188] During the Second Boer War, the British used Maxim machine guns, and the Boers used a modified Maxim, a belt-fed, water-cooled machine gun that fired explosive 1-pound (0.45 kg) rounds (smokeless ammunition) at 450 rounds per minute, which became known as the "pom-pom." Maxim Nordenfelt Guns and Ammunition Company - Wikipedia And: A converted Maxim machine gun, firing a one-pound percussion-fused shell, was adopted for service by the Boers. Known as the "Pom Pom" (from the sound the gun made when firing) it was the first use of automatic light artillery in land warfare. The Pom-Pom was simply a large caliber, belt-fed machine gun firing explosive rounds. It was hard to locate on the battlefield as the smokeless ammunition (some of the first of its kind) produced little or no firing signature. Canvas belts held 25 brass-cased rounds carrying explosive charges with their percussion fuses. First-hand accounts describe the Pom-Pom as being very effective; standard artillery, mostly, could still be avoided by quickly taking cover in the interval between the flash indicating the firing of the shell and its arrival. The Pom-Pom, on the other hand, could keep up a continuous stream of fire, to devastating effect. South Africa's National Museum of Military History (smallarmsreview.com) See also this 15-minute video on the Forgotten Weapons channel describing the weapon. Maxim Pom-Pom 37mm Machine Gun - YouTube
[189] SA Boer-British War Vol 1. Hopkins, 1900
[190] Image source: With the Flag to Pretoria, H. W. Wilson Vol 1 1900 and britishbattles.com PF: From the photographs the interior of the carriage was far more crowded than this sketch indicates.
[191] Black & White Weekly Vol 1-2 1899-1900 (From a sketch by René Bull) "The armoured train disaster at Chieveley, November 15th, where Mr. Winston Churchill was captured."
[192] Image source: With the Flag to Pretoria, H. W. Wilson Vol 1 1900 and britishbattles.com
[193] From a Drawing by René Bull. By permission of "Black and White." From: Winston Spencer Churchill by A. MacCallum Scott, Methuen, London 1905
[194] File:De Albert Medaille met batons VK.jpg - Wikimedia Commons
[195] Black & White Weekly Vol 1-2 1899-1900 (From a sketch by René Bull) "The armoured train disaster at Chieveley, November 15th, where Mr. Winston Churchill was captured."
[196] Black & White Weekly Vol 1-2 1899-1900 Photo by D. Barnett.
[197] Credit: Neville Constantine, Anglo Boer War - Armored Train Ambush 11-15-99 Chuchill & Haldane - Page 2 - Boer War Forum
[198] Credit: Neville Constantine, Anglo Boer War - Armored Train Ambush 11-15-99 Chuchill & Haldane - Page 2 - Boer War Forum
[199] Credit: Neville Constantine, Neville C. at Angloboerwar.com
[200] Note gravesite in the background. The cars left behind have been removed from the tracks. File:Boer War Photograph Taken by B W Caney Q115365.jpg - Wikimedia Commons Author: B.W. Caney.

[201] Navy & Army Illustrated Vol 9 Also: Haldane, How We Escaped from Pretoria.
[202] Credit: Neville Constantine, Neville C. at Angloboerwar.com
[203] Cunliffe, History of the Boer War Vol 1 (NY Library Scan)
[204] Credit: Neville Constantine, Anglo Boer War - Armored Train Ambush 11-15-99 Chuchill & Haldane - Page 2 - Boer War Forum.
[205] Credit: Neville Constantine. Anglo Boer War - Armored Train Ambush 11-15-99 Chuchill & Haldane - Page 2 - Boer War Forum, Taken in 1978.
[206] Credit: Neville Constantine. Anglo Boer War - Armored Train Ambush 11-15-99 Chuchill & Haldane - Page 2 - Boer War Forum, Taken in 1978.
[207] The Biograph in battle: its story in the South African War related with personal experiences. by William Kennedy Laurie Dickson, London: T. Fisher Unwin, 1901. Catalog Search: https://archive.org/search?query=title%3A%28The+Biograph+in+Battle%29
[208] Image source: britishbattles.com
[209] Heroes of the Boer War. Review of Reviews, 1903.
[210] Navy & Army Illustrated Vol 10
[211] With the Flag to Pretoria, H. W. Wilson Vol 2 Dec,. 1899. SA National Archives: http://www.nationalarchives.gov.za/node/1236
[212] Black & White Weekly Vol 1-2 1899-1900
[213] Navy & Army Illustrated Vol 10 and With the Flag to Pretoria, H. W. Wilson Vol 1 1900 Photo by Knight.
[214] File:Boer War Photograph Taken by B W Caney Q115366.jpg - Wikimedia Commons
[215]Black & White Weekly Vol 1-2 1899-1900
[216] War in SA, Harding 1899
[217] Richard Harding Davis With Both Armies in South Africa. Charles Scribner's Sons, New York 1900
[218] SA Boer-British War Vol 1. Hopkins, 1900
[219] Creswicke SA War Vol 2
[220] File:Positions round Ladysmith - November 1899 - Project Gutenberg eText 16466.jpg - Wikimedia Commons
[221] Haldane, How We Escaped from Pretoria. Photo by Stereoscopic Co. and Creswicke, SA War, Vol. 2 Photo by Wilson, Aberdeen.
[222] Item N03503 N03503.jpg (1400×964) (drisa.co.za)
[223] Hillegas, Boers in War, 1900.
[224] Anglo-Boer War Album, Cape Town, D. Edwards & co. [1902?]
[225] Navy & Army Illustrated Vol 9
[226] Davitt, Boer Fight for Freedom, 3rd ed. 1902. (Duke Scan)
[227] Image and heading Credit: Neville Constantine: Anglo Boer War - Artillery and Ammunition - Page 15 - Boer War Forum
[228] Image source: Vera Gallony
[229] South Africa; its History, Mackenzie 1899
[230] History of War SA, Birch, 1899
[231] Black & White Weekly Vol 1-2 1899-1900 Photo by D. Barnett
[232] Haldane, How We Escaped from Pretoria. By: Messrs. Harmsworth
[233] Arthur Mann The Boer in Peace and War, 1900
[234] War with the Boers, H. Brown, London [1902?] Vol 1
[235] Langsnek Tunnel, July 1900. Damaged northern portal after being blown up. - Atom site for DRISA P1628_01.jpg (1400×814) (drisa.co.za)
[236] Booking office for express mail coaches between Johannesburg and Newcastle with coach in foreground. - Atom site for DRISA P0521_05.jpg (1400×936) (drisa.co.za)
[237] Black & White Weekly Vol 1-2 1899-1900
[238] Haldane, How We Escaped from Pretoria. By: Messrs. Harmsworth
[239] Navy & Army Illustrated Vol 9
[240] War in SA, Harding 1899
[241] South Africa; its History, Mackenzie 1899
[242] Navy & Army Illustrated Vol 9
[243] War in SA, Harding 1899

[244] Image Credit Neville Constantine. Anglo Boer War - Armored Train Ambush 11-15-99 Chuchill & Haldane - Page 5 - Boer War Forum
[245] Image Credit Rob Davidson. Anglo Boer War - Armored Train Ambush 11-15-99 Chuchill & Haldane - Page 5 - Boer War Forum at Anglo Boer War
[246] Image source: With the Flag to Pretoria, H. W. Wilson Vol 1 1900 and britishbattles.com
[247] Local Newspaper Publication, Standard & Diggers News, in 1899, Reproduced by: WSC in, My Early Life.
[248] Image credit: Haldane and National Library of Scotland
[249] National Archives Repository Pretoria (Sandys)
[250] Haldane, How We Escaped From Pretoria.
[251] Haldane, How We Escaped from Pretoria.
[252] London to Ladysmith p. 201 also in War Pictures, 1900.
[253] File:Rifles1905-2.jpg - Wikimedia Commons
[254] Davitt, Boer Fight for Freedom, 1902 (U. Cal. Scan)
[255] War with the Boers, H. Brown, London [1902?] Vol 4
[256] Image Credit QSAMike Anglo Boer War - Armored Train Ambush 11-15-99 Chuchill & Haldane - Page 17 - Boer War Forum
[257] From a Sketch by Winston Churchill. By permission of "The Graphic." From: Winston Spencer Churchill by A. MacCallum Scott, Methuen, London 1905
[258] My Escape from the Boers, Strand Magazine Vol 66 p. 537-547. Image by Ernest Prater.
[259] From: Winston Spencer Churchill by A. MacCallum Scott, Methuen, London 1905
[260] WSC, My Early Life Winston Spencer Churchill, My Early Life a Roving Commission, London: Thornton Butterworth, September 1931 My Early Life. A Roving Commission. (fadedpage.com) and Search: fadedpage.com See Sandy's p. 102-104
[261] Image Credit: Francis L. H. de Souza per Candice Millard, Hero of the Empire. Image also appears on page 96 of Celia Sandys, Churchill Wanted Dead or Alive.
[262] Image credit: Francis L. H. de Souza (Mr. de Sozua would sneak in a bottle of whiskey at the bottom of a basket of fruit he would bring in for the prisoners.)
[263] Image source Dr. Jonathan de Souza.
[264] Hofmeyr The Story of My Captivity.
[265] Hofmeyr The Story of My Captivity.
[266] Richard Harding Davis With Both Armies in South Africa. Charles Scribner's Sons, New York 1900
[267] Arthur Mann The Boer in Peace and War, 1900
[268] The Biograph in battle: its story in the South African War related with personal experiences. by William Kennedy Laurie Dickson, London: T. Fisher Unwin, 1901. Catalog Search: https://archive.org/search?query=title%3A%28The+Biograph+in+Battle%29
[269] Haldane, How We Escaped from Pretoria. By: Robert Carnegie
[270] Haldane, How We Escaped from Pretoria. p.94
[271] File:WFReitz CHM VA0957.jpg - Wikimedia Commons
[272] Rosslyn -Twice Captured.
[273] Haldane, How We Escaped from Pretoria. And: Richard Harding Davis With Both Armies in South Africa.
[274] Haldane, How We Escaped from Pretoria.
[275] Haldane, How We Escaped from Pretoria.
[276] Creswicke, Louis South Africa and the Transvaal War, Vol. 1 Photo by Wilson, Aberdeen.
[277] Haldane, How We Escaped from Pretoria. Image credit: Neville Constantine, Neville C. at Angloboerwar.com
[278] Haldane, How We Escaped from Pretoria. By: Major Robert Carnegie.
[279] Haldane, How We Escaped from Pretoria. By Major Robert Carnegie
[280] Haldane, How We Escaped from Pretoria. By: Messrs. Harmsworth
[281] Haldane, How We Escaped from Pretoria. By Major Robert Carnegie
[282] Haldane, How We Escaped from Pretoria. p. 182
[283] Cape cart with horse. - Atom site for DRISA P1205.jpg (1400×959) (drisa.co.za)
[284] Haldane, How We Escaped from Pretoria. p.191
[285] Haldane, How We Escaped from Pretoria. Image credit: Neville Constantine, Neville C. at Angloboerwar.com
[286] Haldane, How We Escaped from Pretoria.
[287] Image source: John Bird.

[288] Image source: John Bird.
[289] With the Central Column in South Africa. From Belmont to Komati-Poort. Compiled from Photographs. The "King and His Navy and Army." By Surgeon-Major Walter Calverley Beevor, 1903 Image and text from: Image per Neville Chamberlain @ Anglo Boer War - Armored Train Ambush 11-15-99 Chuchill & Haldane - Page 17 - Boer War Forum
[290] With the Central Column in South Africa. From Belmont to Komati-Poort. Compiled from Photographs. The "King and His Navy and Army." By Surgeon-Major Walter Calverley Beevor, 1903 Image and text from: Image per Neville Chamberlain @ Anglo Boer War - Armored Train Ambush 11-15-99 Chuchill & Haldane - Page 17 - Boer War Forum
[291] Haldane, How We Escaped from Pretoria.
[292] Head-Gear of the Van Rhyn Gold Mine destroyed by the Boers, January 29th, 1901 Image: After Pretoria: The Guerilla War Vol 3 Jan., 1901.
[293] This a picture of wool that was confiscated by British troops. Navy & Army Vol 13 p. 59 Oct. 5th, 1901.
[294] With the Central Column in South Africa. From Belmont to Komati-Poort. Compiled from Photographs. The "King and His Navy and Army." By Surgeon-Major Walter Calverley Beevor, 1903 Image and text from: Image per Neville Chamberlain @ Anglo Boer War - Armored Train Ambush 11-15-99 Chuchill & Haldane - Page 17 - Boer War Forum
[295] Haldane, How We Escaped from Pretoria.
[296] Haldane, How We Escaped from Pretoria. Image credit: Neville Constantine, Neville C. at Angloboerwar.com
[297] File:Termine cremagliera Saline-Volterra.jpg-Wikimedia Commons
[298] Waterval-Boven, circa 1901. Armoured train during Anglo-Boer War at railway station. - Atom site for DRISA N78689.jpg (1400×1045) (drisa.co.za)
[299] Creswicke Drawing by Donald F. M'Cracken.
[300] Anglo-Boer War Album, Cape Town, D. Edwards & co. [1902?]
[301] Haldane, How We Escaped from Pretoria. p. 222 Note on line 3 the name "Burnham." Had they been discovered, Mr. Burnham would have gone to prison.
[302] File:Flag of Transvaal.svg - Wikimedia Commons Credit: Himasaram
[303] Haldane, How We Escaped from Pretoria. p. 227
[304] Navy & Army Illustrated Vol 10 Photo by Edwards
[305] HMS *Forte* leaving Simonstown, South Africa, c1900 File:The Royal Navy 1899-1902 Q115146.jpg-Wikimedia Commons
[306] File:Arrival of 'Königin Luise' in Fremantle, W.A. - 3 August 1919 (24299353089).jpg - Wikimedia Commons
[307] Haldane, How We Escaped from Pretoria. p 231
[308] Haldane, How We Escaped from Pretoria. p 231
[309] Haldane, How We Escaped from Pretoria. p. 231
[310] Durban. St Pauls Wharf and timber wharf with four sailships and three mule wagons in Durban Harbour. - Atom site for DRISA N06201.jpg (1400×1071) (drisa.co.za)
[311] The Biograph in battle: its story in the South African War related with personal experiences. by William Kennedy Laurie Dickson, London: T. Fisher Unwin, 1901. Catalog Search: https://archive.org/search?query=title%3A%28The+Biograph+in+Battle%29
[312] Le Mesurier as depicted in the Oxford Times, 23rd June 1900 per Neville Chamberlain Anglo Boer War - Armored Train Ambush 11-15-99 Chuchill & Haldane - Page 15 - Boer War Forum
[313] "The Gram" A social Magazine Founded by British Prisoners of War in Pretoria, Facsimile Edition Eyre & Spottiswoode, Her Majesty's Printers, London #1 - The Gram ; a social magazine founded by British prisoners ... (no.1-3). - Full View | HathiTrust Digital Library Scanned from the University of Minnesota Copy
[314] Image Credit QSAMike Anglo Boer War - Armored Train Ambush 11-15-99 Chuchill & Haldane - Page 17 - Boer War Forum
[315] Winston Churchill's Former Apartment in Mayfair, London For Rent (housebeautiful.com) No Author Listed.
[316] File:105 Mount Street Londres.jpg - Wikimedia Commons Author: CVB
[317] File:9 2 258 0008-Staats Model School-Pretoria-s.jpg - Wikimedia Commons
[318] Press Association Photograph. See also Frederick Woods' Young Winston's Wars for citation and reproduction p. 195
[319] File:FWReitz VA0947.jpg - Wikimedia Commons
[320] Deneys Reitz op kommando - Deneys Reitz - Wikipedia
[321] File:St James's Hall.jpg - Wikimedia Commons
[322] Drawn by J. Nash, R.I. [from a sketch by Winston Churchill.] War pictures. v. 1 (no. 1-16) 1900 #1 - War pictures. v. 1 (no. 1-16) 1900. - Full View | HathiTrust Digital Library And: Catalog Record: War pictures | HathiTrust Digital Library

[323] From biography by A. MacCallum Scott of: Winston S. Churchill, Methuen, London 1905. By permission of "The Graphic."
[324] My Escape from the Boers, Strand Magazine Vol 66 p. 537-547. Image by Ernest Prater.
[325] File:1stEarlOfBirkenhead.jpg - Wikimedia Commons
[326] HathiTrust.org 39015005657708 -8-1676306871 University of Michigan Scan.
[327] Portrait by Francis Dodd, 1917 File:Lieut-general J Aylmer L Haldane, Cb, Dso Art.IWMART1806.jpg - Wikimedia Commons.
[328] https://www.librarything.com/author/haldanejamesaylmerlo
[329] Reynolds's Newspaper - Sunday 19 May 1912 p. 7 Col 5 @ bottom. https://www.britishnewspaperarchive.co.uk/viewer/bl/0000101/19120519/146/0007 Same Article: Weekly Dispatch (London) - Sunday 19 May 1912 p.1 https://www.britishnewspaperarchive.co.uk/viewer/bl/0003358/19120519/014/0001.
[330] A Lost Letter of Ancient Rome Blackwood's Magazine, February 1912-02: Vol 191 Iss 1156 Page 270-276 https://archive.org/details/sim_blackwoods-magazine_1912-02_191_1156
[331] History of War SA, Birch, 1899
[332] File:The Second Boer War, 1899-1902 Q72215.jpg - Wikimedia Commons
[333] File:De Albert Medaille met batons VK.jpg - Wikimedia Commons
[334] Cunliffe, History of the Boer War Vol 1 (NY Library Scan)
[335] "Hairy Mary" Locomotive at siege of Ladysmith. "Ladysmith, 1899. Hairy Mary locomotive." - Atom site for DRISA N44246.jpg (1400×1082) (drisa.co.za)
[336] Image Courtesy of Neville Constantine. Anglo Boer War - Armored Train Ambush 11-15-99 Chuchill & Haldane - Page 22 - Boer War Forum
[337] Image thanks to Dougie McMaster / MC Heunis) Anglo Boer War - Armored Train Ambush 11-15-99 Chuchill & Haldane - Page 22 - Boer War Forum
[338] Image credit: Francis L. H. de Souza (Mr. de Sozua would sneak in a bottle of whiskey at the bottom of a basket of fruit he would bring in for the prisoners.)
[339] Image source Dr. Jonathan de Souza
[340] Hofmeyr The Story of My Captivity.
[341] File:TradeCardMasonAndHamlinOrganCoCirca1880s.jpg - Wikimedia Commons
[342] Cunliffe, History of the Boer War Vol 1 (NY Library Scan)
[343] Haldane, How We Escaped from Pretoria. By: Robert Carnegie
[344] Images Credit: Rob Davidson
[345] Anglo-Boer War Album, Cape Town, D. Edwards & co. [1902?]
[346] Haldane, How We Escaped from Pretoria.
[347] Creswicke, Louis South Africa and the Transvaal War, Vol. 1 Photo by Wilson, Aberdeen.
[348] Rosslyn -Twice Captured.
[349] Richard Harding Davis With Both Armies in South Africa. Charles Scribner's Sons, New York 1900
[350] With the Flag to Pretoria, H. W. Wilson Vol 2 June, 1900.
[351] With the Flag to Pretoria, H. W. Wilson Vol 2 June, 1900.
[352] Richard Harding Davis With Both Armies in South Africa. Charles Scribner's Sons, New York 1900
[353] With the Inniskilling Dragoons: the record of a cavalry regiment during the Boer War, 1899-1902 by Lieut.-Colonel J. Watkins Yardley. Longmans, Green, and co., 1904.
[354] Image Credit QSAMike Anglo Boer War - Armored Train Ambush 11-15-99 Chuchill & Haldane - Page 17 - Boer War Forum
[355] Image per Neville Chamberlain. Anglo Boer War - Armored Train Ambush 11-15-99 Chuchill & Haldane - Page 17 - Boer War Forum
[356] Ian Hamilton's March 1900 The Copp, Clark Co. Toronto and online sources. See Bibliography,
[357] File:Dorset Yeomanry Boer War.png - Wikimedia Commons Credit: Axel Ravera. Public Domain.
[358] Navy & Army Illustrated Vol 11 November 17th, 1900
[359] Hillegas, Boers in War, 1900.
[360] With the Flag to Pretoria, H. W. Wilson Vol 2 June, 1900. Photo by J. F. Archibald
[361] Hillegas, Boers in War, 1900.
[362] Richard Harding Davis With Both Armies in South Africa. Charles Scribner's Sons, New York 1900
[363] Hillegas, Boers in War, 1900.

[364] Black & White Weekly Vol 4 1901
[365] With the Flag to Pretoria, H. W. Wilson Vol 2 June, 1900.
[366] Anglo-Boer War Album, Cape Town, D. Edwards & co. [1902?]
[367] After Pretoria: The Guerilla War Vol 3 August, 1900
[368] Richard Harding Davis With Both Armies in South Africa. Charles Scribner's Sons, New York 1900
[369] Richard Harding Davis With Both Armies in South Africa. Charles Scribner's Sons, New York 1900
[370] With the Flag to Pretoria, H. W. Wilson Vol 2 June, 1900.
[371] With the Flag to Pretoria, H. W. Wilson Vol 2 June, 1900.
[372] Richard Harding Davis With Both Armies in South Africa. Charles Scribner's Sons, New York 1900
[373] Library of Congress Prints and Photographs Division Washington, D.C. https://www.loc.gov/pictures/item/00651051/resource/
[374] Anglo-Boer War Album, Cape Town, D. Edwards & co. [1902?]
[375] The Biograph in battle: its story in the South African War related with personal experiences. by William Kennedy Laurie Dickson, London: T. Fisher Unwin, 1901. Catalog Search: https://archive.org/search?query=title%3A%28The+Biograph+in+Battle%29
[376] The Biograph in battle: its story in the South African War related with personal experiences. by William Kennedy Laurie Dickson, London: T. Fisher Unwin, 1901. Catalog Search: https://archive.org/search?query=title%3A%28The+Biograph+in+Battle%29
[377] The Biograph in battle: its story in the South African War related with personal experiences. by William Kennedy Laurie Dickson, London: T. Fisher Unwin, 1901. Catalog Search: https://archive.org/search?query=title%3A%28The+Biograph+in+Battle%29
[378] Image Credit: QSAMIKE: Anglo Boer War - Armored Train Ambush 11-15-99 Chuchill & Haldane - Page 18 - Boer War Forum
[379] Hofmeyr The Story of My Captivity.
[380] Hofmeyr The Story of My Captivity.
[381] Cape waggon crossing Eland's River from: The Queen's Empire. Volume 1. Cassell & Co. London Eland crossing QE1 48 - Elands River (Mpumalanga) - Wikipedia
[382] Arthur Mann The Boer in Peace and War, 1900
[383] Ox wagon stuck in Crocodile River - Atom site for DRISA N00061.jpg (1400×925) (drisa.co.za)
[384] South Africa; its History, Heroes and Wars. Mackenzie [1899]
[385] Author: NOAA John Hayford Album. File:Theb0741 - Flickr - NOAA Photo Library.jpg - Wikimedia Commons
[386] Engineers repairing a Railway destroyed by the Boers near Kroonstadt. File:Reparaties aan de spoorlijn naar Kroonstad, Zuid-Afrika Royal Engineers repairing a Railway destroyed by the Boers - Kroonstadt, South africa (titel op object), RP-F-F09002.jpg - Wikimedia Commons
[387] Heavily loaded ox wagon. - Atom site for DRISA PB1153_001.jpg (1400×922) (drisa.co.za)
[388] South Africa; its History, Heroes and Wars. Mackenzie [1899]
[389] Navy & Army Illustrated Vol 11 p. 123 October 27th, 1900.
[390] Black & White Weekly Vol 1-2 1899-1900
[391] File:TradeCardMasonAndHamlinOrganCoCirca1880s.jpg - Wikimedia Commons
[392] After Pretoria: The Guerilla War Vol 3 p. 52 1900
[393] Anglo-Boer War Album, Cape Town, D. Edwards & co. [1902?]
[394] Anglo-Boer War Album, Cape Town, D. Edwards & co. [1902?]
[395] With the Flag to Pretoria, H. W. Wilson Vol 2 Dec., 1899.
[396] With the Flag to Pretoria, H. W. Wilson Vol 2 Nov. 1899. Photo by Lieut. J. F. Forbes, Gordens.
[397] Hillegas, Boers in War, 1900.
[398] Davitt, Boer Fight for Freedom, 3rd ed. 1902. (Duke Scan)
[399] Unknown Source (Roberts)
[400] Anglo-Boer War Album, Cape Town, D. Edwards & co. [1902?]
[401] Anglo-Boer War Album, Cape Town, D. Edwards & co. [1902?]
[402] Champion Berkshire boar at 2005 Royal Adelaide Show. Source: Photographer: Scott Davis. License under GNU Free Documentation License, Version 1.2 or any later version published by the Free Software Foundation. File:Adelaide champion Berkshire boar 2005.jpg - Wikimedia Commons
[403] File:Viljoensdrif station.jpg - Wikipedia
[404] Times History War in S.A Vol 3 Photo by Mr. Perceval Landon.

[405] Image Source: Exiles | Saint Helena Island Info: All about St Helena, in the South Atlantic Ocean
[406] After Pretoria: The Guerilla War Vol 3 Dec. 1900 Photo by N.P. Edwards.
[407] War with the Boers, H. Brown, London [1902?] Vol 1
[408] Anglo-Boer War Album, Cape Town, D. Edwards & co. [1902?]
[409] With the Flag to Pretoria, H. W. Wilson Vol 1
[410] My Escape from the Boers, Strand Magazine Strand Magazine Vol 67 p. 14-23.
[411] Black & White Weekly Vol 1-2 1899-1900 and Rob Davidson at Anglo Boer War.com
[412] Richard Harding Davis, Real Soldiers of Fortune. Charles Scribner's Sons, New York 1906 : #7 - Real soldiers of fortune, by Richard Harding Davis. - Full View | HathiTrust Digital Library
[413] My Escape from the Boers, Strand Magazine Strand Magazine Vol 67 p. 14-23.
[414] Image Credit Rob Davidson at Anglo Boer War.com Anglo Boer War - Armored Train Ambush 11-15-99 Chuchill & Haldane - Page 11 - Boer War Forum
[415] Hiley and Hassell, The Mobile Boer, Grafton Press, New York, (c. 1902)
[416] Heroes of the Boer War. Review of Reviews, 1903.
[417] Arthur Mann The Boer in Peace and War, 1900
[418] Photo portrait by E. H. Mills 1905 File:W.T. Stead by Mills, 1905.png - Wikimedia Commons
[419] History of War SA, Birch, 1899
[420] War in SA, Harding 1899
[421] Hillegas, Boers in War, 1900.
[422] War with the Boers, H. Brown, London [1902?] Vol 2
[423] Dr. Th. F. Burgers. Provincial Archives of the Free State, Collection Dr. Hendrik Muller, Photo Collection no. VA0957. File:ThFBurgers CHM VA0897.jpg - Wikimedia Commons
[424] File:South Africa 1874 One Pond.jpg - Wikipedia Credit: National Numismatic Collection, National Museum of American History.
[425] File:LL1882 pg008 SIKUKUNI.jpg - Wikimedia Commons
[426] Cunliffe, History of the Boer War Vol 1 (NY Library Scan)
[427] Anglo-Boer War Album, Cape Town, D. Edwards & co. [1902?]
[428] Haldane, How We Escaped from Pretoria. By: Robert Carnegie
[429] Haldane, How We Escaped from Pretoria. And: Richard Harding Davis With Both Armies in South Africa.
[430] Rosslyn -Twice Captured.
[431] With the Flag to Pretoria, H. W. Wilson Vol 2 June, 1900.
[432] Richard Harding Davis With Both Armies in South Africa. Charles Scribner's Sons, New York 1900
[433] With the Flag to Pretoria, H. W. Wilson Vol 2 June, 1900.
[434] Richard Harding Davis With Both Armies in South Africa. Charles Scribner's Sons, New York 1900
[435] Davitt, Boer Fight for Freedom, 3rd ed. 1902. (Duke Scan)
[436] File:WFReitz CHM VA0957.jpg - Wikimedia Commons
[437] Creswicke Drawing by Donald F. M'Cracken.
[438] Anglo-Boer War Album, Cape Town, D. Edwards & co. [1902?]
[439] File:James Francis Harry St Clair-Erskine, 5th Earl Rosslyn,.jpg - Wikimedia Commons
[440] Image per Neville Chamberlain. Anglo Boer War - Armored Train Ambush 11-15-99 Chuchill & Haldane - Page 17 - Boer War Forum
[441] Davitt, Boer Fight for Freedom, 3rd ed. 1902. (Duke Scan)
[442] Haldane, How We Escaped from Pretoria. p. 227
[443] Navy & Army Illustrated Vol 10 Photo by Edwards
[444] File:RMS Umbria.jpg - Wikimedia Commons
[445] Anglo-Boer War Album, Cape Town, D. Edwards & co. [1902?]
[446] Creswicke, Louis South Africa and the Transvaal War, Vol. 1 Photo by Wilson, Aberdeen.
[447] Anglo-Boer War Album, Cape Town, D. Edwards & co. [1902?]
[448] Cape Town. Table Bay Harbour with many sailing vessels and wagonloads of bags in the foreground. - Atom site for DRISA P0692_02G.jpg (1400×1080) (drisa.co.za)
[449] Anglo-Boer War Album, Cape Town, D. Edwards & co. [1902?]
[450] Black & White Weekly Vol 1-2 1899-1900
[451] Anglo-Boer War Album, Cape Town, D. Edwards & co. [1902?]
[452] Anglo-Boer War Album, Cape Town, D. Edwards & co. [1902?]

Made in United States
Orlando, FL
05 November 2026

90230218R00199